CONCEPTS IN THERMAL PHYSICS

Concepts in Thermal Physics

STEPHEN J. BLUNDELL AND
KATHERINE M. BLUNDELL

*Department of Physics,
University of Oxford, UK*

OXFORD
UNIVERSITY PRESS

Great Clarendon Street, Oxford OX2 6DP

Oxford University Press is a department of the University of Oxford.
It furthers the University's objective of excellence in research, scholarship,
and education by publishing worldwide in

Oxford New York

Auckland Cape Town Dar es Salaam Hong Kong Karachi
Kuala Lumpur Madrid Melbourne Mexico City Nairobi
New Delhi Shanghai Taipei Toronto

With offices in

Argentina Austria Brazil Chile Czech Republic France Greece
Guatemala Hungary Italy Japan Poland Portugal Singapore
South Korea Switzerland Thailand Turkey Ukraine Vietnam

Oxford is a registered trade mark of Oxford University Press
in the UK and in certain other countries

Published in the United States
by Oxford University Press Inc., New York

© Stephen J. Blundell and Katherine M. Blundell 2010

The moral rights of the authors have been asserted
Database right Oxford University Press (maker)

First published 2006
Second edition published in 2010

All rights reserved. No part of this publication may be reproduced,
stored in a retrieval system, or transmitted, in any form or by any means,
without the prior permission in writing of Oxford University Press,
or as expressly permitted by law, or under terms agreed with the appropriate
reprographics rights organization. Enquiries concerning reproduction
outside the scope of the above should be sent to the Rights Department,
Oxford University Press, at the address above

You must not circulate this book in any other binding or cover
and you must impose the same condition on any acquirer

British Library Cataloguing in Publication Data

Data available

Library of Congress Cataloging in Publication Data

Data available

Printed in Great Britain
by CPI Group (UK) Ltd, Croydon, CR0 4YY

ISBN 978–0–19–956209–1 (Hbk.)
ISBN 978–0–19–956210–7 (Pbk.)

The manufacturer's authorised representative in the EU for product
safety is Oxford University Press España S.A. of El Parque Empresarial
San Fernando de Henares, Avenida de Castilla, 2 – 28830 Madrid (www.
oup.es/en or product.safety@oup.com).
OUP España S.A. also acts as importer into Spain of products
made by the manufacturer.

To our dear parents

Alan and Daphne Blundell
Alan and Christine Sanders

with love.

Preface

In the beginning was the Word...   (John 1:1, first century AD)

Consider sunbeams. When the sun's rays let in
Pass through the darkness of a shuttered room,
You will see a multitude of tiny bodies
All mingling in a multitude of ways
Inside the sunbeam, moving in the void,
Seeming to be engaged in endless strife,
Battle, and warfare, troop attacking troop,
And never a respite, harried constantly,
With meetings and with partings everywhere.
From this you can imagine what it is
For atoms to be tossed perpetually
In endless motion through the mighty void.

(*On the Nature of Things*, Lucretius, first century BC)

... (we) have borne the burden of the *work* and the *heat* of the day.

(Matthew 20:12, first century AD)

Thermal physics forms a key part of any undergraduate physics course. It includes the fundamentals of *classical thermodynamics* (which was founded largely in the nineteenth century and motivated by a desire to understand the conversion of heat into work using engines) and also *statistical mechanics* (which was founded by Boltzmann and Gibbs, and is concerned with the statistical behaviour of the underlying microstates of the system). Students often find these topics hard, and this problem is not helped by a lack of familiarity with basic concepts in mathematics, particularly in probability and statistics. Moreover, the traditional focus of thermodynamics on steam engines seems remote and largely irrelevant to a twenty-first century student. This is unfortunate since an understanding of thermal physics is crucial to almost all modern physics and to the important technological challenges which face us in this century.

    The aim of this book is to provide an introduction to the key concepts in thermal physics, fleshed out with plenty of modern examples from astrophysics, atmospheric physics, laser physics, condensed matter physics and information theory. The important mathematical principles, particularly concerning probability and statistics, are expounded in some detail. This aims to make up for the material which can no longer be automatically assumed to have been covered in every school mathematics course. In addition, the appendices contain useful mathematics, such as various integrals, mathematical results and identities. There is, unfortunately, no shortcut to mastering the necessary mathematics in studying thermal physics, but the material in the appendix provides a useful *aide-mémoire*.

Many courses on this subject are taught historically: the kinetic theory of gases, then classical thermodynamics are taught first, with statistical mechanics taught last. In other courses, one starts with the principles of classical thermodynamics, followed then by statistical mechanics and kinetic theory is saved until the end. Although there is merit in both approaches, we have aimed at a more integrated treatment. For example, we introduce temperature using a straightforward statistical mechanical argument, rather than on the basis of a somewhat abstract Carnot engine. However, we do postpone detailed consideration of the partition function and statistical mechanics until after we have introduced the functions of state, which manipulation of the partition function so conveniently produces. We present the kinetic theory of gases fairly early on, since it provides a simple, well-defined arena in which to practise simple concepts in probability distributions. This has worked well in the course given in Oxford, but since kinetic theory is only studied at a later stage in courses in other places, we have designed the book so that the kinetic theory chapters can be omitted without causing problems; see Fig. 1.5 on page 10 for details. In addition, some parts of the book contain material that is much more advanced (often placed in boxes, or in the final part of the book), and these can be skipped at first reading.

The book is arranged in a series of short, easily digestible chapters, each one introducing a new concept or illustrating an important application. Most people learn from examples, so plenty of worked examples are given in order that the reader can gain familiarity with the concepts as they are introduced. Exercises are provided at the end of each chapter to allow the students to gain practice in each area.

In choosing which topics to include, and at what level, we have aimed for a balance between pedagogy and rigour, providing a comprehensible introduction with sufficient details to satisfy more advanced readers. We have also tried to balance fundamental principles with practical applications. However, this book does not treat real engines in any engineering depth, nor does it venture into the deep waters of ergodic theory. Nevertheless, we hope that there is enough in this book for a thorough grounding in thermal physics and the recommended further reading gives pointers for additional material. An important theme running through this book is the concept of information, and its connection with entropy. The black hole shown at the start of this preface, with its surface covered in 'bits' of information, is a helpful picture of the deep connection between information, thermodynamics, radiation, and the Universe.

The history of thermal physics is a fascinating one, and we have provided a selection of short biographical sketches of some of the key pioneers in thermal physics. To qualify for inclusion, the person had to have made a particularly important contribution or had a particularly interesting life – and be dead! Therefore one should not conclude from the list of people we have chosen that the subject of thermal physics is in any sense finished, it is just harder to write with the same perspective about current work in this subject. The biographical sketches are necessarily brief, giving only a glimpse of the life-story, so the Bibliography should be consulted for a list of more comprehensive biographies. However, the sketches are designed to provide some light relief in the main narrative and demonstrate that science is a *human* endeavour.

It is a great pleasure to record our gratitude to those who taught us the subject while we were undergraduates in Cambridge, particularly Owen Saxton and Peter Scheuer, and to our friends in Oxford: we have benefitted from many enlightening discussions with colleagues in the physics department, from the intelligent questioning of our Oxford students and from the stimulating environments provided by both Mansfield College and St John's College. In the writing of this book, we have enjoyed the steadfast encouragement of Sönke Adlung and his colleagues at OUP, and in particular Julie Harris' black-belt LaTeX support.

A number of friends and colleagues in Oxford and elsewhere have been kind enough to give their time and read drafts of chapters of this book; they have made numerous helpful comments, which have greatly improved the final result: Fathallah Alouani Bibi, James Analytis, David Andrews, Arzhang Ardavan, Tony Beasley, Michael Bowler, Peter Duffy, Paul Goddard, Stephen Justham, Michael Mackey, Philipp Podsiadlowski, Linda Schmidtobreick, John Singleton and Katrien Steenbrugge. Particular thanks are due to Tom Lancaster, who twice read the entire manuscript at early stages and made many constructive and imaginative suggestions, and to Harvey Brown, whose insights were always stimulating and whose encouragement was always constant. To all these friends, our warmest thanks are due. Errors which we discover after going to press will be posted on the book's website, which may be found at:

http://users.ox.ac.uk/~sjb/ctp

It is our earnest hope that this book will make the study of thermal physics enjoyable and fascinating and that we have managed to communicate something of the enthusiasm we feel for this subject. Moreover, understanding the concepts of thermal physics is vital for humanity's future; the impending energy crisis and the potential consequences of climate change mandate creative, scientific, and technological innovations at the highest levels. This means that thermal physics is a field that some of tomorrow's best minds need to master today.

SJB & KMB
Oxford
June 2006

Preface to the second edition

This new edition keeps the same structure as the first edition but includes additional material on probability, Bayes' theorem, diffusion problems, osmosis, the Ising model, Monte-Carlo simulations, and radiative transfer in atmospheric physics. We have also taken the opportunity to improve the treatment of various topics, including the discussion of constraints and the presentation of the Fermi–Dirac and Bose–Einstein distributions, as well as correcting various errors. We are particularly grateful to the following people who have pointed out errors or omissions and made highly relevant comments: David Andrews, John Aveson, Ryan Buckingham, Radu Coldea, Merlin Cooper, Peter Coulon, Peter Duffy, Ted Einstein, Joe Fallon, Amy Fok, Felix Flicker, William Frass, Andrew Garner, Paul Hennin, Ben Jones, Stephen Justham, Austen Lamacraft, Peter Liley, Gabriel McManus, Adam Micolich, Robin Moss, Alan O'Neill, Elena Nickson, Wilson Poon, Caity Rice, Andrew Steane, Nicola van Leeuwen, Yan Mei Wang, Peter Watson, Helena Wilding, and Michael Williams. We have once again enjoyed the support of the staff of OUP and, in particular, our copy-editor Alison Lees, who trawled through the manuscript with meticulous care, making many important improvements. Myles Allen, David Andrews, and William Ingram gave us very pertinent and instructive comments about the treatment of atmospheric physics and their input has been invaluable. Thanks are also due to Geoff Brooker, who shared his profound insights into the nature of free energies, and Tom Lancaster, who once again made numerous helpful suggestions.

SJB & KMB
Oxford
August 2009

Note added (July 2014; July 2017): We are grateful to the following for pointing out various errors which have been corrected in this reprinting: Cassio Amorim, David Andrews, Piotr Boguslawski, Pablo Gregorian, Robert Jeffrey, Jakob Kastelic, Hong Mao, Marius Schaper, Robert Spry, Olinga Tahzib, Ayngaran Thavanesan, Noah Waterfield Price and particularly Eleftherios N. Economou, Guo-xing Ju and Carl Mungan.

Contents

Preface		vii
Preface to the second edition		x

I Preliminaries   1

1 Introduction   2
- 1.1 What is a mole?   3
- 1.2 The thermodynamic limit   4
- 1.3 The ideal gas   6
- 1.4 Combinatorial problems   7
- 1.5 Plan of the book   9
- Exercises   12

2 Heat   13
- 2.1 A definition of heat   13
- 2.2 Heat capacity   14
- Exercises   17

3 Probability   18
- 3.1 Discrete probability distributions   19
- 3.2 Continuous probability distributions   20
- 3.3 Linear transformation   21
- 3.4 Variance   22
- 3.5 Linear transformation and the variance   23
- 3.6 Independent variables   24
- 3.7 Binomial distribution   26
- Further reading   29
- Exercises   29

4 Temperature and the Boltzmann factor   32
- 4.1 Thermal equilibrium   32
- 4.2 Thermometers   33
- 4.3 The microstates and macrostates   35
- 4.4 A statistical definition of temperature   36
- 4.5 Ensembles   38
- 4.6 Canonical ensemble   38
- 4.7 Applications of the Boltzmann distribution   42
- Further reading   46
- Exercises   46

II Kinetic theory of gases   47

5 The Maxwell–Boltzmann distribution   48
  5.1 The velocity distribution   48
  5.2 The speed distribution   49
  5.3 Experimental justification   51
  Exercises   54

6 Pressure   56
  6.1 Molecular distributions   57
  6.2 The ideal gas law   58
  6.3 Dalton's law   60
  Exercises   61

7 Molecular effusion   64
  7.1 Flux   64
  7.2 Effusion   66
  Exercises   69

8 The mean free path and collisions   70
  8.1 The mean collision time   70
  8.2 The collision cross-section   71
  8.3 The mean free path   73
  Exercises   74

III Transport and thermal diffusion   75

9 Transport properties in gases   76
  9.1 Viscosity   76
  9.2 Thermal conductivity   81
  9.3 Diffusion   83
  9.4 More detailed theory   86
  Further reading   88
  Exercises   89

10 The thermal diffusion equation   90
  10.1 Derivation of the thermal diffusion equation   90
  10.2 The one-dimensional thermal diffusion equation   91
  10.3 The steady state   94
  10.4 The thermal diffusion equation for a sphere   94
  10.5 Newton's law of cooling   99
  10.6 The Prandtl number   100
  10.7 Sources of heat   101
  10.8 Particle diffusion   102
  Exercises   103

IV The first law   107

11 Energy   108
11.1 Some definitions   108
11.2 The first law of thermodynamics   110
11.3 Heat capacity   112
Exercises   115

12 Isothermal and adiabatic processes   118
12.1 Reversibility   118
12.2 Isothermal expansion of an ideal gas   120
12.3 Adiabatic expansion of an ideal gas   121
12.4 Adiabatic atmosphere   121
Exercises   123

V The second law   125

13 Heat engines and the second law   126
13.1 The second law of thermodynamics   126
13.2 The Carnot engine   127
13.3 Carnot's theorem   130
13.4 Equivalence of Clausius' and Kelvin's statements 131
13.5 Examples of heat engines   131
13.6 Heat engines running backwards   133
13.7 Clausius' theorem   134
Further reading   137
Exercises   137

14 Entropy   140
14.1 Definition of entropy   140
14.2 Irreversible change   140
14.3 The first law revisited   142
14.4 The Joule expansion   144
14.5 The statistical basis for entropy   146
14.6 The entropy of mixing   147
14.7 Maxwell's demon   149
14.8 Entropy and probability   150
Exercises   153

15 Information theory   157
15.1 Information and Shannon entropy   157
15.2 Information and thermodynamics   159
15.3 Data compression   160
15.4 Quantum information   162
15.5 Conditional and joint probabilities   165
15.6 Bayes' theorem   165
Further reading   168
Exercises   169

VI Thermodynamics in action — 171

16 Thermodynamic potentials — 172
- 16.1 Internal energy, U — 172
- 16.2 Enthalpy, H — 173
- 16.3 Helmholtz function, F — 174
- 16.4 Gibbs function, G — 175
- 16.5 Constraints — 176
- 16.6 Maxwell's relations — 179
- Exercises — 187

17 Rods, bubbles, and magnets — 191
- 17.1 Elastic rod — 191
- 17.2 Surface tension — 194
- 17.3 Electric and magnetic dipoles — 195
- 17.4 Paramagnetism — 196
- Exercises — 201

18 The third law — 203
- 18.1 Different statements of the third law — 203
- 18.2 Consequences of the third law — 205
- Exercises — 208

VII Statistical mechanics — 209

19 Equipartition of energy — 210
- 19.1 Equipartition theorem — 210
- 19.2 Applications — 213
- 19.3 Assumptions made — 215
- 19.4 Brownian motion — 217
- Exercises — 218

20 The partition function — 219
- 20.1 Writing down the partition function — 220
- 20.2 Obtaining the functions of state — 221
- 20.3 The big idea — 228
- 20.4 Combining partition functions — 228
- Exercises — 232

21 Statistical mechanics of an ideal gas — 233
- 21.1 Density of states — 233
- 21.2 Quantum concentration — 235
- 21.3 Distinguishability — 236
- 21.4 Functions of state of the ideal gas — 237
- 21.5 Gibbs paradox — 240
- 21.6 Heat capacity of a diatomic gas — 241
- Exercises — 243

22 The chemical potential — 244
22.1 A definition of the chemical potential — 244
22.2 The meaning of the chemical potential — 245
22.3 Grand partition function — 247
22.4 Grand potential — 248
22.5 Chemical potential as Gibbs function per particle — 250
22.6 Many types of particle — 250
22.7 Particle number conservation laws — 251
22.8 Chemical potential and chemical reactions — 252
22.9 Osmosis — 257
Further reading — 261
Exercises — 262

23 Photons — 263
23.1 The classical thermodynamics of electromagnetic radiation — 264
23.2 Spectral energy density — 265
23.3 Kirchhoff's law — 266
23.4 Radiation pressure — 268
23.5 The statistical mechanics of the photon gas — 269
23.6 Black-body distribution — 270
23.7 Cosmic microwave background radiation — 273
23.8 The Einstein A and B coefficients — 274
Further reading — 277
Exercises — 278

24 Phonons — 279
24.1 The Einstein model — 279
24.2 The Debye model — 281
24.3 Phonon dispersion — 284
Further reading — 287
Exercises — 287

VIII Beyond the ideal gas — 289

25 Relativistic gases — 290
25.1 Relativistic dispersion relation for massive particles — 290
25.2 The ultrarelativistic gas — 290
25.3 Adiabatic expansion of an ultrarelativistic gas — 293
Exercises — 295

26 Real gases — 296
26.1 The van der Waals gas — 296
26.2 The Dieterici equation — 304
26.3 Virial expansion — 306
26.4 The law of corresponding states — 310
Exercises — 312

27 Cooling real gases — 313
- 27.1 The Joule expansion — 313
- 27.2 Isothermal expansion — 315
- 27.3 Joule–Kelvin expansion — 316
- 27.4 Liquefaction of gases — 318
- Exercises — 320

28 Phase transitions — 321
- 28.1 Latent heat — 321
- 28.2 Chemical potential and phase changes — 324
- 28.3 The Clausius–Clapeyron equation — 324
- 28.4 Stability and metastability — 329
- 28.5 The Gibbs phase rule — 332
- 28.6 Colligative properties — 334
- 28.7 Classification of phase transitions — 335
- 28.8 The Ising model — 338
- Further reading — 343
- Exercises — 343

29 Bose–Einstein and Fermi–Dirac distributions — 345
- 29.1 Exchange and symmetry — 345
- 29.2 Wave functions of identical particles — 346
- 29.3 The statistics of identical particles — 349
- Further reading — 353
- Exercises — 354

30 Quantum gases and condensates — 358
- 30.1 The non-interacting quantum fluid — 358
- 30.2 The Fermi gas — 361
- 30.3 The Bose gas — 366
- 30.4 Bose–Einstein condensation (BEC) — 367
- Further reading — 373
- Exercises — 373

IX Special topics — 375

31 Sound waves — 376
- 31.1 Sound waves under isothermal conditions — 377
- 31.2 Sound waves under adiabatic conditions — 377
- 31.3 Are sound waves in general adiabatic or isothermal? — 378
- 31.4 Derivation of the speed of sound within fluids — 379
- Further reading — 382
- Exercises — 382

32 Shock waves — 383
- 32.1 The Mach number — 383
- 32.2 Structure of shock waves — 383
- 32.3 Shock conservation laws — 385

32.4 The Rankine–Hugoniot conditions	386
Further reading	389
Exercises	389

33 Brownian motion and fluctuations — 390
- 33.1 Brownian motion — 390
- 33.2 Johnson noise — 393
- 33.3 Fluctuations — 394
- 33.4 Fluctuations and the availability — 395
- 33.5 Linear response — 397
- 33.6 Correlation functions — 400
- Further reading — 407
- Exercises — 407

34 Non-equilibrium thermodynamics — 408
- 34.1 Entropy production — 408
- 34.2 The kinetic coefficients — 409
- 34.3 Proof of the Onsager reciprocal relations — 410
- 34.4 Thermoelectricity — 413
- 34.5 Time reversal and the arrow of time — 417
- Further reading — 419
- Exercises — 419

35 Stars — 420
- 35.1 Gravitational interaction — 421
- 35.2 Nuclear reactions — 426
- 35.3 Heat transfer — 427
- Further reading — 434
- Exercises — 434

36 Compact objects — 435
- 36.1 Electron degeneracy pressure — 435
- 36.2 White dwarfs — 437
- 36.3 Neutron stars — 438
- 36.4 Black holes — 440
- 36.5 Accretion — 441
- 36.6 Black holes and entropy — 442
- 36.7 Life, the Universe, and entropy — 443
- Further reading — 445
- Exercises — 445

37 Earth's atmosphere — 446
- 37.1 Solar energy — 446
- 37.2 The temperature profile in the atmosphere — 447
- 37.3 Radiative transfer — 449
- 37.4 The greenhouse effect — 452
- 37.5 Global warming — 456
- Further reading — 460
- Exercises — 460

A	**Fundamental constants**	**461**
B	**Useful formulae**	**462**
C	**Useful mathematics**	**464**
	C.1 The factorial integral	464
	C.2 The Gaussian integral	464
	C.3 Stirling's formula	467
	C.4 Riemann zeta function	469
	C.5 The polylogarithm	470
	C.6 Partial derivatives	471
	C.7 Exact differentials	472
	C.8 Volume of a hypersphere	473
	C.9 Jacobians	473
	C.10 The Dirac delta function	475
	C.11 Fourier transforms	475
	C.12 Solution of the diffusion equation	476
	C.13 Lagrange multipliers	477
D	**The electromagnetic spectrum**	**479**
E	**Some thermodynamical definitions**	**480**
F	**Thermodynamic expansion formulae**	**481**
G	**Reduced mass**	**482**
H	**Glossary of main symbols**	**483**
	Bibliography	**485**
	Index	**489**

Part I

Preliminaries

To explore and understand the rich and beautiful subject that is thermal physics, we need some essential tools in place. Part I provides these, as follows:

- In Chapter 1 we explore the concept of *large numbers*, showing why large numbers appear in thermal physics and explaining how to handle them. Large numbers arise in thermal physics because the number of atoms in the bit of matter under study is usually very large (for example, it can be typically of the order of 10^{23}), but also because many thermal physics problems involve *combinatorial* calculations (and this can produce numbers like $10^{23}!$, where "!" here means a factorial). We introduce *Stirling's approximation*, which is useful for handling expressions, such as $\ln N!$, which frequently appear in thermal physics. We discuss the *thermodynamic limit* and state the *ideal gas equation* (derived later, in Chapter 6, from the kinetic theory of gases).

- In Chapter 2 we explore the concept of *heat*, defining it as "thermal energy in transit", and introduce the idea of a *heat capacity*.

- The ways in which thermal systems behave is determined by the laws of *probability*, so we outline the notion of probability in Chapter 3 and apply it to a number of problems. This chapter may well cover ground that is familiar to some readers, but is a useful introduction to the subject.

- We then use these ideas to define the *temperature* of a system from a statistical perspective and hence derive the *Boltzmann distribution* in Chapter 4. This distribution describes how a thermal system behaves when it is placed in thermal contact with a large *thermal reservoir*. This is a key concept in thermal physics and forms the basis of all that follows.

1 Introduction

1.1 What is a mole? 3
1.2 The thermodynamic limit 4
1.3 The ideal gas 6
1.4 Combinatorial problems 7
1.5 Plan of the book 9
Chapter summary 12
Exercises 12

Some large numbers:

million	10^6
billion	10^9
trillion	10^{12}
quadrillion	10^{15}
quintillion	10^{18}
googol	10^{100}
googolplex	$10^{10^{100}}$

Note: these values assume the US billion, trillion, etc, which are now in general use.

[1] Still more hopeless would be the task of measuring where each molecule is and how fast it is moving in its initial state!

The subject of thermal physics involves studying assemblies of large numbers of atoms. As we will see, it is the large numbers involved in macroscopic systems that allow us to treat some of their properties in a statistical fashion. What do we mean by a large number?

Large numbers turn up in many spheres of life. A book might sell a million (10^6) copies (probably not this one), the Earth's population is (at the time of writing) between six and seven billion people (6–7×10^9), and the US national debt is currently around ten trillion dollars (10^{13} US\$). But even these large numbers pale into insignificance compared with the numbers involved in thermal physics. The number of atoms in an average-sized piece of matter is usually ten to the power of twenty-something, and this puts extreme limits on what sort of calculations we can make to understand them.

Example 1.1

One kilogramme of nitrogen gas contains approximately 2×10^{25} N_2 molecules. Let us see how easy it would be to make predictions about the motion of the molecules in this amount of gas. In one year, there are about 3.2×10^7 seconds, so that a 3 GHz personal computer can count molecules at a rate of roughly 10^{17} year^{-1}, if it counts one molecule every computer clock cycle. Therefore it would take about 0.2 billion years just for this computer to count all the molecules in one kilogramme of nitrogen gas (a time that is roughly a few percent of the age of the Universe!). Counting the molecules is a computationally simpler task than calculating all their movements and collisions with each other. Therefore modelling this quantity of matter by following each and every particle is a hopeless task.[1]

Hence, to make progress in thermal physics it is necessary to make approximations and deal with the statistical properties of molecules, i.e., to study how they behave *on average*. Chapter 3 therefore contains a discussion of probability and statistical methods, which are foundational for understanding thermal physics. In this chapter, we will briefly review the definition of a mole (which will be used throughout the book), consider why very big numbers arise from combinatorial problems in thermal physics and introduce the *thermodynamic limit* and the *ideal gas equation*.

1.1 What is a mole?

A **mole** is, of course, a small burrowing animal, but also a name (first coined about a century ago from the German "Molekül" [molecule]) representing a certain numerical quantity of stuff. It functions in the same way as the word "dozen", which describes a certain number of eggs (12), or "score", which describes a certain number of years (20). It might be easier if we could use the word dozen when describing a certain number of atoms, but a dozen atoms is not many (unless you are building a quantum computer) and since a million, a billion, and even a quadrillion are also too small to be useful, we have ended up with using an even bigger number. Unfortunately, for historical reasons, it isn't a power of ten.

The mole
A **mole** is *defined* as the quantity of matter that contains as many objects (for example, atoms, molecules, formula units, or ions) as the number of atoms in exactly 12 g (= 0.012 kg) of ^{12}C.

A mole is also *approximately* the quantity of matter that contains as many objects (for example, atoms, molecules, formula units, ions) as the number of atoms in exactly 1 g (= 0.001 kg) of ^{1}H, but carbon was chosen as a more convenient international standard since solids are easier to weigh accurately.

A mole of atoms is equivalent to an **Avogadro number** N_A of atoms. The Avogadro number, expressed to four significant figures, is

$$N_A = 6.022 \times 10^{23}. \tag{1.1}$$

One can write N_A as 6.022×10^{23} mol^{-1} as a reminder of its definition, but N_A is dimensionless, as are moles. They are both numbers. By the same logic, one would have to define the 'eggbox number' as 12 dozen^{-1}.

Example 1.2

- 1 mole of carbon is 6.022×10^{23} atoms of carbon.
- 1 mole of benzene is 6.022×10^{23} molecules of benzene.
- 1 mole of NaCl contains 6.022×10^{23} NaCl formula units, etc.

The Avogadro number is an exceedingly large number: a mole of eggs would make an omelette with about half the mass of the Moon!

The **molar mass** of a substance is the mass of one mole of the substance. Thus the molar mass of carbon is 12 g, but the molar mass of water is close to 18 g (because the mass of a water molecule is about $\frac{18}{12}$ times larger than the mass of a carbon atom). The mass m of a single molecule or atom is therefore the molar mass of that substance *divided* by the Avogadro number. Equivalently:

$$\text{molar mass} = mN_A. \tag{1.2}$$

1.2 The thermodynamic limit

In this section, we will explain how the large numbers of molecules in a typical thermodynamic system mean that it is possible to deal with average quantities. Our explanation proceeds using an analogy: imagine that you are sitting inside a tiny hut with a flat roof. It is raining outside, and you can hear the occasional raindrop striking the roof. The raindrops arrive randomly, so sometimes two arrive close together, but sometimes there is quite a long gap between raindrops. Each raindrop transfers its momentum to the roof and exerts an impulse[2] on it. If you knew the mass and terminal velocity of a raindrop, you could estimate the force on the roof of the hut. The force as a function of time would look like that shown in Fig. 1.1(a), each little blip corresponding to the impulse from one raindrop.

Now imagine that you are sitting inside a much bigger hut with a flat roof a thousand times the area of the first roof. Many more raindrops will now be falling on the larger roof area and the force as a function of time would look like that shown in Fig. 1.1(b). Now scale up the area of the flat roof by a further factor of one hundred and the force would look like that shown in Fig. 1.1(c). Notice two key things about these graphs:

(1) The force, on average, gets bigger as the area of the roof gets bigger. This is not surprising because a bigger roof catches more raindrops.

(2) The *fluctuations* in the force get smoothed out and the force looks like it stays much closer to its average value. In fact, the fluctuations are still big but, as the area of the roof increases, they grow more slowly than the average force does.

The force grows with area, so it is useful to consider the **pressure**, which is defined as

$$\text{pressure} = \frac{\text{force}}{\text{area}}. \tag{1.3}$$

The average pressure due to the falling raindrops will not change as the area of the roof increases, but the fluctuations in the pressure will decrease. In fact, we can completely ignore the fluctuations in the pressure in the limit that the area of the roof *grows to infinity*. This is precisely analogous to the limit we refer to as the **thermodynamic limit**.

Consider now the molecules of a gas which are bouncing around in a container. Each time the molecules bounce off the walls of the container, they exert an impulse on the walls. The net effect of all these impulses is a pressure, a force per unit area, exerted on the walls of the container. If the container were very small, we would have to worry about fluctuations in the pressure (the random arrival of individual molecules on the wall, much like the raindrops in Fig. 1.1(a)). However, in most cases that one meets, the number of molecules in a container of gas is extremely large, so these fluctuations can be ignored and the pressure of the gas appears to be completely uniform. Again, our description of the pressure of this

---

[2] An impulse is the product of force and a time interval. The impulse is equal to the change of momentum.

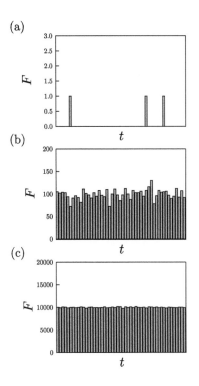

Fig. 1.1 Graphs of the force on a roof as a function of time due to falling rain drops.

system can be said to be "in the thermodynamic limit", where we have let the number of molecules be regarded as tending to infinity in such a way that the density of the gas is a constant.

Suppose that the container of gas has volume V, that the temperature is T, the pressure is p, and the kinetic energy of all the gas molecules adds up to U. Imagine slicing the container of gas in half with an imaginary plane, and now just focus your attention on the gas on one side of the plane. The volume of this half of the gas, let's call it V^*, is by definition half that of the original container, i.e.,

$$V^* = \frac{V}{2}. \tag{1.4}$$

The kinetic energy of this half of the gas, let's call it U^*, is clearly half that of the total kinetic energy, i.e.,

$$U^* = \frac{U}{2}. \tag{1.5}$$

However, the pressure p^* and the temperature T^* of this half of the gas are the same as for the whole container of gas, so that

$$p^* = p, \tag{1.6}$$
$$T^* = T. \tag{1.7}$$

Variables which scale with the system size, like V and U, are called **extensive variables**. Those which are independent of system size, like p and T, are called **intensive variables**.

Thermal physics evolved in various stages and has left us with various approaches to the subject:

- The subject of **classical thermodynamics** deals with macroscopic properties, such as pressure, volume, and temperature, without worrying about the underlying microscopic physics. It applies to systems that are sufficiently large that microscopic fluctuations can be ignored, and it does not assume that there is an underlying atomic structure to matter.

- The **kinetic theory of gases** tries to determine the properties of gases by considering probability distributions associated with the motions of individual molecules. This was initially somewhat controversial since the existence of atoms and molecules was doubted by many until the late nineteenth and early twentieth centuries.

- The realization that atoms and molecules exist led to the development of **statistical mechanics**. Rather than starting with descriptions of macroscopic properties (as in thermodynamics) this approach begins with trying to describe the individual microscopic states of a system and then uses statistical methods to derive the macroscopic properties from them. This approach received an additional impetus with the development of **quantum theory**, which showed explicitly how to describe the microscopic quantum states of different systems. The thermodynamic behaviour of a system is then asymptotically approximated by the results of statistical mechanics in the *thermodynamic limit*, i.e., as the number of particles tends to infinity (with intensive quantities such as pressure and density remaining finite).

In the next section, we will state the *ideal gas law*, which was first found experimentally but can be deduced from the kinetic theory of gases (see Chapter 6).

1.3 The ideal gas

Experiments on gases show that the pressure p of a volume V of gas depends on its temperature T. For example, a fixed amount of gas at constant temperature obeys

$$p \propto 1/V, \tag{1.8}$$

a result which is known as **Boyle's law** (sometimes as the Boyle–Mariotte law); it was discovered experimentally by Robert Boyle (1627–1691) in 1662 and independently by Edmé Mariotte (1620–1684) in 1676. At constant pressure, the gas also obeys

$$V \propto T, \tag{1.9}$$

where T is measured in kelvin. This is known as **Charles' law** and was discovered experimentally, in a crude fashion, by Jacques Charles (1746–1823) in 1787, and more completely by Joseph Louis Gay-Lussac (1778–1850) in 1802, though their work was partly anticipated by Guillaume Amontons (1663–1705) in 1699, who also noticed that a fixed volume of gas obeys

$$p \propto T, \tag{1.10}$$

a result that Gay-Lussac himself found independently in 1809 and is often known as **Gay-Lussac's law**.[3]

These three empirical laws can be combined to give

$$pV \propto T. \tag{1.11}$$

It turns out that, if there are N molecules in the gas, this finding can be expressed as follows:

$$\boxed{pV = Nk_\text{B}T.} \tag{1.12}$$

This is known as the **ideal gas equation**, and the constant k_B is known as the **Boltzmann constant**.[4] We now make some comments about the ideal gas equation.

- We have stated this law purely as an empirical law, observed in experiment. We will derive it from first principles using the kinetic theory of gases in Chapter 6. This theory assumes that a gas can be modelled as a collection of individual tiny particles which can bounce off the walls of the container, and each other (see Fig. 1.2).

[3] Note that none of these scientists expressed temperature in this way, since the kelvin scale and absolute zero had yet to be invented. For example, Gay-Lussac found merely that $V = V_0(1 + \alpha\tilde{T})$, where V_0 and α are constants and $\tilde{T}$ is temperature in his scale.

[4] It takes the numerical value $k_\text{B} = 1.3807 \times 10^{-23}$ J K^{-1}. We will meet this constant again in eqn 4.7.

- Why do we call it "ideal"? The microscopic justification that we will present in Chapter 6 proceeds under various assumptions: (i) we assume that there are no intermolecular forces, so that the molecules are not attracted to each other; (ii) we assume that molecules are point-like and have zero size. These are idealized assumptions and so we do not expect the ideal gas model to describe real gases under all circumstances. However, it does have the virtue of simplicity: eqn 1.12 is simple to write down and remember. Perhaps more importantly, it does describe gases quite well under quite a wide range of conditions.
- The ideal gas equation forms the basis of much of our study of classical thermodynamics. Gases are common in nature: they are encountered in astrophysics and atmospheric physics; it is gases which are used to drive engines, and thermodynamics was invented to try and understand engines. Therefore this equation is fundamental in our treatment of thermodynamics and should be memorized.
- The ideal gas law, however, doesn't describe all important gases, and several chapters in this book are devoted to seeing what happens when various assumptions fail. For example, the ideal gas equation assumes that the gas molecules move non-relativistically. When this is not the case, we have to develop a model of relativistic gases (see Chapter 25). At low temperatures and high densities, gas molecules do attract one another (this must occur for liquids and solids to form) and this is considered in Chapters 26, 27, and 28. Furthermore, when quantum effects are important we need a model of quantum gases, and this is outlined in Chapter 30.
- Of course, thermodynamics applies also to systems which are not gaseous (so the ideal gas equation, though useful, is not a cure for all ills), and we will look at the thermodynamics of rods, bubbles, and magnets in Chapter 17.

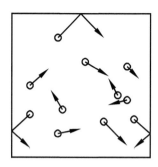

Fig. 1.2 In the kinetic theory of gases, a gas is modelled as a number of individual tiny particles which can bounce off the walls of the container, and each other.

1.4 Combinatorial problems

Even larger numbers than N_A occur in problems involving combinations, and these turn out to be very important in thermal physics. The following example illustrates a simple combinatorial problem which captures the essence of what we are going to have to deal with.

Example 1.3

Let us imagine that a certain system contains ten atoms. Each of these atoms can exist in one of two states, according to whether it has zero units or one unit of energy. These "units" of energy are called **quanta** of energy. How many distinct arrangements of quanta are possible for this system if you have at your disposal (a) ten quanta of energy; (b) four quanta of energy?

Fig. 1.3 Ten atoms that can accommodate four quanta of energy. An atom with a single quantum of energy is shown as a filled circle, otherwise it is shown as an empty circle. One configuration is shown here.

Solution:
We can represent the ten atoms by drawing ten boxes; an empty box signifies an atom with zero quanta of energy; a filled box signifies an atom with one quantum of energy (see Fig. 1.3). We give two methods for calculating the number of ways of arranging r quanta among n atoms:

(1) In the first method, we realize that the first quantum can be assigned to any of the n atoms, the second quantum can be assigned to any of the remaining atoms (there are $n-1$ of them), and so on until the rth quantum can be assigned to any of the remaining $n-r+1$ atoms. Thus our first guess for the number of possible arrangements of the r quanta we have assigned is $\Omega_{\text{guess}} = n \times (n-1) \times (n-2) \times \ldots \times (n-r+1)$. This can be simplified as follows:

$$\Omega_{\text{guess}} = \frac{n \times (n-1) \times (n-2) \times \ldots \times 1}{(n-r) \times (n-r-1) \times \ldots \times 1} = \frac{n!}{(n-r)!}. \quad (1.13)$$

However, this assumes that we have labelled the quanta as "the first quantum", "the second quantum" etc. In fact, we don't care which quantum is which because they are indistinguishable. We can rearrange the r quanta in any one of $r!$ arrangements. Hence our answer Ω_{guess} needs to be divided by $r!$, so that the number Ω of unique arrangements is

$$\Omega = \frac{n!}{(n-r)!\,r!} \equiv {}^nC_r, \quad (1.14)$$

where ${}^nC_r$ is the symbol for a **combination**.[5]

[5]Other symbols sometimes used for ${}^nC_r$ include ${}^n_rC$ and $\binom{n}{r}$.

(2) In the second method, we recognize that there are r atoms each with one quantum and $n-r$ atoms with zero quanta. The number of arrangements is then simply the number of ways of arranging r ones and $n-r$ zeros. There are $n!$ ways of arranging a sequence of n distinguishable symbols. If r of these symbols are the same (all ones), there are $r!$ ways of arranging these without changing the pattern. If the remaining $n-r$ symbols are all the same (all zeros), there are $(n-r)!$ ways of arranging these without changing the pattern. Hence we again find that

$$\Omega = \frac{n!}{(n-r)!\,r!}. \quad (1.15)$$

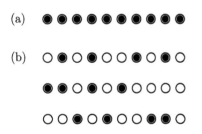

Fig. 1.4 Each row shows the ten atoms that can accommodate r quanta of energy. An atom with a single quantum of energy is shown as a filled circle, otherwise it is shown as an empty circle. (a) For $r=10$ there is only one possible configuration. (b) For $r=4$, there are 210 possibilities, of which three are shown.

For the specific cases shown in Fig. 1.4:
(a) $n=10$, $r=10$, so $\Omega = 10!/(10! \times 0!) = 1$. This one possibility, with each atom having a quantum of energy, is shown in Fig. 1.4(a).
(b) $n=10$, $r=4$, so $\Omega = 10!/(6! \times 4!) = 210$. A few of these possibilities are shown in Fig. 1.4(b).

If instead we had chosen ten times as many atoms (so $n=100$) and ten times as many quanta, the numbers for (b) would have come out much much bigger. In this case, we would have $r=40$, $\Omega \sim 10^{28}$. A further factor of ten sends these numbers up much further, so for $n=1000$ and $r=400$, $\Omega \sim 10^{290}$ – a staggeringly large number.

The numbers in the above example are so large because factorials increase very quickly. In our example we treated 10 atoms; we are clearly going to run into trouble when we attempt to deal with a mole of atoms, i.e., when $n = 6 \times 10^{23}$.

One way of bringing large numbers down to size is to look at their logarithms.[6] Thus, if Ω is given by eqn 1.15, we could calculate

$$\ln \Omega = \ln(n!) - \ln((n-r)!) - \ln(r!). \tag{1.16}$$

[6] We will use "ln" to signify log to the base e, i.e., $\ln = \log_e$. This is known as the natural logarithm.

This expression involves the logarithm of a factorial, and it is going to be very useful to be able to evaluate this. Most pocket calculators have difficulty in evaluating factorials above 69! (because $70! > 10^{100}$ and many pocket calculators give an overflow error for numbers above 9.999×10^{99}), so some low cunning will be needed to overcome this. Such low cunning is provided by an expression termed **Stirling's formula**:

$$\boxed{\ln n! \approx n \ln n - n.} \tag{1.17}$$

This expression[7] is derived in Appendix C.3.

[7] As shown in Appendix C.3, it is slightly more accurate to use the formula $\ln n! \approx n \ln n - n + \frac{1}{2} \ln 2\pi n$, but this only gives a significant advantage when n is not too large.

Example 1.4

Estimate the order of magnitude of $10^{23}!$.
Solution:
Using Stirling's formula, we can estimate

$$\ln 10^{23}! \approx 10^{23} \ln 10^{23} - 10^{23} = 5.2 \times 10^{24}, \tag{1.18}$$

and hence

$$10^{23}! = \exp(\ln 10^{23}!) \approx \exp(5.20 \times 10^{24}). \tag{1.19}$$

We have our answer in the form e^x, but we would really like it as ten to some power. Now if $e^x = 10^y$, then $y = x/\ln 10$ and hence

$$10^{23}! \approx 10^{2.26 \times 10^{24}}. \tag{1.20}$$

Just pause for a moment to take in how big this number is. It is roughly one followed by about 2.26×10^{24} zeros! Our claim that combinatorial numbers are big seems to be justified!

1.5 Plan of the book

This book aims to introduce the concepts of thermal physics one by one, steadily building up the techniques and ideas that make up the subject. Part I contains various preliminary topics. In Chapter 2 we define heat and introduce the idea of heat capacity. In Chapter 3, the ideas of probability are presented for discrete and continuous distributions. (For

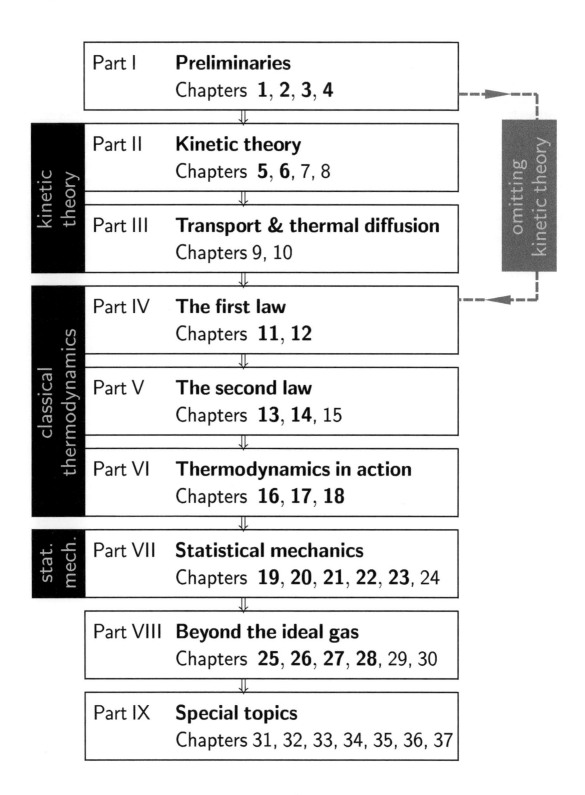

Fig. 1.5 Organization of the book. The dashed line shows a possible route through the material that avoids the kinetic theory of gases. The numbers of the core chapters are given in bold type. The other chapters can be omitted on a first reading, or for a reduced-content course.

a reader familiar with probability theory, this chapter can be omitted.) We then define temperature in Chapter 4, and this allows us to introduce the Boltzmann distribution, which is the probability distribution for systems in contact with a thermal reservoir.

The plan for the remaining parts of the book is sketched in Fig. 1.5. The following two parts contain a presentation of the kinetic theory of gases, which justifies the ideal gas equation from a microscopic model. Part II presents the Maxwell–Boltzmann distribution of molecular speeds in a gas and the derivation of formulae for pressure, molecular effusion, and mean free path. Part III concentrates on transport and thermal diffusion. Parts II and III can be omitted in courses in which kinetic theory is treated at a later stage.

In Part IV, we begin our introduction to mainstream thermodynamics. The concept of energy is covered in Chapter 11, along with the zeroth and first laws of thermodynamics. These are applied to isothermal and adiabatic processes in Chapter 12.

Part V contains the crucial second law of thermodynamics. The idea of a heat engine is introduced in Chapter 13, which leads to various statements of the second law of thermodynamics. Hence the important concept of entropy is presented in Chapter 14 and its application to information theory is discussed in Chapter 15.

Part VI introduces the rest of the machinery of thermodynamics. Various thermodynamic potentials, such as the enthalpy, Helmholtz function, and Gibbs function, are introduced in Chapter 16, and their usage illustrated. Thermal systems include not only gases, and Chapter 17 looks at other possible systems, such as elastic rods and magnetic systems. The third law of thermodynamics is described in Chapter 18 and provides a deeper understanding of how entropy behaves as the temperature is reduced to absolute zero.

Part VII focuses on statistical mechanics. Following a discussion of the equipartition of energy in Chapter 19, so useful for understanding high temperature limits, the concept of the partition function is presented in some detail in Chapter 20, which is foundational for understanding statistical mechanics. The idea is applied to the ideal gas in Chapter 21. Particle number becomes important when considering different types of particle, so the chemical potential and grand partition function are presented in Chapter 22. Two simple applications where the chemical potential is zero are photons and phonons, discussed in Chapters 23 and 24 respectively.

The discussion up to this point has concentrated on the ideal gas model and we go beyond this in Part VIII: Chapter 25 discusses the effect of relativistic velocities and Chapters 26 and 27 discuss the effect of intermolecular interactions, while phase transitions are discussed in Chapter 28, where the important Clausius–Clapeyron equation for a phase boundary is derived. Another quantum mechanical implication is the existence of identical particles and the difference between fermions and bosons, discussed in Chapter 29; the consequences for the properties of quantum gases are presented in Chapter 30.

The remainder of the book, Part IX, contains more detailed information on various special topics which allow the power of thermal physics to be demonstrated. In Chapters 31 and 32 we describe sound waves and shock waves in fluids. We draw some of the statistical ideas of the book together in Chapter 33 and discuss non-equilibrium thermodynamics and the arrow of time in Chapter 34. Applications of the concepts in the book to astrophysics are described in Chapters 35 and 36 and to atmospheric physics in Chapter 37.

Chapter summary

- In this chapter, the idea of big numbers has been introduced. These arise in thermal physics for two main reasons:
    (1) The number of atoms in a typical macroscopic lump of matter is large. It is measured in the units of the mole. One mole of atoms contains N_A atoms, where $N_A = 6.022 \times 10^{23}$.
    (2) Combinatorial problems generate very large numbers. To make these numbers manageable, we often consider their logarithms and use Stirling's approximation: $\ln n! \approx n \ln n - n$.

Exercises

(1.1) What is the mass of 3 moles of carbon dioxide (CO_2)? (1 mole of oxygen atoms has a mass of 16 g.)

(1.2) A typical bacterium has a mass of 10^{-12} g. Calculate the mass of a mole of bacteria. (Interestingly, this is about the total number of bacteria living in the guts of all humans resident on planet Earth.) Give your answer in units of elephant-masses (elephants have a mass ≈ 5000 kg).

(1.3) (a) How many water molecules are there in your body? (Assume that you are nearly all water.)
(b) How many drops of water are there in all the oceans of the world? (The mass of the world's oceans is about 10^{21} kg. Estimate the size of a typical drop of water.)
(c) Which of these two numbers from (a) and (b) is the larger?

(1.4) A system contains n atoms, each of which can only have zero or one quanta of energy. How many ways can you arrange r quanta of energy when (a) $n = 2$, $r = 1$; (b) $n = 20$, $r = 10$; (c) $n = 2 \times 10^{23}$, $r = 10^{23}$?

(1.5) What fractional error do you make when using Stirling's approximation (in the form $\ln n! \approx n \ln n - n$) to evaluate

  (a) $\ln 10!$,
  (b) $\ln 100!$, and
  (c) $\ln 1000!$?

(1.6) Show that eqn C.19 is equivalent to writing

$$n! \approx n^n e^{-n} \sqrt{2\pi n}, \qquad (1.21)$$

and

$$n! \approx \sqrt{2\pi} n^{n+\frac{1}{2}} e^{-n}. \qquad (1.22)$$

Heat

2

In this chapter, we will introduce the concepts of heat and heat capacity.

2.1 A definition of heat	13
2.2 Heat capacity	14
Chapter summary	17
Exercises	17

2.1 A definition of heat

We all have an intuitive notion of what heat is: sitting next to a roaring fire in winter, we *feel* its heat warming us up, increasing our temperature; lying outside in the sunshine on a warm day, we *feel* the Sun's heat warming us up. In contrast, holding a snowball, we feel heat leaving our hand and transferring to the snowball, making our hand feel cold. Heat seems to be some sort of energy transferred from hot things to cold things when they come into contact. We therefore make the following definition:

> **heat** is *thermal energy in transit.*

We now stress a couple of important points about this definition.

(1) Experiments suggest that heat spontaneously transfers from a hotter body to a colder body when they are in contact, and not in the reverse direction. However, there are circumstances when it is possible for heat to go in the reverse direction. A good example of this is a kitchen freezer: you place food, initially at room temperature, into the freezer and shut the door; the freezer then sucks heat out of the food and cools the food down to below freezing point. Heat is being transferred from your colder freezer to the warmer kitchen, apparently in the "wrong" direction. Of course, to achieve this, you have to be paying your electricity bill and therefore be putting energy in to your freezer. If there is a power cut, heat will slowly leak back into the freezer from the warmer kitchen and thaw out all your frozen food. This shows that it is possible to reverse the direction of heat flow, but only if you intervene by putting additional energy in. We will return to this point in Section 13.5 when we consider refrigerators, but for now let us note that we are defining heat as *thermal energy in transit* and not hard-wiring into the definition anything about which direction it goes.

(2) The "in transit" part of our definition is very important. Though you can add heat to an object, you *cannot* say that "an object contains a certain quantity of heat." This is very different from

the case of the fuel in your car: you can add fuel to your car, and you are quite entitled to say that your car "contains a certain quantity of fuel". You even have a gauge for measuring it! But heat is quite different. Objects do not and cannot have gauges which read out how much heat they contain, because heat only makes sense when it is "in transit".[1]

To see this, consider your cold hands on a chilly winter day. You can increase the temperature of your hands in two different ways: (i) by adding heat, for example by putting your hands close to something hot, like a roaring fire; (ii) by rubbing your hands together. In one case you have added heat from the outside, in the other case you have not added any heat but have done some work.[2] In both cases, you end up with the same final situation: hands that have increased in temperature. There is no physical difference between hands that have been warmed by heat and hands that have been warmed by work.[3]

Heat is measured in joules (J). The rate of heating has the units of watts (W), where $1\,\text{W} = 1\,\text{J}\,\text{s}^{-1}$ (i.e., 1 watt=1 joule per second).

[1] We will see later that objects can contain a certain quantity of *energy*, so it is possible, at least in principle, to have a gauge that reads out how much energy is contained.

[2] Work is also a type of energy in transit, since you always do work *on* something. For example you do work on a mass by lifting it a height h. We could define work as "mechanical energy in transit". We will explore how work and heat can be interchanged in Chapter 13.

[3] We have made this point by giving a plausible example, but in Chapter 11 we will show using more mathematical arguments that heat only makes sense as energy "in transit".

Example 2.1

A 1 kW electric heater is switched on for ten minutes. How much heat does it produce?
Solution:
Ten minutes equals 600 s, so the heat Q is given by

$$Q = 1\,\text{kW} \times 600\,\text{s} = 600\,\text{kJ}. \tag{2.1}$$

Notice in this last example that the power in the heater is supplied by electrical work. Thus it is possible to produce heat by doing work. We will return to the question of whether one can produce work from heat in Chapter 13.

2.2 Heat capacity

In the previous section, we explained that it is not possible for an object to contain a certain quantity of heat, because heat is defined as "thermal energy in transit". It is therefore with a somewhat heavy heart that we turn to the topic of "heat capacity", since we have argued that objects have no capacity for heat! (This is one of those occasions in physics when decades of use of a name have made it completely standard, even though it is really a misleading name to use.) What we are going to derive in this section might be better termed "energy capacity", but to do this would put us at odds with common usage throughout physics. All of this being said, we can proceed quite legitimately by asking the following simple question:

How much heat needs to be supplied to an object to raise its temperature by a small amount dT?

The answer to this question is the heat $dQ = C\,dT$, where we define the **heat capacity** C of an object using

$$C = \frac{dQ}{dT}. \tag{2.2}$$

As long as we remember that heat capacity tells us simply how much heat is needed to warm an object (and is nothing about the capacity of an object for heat) we shall be on safe ground. As can be inferred from eqn 2.2, the heat capacity C has units $J\,K^{-1}$.

As shown in the following example, although objects have a heat capacity, one can also express the heat capacity of a particular substance *per unit mass*, or *per unit volume*.[4]

[4] We will use the symbol C to represent a heat capacity, whether of an object, or per unit volume, or per mole. We will always state which is being used. The heat capacity per unit mass is distinguished by the use of the lower-case symbol c. We will usually reserve the use of subscripts on the heat capacity to denote the constraint being applied (see eqns 2.6 and 2.7).

Example 2.2
The heat capacity of $0.125\,\text{kg}$ of water is measured to be $523\,\text{J}\,\text{K}^{-1}$ at room temperature. Hence calculate the heat capacity of water (a) per unit mass and (b) per unit volume.
Solution:
(a) The heat capacity per unit mass c is given by dividing the heat capacity by the mass, and hence

$$c = \frac{523\,\text{J}\,\text{K}^{-1}}{0.125\,\text{kg}} = 4.184 \times 10^3\,\text{J}\,\text{K}^{-1}\,\text{kg}^{-1}. \tag{2.3}$$

(b) The heat capacity per unit volume C is obtained by multiplying the previous answer by the density of water, namely $1000\,\text{kg}\,\text{m}^{-3}$, so that

$$C = 4.184 \times 10^3\,\text{J}\,\text{K}^{-1}\,\text{kg}^{-1} \times 1000\,\text{kg}\,\text{m}^{-3} = 4.184 \times 10^6\,\text{J}\,\text{K}^{-1}\,\text{m}^{-3}. \tag{2.4}$$

The heat capacity per unit mass c occurs quite frequently, and it is given a special name: the **specific heat capacity**.

Example 2.3
Calculate the specific heat capacity of water.
Solution:
This is given in answer (a) from the previous example: the specific heat capacity of water is $4.184 \times 10^3\,\text{J}\,\text{K}^{-1}\,\text{kg}^{-1}$.

Also useful is the **molar heat capacity**, which is the heat capacity of one mole of the substance.

Example 2.4

Calculate the molar heat capacity of water. (The molar mass of water is 18 g.)
Solution:
The molar heat capacity is obtained by multiplying the specific heat capacity by the molar mass, and hence

$$C = 4.184 \times 10^3 \,\mathrm{J\,K^{-1}\,kg^{-1}} \times 0.018\,\mathrm{kg} = 75.2\,\mathrm{J\,K^{-1}\,mol^{-1}}. \quad (2.5)$$

When we think about the heat capacity of a gas, there is a further complication.[5] We are trying to ask the question: how much heat should you add to raise the temperature of our gas by one kelvin? But we can imagine doing the experiment in two ways (see also Fig. 2.1):

[5] This complication is there for liquids and solids, but doesn't make such a big difference.

(1) Place our gas in a sealed box and add heat (Fig. 2.1(a)). As the temperature rises, the gas will not be allowed to expand because its volume is fixed, so its pressure will increase. This method is known as heating *at constant volume*.

(2) Place our gas in a chamber connected to a piston and heat it (Fig. 2.1(b)). The piston is well lubricated, and so will slide in and out to maintain the pressure in the chamber to be identical to that in the lab. As the temperature rises, the piston is forced out (doing work against the atmosphere) and the gas is allowed to expand, keeping its pressure constant. This method is known as heating *at constant pressure*.

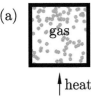

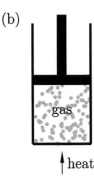

Fig. 2.1 Two methods of heating a gas: (a) constant volume, (b) constant pressure.

In both cases, we are applying a **constraint** to the system, either constraining the volume of the gas to be fixed, or constraining the pressure of the gas to be fixed. We need to modify our definition of heat capacity given in eqn 2.2, and hence we define two new quantities: C_V is the heat capacity *at constant volume* and C_p is the heat capacity *at constant pressure*. We can write them using partial differentials as follows:

$$C_V = \left(\frac{\partial Q}{\partial T}\right)_V, \quad (2.6)$$

$$C_p = \left(\frac{\partial Q}{\partial T}\right)_p. \quad (2.7)$$

We expect that C_p will be bigger than C_V for the simple reason that more heat will need to be added when heating at constant pressure than when heating at constant volume. This is because in the former case additional energy will be expended on doing work on the atmosphere as the gas expands. It turns out that indeed C_p is bigger than C_V in practice.[6]

[6] We will *calculate* the relative sizes of C_V and C_p in Section 11.3.

Example 2.5

The specific heat capacity of helium gas is measured to be $3.12\,\mathrm{kJ\,K^{-1}\,kg^{-1}}$ at constant volume and $5.19\,\mathrm{kJ\,K^{-1}\,kg^{-1}}$ at constant pressure. Calculate the molar heat capacities. (The molar mass of helium is $4\,\mathrm{g}$.)

Solution:
The molar heat capacity is obtained by multiplying the specific heat capacity by the molar mass, and hence

$$C_V = 12.48\,\mathrm{J\,K^{-1}\,mol^{-1}}, \tag{2.8}$$
$$C_p = 20.76\,\mathrm{J\,K^{-1}\,mol^{-1}}. \tag{2.9}$$

(Interestingly, these answers are almost exactly $\frac{3}{2}R$ and $\frac{5}{2}R$ where R is the gas constant.[7] We will see why in Section 11.3.)

[7] $R = 8.31447\,\mathrm{J\,K^{-1}\,mol^{-1}}$ is known as the gas constant and is equal to the product of the Avogadro number N_A and the Boltzmann constant k_B (see Section 6.2).

Chapter summary

- In this chapter, the concepts of heat and heat capacity have been introduced.
- Heat is "thermal energy in transit".
- The heat capacity C of an object is given by $C = \mathrm{d}Q/\mathrm{d}T$. The heat capacity of a substance can also be expressed per unit volume or per unit mass (in the latter case it is called *specific heat capacity*).

Exercises

(2.1) Using data from this chapter, estimate the energy needed to (a) boil enough tap water to make a cup of tea, (b) heat the water for a bath.

(2.2) The world's oceans contain approximately $10^{21}\,\mathrm{kg}$ of water. Estimate the total heat capacity of the world's oceans.

(2.3) The world's power consumption is currently about $13\,\mathrm{TW}$, and growing! ($1\,\mathrm{TW} = 10^{12}\,\mathrm{W}$.) Burning one ton of crude oil (which is nearly seven barrels worth) produces about $42\,\mathrm{GJ}$ ($1\,\mathrm{GJ} = 10^9\,\mathrm{J}$). If the world's total power needs were to come from burning oil (a large fraction currently does), how much oil would we be burning per second?

(2.4) The molar heat capacity of gold is $25.4\,\mathrm{J\,mol^{-1}\,K^{-1}}$. Its density is $19.3\times10^3\,\mathrm{kg\,m^{-3}}$. Calculate the specific heat capacity of gold and the heat capacity per unit volume. What is the heat capacity of $4\times 10^6\,\mathrm{kg}$ of gold? (This is roughly the holdings of Fort Knox.)

(2.5) Two bodies, with heat capacities C_1 and C_2 (assumed independent of temperature) and initial temperatures T_1 and T_2 respectively, are placed in thermal contact. Show that their final temperature T_f is given by $T_\mathrm{f} = (C_1 T_1 + C_2 T_2)/(C_1 + C_2)$. If C_1 is much larger than C_2, show that $T_\mathrm{f} \approx T_1 + C_2(T_2 - T_1)/C_1$.

3 Probability

- 3.1 Discrete probability distributions 19
- 3.2 Continuous probability distributions 20
- 3.3 Linear transformation 21
- 3.4 Variance 22
- 3.5 Linear transformation and the variance 23
- 3.6 Independent variables 24
- 3.7 Binomial distribution 26
- Chapter summary 28
- Further reading 29
- Exercises 29

Life is full of uncertainties, and has to be lived according to our best guesses based on the information available to us. This is because the chain of events that lead to various outcomes can be so complex that the exact outcomes are unpredictable. Nevertheless, things can still be said even in an uncertain world: for example, it is more helpful to know that there is a 20% chance of rain tomorrow than that the weather forecaster has absolutely no idea; or worse still that he or she claims that there will definitely be no rain, when there might be! Probability is therefore an enormously useful and powerful subject, since it can be used to *quantify* uncertainty.

The foundations of probability theory were laid by the French mathematicians Pierre de Fermat (1601–1665) and Blaise Pascal (1623–1662), through their correspondence in 1654, which originated from a problem set to them by a gentleman gambler. The ideas proved to be intellectually infectious and the first probability textbook was written by the Dutch physicist Christian Huygens (1629–1695) in 1657, who applied it to the working out of life expectancy. Probability was thought to be useful only for determining possible outcomes in situations in which we lacked complete knowledge. The supposition was that if we could know the motions of all particles at the microscopic level, we could determine every outcome precisely. In the twentieth century, the discovery of quantum theory has led to the understanding that, at the microscopic level, outcomes are purely probabilistic.

Probability has had a huge impact on thermal physics. This is because we are often interested in systems containing huge numbers of particles, so that predictions based on probability turn out to be precise enough for most purposes. In a thermal physics problem, one is often interested in the values of quantities that are the sum of many small contributions from individual atoms. Though each atom behaves differently, the *average* behaviour is what comes through, and therefore it becomes necessary to be able to extract average values from probability distributions.

In this chapter, we will define some basic concepts in probability theory. Let us begin by stating that the probability of occurrence of a particular event, taken from a finite set of possible events, is zero if that event is impossible, is one if that event is certain, and takes a value somewhere in between zero and one if that event is possible but not certain. We begin by considering two different types of probability distribution: *discrete* and *continuous*.

3.1 Discrete probability distributions

Discrete random variables can only take a finite number of values. Examples include the number obtained when throwing a die (1, 2, 3, 4, 5, or 6), the number of children in each family (0, 1, 2, ...), and the number of people killed per year in the UK in bizarre gardening accidents (0, 1, 2, ...). Let x be a **discrete random variable** which takes values x_i with probability P_i. We require that the sum of the probabilities of every possible outcome adds up to one. This may be written

$$\sum_i P_i = 1. \tag{3.1}$$

We define the **mean** (or **average** or **expected value**) of x to be

$$\langle x \rangle = \sum_i x_i P_i. \tag{3.2}$$

The idea is that you weight by its probability each value taken by the random variable x.

Alternative notations for the mean of x include $\bar{x}$ and $E(x)$. We prefer the one given in the main text since it is easier to distinguish quantities such as $\langle x^2 \rangle$ and $\langle x \rangle^2$ with this notation, particularly when writing quickly.

Example 3.1

Note that the mean, $\langle x \rangle$, may be a value that x cannot actually take. A common example of this is the number of children in families, which is often quoted as 2.4. Any individual couple can only have an integer number of children. Thus the expected value of x is actually an impossibility!

It is also possible to define the mean squared value of x using

$$\langle x^2 \rangle = \sum_i x_i^2 P_i. \tag{3.3}$$

In fact, any function of x can be averaged, using (by analogy)

$$\langle f(x) \rangle = \sum_i f(x_i) P_i. \tag{3.4}$$

Now let us actually evaluate the mean of x for a particular discrete distribution.

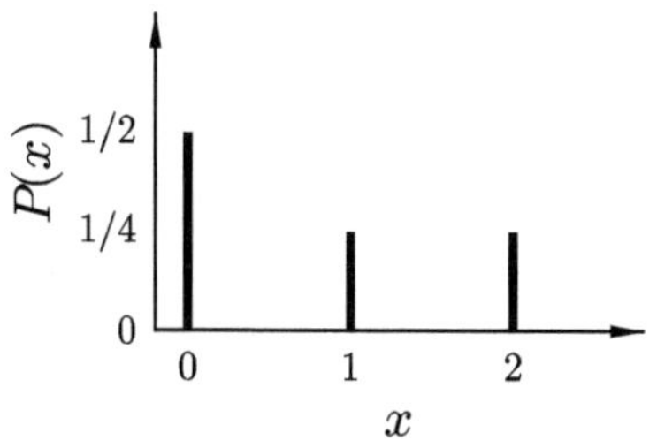

Example 3.2

Let x take values 0, 1, and 2 with probabilities $\frac{1}{2}$, $\frac{1}{4}$, and $\frac{1}{4}$ respectively. This distribution is shown in Fig. 3.1. Calculate $\langle x \rangle$ and $\langle x^2 \rangle$.

Fig. 3.1 An example of a discrete probability distribution.

Solution:
First check that $\sum P_i = 1$. Since $\frac{1}{2} + \frac{1}{4} + \frac{1}{4} = 1$, this is fine. Now we can calculate the averages as follows:

$$\begin{aligned} \langle x \rangle &= \sum_i x_i P_i \\ &= 0 \cdot \frac{1}{2} + 1 \cdot \frac{1}{4} + 2 \cdot \frac{1}{4} \\ &= \frac{3}{4}. \end{aligned} \quad (3.5)$$

Again, we find that the mean $\langle x \rangle$ is not actually one of the possible values of x. We can now calculate the value of $\langle x^2 \rangle$ as follows:

$$\begin{aligned} \langle x^2 \rangle &= \sum_i x_i^2 P_i \\ &= 0 \cdot \frac{1}{2} + 1 \cdot \frac{1}{4} + 4 \cdot \frac{1}{4} \\ &= \frac{5}{4}. \end{aligned} \quad (3.6)$$

3.2 Continuous probability distributions

Let x now be a **continuous random variable**[1] which has a probability $P(x)\,\mathrm{d}x$ of having a value between x and $x + \mathrm{d}x$. Continuous random variables can take a range of possible values. Examples include the height of children in a class, the length of time spent in a waiting room, and the amount a person's blood pressure increases when reading their mobile-phone bill. These quantities are not restricted to any finite set of values, but can take a continuous set of values.

As before, we require that the total probability of all possible outcomes is one. Because we are dealing with continuous distributions, the sums become integrals, and we have

$$\int P(x)\,\mathrm{d}x = 1. \quad (3.7)$$

The mean is defined as

$$\langle x \rangle = \int x\,P(x)\,\mathrm{d}x. \quad (3.8)$$

Similarly, the mean square value is defined as

$$\langle x^2 \rangle = \int x^2\,P(x)\,\mathrm{d}x, \quad (3.9)$$

and the mean of any function of x, $f(x)$, can be defined as

$$\langle f(x) \rangle = \int f(x)\,P(x)\,\mathrm{d}x. \quad (3.10)$$

[1] For a continuous random variable, there are an infinite number of possible values it can take, so the probability of any one of them occurring is zero! Hence we talk about the probability of the variable lying in some range, such as "between x and $x + \mathrm{d}x$".

Example 3.3

Let $P(x) = Ce^{-x^2/2a^2}$ where C and a are constants. This probability is illustrated in Fig. 3.2 and this curve is known as a **Gaussian**.[2] Calculate $\langle x \rangle$ and $\langle x^2 \rangle$ given this probability distribution.

[2]See Appendix C.2.

Solution:

The first thing to do is to normalize the probability distribution (i.e., to ensure that the sum over all probabilities is one). This allows us to find the constant C using eqn C.3 to evaluate the integral:

$$\begin{aligned} 1 = \int_{-\infty}^{\infty} P(x)\,dx &= C\int_{-\infty}^{\infty} e^{-x^2/2a^2}\,dx \\ &= C\sqrt{2\pi a^2}, \end{aligned} \quad (3.11)$$

so we find that $C = 1/\sqrt{2\pi a^2}$, which gives

$$P(x) = \frac{1}{\sqrt{2\pi a^2}} e^{-x^2/2a^2}. \quad (3.12)$$

The mean of x can then be evaluated using

$$\begin{aligned} \langle x \rangle &= \frac{1}{\sqrt{2\pi a^2}} \int_{-\infty}^{\infty} x\, e^{-x^2/2a^2}\,dx \\ &= 0, \end{aligned} \quad (3.13)$$

because the integrand is an odd function. The mean of x^2 can also be evaluated as follows:

$$\begin{aligned} \langle x^2 \rangle &= \frac{1}{\sqrt{2\pi a^2}} \int_{-\infty}^{\infty} x^2\, e^{-x^2/2a^2}\,dx \\ &= \frac{1}{\sqrt{2\pi a^2}} \frac{1}{2}\sqrt{8\pi a^6} \\ &= a^2, \end{aligned} \quad (3.14)$$

where the integrals are performed as described in Appendix C.2.

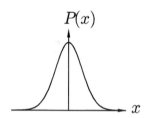

Fig. 3.2 An example continuous probability distribution.

3.3 Linear transformation

Sometimes one has a random variable, and one wants to make a second random variable by performing a linear transformation on the first one. If y is a random variable, which is related to the random variable x by the equation

$$y = ax + b, \quad (3.15)$$

where a and b are constants, then the average value of y is given by

$$\langle y \rangle = \langle ax + b \rangle = a\langle x \rangle + b. \quad (3.16)$$

The proof of this result is straightforward and is left as an exercise.

Example 3.4

Temperatures in degrees Celsius and degrees Fahrenheit are related by the simple formula $C = \frac{5}{9}(F - 32)$, where C is the temperature in degrees Celsius and F the temperature in degrees Fahrenheit. Hence the average temperature of a particular temperature distribution is $\langle C \rangle = \frac{5}{9}(\langle F \rangle - 32)$. The average annual temperature in New York Central Park is $54\,°\text{F}$. One can convert this to Celsius using the formula above to get $\approx 12\,°\text{C}$.

3.4 Variance

We now know how to calculate the average of a set of values, but what about the spread in the values? The first idea one might have to quantify the spread of values in a distribution is to consider the **deviation** from the mean for a particular value of x. This is defined by

$$x - \langle x \rangle. \tag{3.17}$$

This quantity tells you by how much a particular value is above or below the mean value. We can work out the average of the deviation (averaging over all values of x) as follows:

$$\langle x - \langle x \rangle \rangle = \langle x \rangle - \langle x \rangle = 0, \tag{3.18}$$

which follows from the equation for linear transformation (eqn 3.16). Thus the average deviation is not going to be a very helpful indicator! Of course, the problem is that the deviation is sometimes positive and sometimes negative, and the positive and negative deviations cancel out. A more useful quantity would be the modulus of the deviation,

$$|x - \langle x \rangle|, \tag{3.19}$$

which is always positive, but this will suffer from the disadvantage that modulus signs in algebra can be both confusing and tedious. Therefore, another approach is to use a different quantity which is always positive, the square of the deviation, $(x - \langle x \rangle)^2$. This quantity is what we need: always positive and easy to manipulate algebraically. Hence, its average is given a special name, the **variance**. Consequently the variance of x, written as σ_x^2, is defined as the *mean squared deviation*:[3]

$$\sigma_x^2 = \langle (x - \langle x \rangle)^2 \rangle. \tag{3.20}$$

We will call σ_x the **standard deviation**, and it is defined as the square root of the variance:

$$\sigma_x = \sqrt{\langle (x - \langle x \rangle)^2 \rangle}. \tag{3.21}$$

---

[3]In fact, in general we can define the kth **moment** about the mean as $\langle (x - \langle x \rangle)^k \rangle$. The first moment about the mean is the mean deviation, and it is zero, as we have seen. The second moment about the mean is the variance. The third moment about the mean is known as the skewness parameter, and sometimes turns out to be useful. The fourth moment about the mean is called the kurtosis.

The standard deviation represents the "root mean square" (known as the "rms") scatter or spread in the data.

The following identity is extremely useful:

$$\begin{aligned} \sigma_x^2 &= \langle (x - \langle x \rangle)^2 \rangle \\ &= \langle x^2 - 2x\langle x \rangle + \langle x \rangle^2 \rangle \\ &= \langle x^2 \rangle - 2\langle x \rangle \langle x \rangle + \langle x \rangle^2 \\ &= \langle x^2 \rangle - \langle x \rangle^2. \end{aligned} \quad (3.22)$$

Example 3.5

For Examples 3.2 and 3.3 above, work out σ_x^2, the variance of the distribution, in each case.

Solution:

For Example 3.2

$$\sigma_x^2 = \langle x^2 \rangle - \langle x \rangle^2 = \frac{5}{4} - \frac{9}{16} = \frac{11}{16}. \quad (3.23)$$

For Example 3.3

$$\sigma_x^2 = \langle x^2 \rangle - \langle x \rangle^2 = a^2 - 0 = a^2. \quad (3.24)$$

3.5 Linear transformation and the variance

We return to the problem of a linear transformation of a random variable. What happens to the variance in this case?

If y is a random variable which is related to the random variable x by the equation

$$y = ax + b, \quad (3.25)$$

where a and b are constants, then we have seen that

$$\langle y \rangle = \langle ax + b \rangle = a\langle x \rangle + b. \quad (3.26)$$

Hence, we can work out $\langle y^2 \rangle$, which is

$$\begin{aligned} \langle y^2 \rangle &= \langle (ax + b)^2 \rangle \\ &= \langle a^2 x^2 + 2abx + b^2 \rangle \\ &= a^2 \langle x^2 \rangle + 2ab\langle x \rangle + b^2. \end{aligned} \quad (3.27)$$

Also, we can work out $\langle y \rangle^2$, which is

$$\langle y \rangle^2 = (a\langle x \rangle + b)^2 = a^2 \langle x \rangle^2 + 2ab\langle x \rangle + b^2. \quad (3.28)$$

Hence, using eqn 3.22, the variance in y is given by eqn 3.27 minus eqn 3.28, i.e.

$$\begin{aligned}\sigma_y^2 &= \langle y^2\rangle - \langle y\rangle^2 \\ &= a^2\langle x^2\rangle - a^2\langle x\rangle^2 \\ &= a^2\sigma_x^2.\end{aligned} \qquad (3.29)$$

Notice that the variance depends on a but not on b. This makes sense because the variance tells us about the width of a distribution, and nothing about its absolute position. The standard deviation of y is therefore given by

$$\sigma_y = a\sigma_x. \qquad (3.30)$$

Example 3.6
The average temperature in a town in the USA in January is $23\,°\mathrm{F}$ and the standard deviation is $9\,°\mathrm{F}$. Convert these figures into degrees Celsius using the relation in Example 3.4.
Solution:
The average temperature in degrees Celsius is given by

$$\langle C\rangle = \frac{5}{9}(\langle F\rangle - 32) = \frac{5}{9}(23 - 32) = -5\,°\mathrm{C}, \qquad (3.31)$$

and the standard deviation is given by $\frac{5}{9}\times 9 = 5\,°\mathrm{C}$.

3.6 Independent variables

If u and v are **independent random variables**[4] the probability that u is in the range from u to $u+\mathrm{d}u$ and v is in the range from v to $v+\mathrm{d}v$ is given by the product

$$P_u(u)\mathrm{d}u\, P_v(v)\mathrm{d}v. \qquad (3.32)$$

Hence, the average value of the product of u and v is

$$\begin{aligned}\langle uv\rangle &= \iint uv P_u(u)P_v(v)\,\mathrm{d}u\,\mathrm{d}v \\ &= \int u P_u(u)\,\mathrm{d}u \int v P_v(v)\,\mathrm{d}v \\ &= \langle u\rangle\langle v\rangle,\end{aligned} \qquad (3.33)$$

because the integrals separate for *independent* random variables. This implies that the average value of the product of u and v is equal to the product of their average values.

---

[4] Two random variables are independent if knowing the value of one of them yields no information about the value of the other. For example, the height of a person chosen at random from a city and the number of hours of rainfall in that city on the first Tuesday of September are two independent random variables.

Example 3.7

Suppose that there are n independent random variables, X_i, each with the same mean $\langle X \rangle$ and variance σ_X^2. Let Y be the sum of the random variables, so that $Y = X_1 + X_2 + \cdots + X_n$. Find the mean and variance of Y.

Solution:
The mean of Y is simply

$$\langle Y \rangle = \langle X_1 \rangle + \langle X_2 \rangle + \cdots + \langle X_n \rangle, \qquad (3.34)$$

but since all the X_i have the same mean $\langle X \rangle$ this can be written

$$\langle Y \rangle = n \langle X \rangle. \qquad (3.35)$$

Hence the mean of Y is n times the mean of the X_i. To find the variance of Y, we can use the formula

$$\sigma_Y^2 = \langle Y^2 \rangle - \langle Y \rangle^2. \qquad (3.36)$$

Hence

$$\begin{aligned}\langle Y^2 \rangle &= \langle X_1^2 + \cdots + X_N^2 + X_1 X_2 + X_2 X_1 + X_1 X_3 + \cdots \rangle \qquad (3.37) \\ &= \langle X_1^2 \rangle + \cdots + \langle X_N^2 \rangle + \langle X_1 X_2 \rangle + \langle X_2 X_1 \rangle + \langle X_1 X_3 \rangle + \cdots \end{aligned}$$

There are n terms like $\langle X_1^2 \rangle$ on the right-hand side, and $n(n-1)$ terms like $\langle X_1 X_2 \rangle$. The former terms take the value $\langle X^2 \rangle$ and the latter terms (because they are the product of two independent random variables) take the value $\langle X \rangle \langle X \rangle = \langle X \rangle^2$. Hence, using eqn 3.35,

$$\langle Y^2 \rangle = n \langle X^2 \rangle + n(n-1) \langle X \rangle^2, \qquad (3.38)$$

so that

$$\begin{aligned}\sigma_Y^2 &= \langle Y^2 \rangle - \langle Y \rangle^2 \\ &= n \langle X^2 \rangle - n \langle X \rangle^2 \\ &= n \sigma_X^2. \qquad (3.39)\end{aligned}$$

The results proved in this last example have some interesting applications. The first concerns experimental measurements. Imagine that a quantity X is measured n times, each time with an independent error, which we call σ_X. If you add up the results of the measurements to make $Y = \sum X_i$, then the rms error in Y is only $\sqrt{n}$ times the rms error of a single X. Hence if you try and get a good estimate of X by calculating $(\sum X_i)/n$, the error in this quantity is equal to $\sigma_X/\sqrt{n}$. Thus, for example, if you make four measurements of a quantity and average your results, the random error in your average is half of what it would be if you'd just taken a single measurement. Of course, you may still have *systematic* errors in your experiment. If you are consistently overestimating your quantity by an error in your experimental setup, that error won't reduce by repeated measurement!

A second application is in the theory of **random walks**. Imagine a drunken person staggering out of a pub and attempting to walk along a narrow street (which confines him or her to motion in one dimension). Let's pretend that with each inebriated step, the drunken person is equally likely to travel one step forwards or one step backwards. The effects of intoxication are such that each step is uncorrelated with the previous one. Thus the average distance travelled in a single step is $\langle X \rangle = 0$. After n such steps, we would have an expected total distance travelled of $\langle Y \rangle = \sum \langle X_i \rangle = 0$. However, in this case the root mean squared distance is more revealing. In this case $\langle Y^2 \rangle = n \langle X^2 \rangle$, so that the rms length of a random walk of n steps is $\sqrt{n}$ times the length of a single step. This result will be useful in considering Brownian motion in Chapter 33.

3.7 Binomial distribution

A probability distribution, which is very important in thermal physics, is based on what is called a **Bernoulli trial**,[5] an "experiment" with two possible outcomes. One outcome (which we will call "success") occurs with probability p and the other outcome (which we will call "failure") occurs with probability $1 - p$. An example of a Bernoulli trial is the tossing of a coin: one outcome is "heads", the other is "tails".

[5] Jacob Bernoulli (1654–1705).

Example 3.8

Let x be a random variable which takes the value 1 for success and 0 for failure. Then, assuming p to be the probability of success and using eqns 3.2, 3.3 and 3.21

$$\langle x \rangle = 0 \times (1-p) + 1 \times p = p \quad (3.40)$$
$$\langle x^2 \rangle = 0^2 \times (1-p) + 1^2 \times p = p \quad (3.41)$$
$$\sigma_x = \sqrt{\langle x^2 \rangle - \langle x \rangle^2} = \sqrt{p(1-p)}. \quad (3.42)$$

The **binomial distribution** is the discrete probability distribution $P(n, k)$ of getting k successes from n independent Bernoulli trials. The function $P(n, k)$ can be worked out by realizing that (a) the probability of a particular series of k successes and $n - k$ failures is $p^k(1-p)^{n-k}$ and (b) that there are ^nC_k ways of arranging k successes and $n - k$ failures in a sequence. Thus $P(n, k)$ is a product of these factors and hence

$$P(n, k) = {}^nC_k \, p^k (1-p)^{n-k}. \quad (3.43)$$

The **binomial theorem** of elementary algebra states that

$$(x+y)^n = \sum_{k=0}^{n} {}^nC_k \, x^k y^{n-k}. \quad (3.44)$$

Hence by writing $x = p$ and $y = 1 - p$ we can easily show that

$$\sum_{k=1}^{n} P(n,k) = 1, \quad (3.45)$$

as required for a well-behaved probability distribution. Since the binomial distribution is the sum of n *independent* Bernoulli trials, then

$$\langle k \rangle = np \quad (3.46)$$
$$\sigma_k^2 = np(1-p). \quad (3.47)$$

The **fractional width** of the distribution[6] is obtained by dividing the standard deviation by the mean and is given by $\sigma_k/\langle k \rangle = \sqrt{(1-p)/np}$, which is proportional to $1/\sqrt{n}$, and therefore decreases as n increases. This causes the binomial distribution to become more sharply peaked near the mean value as n increases, as shown in Fig. 3.3.

[6]The mean, $\langle k \rangle$ is proportional to n. The standard deviation σ_k is proportional to $\sqrt{n}$. Both quantities increase with n, but the mean increases faster. The fractional width is the width of the distribution (the standard deviation) divided by the mean, and so decreases with n because the mean increases faster than the standard deviation.

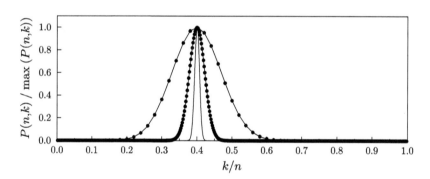

Fig. 3.3 Binomial probability for $p = 0.4$. The three plots are for $n = 50$ (outermost), $n = 500$ and $n = 5000$ (innermost) and are scaled so that their maximum amplitudes are the same. This demonstrates that as n increases, the *fractional width* decreases.

Example 3.9

Coin tossing with a fair coin. In this case, $p = \frac{1}{2}$.

- For $n = 16$ tosses, the expected number of heads is $np = 8$. The standard deviation is $\sqrt{np(1-p)} = 2$, a quarter of the expected number.
- For $n = 10^{20}$ tosses, the expected number of heads is $np = 5 \times 10^{19}$. The standard deviation is $\sqrt{np(1-p)} = 5 \times 10^9$, ten orders of magnitude smaller than the expected number.

Example 3.10

A one-dimensional random walk can be considered as a succession of n Bernoulli trials in which the choice is either a step forwards $+L$ or a step backwards $-L$, each with equal probability (so $p = \frac{1}{2}$). If there are n steps, k of which are forwards, the distance travelled is $x = kL - (n-k)L = (2k-n)L$. For a binomial distribution with $p = \frac{1}{2}$, $\langle k \rangle = \frac{n}{2}$, and $\sigma_k^2 = \langle k^2 \rangle - \langle k \rangle^2 = np(1-p) = \frac{n}{4}$. This implies that $\langle k^2 \rangle = \frac{n}{4} + \frac{n^2}{4}$. Hence, the mean distance travelled is

$$\langle x \rangle = (2\langle k \rangle - n)L = 0, \qquad (3.48)$$

as expected, since the random walker is just as likely to travel forwards as backwards. The mean squared distance travelled, $\langle x^2 \rangle$, is

$$\langle x^2 \rangle = (4\langle k^2 \rangle - 4\langle k \rangle n + n^2)L^2 = nL^2, \qquad (3.49)$$

and hence $\sigma_x = \sqrt{\langle x^2 \rangle - \langle x \rangle^2} = \sqrt{n}L$, in agreement with Section 3.6.

Chapter summary

- In this chapter, several introductory concepts in probability theory have been introduced.
- The **mean** of a discrete probability distribution is given by

$$\langle x \rangle = \sum_i x_i P_i,$$

and the mean of a continuous probability distribution is given by

$$\langle x \rangle = \int x P(x) \, dx.$$

- The **variance** is given by

$$\sigma_x^2 = \langle (x - \langle x \rangle)^2 \rangle,$$

where σ_x is the standard deviation.
- If $y = ax + b$, then $\langle y \rangle = a\langle x \rangle + b$ and $\sigma_y = a\sigma_x$.
- If u and v are **independent** random variables, then $\langle uv \rangle = \langle u \rangle \langle v \rangle$. In particular, if $Y = X_1 + X_2 + \cdots + X_n$, where the X_i are all from the same distribution, $\langle Y \rangle = n\langle x \rangle$ and $\sigma_Y = \sqrt{n}\,\sigma_X$.
- The **binomial distribution** describes the probability of getting k successes from n independent **Bernoulli trials**. The mean of this distribution is $\langle k \rangle = np$ and the variance is $\sigma_k^2 = np(1-p)$.

Further reading

There are many good books on probability theory and statistics. Recommended ones include Papoulis (1984), Saha (2003), Wall and Jenkins (2003), and Sivia and Skilling (2006).

Exercises

(3.1) A throw of a regular die yields the numbers 1, 2, ..., 6, each with probability 1/6. Find the mean, variance, and standard deviation of the numbers obtained.

(3.2) The mean birth-weight of babies in the UK is about 3.2 kg with a standard deviation of 0.5 kg. Convert these figures into pounds (lb), given that 1 kg = 2.2 lb.

(3.3) This question is about a *discrete* probability distribution known as the **Poisson distribution**. Let x be a discrete random variable that can take the values $0, 1, 2, \ldots$ A quantity x is said to be Poisson distributed if the probability $P(x)$ of obtaining x is

$$P(x) = \frac{e^{-m} m^x}{x!},$$

where m is a particular number (which we will show in part (b) of this exercise is the mean value of x).

(a) Show that $P(x)$ is a well-behaved probability distribution in the sense that

$$\sum_{x=0}^{\infty} P(x) = 1.$$

(Why is this condition important?)

(b) Show that the mean value of the probability distribution is $\langle x \rangle = \sum_{x=0}^{\infty} x P(x) = m$.

(c) The Poisson distribution is useful for describing very rare events, which occur independently and whose average rate does not change over the period of interest. Examples include birth defects measured per year, traffic accidents at a particular junction per year, numbers of typographical errors on a page, and the number of activations of a Geiger counter per minute. The first recorded example of a Poisson distribution, the one which in fact motivated Poisson, was connected with the rare event of someone being kicked to death by a horse in the Prussian army. The number of horse-kick deaths of Prussian military personnel was recorded for each of 10 corps in each of 20 years from 1875–1894 and the following data recorded:

Number of deaths per year, per corps	Observed frequency
0	109
1	65
2	22
3	3
4	1
≥ 5	0
Total	200

Calculate the mean number of deaths per year per corps. Compare the observed frequency with a calculated frequency assuming the number of deaths per year per corps are Poisson distributed with this mean.

(3.4) This question is about a *continuous* probability distribution known as the **exponential distribution**. Let x be a continuous random variable that can take any value $x \geq 0$. A quantity is said to be exponentially distributed if it takes values between x and $x + \mathrm{d}x$ with probability

$$P(x)\,\mathrm{d}x = A e^{-x/\lambda}\,\mathrm{d}x,$$

where λ and A are constants.

(a) Find the value of A that makes $P(x)$ a well-defined continuous probability distribution so that $\int_0^\infty P(x)\,dx = 1$.

(b) Show that the mean value of the probability distribution is $\langle x \rangle = \int_0^\infty x P(x)\,dx = \lambda$.

(c) Find the variance and standard deviation of this probability distribution. Both the exponential distribution and the Poisson distribution are used to describe similar processes, but for the exponential distribution x is the *actual time* between, for example, successive radioactive decays, successive molecular collisions, or successive horse-kicking incidents (rather than, as with the Poisson distribution, x being simply *the number* of such events in a specified interval).

(3.5) If θ is a continuous random variable which is uniformly distributed between 0 and π, write down an expression for $P(\theta)$. Hence find the value of the following averages:

(a) $\langle \theta \rangle$;
(b) $\langle \theta - \frac{\pi}{2} \rangle$;
(c) $\langle \theta^2 \rangle$;
(d) $\langle \theta^n \rangle$ (for the case $n \geq 0$);
(e) $\langle \cos\theta \rangle$;
(f) $\langle \sin\theta \rangle$;
(g) $\langle |\cos\theta| \rangle$;
(h) $\langle \cos^2\theta \rangle$;
(i) $\langle \sin^2\theta \rangle$;
(j) $\langle \cos^2\theta + \sin^2\theta \rangle$.

Check that your answers are what you expect.

(3.6) In experimental physics, it is important to repeat measurements. Assuming that errors are random, show that if the error in making a single measurement of a quantity X is Δ, the error obtained after using n measurements is $\Delta/\sqrt{n}$. (Hint: after n measurements, the procedure would be to take the n results and average them. So you require the standard deviation of the quantity $Y = (X_1+X_2+\cdots+X_n)/n$ where $X_1, X_2, \ldots, X_n$ can be assumed to be independent, and each has standard deviation Δ.)

(3.7) (a) Show that the binomial distribution can be approximated by a Poisson distribution with mean np when $n \gg 1$ but np remains small. (This therefore represents the case when $p \ll 1$ so that "success" is a rare event.)

(b) A harder problem is to show that when $n \gg 1$ and also $np(1-p) \gg 1$ the binomial distribution can be approximated by a Gaussian distribution with mean np and variance $np(1-p)$. Assuming this to be the case, revisit the one-dimensional random walk in Example 3.10 and assume that the walker takes a step when time $t = n\tau$, where n is an integer. Writing $D = L^2/2\tau$ and using eqns 3.48 and 3.49 show that when $t \gg \tau$ the probability of finding the particle between x and $x + dx$ is

$$P(x)\,dx = \frac{1}{\sqrt{4\pi Dt}} e^{-x^2/4Dt}\,dx. \qquad (3.50)$$

[See also Appendix C.12 for an alternative derivation of eqn 3.50.]

(c) Show that the standard deviation of the distribution in eqn 3.50 is given by $\sigma_x = \sqrt{2Dt}$. As the random walker "diffuses" backwards and forwards, you could try and define its diffusion speed by σ_x/t. This gives a speed that is proportional to $t^{-1/2}$ and is clearly nonsense. The point about diffusion (the behaviour of random walkers) is that since $\sigma_x \propto t^{1/2}$ you need 100 times as much time to diffuse a distance 10 times as big. A small molecule in water diffuses at a rate governed by $D = 10^{-9}\,\mathrm{m^2\,s^{-1}}$. Estimate the time needed for this molecule to diffuse about (i) $1\,\mu\mathrm{m}$ (the width of a bacterium) and (ii) $1\,\mathrm{cm}$ (the width of a test tube).

(3.8) This question introduces a rather efficient method for calculating the mean and variance of probability distributions. We define the **moment generating function** $M(t)$ for a random variable x by

$$M(t) = \langle e^{tx} \rangle. \qquad (3.51)$$

Show that this definition implies that

$$\langle x^n \rangle = M^{(n)}(0), \qquad (3.52)$$

where $M^{(n)}(t) = d^n M/dt^n$ and further that the mean $\langle x \rangle = M^{(1)}(0)$ and the variance $\sigma_x = M^{(2)}(0) - [M^{(1)}(0)]^2$. Hence show that:

(a) for a single Bernoulli trial,

$$M(t) = pe^t + 1 - p; \qquad (3.53)$$

(b) for the binomial distribution,

$$M(t) = (pe^t + 1 - p)^n; \qquad (3.54)$$

(c) for the Poisson distribution,

$$M(t) = e^{m(e^t - 1)}; \qquad (3.55)$$

(d) for the exponential distribution,

$$M(t) = \frac{1}{1 - \lambda t}. \qquad (3.56)$$

Hence derive the mean and variance in each case and show that they agree with the results derived earlier.

Ludwig Boltzmann (1844–1906)

Ludwig Boltzmann made major contributions to the applications of probability to thermal physics. He worked out much of the kinetic theory of gases independently of Maxwell, and together they share the credit for the Maxwell–Boltzmann distribution (see Chapter 5). Boltzmann was very much in awe of Maxwell all his life, and was one of the first to see the significance of Maxwell's theory of electromagnetism. "Was it a god who wrote these lines?" was Boltzmann's comment (quoting Goethe) on Maxwell's work. Boltzmann's great insight was to recognize the statistical connection between thermodynamic entropy and the number of microstates, and through a series of technical papers was able to put the subject of statistical mechanics on a firm footing (his work was, independently, substantially extended by the American physicist Gibbs). Boltzmann was able to show that the second law of thermodynamics (considered in Part IV of this book) could be derived from the principles of classical mechanics, although the fact that classical mechanics makes no distinction between the direction of time meant that he had to smuggle in some assumptions, which mired his approach in some controversy. However, his derivation of what is known as the Boltzmann transport equation, which extends the ideas of the kinetic theory of gases, led to important developments in the electron transport theory of metals and in plasma physics.

Boltzmann also showed how to derive from the principles of thermodynamics the empirical law discovered by his teacher, Josef Stefan, which stated that the total radiation from a hot body was proportional to the fourth power of its absolute temperature (see Chapter 23).

Boltzmann was born in Vienna and did his doctorate in the kinetic theory of gases at the University of Vienna under the supervision of Stefan. His subsequent career took him to Graz, Heidelberg, Berlin, then Vienna again, back to Graz, then Vienna, Leipzig, and finally back to Vienna. His own temperament was in accord with this physical restlessness and lack of stability. The moving around was also partly due to his difficult relationships with various other physicists, particularly Ernst Mach, who was appointed to a chair in Vienna (which occasioned Boltzmann's move to Leipzig in 1900), and Wilhelm Ostwald (whose opposition in Leipzig, together with Mach's retirement in 1901, motivated Boltzmann's return to Vienna in 1902, although not before Boltzmann had attempted suicide).

Fig. 3.4 Ludwig Boltzmann

The notions of irreversibility inherent in thermodynamics led to some controversial implications, particularly to a Universe based on Newtonian mechanics, which are reversible in time. Boltzmann's approach used probability to understand how the behaviour of atoms determined the properties of matter. Ostwald, a physical chemist, who had himself recognized the importance of Gibbs' work (see Chapters 16, 20, and 22) to the extent that he had translated Gibbs' papers into German, was nevertheless a vigorous opponent of theories that involved what he saw as unmeasurable quantities. Ostwald was one of the last opponents of atomism, and became a dedicated opponent of Boltzmann. Ostwald himself was finally convinced of the validity of atoms nearly a decade after Boltzmann's death, by which time Ostwald had been awarded a Nobel Prize, in 1909, for his work on catalysis.

Boltzmann died just before his atomistic viewpoint became obviously vindicated and universally accepted. Boltzmann had suffered from depression and mood swings throughout his life. On holiday in Italy in 1906, Ludwig Boltzmann hanged himself while his wife and daughter were swimming. His famous equation relating entropy S with number of microstates W (Ω in this book) is

$$S = k \log W \qquad (3.57)$$

and is engraved on his tombstone in Vienna. The constant k is called the Boltzmann constant, and is written as k_B in this book.

4 Temperature and the Boltzmann factor

4.1 Thermal equilibrium	32
4.2 Thermometers	33
4.3 The microstates and macrostates	35
4.4 A statistical definition of temperature	36
4.5 Ensembles	38
4.6 Canonical ensemble	38
4.7 Applications of the Boltzmann distribution	42
Chapter summary	45
Further reading	46
Exercises	46

In this chapter, we will explore the concept of *temperature* and show how it can be defined in a statistical manner. This leads to the idea of a *Boltzmann distribution* and a *Boltzmann factor*. Now of course the concept of **temperature** seems such an intuitively obvious one that you might wonder why we need a whole chapter to discuss it. Temperature is simply a measure of "hotness" or "coldness", so that we say that a hot body has a higher temperature than a cold one. For example, as shown in Fig. 4.1(a) if an object has temperature T_1 and is hotter than a second body with temperature T_2, we expect that $T_1 > T_2$. But what do these numbers T_1 and T_2 signify? What does temperature actually mean?

4.1 Thermal equilibrium

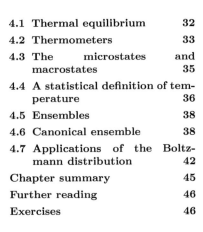

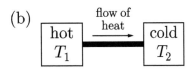
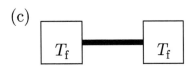

Fig. 4.1 (a) Two objects at different temperatures. (b) The objects are now placed in thermal contact and heat flows from the hot object to the cold object. (c) After a long time, the two objects have the same final temperature T_f.

To begin to answer these questions, let us consider what happens if our hot and cold bodies are placed in **thermal contact** which means that they are able to exchange energy. As described in Chapter 2, heat is "thermal energy in transit" and experiment suggests that, if nothing else is going on,[1] heat will always flow from the hotter body to the colder body, as shown in Fig. 4.1(b). This is backed up by our experience of the world: we always seem to burn ourselves when we touch something very hot (heat flows into us from the hot object) and become very chilled when we touch something very cold (heat flows out of us into the cold object). As heat flows from the hotter body to the colder body, we expect that the energy content and the temperatures of the two bodies will each change with time.

After some time being in thermal contact, we reach the situation in Fig. 4.1(c). The macroscopic properties of the two bodies are now no longer changing with time. If any energy flows from the first body to the second body, this is equal to the energy flowing from the second body to the first body; thus, there is *no net heat flow* between the two bodies. The two bodies are said to be in **thermal equilibrium**, which

---

[1]This is assuming that no additional power is being fed into the systems, such as occurs in the operation of a refrigerator, which sucks heat out of the cold interior and dumps it into your warmer kitchen, but only because you are supplying electrical power.

is defined by saying that the energy content and the temperatures of the two bodies will no longer be changing with time. We would expect that the two bodies in thermal equilibrium are now at the *same temperature*.

It seems that something irreversible has happened. Once the two bodies are put in thermal contact, the change from Fig. 4.1(b) to Fig. 4.1(c) proceeds inevitably. However, if we started with two bodies at the same temperature and placed them in thermal contact as in Fig. 4.1(c), the reverse process, i.e., ending up with Fig. 4.1(b), would not occur.[2] Thus as a function of time, systems in thermal contact tend towards thermal equilibrium, rather than away from it. The process that leads to thermal equilibrium is called **thermalization**.

If various bodies are all in thermal equilibrium with each other, then we would expect that their temperatures should be the same. This idea is encapsulated in the **zeroth law of thermodynamics**:

[2]Thermal processes thus define an arrow of time. We will return to this point later in Section 34.5.

Zeroth law of thermodynamics
Two systems, each separately in thermal equilibrium with a third, are in equilibrium with each other.

You can tell by the numbering of the law that although it is an assumption that comes before the other laws of thermodynamics, it was added after the first three laws had been formulated. Early workers in thermodynamics took the content of the zeroth law as so obvious it hardly needed stating, and you might well agree with them! Nevertheless, the zeroth law gives us some justification for how to actually measure temperature: we place the body whose temperature needs to be measured in thermal contact with a second body, which displays some property that has a well-known dependence on temperature, and wait for them to come into thermal equilibrium. The second body is called a **thermometer**. The zeroth law then guarantees that if we have calibrated this second body against any other standard thermometer, we should always get consistent results. Thus, a more succinct statement of the zeroth law[3] is: "thermometers work".

[3]This version is from our colleague M. G. Bowler.

4.2 Thermometers

We now make some remarks concerning thermometers.

- For a thermometer to work well, its heat capacity must be much lower than that of the object whose temperature one wants to measure. If this is not the case, the action of measurement (placing the thermometer in thermal contact with the object) could alter the temperature of the object.
- A common type of thermometer utilizes the fact that liquids expand when they are heated. Galileo Galilei used a water thermometer based on this principle in 1593, but it was Daniel Gabriel Fahrenheit (1686–1736) who devised thermometers based on al-

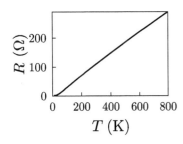

Fig. 4.2 The temperature dependence of the resistance of a typical platinum sensor.

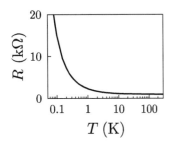

Fig. 4.3 The temperature dependence of the resistance of a typical RuO$_2$ sensor.

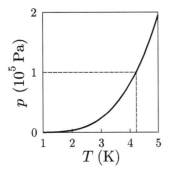

Fig. 4.4 The vapour pressure of ^{4}He as a function of temperature. The dashed line labels atmospheric pressure and the corresponding boiling point for liquid ^{4}He.

[4] We will introduce the Carnot engine in Section 13.2. The definition of temperature that arises from this is based on eqn 13.7 and states that the ratio of the temperature of a body to the heat flow from it is a constant in a reversible Carnot cycle.

cohol (1709) and mercury (1714) that bear most resemblance to modern household thermometers. He introduced his famous temperature scale, which was then superseded by the more logical scheme devised by Anders Celsius (1701–1744).

- Another method is to measure the electrical resistance of a material which has a well-known dependence of resistance on temperature. Platinum is a popular choice since it is chemically resistant, ductile (so can be easily drawn into wires) and has a large temperature-coefficient of resistance; see Fig. 4.2. Other commonly used thermometers are based on doped germanium (a semiconductor that is very stable after repeated thermal cycling), carbon sensors and RuO$_2$ (in contrast with platinum, the electrical resistance of these thermometers increases as they are cooled; see Fig. 4.3).

- Using the ideal gas equation (eqn 1.12), one can measure the temperature of a gas by measuring its pressure with its volume fixed (or by measuring its volume with its pressure fixed). This works well as far as the ideal gas equation works, although at very low temperature gases liquefy and show departures from the ideal gas equation.

- Another method, which is useful in cryogenics, is to have a liquid coexisting with its vapour and to measure the vapour pressure. For example, liquid helium (^{4}He, the most common isotope) has the vapour pressure dependence on temperature shown in Fig. 4.4.

All of these methods use some measurable property, like resistance or pressure, which depends in some, sometimes complicated, manner on temperature. However, none of them is completely linear across the entire temperature range of interest: mercury solidifies at very low temperature and becomes gaseous at very high temperature, the resistance of platinum saturates at very low temperature and platinum wire melts at very high temperature, etc. However, against what standard thermometer can one possibly assess the relative merits of these different thermometers? Which thermometer is perfect and gives the real thing, against which all other thermometers should be judged?

It is clear that we need some absolute definition of temperature based on fundamental physics. In the nineteenth century, one such definition was found, and it was based on a hypothetical machine, which has never been built, called a *Carnot engine*.[4] Subsequently, it was found that temperature could be defined in terms of a purely statistical argument using ideas from probability theory, and this is the definition we will use, which we introduce in Section 4.4. In the following section we will introduce the terminology of *microstates* and *macrostates* that will be needed for this argument.

4.3 The microstates and macrostates

To make the distinction between microstates and macrostates, consider the following example.

Example 4.1

Imagine that you have a large box containing 100 identical coins. With the lid on the box, you give it a really good long and hard shake, so that you can hear the coins flipping, rattling, and being generally tossed around. Now you open the lid and look inside the box. Some of the coins will be lying with heads facing up and some with tails facing up. There are lots of possible configurations that one could achieve (2^{100} to be precise, which is approximately 10^{30}) and we will assume that *each of these different configurations is equally likely*. Each possible configuration therefore has a probability of approximately 10^{-30}. We will call each particular configuration a **microstate** of this system. An example of one of these microstates would be: "Coin number 1 is heads, coin number 2 is heads, coin number 3 is tails, etc". To identify a microstate, you would somehow need to identify each coin individually, which would be a bit of a bore. However, probably the way you would categorize the outcome of this experiment is by simply counting the number of coins which are heads and the number which are tails (e.g., 53 heads and 47 tails). This sort of categorization we call a **macrostate** of this system. The macrostates are not equally likely. For example, of the $\approx 10^{30}$ possible individual configurations (microstates),

$$\begin{aligned}
\text{the number with 50 heads and 50 tails} &= \tfrac{100!}{(50!)^2} \approx 1 \times 10^{29}, \\
\text{the number with 53 heads and 47 tails} &= \tfrac{100!}{53!47!} \approx 8 \times 10^{28}, \\
\text{the number with 90 heads and 10 tails} &= \tfrac{100!}{90!10!} \approx 2 \times 10^{13}, \text{ and} \\
\text{the number with 100 heads and 0 tails} &= 1.
\end{aligned}$$

Thus, the outcome with all 100 coins with their heads facing up is a very unlikely outcome. This macrostate contains only a single microstate. If that were the result of the experiment, you would probably conclude that (i) your shaking had not been very vigorous and that (ii) someone had carefully prepared the coins to be lying heads up at the start of the experiment. Of course, a *particular* microstate with 53 heads and 47 tails is just as unlikely; it is just that there are about 8×10^{28} other microstates having 53 heads and 47 tails that look extremely similar.

This simple example shows two crucial points:

- The system could be described by a very large number of *equally likely* microstates.
- What you actually measure[5] is a property of the macrostate of the

[5] In our example, the measurement was opening the large box and counting the number of coins that were heads and those that were tails.

system. The macrostates are *not* equally likely, because different macrostates correspond to different numbers of microstates.

The most likely macrostate that the system will find itself in is the one that corresponds to the largest number of microstates.

Thermal systems behave in a very similar way to the example we have just considered. To specify a microstate for a thermal system, you would need to give the microscopic configurations (perhaps position and velocity, or perhaps energy) of each and every atom in the system. In general it is impossible to measure which microstate the system is in. The macrostate of a thermal system on the other hand would be specified only by giving the macroscopic properties of the system, such as the pressure, the total energy, or the volume. A macroscopic configuration, such as a gas with pressure 10^5 Pa in a volume $1\,\text{m}^3$, would be associated with an enormous number of microstates. In the next section, we are going to give a statistical definition of temperature, which is based on the idea that a thermal system can have a large number of equally likely microstates, but you are only able to measure the macrostate of the system. At this stage, we are *not* going to worry about what the microstates of the system actually are; we are simply going to posit their existence and say that if the system has energy E, then it could be in any one of $\Omega(E)$ equally likely microstates, where $\Omega(E)$ is some enormous number.

4.4 A statistical definition of temperature

We return to our example of Section 4.1 and consider two large systems that can exchange energy with each other, but not with anything else (Fig. 4.5). In other words, the two systems are in thermal contact with each other, but thermally isolated from their surroundings. The first system has energy E_1 and the second system has energy E_2. The total energy $E = E_1 + E_2$ is therefore assumed fixed since the two systems cannot exchange energy with anything else. Hence the value of E_1 is enough to determine the macrostate of this joint system. Each of these systems can be in a number of possible microstates. This number of possible microstates could in principle be calculated as in Section 1.4 (and in particular, Example 1.3) and will be a very large, combinatorial number, but we will not worry about the details of this. Let us assume that the first system can be in any one of $\Omega_1(E_1)$ microstates and the second system can be in any one of $\Omega_2(E_2)$ microstates. Thus the whole system can be in any one of $\Omega_1(E_1)\Omega_2(E_2)$ microstates.[6]

The systems are able to exchange energy with each other, and we will assume that they have been left in the condition of being joined together for a sufficiently long time that they have come into *thermal equilibrium*. This means that E_1 and E_2 have come to fixed values. The crucial insight which we must make is that *a system will appear to choose a macroscopic configuration that maximizes the number of microstates*. This idea is based on the following assumptions:

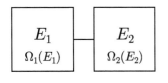

Fig. 4.5 Two systems able only to exchange energy between themselves.

[6] We use the product of the two quantities, $\Omega_1(E_1)$ and $\Omega_2(E_2)$, because for each of the $\Omega_1(E_1)$ states of the first system, the second system can be in any of its $\Omega_2(E_2)$ different states. Hence the total number of possible combined states is the product of $\Omega_1(E_1)$ and $\Omega_2(E_2)$.

(1) Each one of the possible microstates of a system is equally likely to occur;

(2) The system's internal dynamics are such that the microstates of the system are continually changing;

(3) Given enough time, the system will explore all possible microstates and spend an equal time in each of them.[7]

[7]This is the so-called ergodic hypothesis.

These assumptions imply that the system will most likely be found in a configuration that is represented by the most microstates. For a large system our phrase "most likely" becomes "absolutely, overwhelmingly likely"; what appears at first sight to be a somewhat weak, probabilistic statement (perhaps on the same level as a five-day weather forecast) becomes an utterly reliable prediction on whose basis you can design an aircraft engine and trust your life to it!

For our problem of two connected systems, the most probable division of energy between the two systems is the one that maximizes $\Omega_1(E_1)\Omega_2(E_2)$, because this will correspond to the greatest number of possible microstates. Our systems are large and hence we can use calculus to study their properties; we can therefore consider making infinitesimal changes to the energy of one of the systems and seeing what happens. Therefore, we can maximize this expression with respect to E_1 by writing

$$\frac{\mathrm{d}}{\mathrm{d}E_1}(\Omega_1(E_1)\Omega_2(E_2)) = 0 \tag{4.1}$$

and hence, using standard rules for the differentiation of a product,

$$\Omega_2(E_2)\frac{\mathrm{d}\Omega_1(E_1)}{\mathrm{d}E_1} + \Omega_1(E_1)\frac{\mathrm{d}\Omega_2(E_2)}{\mathrm{d}E_2}\frac{\mathrm{d}E_2}{\mathrm{d}E_1} = 0. \tag{4.2}$$

Since the total energy $E = E_1 + E_2$ is assumed fixed, this implies that

$$\mathrm{d}E_1 = -\mathrm{d}E_2, \tag{4.3}$$

and hence

$$\frac{\mathrm{d}E_2}{\mathrm{d}E_1} = -1, \tag{4.4}$$

so that eqn 4.2 becomes

$$\frac{1}{\Omega_1}\frac{\mathrm{d}\Omega_1}{\mathrm{d}E_1} - \frac{1}{\Omega_2}\frac{\mathrm{d}\Omega_2}{\mathrm{d}E_2} = 0, \tag{4.5}$$

and hence

$$\frac{\mathrm{d}\ln\Omega_1}{\mathrm{d}E_1} = \frac{\mathrm{d}\ln\Omega_2}{\mathrm{d}E_2}. \tag{4.6}$$

This condition defines the most likely division of energy between the two systems if they are allowed to exchange energy since it maximizes the total number of microstates. This division of energy is, of course, more usually called "being at the same temperature", and so we identify $\mathrm{d}\ln\Omega/\mathrm{d}E$ with the temperature T (so that $T_1 = T_2$). We will define the **temperature** T by

$$\frac{1}{k_\mathrm{B}T} = \frac{\mathrm{d}\ln\Omega}{\mathrm{d}E}, \tag{4.7}$$

We will see later (Section 14.5) that in statistical mechanics, the quantity $k_B \ln \Omega$ is called the **entropy**, S, and hence eqn 4.7 is equivalent to

$$\frac{1}{T} = \frac{dS}{dE}.$$

where k_B is the **Boltzmann constant**, which is given by

$$k_B = 1.3807 \times 10^{-23} \, \text{J K}^{-1}. \tag{4.8}$$

With this choice of constant, T has its usual interpretation and is measured in kelvin. We will show in later chapters that this choice of definition leads to experimentally verifiable consequences, such as the correct expression for the pressure of a gas.

4.5 Ensembles

We are using probability to describe thermal systems and our approach is to imagine repeating an experiment to measure a property of a system again and again because we cannot control the microscopic properties (as described by the system's microstates). In an attempt to formalize this, Josiah Willard Gibbs in 1878 introduced a concept known as an **ensemble**. This is an idealization in which one considers making a large number of mental "photocopies" of the system, each one of which represents a possible state the system could be in. There are three main ensembles that tend to be used in thermal physics:

(1) The **microcanonical ensemble**: an ensemble of systems that each have the same fixed energy.

(2) The **canonical ensemble**: an ensemble of systems, each of which can exchange its energy with a large reservoir of heat. As we shall see, this fixes (and defines) the temperature of the system.

(3) The **grand canonical ensemble**: an ensemble of systems, each of which can exchange both energy and particles with a large reservoir. (This fixes the system's temperature and a quantity known as the system's chemical potential. We will not consider this again until Chapter 22 and it can be ignored for the present.)

In the next section we will consider the canonical ensemble in more detail and use it to derive the probability of a system at a fixed temperature being in a particular microstate.

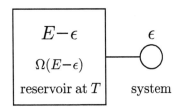

Fig. 4.6 A large reservoir (or heat bath) at temperature T connected to a small system.

4.6 Canonical ensemble

We now consider two systems coupled as before in such a way that they can exchange energy (Fig. 4.6). This time, we will make one of them enormous, and call it the **reservoir** (also known as a **heat bath**). It is so large that you can take quite a lot of energy out of it and yet it can remain at essentially the same temperature. In the same way, if you stand on the seashore and take an eggcupful of water out of the ocean, you do not notice the level of the ocean going down (although it does in fact go down, but by an unmeasurably small amount). The number of ways of arranging the quanta of energy of the reservoir will therefore be colossal. The other system is small and will be known as the **system**.

We will assume that for each allowed energy of the system there is only a single microstate, and therefore the system always has a value of Ω equal to one. Once again, we fix[8] the total energy of the system plus reservoir to be E. The energy of the reservoir is taken to be $E - \epsilon$ while the energy of the system is taken to be ϵ. This situation of a system in thermal contact with a large reservoir is very important and is known as the **canonical ensemble**.[9]

The probability $P(\epsilon)$ that the system has energy ϵ is proportional to the number of microstates that are accessible to the reservoir multiplied by the number of microstates that are accessible to the system. This is therefore

$$P(\epsilon) \propto \Omega(E - \epsilon) \times 1. \tag{4.9}$$

Since we have an expression for temperature in terms of the logarithm of Ω (eqn 4.7), and since $\epsilon \ll E$, we can perform a Taylor expansion[10] of $\ln \Omega(E - \epsilon)$ around $\epsilon = 0$, so that

$$\ln \Omega(E - \epsilon) = \ln \Omega(E) - \frac{d \ln \Omega(E)}{dE} \epsilon + \cdots \tag{4.10}$$

and so now using eqn. 4.7, we have

$$\ln \Omega(E - \epsilon) = \ln \Omega(E) - \frac{\epsilon}{k_B T} + \cdots, \tag{4.11}$$

where T is the temperature of the reservoir. In fact, we can neglect the further terms in the Taylor expansion (see Exercise 4.4) and hence eqn 4.11 becomes

$$\Omega(E - \epsilon) = \Omega(E) \, e^{-\epsilon/k_B T}. \tag{4.12}$$

Using eqn 4.9 we thus arrive at the following result for the probability distribution describing the system, which is given by

$$P(\epsilon) \propto e^{-\epsilon/k_B T}. \tag{4.13}$$

Since the system is now in equilibrium with the reservoir, it must also have the same temperature as the reservoir. But notice that although the system therefore has fixed temperature T, its energy ϵ is not a constant but is governed by the probability distribution in eqn 4.13 (and is plotted in Fig. 4.7). This is known as the **Boltzmann distribution** and also as the **canonical distribution**. The term $e^{-\epsilon/k_B T}$ is known as a **Boltzmann factor**.

We now have a probability distribution that describes exactly how a small system behaves when coupled to a large reservoir at temperature T. The system has a reasonable chance of achieving an energy ϵ that is less than $k_B T$, but the exponential in the Boltzmann distribution quickly begins to reduce the probability of achieving an energy much greater than $k_B T$. However, to quantify this properly we need to normalize the probability distribution. If a system is in contact with a reservoir and has a microstate r with energy E_r, then

$$P(\text{microstate } r) = \frac{e^{-E_r/k_B T}}{\sum_i e^{-E_i/k_B T}}, \tag{4.14}$$

[8] In this respect, the system plus reservoir as a whole can be considered as being in the microcanonical ensemble, which has fixed energy, with each of the microstates of the combined entity being equally likely.

[9] "Canonical" means part of the "canon", the store of generally accepted things one should know. It's an odd word, but we're stuck with it. Focussing on a system whose energy is *not* fixed, but which can exchange energy with a big reservoir, is something we do a lot in thermal physics and is therefore in some sense canonical.

[10] See Appendix B.

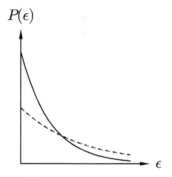

Fig. 4.7 The Boltzmann distribution. The dashed curve corresponds to a higher temperature than the solid curve.

The partition function is the subject of Chapter 20.

where the sum in the denominator makes sure that the probability is normalized. The sum in the denominator is called the **partition function** and is given the symbol Z.

We have derived the Boltzmann distribution on the basis of statistical arguments that show that this distribution of energy maximizes the number of microstates. It is instructive to verify this for a small system, so the following example presents the results of a computer experiment to demonstrate the validity of the Boltzmann distribution.

Example 4.2

To illustrate the statistical nature of the Boltzmann distribution, let us play a game in which quanta of energy are distributed in a lattice. We choose a lattice of 400 sites, arranged for convenience on a 20×20 grid. Each site initially contains a single energy quantum, as shown in Fig. 4.8(a). The adjacent histogram shows that there are 400 sites with one quantum on each. We now choose a site at random and remove the quantum from that site and place it on a second, randomly chosen site. The resulting distribution is shown in Fig. 4.8(b), and the histogram shows that we now have 398 sites each with 1 quantum, 1 site with no quanta and 1 site with two quanta. This redistribution process is repeated many times and the resulting distribution is as shown in Fig. 4.8(c). The histogram describing this looks very much like a Boltzmann exponential distribution.

The initial distribution shown in Fig. 4.8(a) is very equitable and gives a distribution of energy quanta between sites of which Karl Marx would have been proud. It is however very *statistically unlikely* because it is associated with only a single microstate, i.e., $\Omega = 1$. There are many more microstates associated with other macrostates, as we shall now show. For example, the state obtained after a single iteration, such as the one shown in Fig. 4.8(b), is much more likely, since there are 400 ways to choose the site from which a quantum has been removed, and then 399 ways to choose the site to which a quantum is added; hence $\Omega = 400 \times 399 = 19600$ for this histogram (which contains 398 singly occupied sites, one site with zero quanta and one site with two quanta). The state obtained after many iterations in Fig. 4.8(c) is much, much more likely to occur if quanta are allowed to rearrange randomly as the number of microstates associated with the Boltzmann distribution is absolutely enormous. The Boltzmann distribution is simply a matter of probability.

In the model considered in this example, the rôle of temperature is played by the total number of energy quanta in play. So, for example, if instead the initial arrangement had been two quanta per site rather than one quantum per site, then after many iterations one would obtain the arrangement shown in Fig. 4.8(d). Since the initial arrangement has more energy, the final state is a Boltzmann distribution with a higher temperature (leading to more sites with more energy quanta).

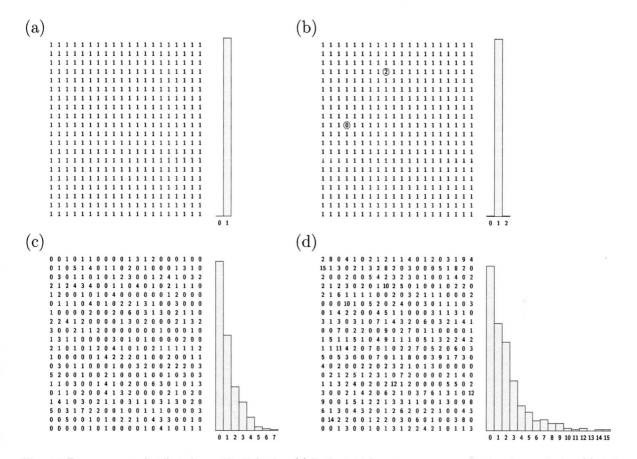

Fig. 4.8 Energy quanta distributed on a 20×20 lattice. (a) In the initial state, one quantum is placed on each site. (b) A site is chosen at random and a quantum is removed from that site and placed on a second randomly chosen site. (c) After many repetitions of this process, the resulting distribution resembles a Boltzmann distribution. (d) The analogous final distribution following redistribution from an initial state with two quanta per site. The adjacent histogram in each case shows how many quanta are placed on each site.

Let us now start with a bigger lattice, now containing 10^6 sites, and place a quantum of energy on each site. We randomly move quanta from site to site as before, and in our computer program we let this proceed for a large number of iterations (in this case 10^{10}). The resulting distribution is shown in Fig. 4.9, which displays a graph on a logarithmic scale of the number of sites N with n quanta. The straight line is a fit to the expected Boltzmann distribution. This example is considered in more detail in the exercises.

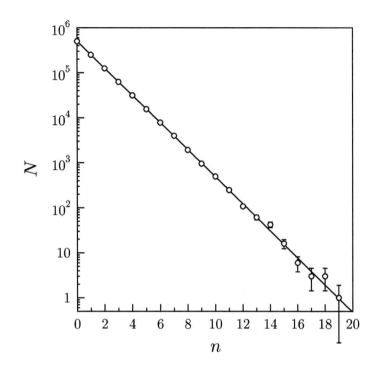

Fig. 4.9 The final distribution for a lattice of size 1000×1000 with one quantum of energy initially placed on each site. The error bars are calculated by assuming Poisson statistics and have length $\sqrt{N}$, where N is the number of sites having n quanta.

4.7 Applications of the Boltzmann distribution

To illustrate the application of the Boltzmann distribution, we now conclude this chapter with some examples. These examples involve little more than a simple application of the Boltzmann distribution, but they have important consequences.

Before we do so, let us introduce a piece of shorthand. Since we will often need to write the quantity $1/k_\mathrm{B}T$, we will use the shorthand

$$\beta \equiv \frac{1}{k_\mathrm{B}T}, \tag{4.15}$$

so that the Boltzmann factor becomes simply $\mathrm{e}^{-\beta E}$. Using this shorthand, we can also write eqn 4.7 as

$$\beta = \frac{\mathrm{d}\ln\Omega}{\mathrm{d}E}. \tag{4.16}$$

$k_\mathrm{B}T$ has units of energy. It is often helpful to remember that $T = 300\,\mathrm{K}$ corresponds to an energy of about 25 meV.

Example 4.3

The two state system
The first example is one of the simplest one can think of. In a two-state system, there are only two states, one with energy 0 and the other with energy $\epsilon > 0$. What is the average energy of the system?

4.7 Applications of the Boltzmann distribution

Solution:
The probability of being in the lower state is given by eqn 4.14, so we have
$$P(0) = \frac{1}{1+e^{-\beta\epsilon}}. \tag{4.17}$$

Similarly, the probability of being in the upper state is
$$P(\epsilon) = \frac{e^{-\beta\epsilon}}{1+e^{-\beta\epsilon}}. \tag{4.18}$$

The average energy $\langle E \rangle$ of the system is then
$$\begin{aligned}\langle E \rangle &= 0 \cdot P(0) + \epsilon \cdot P(\epsilon) \\ &= \epsilon \frac{e^{-\beta\epsilon}}{1+e^{-\beta\epsilon}} \\ &= \frac{\epsilon}{e^{\beta\epsilon}+1}.\end{aligned} \tag{4.19}$$

This expression (plotted in Fig. 4.10) behaves as expected: when T is very low, $k_B T \ll \epsilon$, and so $\beta\epsilon \gg 1$ and $\langle E \rangle \to 0$ (the system is in the ground state). When T is very high, $k_B T \gg \epsilon$, and so $\beta\epsilon \ll 1$ and $\langle E \rangle \to \epsilon/2$ (both levels are equally occupied on average).

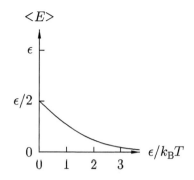

Fig. 4.10 The value of $\langle E \rangle$ as a function of $\epsilon/k_B T = \beta\epsilon$, following eqn 4.19. As $T \to \infty$, each energy level is equally likely to be occupied and so $\langle E \rangle = \epsilon/2$. When $T \to 0$, only the lower level is occupied and $\langle E \rangle = 0$.

Example 4.4

Isothermal atmosphere
Estimate the number of molecules in an isothermal[11] atmosphere as a function of height.

[11]"Isothermal" means constant temperature. A more sophisticated treatment of the atmosphere is postponed until Section 12.4; see also Chapter 37.

Solution:
This is our first attempt at modelling the atmosphere, where we make the rather naive assumption that the temperature of the atmosphere is constant. Consider a molecule in an ideal gas at temperature T in the presence of gravity. The probability $P(z)$ of the molecule of mass m being at height z is given by
$$P(z) \propto e^{-mgz/k_B T}, \tag{4.20}$$

because its potential energy is mgz. Hence, the number density[12] of molecules $n(z)$ at height z, which will be proportional to the probability function $P(z)$ of finding a molecule at height z, is given by

[12]Number density means number per unit volume.

$$n(z) = n(0)e^{-mgz/k_B T}. \tag{4.21}$$

This result (plotted in Fig. 4.11) agrees with a more pedestrian derivation, which goes as follows: consider a layer of gas between height z and $z+dz$. There are $n\,dz$ molecules per unit area in this layer, and therefore they exert a pressure (force per unit area)
$$dp = -n\,dz \cdot mg \tag{4.22}$$

downwards (because each molecule has weight mg). We note in passing that eqn 4.22 can be rearranged using $\rho = nm$ to show that

$$dp = -\rho g\, dz, \tag{4.23}$$

which is known as the **hydrostatic equation**. Using the ideal gas law (in the form derived in Chapter 6), which is $p = nk_B T$, we have that

$$\frac{dn}{n} = -\frac{mg}{k_B T} dz, \tag{4.24}$$

which is a simple differential equation yielding

$$\ln n(z) - \ln n(0) = -\frac{mg}{k_B T} z, \tag{4.25}$$

so that, again, we have

$$n(z) = n(0)e^{-mgz/k_B T}. \tag{4.26}$$

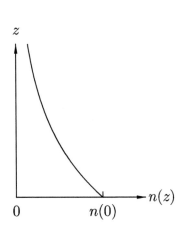

Fig. 4.11 The number density $n(z)$ of molecules at height z for an isothermal atmosphere.

Our prediction is that the number density falls off exponentially with height, but the reality is somewhat different. Our assumption of constant T is at fault (the temperature falls as the altitude increases, at least initially) and we will return to this problem in Section 12.4, and also in Chapter 37.

Example 4.5

Chemical reactions

Many chemical reactions have an activation energy E_{act} of about $\frac{1}{2}$ eV. At $T = 300$ K, which is about room temperature, the probability that a particular reaction occurs is proportional to

$$\exp(-E_{\text{act}}/(k_B T)). \tag{4.27}$$

If the temperature is increased to $T + \Delta T = 310$ K, the probability increases to

$$\exp(-E_{\text{act}}/(k_B (T + \Delta T))), \tag{4.28}$$

which is larger by a factor

$$\begin{aligned}
\frac{\exp(-E_{\text{act}}/(k_B(T+\Delta T)))}{\exp(-E_{\text{act}}/(k_B T))} &= \exp\left(-\frac{E_{\text{act}}}{k_B}[(T+\Delta T)^{-1} - T^{-1}]\right) \\
&\approx \exp\left(\frac{E_{\text{act}}}{k_B T}\frac{\Delta T}{T}\right) \\
&\approx 2.
\end{aligned} \tag{4.29}$$

Hence many chemical reactions roughly double in speed when the temperature is increased by about 10 K.

Example 4.6
The Sun

The main fusion reaction in the Sun[13] is

$$p^+ + p^+ \rightarrow d^+ + e^+ + \bar{\nu} \tag{4.30}$$

but the main barrier to this occuring is the electrostatic repulsion of the two protons coming together in the first place. This energy is

$$E = \frac{e^2}{4\pi\epsilon_0 r}, \tag{4.31}$$

which for $r = 10^{-15}\,\text{m}$, the distance which they must approach each other for fusion to occur, E is about $1\,\text{MeV}$. The Boltzmann factor for this process at a temperature of $T \approx 10^7\,\text{K}$ (at the centre of the Sun) is

$$e^{-E/k_B T} \approx 10^{-1200}. \tag{4.32}$$

This is extremely small, suggesting that the Sun is unlikely to undergo fusion. However, our lazy sunny afternoons are saved by the fact that quantum mechanical tunnelling allows the protons to pass *through* this barrier vastly more often than this calculation predicts that they could pass *over the top* of it.

[13]p^+ is a proton, d^+ is a deuteron (a proton and a neutron), e^+ is a positron and $\bar{\nu}$ is a neutrino. This reaction and its consequences are explored more fully in Section 35.2.

Chapter summary

- The temperature T of a system is given by

$$\beta \equiv \frac{1}{k_B T} = \frac{\mathrm{d}\ln\Omega}{\mathrm{d}E},$$

where k_B is the Boltzmann constant, E is its energy, and Ω is the number of microstates (i.e., the number of ways of arranging the quanta of energy in the system).

- The *microcanonical ensemble* is an idealized collection of systems that each have the same fixed energy.

- The *canonical ensemble* is an idealized collection of systems, each of which can exchange its energy with a large reservoir of heat.

- For the canonical ensemble, the probability that a particular system has energy ϵ is given by

$$P(\epsilon) \propto e^{-\beta\epsilon}$$

(Boltzmann distribution), and the factor $e^{-\beta\epsilon}$ is known as the *Boltzmann factor*. Its use has been illustrated for a number of physical situations.

Further reading

Methods of measuring temperature are described in Pobell (1996) and White and Meeson (2002).

Exercises

(4.1) Check that the probability in eqn 4.14 is normalized, so that the sum of all possible probabilities is one.

(4.2) For the two-state system described in Example 4.3, derive an expression for the variance of the energy.

(4.3) A system comprises N states, which can have energy 0 or Δ. Show that the number of ways $\Omega(E)$ of arranging the total system to have energy $E = r\Delta$ (where r is an integer) is given by

$$\Omega(E) = \frac{N!}{r!(N-r)!}. \tag{4.33}$$

Now remove a small amount of energy $s\Delta$ from the system, where $s \ll r$. Show that

$$\Omega(E-\epsilon) \approx \Omega(E)\frac{r^s}{(N-r)^s}, \tag{4.34}$$

and hence show that the system has temperature T given by

$$\frac{1}{k_\mathrm{B} T} = \frac{1}{\Delta}\ln\left(\frac{N-r}{r}\right). \tag{4.35}$$

Sketch $k_\mathrm{B} T$ as a function of r from $r=0$ to $r=N$ and explain the result.

(4.4) In eqn 4.10, we neglected the next term in the Taylor expansion, which is

$$\frac{1}{2}\frac{\mathrm{d}^2 \ln \Omega}{\mathrm{d}E^2}\epsilon^2. \tag{4.36}$$

Show that this term equals

$$-\frac{\epsilon^2}{2k_\mathrm{B} T^2}\frac{\mathrm{d}T}{\mathrm{d}E}, \tag{4.37}$$

and hence show that it can be neglected compared with the first two terms if the reservoir is large. (Hint: how much should the temperature of the reservoir change when you change its energy by order ϵ?)

(4.5) A photon of visible light with energy 2 eV is absorbed by a macroscopic body held at room temperature. By what factor does Ω for the macroscopic body change? Repeat the calculation for a photon that originated from an FM radio transmitter.

(4.6) Figure 4.10 is a plot of $\langle E \rangle$ as a function of $\beta\epsilon$. Sketch $\langle E \rangle$ as a function of temperature T (measured in units of ϵ/k_B).

(4.7) Find the average energy $\langle E \rangle$ for
(a) An n-state system, in which a given state can have energy $0, \epsilon, 2\epsilon, \ldots, n\epsilon$.
(b) A harmonic oscillator, in which a given state can have energy $0, \epsilon, 2\epsilon, \ldots$ (i.e., with no upper limit).

(4.8) Estimate $k_\mathrm{B} T$ at room temperature, and convert this energy into electronvolts (eV). Using this result, answer the following:
(a) Would you expect hydrogen atoms to be ionized at room temperature? (The binding energy of an electron in a hydrogen atom is 13.6 eV.)
(b) Would you expect the rotational energy levels of diatomic molecules to be excited at room temperature? (It costs about 10^{-4} eV to promote such a system to an excited rotational energy level.)

(4.9) Write a computer program to reproduce the results in Example 4.2. For the case of $\mathcal{N} \gg 1$ sites with initially one quantum per site, show that after many iterations you would expect there to be $N(n)$ sites with n quanta, where

$$N(n) \approx 2^{-n}\mathcal{N}, \tag{4.38}$$

and explain why this is a Boltzmann distribution. Generalize your results for $\mathcal{Q} \gg 1$ quanta distributed on $\mathcal{N} \gg 1$ sites.

Part II

Kinetic theory of gases

In the second part of this book, we apply the results of Part I to the properties of gases. This is the **kinetic theory of gases**, in which it is the motion of individual gas atoms, behaving according to the Boltzmann distribution, that determines quantities such as the pressure of a gas, or the rate of effusion. This part is structured as follows:

- In Chapter 5, we show that the Boltzmann distribution applied to gases gives rise to a speed distribution known as the *Maxwell–Boltzmann distribution*. We show how this can be measured experimentally.
- A treatment of pressure in Chapter 6 using the results so far developed allows us to derive *Boyle's law* and the *ideal gas law*.
- We are then able to treat the *effusion of gases* through small holes in Chapter 7, which also introduces the concept of *flux*.
- Chapter 8 considers the nature of molecular collisions and introduces the concepts of the *mean scattering time*, the *collision cross-section* and the *mean free path*.

5 The Maxwell–Boltzmann distribution

5.1 The velocity distribution 48
5.2 The speed distribution 49
5.3 Experimental justification 51
Chapter summary 54
Exercises 54

In this chapter we will apply the results of the Boltzmann distribution (eqn 4.13) to the problem of the motion of molecules in a gas. For the present, we will neglect any rotational or vibrational motion of the molecules and consider only translational motion (so these results are strictly applicable only to a monatomic gas). In this case the energy of a molecule is given by

$$\frac{1}{2}mv_x^2 + \frac{1}{2}mv_y^2 + \frac{1}{2}mv_z^2 = \frac{1}{2}mv^2, \tag{5.1}$$

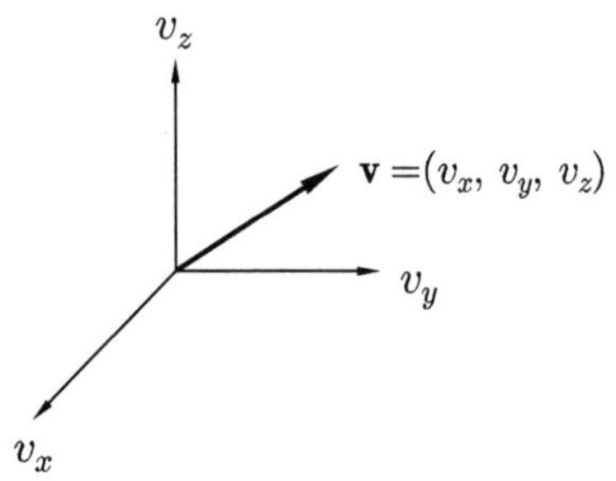

where $\boldsymbol{v} = (v_x, v_y, v_z)$ is the molecular velocity, and $v = |\boldsymbol{v}|$ is the molecular speed. This molecular velocity can be represented in velocity space (see Fig. 5.1). The aim is to determine the distribution of molecular velocities and to determine the distribution of molecular speeds. This we will do in the next two sections. To make some progress, we will make a couple of assumptions: first, that the molecular size is much less than the intermolecular separation, so that we assume that molecules spend most of their time whizzing around and only rarely bumping into each other; second, we will ignore any intermolecular forces. Molecules can exchange energy with each other due to collisions, but everything remains in equilibrium. Each molecule therefore behaves like a small system connected to a heat reservoir at temperature T, where the heat reservoir is "all the other molecules in the gas". Hence the results of the Boltzmann distribution of energies (described in the previous chapter) will hold.

Fig. 5.1 The velocity of a molecule is shown as a vector in velocity space.

5.1 The velocity distribution

To work out the velocity distribution of molecules in a gas, we must first choose a given direction and see how many molecules have particular components of velocity along it. We define the **velocity distribution function** as the fraction of molecules with velocities in, say, the x-direction,[1] between v_x and $v_x + dv_x$, as $g(v_x) \, dv_x$. The velocity distribution function is proportional to a Boltzmann factor, namely e to the power of the relevant energy, in this case $\frac{1}{2}mv_x^2$, divided by $k_\mathrm{B}T$. Hence

$$\boxed{g(v_x) \propto e^{-mv_x^2/2k_\mathrm{B}T}.} \tag{5.2}$$

[1]But we could choose any direction of motion we like!

This velocity distribution function is sketched in Fig. 5.2. To normalize this function, so that $\int_{-\infty}^{\infty} g(v_x)\,dv_x = 1$, we need to evaluate the integral[2]

$$\int_{-\infty}^{\infty} e^{-mv_x^2/2k_BT}\,dv_x = \sqrt{\frac{\pi}{m/2k_BT}} = \sqrt{\frac{2\pi k_BT}{m}}, \quad (5.3)$$

so that

$$g(v_x) = \sqrt{\frac{m}{2\pi k_BT}} e^{-mv_x^2/2k_BT}. \quad (5.4)$$

It is then possible to find the following expected values of this distribution (using the integrals in Appendix C.2):

$$\langle v_x \rangle = \int_{-\infty}^{\infty} v_x g(v_x)\,dv_x = 0, \quad (5.5)$$

$$\langle |v_x| \rangle = 2\int_0^{\infty} v_x g(v_x)\,dv_x = \sqrt{\frac{2k_BT}{\pi m}}, \quad (5.6)$$

$$\langle v_x^2 \rangle = \int_{-\infty}^{\infty} v_x^2 g(v_x)\,dv_x = \frac{k_BT}{m}. \quad (5.7)$$

Of course, it does not matter which component of the velocity was initially chosen. Identical results would have been obtained for v_y and v_z. Hence the fraction of molecules with velocities between (v_x, v_y, v_z) and $(v_x + dv_x, v_y + dv_y, v_z + dv_z)$ is given by

$$g(v_x)dv_x\, g(v_y)dv_y\, g(v_z)dv_z$$
$$\propto e^{-mv_x^2/2k_BT}dv_x\, e^{-mv_y^2/2k_BT}dv_y\, e^{-mv_z^2/2k_BT}dv_z$$
$$= e^{-mv^2/2k_BT}\, dv_x\, dv_y\, dv_z. \quad (5.8)$$

5.2 The speed distribution

We now wish to turn to the problem of working out the distribution of molecular speeds in a gas. We want the fraction of molecules which are travelling with speeds between $v = |\boldsymbol{v}|$ and $v + dv$, and this corresponds to a spherical shell in velocity space of radius v and thickness dv (see Fig. 5.3). The volume of velocity space corresponding to speeds between v and $v + dv$ is therefore equal to

$$4\pi v^2\, dv, \quad (5.9)$$

so that the fraction of molecules with speeds between v and $v + dv$ can be defined as $f(v)\,dv$, where $f(v)$ is given by

$$\boxed{f(v)\,dv \propto v^2\,dv\, e^{-mv^2/2k_BT}.} \quad (5.10)$$

In this expression the 4π factor has been absorbed in the proportionality sign.

[2] The integral may be evaluated using eqn C.3.

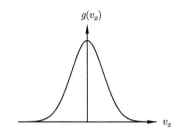

Fig. 5.2 $g(v_x)$, the distribution function for a particular component of molecular velocity (which is a Gaussian distribution).

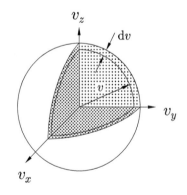

Fig. 5.3 Molecules with speeds between v and $v + dv$ occupy a volume of velocity space inside a spherical shell of radius v and thickness dv. (An octant of this sphere is shown cut-away.)

[3] We integrate between 0 and ∞, not between $-\infty$ and ∞, because the speed $v = |\mathbf{v}|$ is a positive quantity.

To normalize[3] this function, so that $\int_0^\infty f(v)\,dv = 1$, we must evaluate the integral (using eqn C.3)

$$\int_0^\infty v^2 e^{-mv^2/2k_BT}\,dv = \frac{1}{4}\sqrt{\frac{\pi}{(m/2k_BT)^3}}, \quad (5.11)$$

so that

$$\boxed{f(v)\,dv = \frac{4}{\sqrt{\pi}}\left(\frac{m}{2k_BT}\right)^{3/2} v^2\,dv\, e^{-mv^2/2k_BT}.} \quad (5.12)$$

This speed distribution function is known as the **Maxwell–Boltzmann speed distribution**, or sometimes simply as a **Maxwellian distribution** and is plotted in Fig. 5.4. Having derived the Maxwell–Boltzmann distribution function (eqn 5.10) we are now in a position to derive some of its properties.

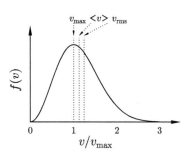

Fig. 5.4 $f(v)$, the distribution function for molecular speeds (Maxwell–Boltzmann distribution).

5.2.1 $\langle v \rangle$ and $\langle v^2 \rangle$

It is straightforward to find the following expected values of the Maxwell–Boltzmann distribution using standard integrals:

$$\langle v \rangle = \int_0^\infty v f(v)\,dv = \sqrt{\frac{8k_BT}{\pi m}}, \quad (5.13)$$

$$\langle v^2 \rangle = \int_0^\infty v^2 f(v)\,dv = \frac{3k_BT}{m}. \quad (5.14)$$

Note that using eqns 5.7 and 5.14 we can write

$$\langle v_x^2 \rangle + \langle v_y^2 \rangle + \langle v_z^2 \rangle = \frac{k_BT}{m} + \frac{k_BT}{m} + \frac{k_BT}{m} = \frac{3k_BT}{m} = \langle v^2 \rangle \quad (5.15)$$

as expected.

Note also that the root mean squared speed of a molecule

$$v_{\text{rms}} = \sqrt{\langle v^2 \rangle} = \sqrt{\frac{3k_BT}{m}} \quad (5.16)$$

is proportional to $m^{-1/2}$.

5.2.2 The mean kinetic energy of a gas molecule

The mean kinetic energy of a gas molecule is given by

$$\langle E_{\text{KE}} \rangle = \frac{1}{2}m\langle v^2 \rangle = \frac{3}{2}k_BT. \quad (5.17)$$

This is an important result, and we will later derive it again by a different route (see Section 19.2.1). It demonstrates that the average energy of a molecule in a gas depends only on temperature.

5.2.3 The maximum of $f(v)$

The maximum value of $f(v)$ is found by setting

$$\frac{\mathrm{d}f}{\mathrm{d}v} = 0, \tag{5.18}$$

and straightforward differentiation of eqn 5.10 yields

$$v_{\max} = \sqrt{\frac{2k_\mathrm{B}T}{m}}. \tag{5.19}$$

Since

$$\sqrt{2} < \sqrt{\frac{8}{\pi}} < \sqrt{3}, \tag{5.20}$$

we have that

$$v_{\max} < \langle v \rangle < v_{\mathrm{rms}} \tag{5.21}$$

and hence the points marked on Fig. 5.4 are in the order drawn. The mean speed of the Maxwell–Boltzmann distribution is higher than the value of the speed corresponding to the maximum in the distribution since the shape of $f(v)$ is such that the tail to the right is very long.

Example 5.1

Calculate the rms speed of a nitrogen (N_2) molecule at room temperature. [One mole of N_2 has a mass of 28 g.]

Solution:
For nitrogen at room temperature, $m = (0.028\,\mathrm{kg})/(6.022 \times 10^{23})$ and so $v_\mathrm{rms} \approx 500\,\mathrm{m\,s^{-1}}$. This is about 1100 miles per hour, and is the same order of magnitude as the speed of sound.

5.3 Experimental justification

How do you demonstrate that the velocity distribution in a gas obeys the Maxwell–Boltzmann distribution? A possible experimental apparatus is shown in Fig. 5.5. This consists of an oven, a velocity selector, and a detector, which are mounted on an optical bench. Hot gas atoms emerge from the oven and pass through a collimating slit. Velocity selection of molecules is achieved using discs with slits cut into them, which are rotated at high angular speed by a motor. A phase shifter varies the phase of the voltage fed to the motor for one disc relative to that of the other, so that the angle between the slits on the two discs can be continuously adjusted. Thus only molecules travelling with a particular speed from the oven will pass through the slits in both discs. A beam of light can be used to determine when the velocity selector is set for zero transit time. This beam is produced by a small light source near

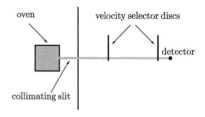

Fig. 5.5 The experimental apparatus that can be used to measure the Maxwell–Boltzmann distribution.

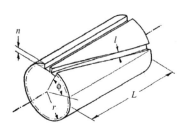

Fig. 5.6 Diagram of the velocity selector. (After R. C. Miller and P. Kusch, Phys. Rev. **99**, 1314 (1955).) copyright (1955) by the American Physical Society.

Fig. 5.7 Intensity data measured for potassium atoms using the velocity selector shown in Fig. 5.6 (from R. C. Miller and P. Kusch, Phys. Rev. **99**, 1314 (1955), Copyright (1955) by the American Physical Society). The line shows the best fit to an expression of the form $v^4 e^{-mv^2/2k_\mathrm{B}T}$ (see text).

one disc and passes through the velocity selector and is detected by a photocell near the other disc.

Another way of selecting the velocity is shown in Fig. 5.6. This consists of a solid surface on whose surface is cut a helical slot, and which is capable of rotation around the cylinder's axis at a rate ω. A molecule of velocity v which goes through the slot without changing its position relative to the sides of the slot will satisfy the equation

$$v = \frac{\omega L}{\phi}, \tag{5.22}$$

in which ϕ and L are the fixed angle and length shown in Fig. 5.6. Tuning ω allows you to tune the selected velocity v.

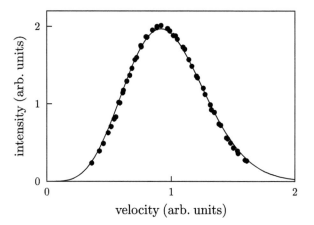

Data from this experiment are shown in Fig. 5.7. In fact, the intensity as a function of velocity v does not follow the expected $v^2 e^{-mv^2/2k_\mathrm{B}T}$ distribution but instead fits to $v^4 e^{-mv^2/2k_\mathrm{B}T}$. What has gone wrong?

Nothing has gone wrong, but there are two factors of v that must be included for two different reasons. One factor of v comes from the fact that the gas atoms emerging through the small aperture in the wall of the oven are not completely representative of the atoms inside the oven. This effect will be analysed in Chapter 7. The other factor of v comes from the fact that the range Δv of molecular velocities transmitted by the velocity selector also depends on ω. This can be understood in detail as follows. Because of the finite width of the slit, the velocity selector selects molecules with a range of velocities. The limiting velocities correspond to molecules that enter the slot at one wall and leave the slot at the opposite wall. This leads to velocities that range all the way from $\omega L/\phi_-$ to $\omega L/\phi_+$, where $\phi_\pm = \phi \pm l/r$ and l and r are as defined in Fig. 5.6. Thus the range, Δv, of velocities transmitted is given by

$$\Delta v = \omega L \left(\frac{1}{\phi_-} - \frac{1}{\phi_+} \right) \approx \frac{2l}{\phi r} v, \tag{5.23}$$

and thus increases as the selected velocity increases. This gives rise to the second additional factor of v.

Another way to justify the treatment in this chapter experimentally is to look at spectral lines of hot gas atoms. The limit on resolution is often set by **Doppler broadening**: those atoms travelling with a component of velocity v_x towards the detector will have transition frequencies that differ from those of atoms at rest due to the Doppler shift. A spectral line with frequency ω_0 (and wavelength $\lambda_0 = 2\pi c/\omega_0$, where c is the speed of light) will be Doppler-shifted to a frequency $\omega_0(1 \pm v_x/c)$ and the $\pm$ sign reflects molecules travelling towards or away from the detector. The Gaussian distribution of velocities given by eqn 5.2 now gives rise to a Gaussian shape of the spectral line $I(\omega)$ (see Fig. 5.8), which is given by

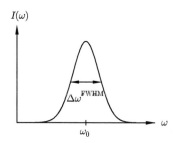

Fig. 5.8 The intensity of a Doppler-broadened spectral line.

$$I(\omega) \propto \exp\left(-\frac{mc^2(\omega_0 - \omega)^2}{2k_B T \omega_0^2}\right), \tag{5.24}$$

and the full-width at half-maximum (FWHM) of this spectral line, written as $\Delta\omega^{\text{FWHM}}$ (or in wavelength units: $\Delta\lambda^{\text{FWHM}}$), is given by

$$\frac{I(\omega_0 + \Delta\omega^{\text{FWHM}}/2)}{I(\omega_0)} = \frac{1}{2} \tag{5.25}$$

so that

$$\frac{\Delta\omega^{\text{FWHM}}}{\omega_0} = \frac{\Delta\lambda^{\text{FWHM}}}{\lambda_0} = 2\sqrt{2\ln 2 \frac{k_B T}{mc^2}}. \tag{5.26}$$

Another source of broadening of spectral lines arises from molecular collisions. This is called **collisional broadening** or sometimes **pressure broadening** (since collisions are more frequent in a gas when the pressure is higher, see Section 8.1). Doppler broadening is therefore most important in low-pressure gases.

Chapter summary

- A physical situation that is very important in kinetic theory is the translational motion of atoms or molecules in a gas. The probability distribution for a given component of velocity is given by

$$g(v_x) \propto e^{-mv_x^2/2k_BT}.$$

- We have shown that the corresponding expression for the probability distribution of molecular speeds is given by

$$\boxed{f(v) \propto v^2 e^{-mv^2/2k_BT}.}$$

This is known as a **Maxwell–Boltzmann** distribution, or sometimes as a **Maxwellian** distribution.

- Two important average values of the Maxwell–Boltzmann distribution are:

$$\langle v \rangle = \sqrt{\frac{8k_BT}{\pi m}}, \qquad \langle v^2 \rangle = \frac{3k_BT}{m}.$$

Exercises

(5.1) Evaluate the integrals in eqns 5.5–5.7 and eqns 5.13 and 5.14, and check that you get the same answers.

(5.2) Calculate the rms speed of hydrogen (H$_2$), helium (He) and oxygen (O$_2$) at room temperature. [The atomic masses of H, He, and O are 1, 4, and 16 respectively.] Compare these speeds with the escape velocity on the surface of (i) the Earth, (ii) the Sun.

(5.3) What fractional error do you make if you approximate $\sqrt{\langle v^2 \rangle}$ by $\langle v \rangle$ for a Maxwell–Boltzmann gas?

(5.4) A Maxwell–Boltzmann distribution implies that a given molecule (mass m) will have a speed between v and $v+dv$ with probability equal to $f(v)\,dv$ where

$$f(v) \propto v^2 e^{-mv^2/2k_BT},$$

and the proportionality sign is used because a normalization constant has been omitted. (You can correct for this by dividing any averages you work out by $\int_0^\infty f(v)\,dv$.) For this distribution, calculate the mean speed $\langle v \rangle$ and the mean inverse speed $\langle 1/v \rangle$. Show that

$$\langle v \rangle \langle 1/v \rangle = \frac{4}{\pi}.$$

(5.5) The width of a spectral line (FWHM) is often quoted as

$$\Delta\lambda^{\text{FWHM}} = 7.16 \times 10^{-7}\lambda_0\sqrt{\frac{T}{m}}, \qquad (5.27)$$

where T is the temperature in kelvin, λ_0 is the wavelength at the centre of the spectral line in the rest frame and m is the atomic mass of the gas measured in atomic mass units (i.e., multiples of the mass of a proton). Does this formula make sense?

(5.6) What is the Doppler broadening of the 21 cm line in an interstellar gas cloud (temperature 100 K) composed of neutral hydrogen (i.e., non-ionized atomic hydrogen)? Express your answer in kHz.

(5.7) Calculate the rms speed of a sodium atom in the solar atmosphere at 6000 K. (The atomic mass of sodium is 23.) The sodium D lines ($\lambda = 5900$ Å) are observed in a solar spectrum. Estimate the Doppler broadening in GHz.

James Clerk Maxwell (1831–1879)

Born in Edinburgh, James Clerk Maxwell was brought up in the Scottish countryside at Glenair. He was educated at home until, at the age of ten, he was sent to the Edinburgh Academy where his unusual homemade clothes and distracted air earned him the nickname "Dafty".

Fig. 5.9 James Clerk Maxwell

But a lot was going on in his head and he wrote his first scientific paper aged 14. Maxwell went to Peterhouse, Cambridge in 1850 but then moved to Trinity College, where he gained a fellowship in 1854. There he worked on the perception of colour, and also put Michael Faraday's ideas of lines of electrical force onto a sound mathematical basis. In 1856 he took up a chair in Natural Philosophy in Aberdeen where he worked on a theory of the rings of Saturn (confirmed by the Voyager spacecraft visits of the 1980's) and, in 1858, married the College Principal's daughter, Katherine Mary Dewar.

In 1859, he was inspired by a paper of Clausius on diffusion in gases to conceive of his theory of speed distributions in gases, outlined in Chapter 5, which, with its subsequent elaborations by Boltzmann, is known as the Maxwell–Boltzmann distribution. These triumphs were not enough to preserve him from the consequences of the merging of Aberdeen's two Universities in 1860 when, incredibly, the powers that be decided that it was Maxwell out of the two Professors of Natural Philosophy who should be made redundant. He failed to obtain a chair at Edinburgh (losing out to Tait) but instead moved to King's College, London. There, he produced the world's first colour photograph, came up with his theory of electromagnetism that proposed that light was an electromagnetic wave and explained its speed in terms of electrical properties, and chaired a committee to decide on a new system of units to incorporate the new understanding of the link between electricity and magnetism (and which became known as the "Gaussian", or cgs, system – though "Maxwellian system" would have been more appropriate). He also constructed his apparatus for measuring the viscosity of gases (see Chapter 9), verifying some of his predictions, but not others.

In 1865, he resigned his chair at King's and moved full time to Glenair, where he wrote his *Theory of Heat* which introduced what are now known as Maxwell relations (Chapter 16) and the concept of the Maxwell's demon (Section 14.7). He applied for, but did not get, the position of Principal of St Andrews' University, but in 1871 was appointed to the newly-established Professorship of Experimental Physics in Cambridge (after William Thomson and Hermann Helmholtz both turned the job down). There he supervised the building of the Cavendish Laboratory and wrote his celebrated *A Treatise on Electricity and Magnetism* (1873) where his four electromagnetic equations ("Maxwell's equations") first appear. In 1877 he was diagnosed with abdominal cancer and died in Cambridge in 1879.

In his short life Maxwell had been one of the most prolific, inspirational, and creative scientists who has ever lived. His work has had far-reaching implications in much of physics, not just in thermodynamics. He had also lived a devout and contemplative life in which he had been free of pride, selfishness, and ego, always generous and courteous to everyone. The doctor who tended him in his last days wrote:

> I must say that he is one of the best men I have ever met, and a greater merit than his scientific achievements is his being, so far as human judgement can discern, a most perfect example of a Christian gentleman.

Maxwell summed up his own philosophy as follows:

> Happy is the man who can recognize in the work of Today a connected portion of the work of life, and an embodiment of the work of Eternity. The foundations of his confidence are unchangeable, for he has been made a partaker of Infinity.

6 Pressure

6.1 Molecular distributions   57
6.2 The ideal gas law   58
6.3 Dalton's law   60
Chapter summary   61
Exercises   61

One of the most fundamental variables in the study of gases is pressure. The pressure p due to a gas (or in fact any fluid) is defined as the ratio of the perpendicular contact force to the area of contact. The unit is therefore that of force (N) divided by that of area (m^2) and is called the pascal (Pa = Nm^{-2}). The direction in which pressure acts is always at right angles to the surface upon which it is acting.

> *Other units* for measuring pressure are sometimes encountered, such as the bar (1 bar = 10^5 Pa) and the almost equivalent atmosphere (1 atm = 1.01325×10^5 Pa). The pressure of the atmosphere at sea level actually varies depending on the weather by approximately ± 50 mbar around the standard atmosphere of 1013.25 mbar, though pressures (adjusted for sea level) as low as 882 mbar and as high as 1084 mbar have been recorded. An archaic unit is the torr, which is equal to a millimetre of mercury (Hg): 1 torr = 133.32 Pa.

Example 6.1

Air has a density of about $1.29 \, \mathrm{kg\,m^{-3}}$. Give a rough estimate of the height of the atmosphere assuming that the density of air in the atmosphere is uniform.
Solution:
Atmospheric pressure $p \approx 10^5$ Pa is due to the weight of air $\rho g h$ in the atmosphere (with assumed height h and uniform density ρ) pressing down on each square metre. Hence $h = p/\rho g \approx 10^4$ m (which is about the cruising altitude of planes). Of course, in reality the density of the atmosphere falls off with increasing height (see Chapter 37).

The pressure p of a volume V of gas (comprising N molecules) depends on its temperature T via an **equation of state**, which is an expression of the form

$$p = f(T, V, N), \tag{6.1}$$

where f is some function. One example of an equation of state is that for an **ideal gas**, which was given in eqn 1.12:

$$pV = Nk_\mathrm{B}T. \tag{6.2}$$

Daniel Bernoulli (1700–1782) attempted an explanation of Boyle's law ($p \propto 1/V$) by assuming (controversially at the time) that gases were composed of a vast number of tiny particles (see Fig. 6.1). This was the first serious attempt at a kinetic theory of gases of the sort that we will describe in this chapter to derive the ideal gas equation.

6.1 Molecular distributions

In the previous chapter we derived the Maxwell–Boltzmann speed distribution function $f(v)$. We denote the total number of molecules per unit volume by the symbol n. The number of molecules per unit volume travelling with speeds between v and $v + dv$ is then given by $nf(v)\,dv$. We now seek to determine the distribution function of molecules travelling in different directions.

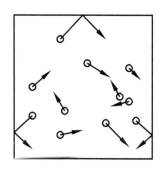

Fig. 6.1 In the kinetic theory of gases, a gas is modelled as a number of individual tiny particles (atoms or molecules), which can bounce off the walls of the container and each other.

6.1.1 Solid angles

Recall that an angle θ in a circle is defined by dividing the arc length s which the angle subtends by the radius r (see Fig. 6.2), so that

$$\theta = \frac{s}{r}. \tag{6.3}$$

The angle is measured in radians. The angle subtended by the whole circle at its centre is then

$$\frac{2\pi r}{r} = 2\pi. \tag{6.4}$$

By analogy, a **solid angle** Ω in a sphere (see Fig. 6.3) is defined by dividing the surface area A which the solid angle subtends by the radius squared, so that

$$\Omega = \frac{A}{r^2}. \tag{6.5}$$

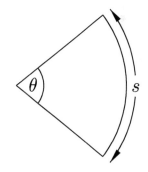

Fig. 6.2 The definition of angle θ in terms of the arc length.

The solid angle is measured in **steradians**. The solid angle subtended by a whole sphere at its centre is then

$$\frac{4\pi r^2}{r^2} = 4\pi. \tag{6.6}$$

6.1.2 The number of molecules travelling in a certain direction at a certain speed

If all molecules are equally likely to be travelling in any direction, the fraction whose trajectories lie in an elemental solid angle $d\Omega$ is

$$\frac{d\Omega}{4\pi}. \tag{6.7}$$

If we choose a particular direction, then the solid angle $d\Omega$ corresponding to molecules travelling at angles between θ and $\theta + d\theta$ to that direction is

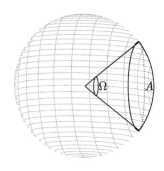

Fig. 6.3 The definition of solid angle $\Omega = A/r^2$ where r is the radius of the sphere and A is the surface area over the region of the sphere indicated.

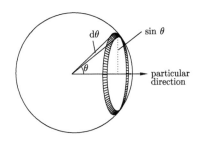

Fig. 6.4 The area of the shaded region on this sphere of unit radius is equal to the circumference of a circle of radius $\sin\theta$ multiplied by the width $d\theta$ and is hence given by $2\pi \sin\theta\, d\theta$.

equal to the area of the annular region shown shaded in the unit-radius sphere of Fig. 6.4 which is given by

$$d\Omega = 2\pi \sin\theta\, d\theta, \tag{6.8}$$

so that

$$\frac{d\Omega}{4\pi} = \frac{1}{2}\sin\theta\, d\theta. \tag{6.9}$$

Therefore, a number of molecules per unit volume given by

$$\boxed{n\, f(v)\, dv\, \tfrac{1}{2}\sin\theta\, d\theta} \tag{6.10}$$

have speeds between v and $v+dv$ and are travelling at angles between θ and $\theta + d\theta$ to the chosen direction, where $f(v)$ is the speed distribution function.

6.1.3 The number of molecules hitting a wall

We now let our particular direction, up until now arbitarily chosen, lie perpendicular to a wall of area A (see Fig. 6.5). In a small time dt, the molecules travelling at angle θ to the normal to the wall sweep out a volume

$$A\,v\,dt\,\cos\theta. \tag{6.11}$$

Multiplying this volume by the number in expression 6.10 implies that in time dt, the number of molecules hitting a wall of area A is

$$A\,v\,dt\,\cos\theta\, n\, f(v)\, dv\, \frac{1}{2}\sin\theta\, d\theta. \tag{6.12}$$

Hence, the number of molecules hitting unit area of wall in unit time, and having speeds between v and $v+dv$ and travelling at angles between θ and $\theta + d\theta$, is given by

$$\boxed{v\cos\theta\, n\, f(v)\, dv\, \tfrac{1}{2}\sin\theta\, d\theta.} \tag{6.13}$$

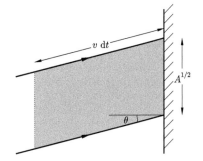

Fig. 6.5 Molecules hit a region of wall (of cross-sectional area $A^{1/2} \times A^{1/2} = A$) at an angle θ. The number hitting in time dt is the volume of the shaded region ($A\,v\,dt\,\cos\theta$) multiplied by $n\,f(v)\,dv\,\tfrac{1}{2}\sin\theta$.

The angle θ runs from 0 to π for all molecules. However, in the integral we only consider molecules with θ from 0 to $\pi/2$ because these are the ones which will hit the wall. Those with $\pi/2 < \theta < \pi$ are travelling *away* from the wall.

6.2 The ideal gas law

We are now in a position to calculate the pressure of a gas on its container. Each molecule hitting the wall of the container has a momentum change of $2mv\cos\theta$, which is perpendicular to the wall. This change of momentum is equivalent to an impulse. Hence, if we multiply $2mv\cos\theta$ (the momentum change arising from one molecule hitting the container walls) by the number of molecules hitting unit area per unit time, and having speeds between v and $v + dv$ and angles between θ and $\theta + d\theta$ (which we derived in eqn 6.13), and then integrating over θ and v, we should get the pressure p. Thus

$$\begin{aligned} p &= \int_0^\infty \int_0^{\pi/2} (2mv\cos\theta)\left(v\cos\theta\, n\, f(v)\, dv\, \frac{1}{2}\sin\theta\, d\theta\right) \\ &= mn \int_0^\infty dv\, v^2 f(v) \int_0^{\pi/2} \cos^2\theta \sin\theta\, d\theta, \end{aligned} \tag{6.14}$$

and using the integral $\int_0^{\pi/2} \cos^2\theta \sin\theta \, d\theta = \frac{1}{3}$, we have that

$$p = \tfrac{1}{3} nm \langle v^2 \rangle. \qquad (6.15)$$

If we write the total number of molecules N in volume V as

$$N = nV, \qquad (6.16)$$

then this equation can be written as

$$pV = \frac{1}{3} Nm \langle v^2 \rangle. \qquad (6.17)$$

Using $\langle v^2 \rangle = 3k_\mathrm{B} T/m$, this can be rewritten as

$$pV = Nk_\mathrm{B} T, \qquad (6.18)$$

which is the **ideal gas equation** we met in eqn 1.12. This completes the kinetic theory derivation of the ideal gas law.

Equivalent forms of the ideal gas law:

- The form given in eqn 6.18 is

$$pV = Nk_\mathrm{B} T,$$

  and contains an "N", which we reiterate is the total number of molecules in the gas.

- An equivalent form of the ideal gas equation can be derived by dividing both sides of eqn 6.18 by volume, so that

$$p = nk_\mathrm{B} T, \qquad (6.19)$$

  where $n = N/V$ is the number of molecules per unit volume.

- Another form of the ideal gas law can be obtained by writing the number of molecules $N = n_\mathrm{m} N_\mathrm{A}$, where n_m is the number of moles and N_A is the Avogadro number (the number of molecules in a mole, see Section 1.1). In this case, eqn 6.18 becomes

$$pV = n_\mathrm{m} RT, \qquad (6.20)$$

  where

$$R = N_\mathrm{A} k_\mathrm{B} \qquad (6.21)$$

  is the **gas constant** ($R = 8.31447 \, \mathrm{J\,K^{-1}\,mol^{-1}}$).

The ideal gas law ($p = nk_\mathrm{B} T$) expresses the important point that the pressure of an ideal gas does not depend on the mass m of the molecules. Although more massive molecules transfer greater momentum to the container walls than light molecules, their mean velocity is lower and so they make fewer collisions with the walls. Therefore the pressure is the same for a gas of light or massive molecules; it depends only on n, the number per unit volume, and the temperature.

Example 6.2

What is the volume occupied by one mole of ideal gas at **standard temperature and pressure** (STP, defined as $0\,°\mathrm{C}$ and 1 atm)?

Solution:
At $p = 1.01325 \times 10^5\,\mathrm{Pa}$ and $T = 273.15\,\mathrm{K}$, the **molar volume** V_m can be obtained from eqn 6.20 as

$$V_\mathrm{m} = \frac{RT}{p} = 0.022414\,\mathrm{m^3} = 22.414\,\mathrm{litres}. \tag{6.22}$$

Example 6.3

What is the connection between pressure and kinetic energy density?

Solution:
The kinetic energy of a gas molecule moving with speed v is

$$\frac{1}{2}mv^2. \tag{6.23}$$

The total kinetic energy of the molecules of a gas per unit volume, i.e., the kinetic energy density, which we will call u, is therefore given by

$$u = n \int_0^\infty \frac{1}{2}mv^2 f(v)\,\mathrm{d}v = \frac{1}{2}nm\langle v^2 \rangle, \tag{6.24}$$

so that comparing with eqn 6.15 we have that

$$p = \frac{2}{3}u. \tag{6.25}$$

This expression is true for a non-relativistic gas of particles. For an ultrarelativistic gas, the correct expression is given in eqn 25.21.

6.3 Dalton's law

If one has a mixture of gases in thermal equilibrium, then the total pressure $p = nk_\mathrm{B}T$ is simply the sum of the pressures due to each component of the mixture. We can write n as

$$n = \sum_i n_i, \tag{6.26}$$

where n_i is the number density of the ith species. Therefore

$$p = \left(\sum_i n_i\right) k_\mathrm{B}T = \sum_i p_i, \tag{6.27}$$

where $p_i = n_i k_\mathrm{B}T$ is known as the **partial pressure** of the ith species. The observation that $p = \sum_i p_i$ is known as **Dalton's law**, after the British chemist John Dalton (1766–1844), who was a pioneer of the atomic theory.

Example 6.4

Air is 75.5% N_2, 23.2% O_2, 1.3% Ar, and 0.05% CO_2 by mass. Calculate the partial pressure of CO_2 in air at atmospheric pressure.

Solution:
Dalton's law states that the partial pressure is proportional to the number density. The number density is proportional to the mass fraction divided by the molar mass. The molar masses of the species (in grammes) are 28 (N_2), 32 (O_2), 40 (Ar), and 44 (CO_2). Hence, the partial pressure of CO_2 is

$$p_{CO_2} = \frac{\frac{0.05}{44} \times 1 \text{ atm}}{\frac{75.5}{28} + \frac{23.2}{32} + \frac{1.3}{40} + \frac{0.05}{44}} = 0.00033 \text{ atm}. \qquad (6.28)$$

Chapter summary

- The pressure, p, is given by

$$p = \frac{1}{3} nm \langle v^2 \rangle,$$

where n is the number of molecules per unit volume, m is the molecular mass, and v is the molecular velocity.

- This expression agrees with the ideal gas equation,

$$p = nk_B T,$$

where T is the temperature and k_B is the Boltzmann constant.

Exercises

(6.1) What is the volume occupied by 1 mole of gas at 10^{-10} torr, the pressure inside an "ultra high vacuum" (UHV) chamber.

(6.2) Calculate u, the kinetic energy density, for air at atmospheric pressure.

(6.3) Mr Fourier sits in his living room at 18 °C. He decides he is rather cold and turns the heating up so that the temperature is 25 °C. What happens to the total energy of the air in his living room? [Hint: what controls the pressure in the room?]

(6.4) A diffuse cloud of neutral hydrogen atoms (known as HI) in space has a temperature of 50 K and number density 500 cm^{-3}. Calculate the pressure (in Pa) and the volume (in cubic light years) occupied by the cloud if its mass is $100 M_\odot$. ($M_\odot$ is the symbol for the mass of the Sun, see Appendix A.)

(6.5) (a) Given that the number of molecules hitting unit area of a surface per second with speeds between v and $v + dv$ and angles between θ and $\theta + d\theta$ to the normal is

$$\frac{1}{2} v n f(v) dv \sin\theta \cos\theta \, d\theta,$$

show that the average value of $\cos\theta$ for these molecules is $\frac{2}{3}$.

(b) Using the results above, show that for a gas obeying the Maxwellian distribution (i.e., $f(v) \propto v^2 e^{-mv^2/2k_B T}$) the average energy of all the molecules is $\frac{3}{2} k_B T$, but the average energy of those hitting the surface is $2 k_B T$.

(6.6) The molecules in a gas travel with different velocities. A particular molecule will have velocity $\boldsymbol{v}$ and speed $v = |\boldsymbol{v}|$ and will move at an angle θ to some chosen fixed axis. We have shown that the number of molecules in a gas with speeds between v and $v + dv$, and moving at angles between θ and $\theta + d\theta$ to any chosen axis is given by

$$\frac{1}{2} n f(v) \, dv \sin\theta \, d\theta,$$

where n is the number of molecules per unit volume and $f(v)$ is some function of v only. [$f(v)$ could be the Maxwellian distribution given above; however you should not assume this but rather calculate the general case.] Hence show by integration that:
(a) $\langle u \rangle = 0$,
(b) $\langle u^2 \rangle = \frac{1}{3} \langle v^2 \rangle$,
(c) $\langle |u| \rangle = \frac{1}{2} \langle v \rangle$,
where u is any one Cartesian component of $\boldsymbol{v}$, i.e., v_x, v_y, or v_z.
[Hint: You can take u as the z-component of $\boldsymbol{v}$ without loss of generality. Why? Then express u in terms of v and θ and average over v and θ. You can use expressions such as

$$\langle v \rangle = \frac{\int_0^\infty v f(v) \, dv}{\int_0^\infty f(v) \, dv}$$

and similarly for $\langle v^2 \rangle$. Make sure you understand why.]

(6.7) If v_1, v_2, v_3 are three Cartesian components of $\boldsymbol{v}$, what value do you expect for $\langle v_1 v_2 \rangle$, $\langle v_1 v_3 \rangle$, and $\langle v_2 v_3 \rangle$? Evaluate one of them by integration to check your deduction.

(6.8) Calculate the partial pressure of O_2 in air at atmospheric pressure.

(6.9) This question provides an alternative derivation of the formula for pressure. Without loss of generality, let us consider molecules travelling towards a wall which lies in the yz plane. The momentum change of a molecule of mass m and velocity $\boldsymbol{v} = (v_x, v_y, v_z)$ bouncing off the wall will be $2m v_x$. Explain why the pressure p on the wall is given by

$$p = \int_0^\infty (2m v_x) v_x n g(v_x) \, dv_x, \qquad (6.29)$$

where $g(v_x)$ is the function given in eqn 5.2. Hence show that $p = n k_B T$. Using the same approach show that Φ, the number of molecules hitting unit area of the wall per second is given by

$$\Phi = \int_0^\infty v_x n g(v_x) \, dv_x = n \sqrt{\frac{k_B T}{\pi m}} = \frac{1}{4} n \langle v \rangle. \qquad (6.30)$$

This result will be derived using a different method in the next chapter.

Robert Boyle (1627–1691)

Robert Boyle was born into wealth. His father was a self-made man of humble yeoman stock who, at the age of 22, had left England for Ireland to seek his fortune. This his father found or, possibly more accurately, "grabbed" and through rapid land acquisition of a rather dubious nature Boyle senior became one of England's richest men and the Earl of Cork to boot. Robert was born when his father was in his sixties and was the last but one of his father's sixteen children. His father, as a new member of the aristocracy, believed in the best education for his children, and Robert was duly packed off to Eton and then, at the age of 12, sent off for a European Grand Tour, taking in Geneva, Venice, and Florence. Boyle studied the works of Galileo, who died in Florence while Boyle was staying in the city. Meanwhile, his father was getting into a spot of bother with the Irish rebellion of 1641–1642, resulting in the loss of the rents that kept him and his family in the manner to which they had become accustomed, and hence also causing Robert Boyle some financial difficulty. He was almost married off at this time to a wealthy heiress, but Boyle managed to escape this fate and remained unmarried for the rest of his life. His father died in 1643 and Boyle returned to England the following year, inheriting his father's Dorset estate.

Fig. 6.6 Robert Boyle

However, by this time the Civil War (which had started in 1642) was in full swing and Boyle tried hard not to take sides. He kept his head down, devoting his time to study, building a chemical laboratory in his house and worked on moral and theological essays. Cromwell's defeat of the Irish in 1652 worked well for Boyle as many Irish lands were handed over to the English colonists. Financially, Boyle was now secure and ready to live the life of a gentleman. In London, he had met John Wilkins, who had founded an intellectual society, which he called "The Invisible College" and which suddenly brought Boyle into contact with the leading thinkers of the day. When Wilkins was appointed Warden of Wadham College, Oxford, Boyle decided to move to Oxford and set up a laboratory there. He set up an air pump and, together with a number of talented assistants (the most famous of which was Robert Hooke, later to discover his law of springs and to observe a cell with a microscope, in addition to numerous other discoveries) Boyle and his team conducted a large number of elaborate experiments in this new vacuum. They showed that sound did not travel in a vacuum, and that flames and living organisms could not be sustained, and discovered the "spring of air", namely that compressing air resulted in its pressure increasing, and that the pressure of a gas and its volume were in inverse proportion.

Boyle was much taken with the atomistic viewpoint as described by the French philosopher Pierre Gassendi (1592–1655), which seems particularly appropriate for someone whose work led to the path for the development of the kinetic theory of gases. His greatest legacy was in his reliance on experiment as a means of determining scientific truth. He was, however, also someone who often worked vicariously through a band of assistants, citing his weakness of health and of eyesight as a reason for failing to write his papers as he wished to and to have read other peoples' works as he ought; his writings are, however, full of criticisms of his assistants for making mistakes, failing to record data, and generally slowing down his research endeavours.

With the restoration of the monarchy in 1660, the Invisible College, which had been meeting for several years in Gresham College, London, sought the blessing of the newly crowned Charles II and became the Royal Society, which has existed ever since as a thriving scientific society. In 1680, Boyle (who had been a founding fellow of the Royal Society) was elected President of the Royal Society, but declined to hold the office, citing an unwillingness to take the necessary oaths. Boyle retained a strong Christian faith throughout his life, and prided himself on his honesty and pure seeking of the truth. In 1670, Boyle suffered a stroke but made a good recovery, staying active in research until the mid-1680's. He died in 1691, shortly after the death of his sister Katherine to whom he had been extremely close.

7 Molecular effusion

7.1 Flux   64
7.2 Effusion   66
Chapter summary   69
Exercises   69

Effusion is the process by which a gas escapes from a very small hole. The empirical relation known as **Graham's law of effusion** [after Thomas Graham (1805–1869)] states that the rate of effusion is inversely proportional to the square root of the mass of the effusing molecule.

Example 7.1

Effusion can be used to separate different isotopes of a gas (which cannot be separated chemically). For example, in the separation of $^{235}\mathrm{UF}_6$ and $^{238}\mathrm{UF}_6$ the ratio of the effusion rates of the two gases is equal to

$$\sqrt{\frac{\text{mass of }{}^{238}\mathrm{UF}_6}{\text{mass of }{}^{235}\mathrm{UF}_6}} = \sqrt{\frac{352.0412}{348.0343}} = 1.00574, \quad (7.1)$$

which, although small, was enough for many kilogrammes of $^{235}\mathrm{UF}_6$ to be extracted for the Manhattan project in 1945 to produce the first uranium atom bomb, which was subsequently dropped on Hiroshima.

Isotopes (the word means "same place") are atoms of a chemical element with the same atomic number Z (and hence number of protons in the nucleus) but different atomic weights A (and hence different number of neutrons in the nucleus).

Example 7.2

How much faster does helium gas effuse out of a small hole than N_2?
Solution:

$$\sqrt{\frac{\text{mass of }\mathrm{N}_2}{\text{mass of He}}} = \sqrt{\frac{28}{4}} = 2.6. \quad (7.2)$$

In this chapter, we will discover where Graham's law comes from. We begin by evaluating the flux of particles hitting the inside walls of the container of a gas.

7.1 Flux

The concept of **flux** is a very important one in thermal physics. It quantifies the flow of particles or the flow of energy or even the flow of momentum. Of relevance to this chapter is the molecular flux, Φ, which is defined to be the number of molecules striking unit area per second. Thus

$$\text{molecular flux} = \frac{\text{number of molecules}}{\text{area} \times \text{time}}. \tag{7.3}$$

The units of molecular flux are therefore $\mathrm{m^{-2}\,s^{-1}}$. We can also define heat flux using

$$\text{heat flux} = \frac{\text{amount of heat}}{\text{area} \times \text{time}}. \tag{7.4}$$

The units of heat flux are therefore $\mathrm{J\,m^{-2}\,s^{-1}}$. In Section 9.1, we will also come across a flux of momentum.

Returning to the effusion problem, we note that the flux of molecules in a gas can be evaluated by integrating expression 6.13 over all v and θ, so that

$$\begin{aligned}
\Phi &= \int_0^\infty \int_0^{\pi/2} v \cos\theta \, n \, f(v) \, dv \, \frac{1}{2}\sin\theta \, d\theta \\
&= \frac{n}{2} \int_0^\infty dv \, v \, f(v) \int_0^{\pi/2} d\theta \, \cos\theta \sin\theta \tag{7.5}
\end{aligned}$$

so that

$$\boxed{\Phi = \tfrac{1}{4} n \langle v \rangle.} \tag{7.6}$$

An alternative expression for Φ can be found as follows: rearranging the ideal gas law $p = n k_\mathrm{B} T$, we can write

$$n = \frac{p}{k_\mathrm{B} T}, \tag{7.7}$$

and using the expression for the average speed of molecules in a gas from eqn 5.13

$$\langle v \rangle = \sqrt{\frac{8 k_\mathrm{B} T}{\pi m}}, \tag{7.8}$$

we can substitute these expressions into eqn 7.6 and obtain

$$\boxed{\Phi = \frac{p}{\sqrt{2\pi m k_\mathrm{B} T}}.} \tag{7.9}$$

Note that consideration of eqn 7.9 shows us that the effusion rate depends inversely on the square root of the mass, in agreement with Graham's law.

> In the integral we only consider molecules with θ from 0 to $\pi/2$ because these are the ones which will contribute to the flux through a particular area. Those with $\pi/2 < \theta < \pi$ are travelling away from the area.
>
> We have used $\int_0^{\pi/2} d\theta \, \cos\theta \sin\theta = \tfrac{1}{2}$. (Hint: substitute $u = \sin\theta$, $du = \cos\theta \, d\theta$, so that the integral becomes $\int_0^1 u \, du = \tfrac{1}{2}$.)

Example 7.3

Calculate the particle flux from N_2 gas at STP (standard temperature and pressure, i.e., 1 atm and 0°C).
Solution:

$$\begin{aligned}
\Phi &= \frac{1.01325 \times 10^5 \,\mathrm{Pa}}{\sqrt{2\pi \times (28 \times 1.67 \times 10^{-27}\,\mathrm{kg}) \times 1.38 \times 10^{-23}\,\mathrm{J\,K^{-1}} \times 273\,\mathrm{K}}} \\
&\approx 3 \times 10^{27}\,\mathrm{m^{-2}\,s^{-1}}. \tag{7.10}
\end{aligned}$$

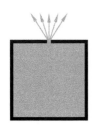

Fig. 7.1 A gas effuses from a small hole in its container.

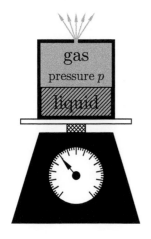

Fig. 7.2 The Knudsen method.

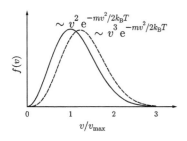

Fig. 7.3 The distribution function for molecular speeds (Maxwell–Boltzmann distribution) in a gas is proportional to $v^2 e^{-mv^2/2k_B T}$ (solid line) but the gas effusing from a small hole has a distribution function that is proportional to $v^3 e^{-mv^2/2k_B T}$ (dashed line). The distinction between the two situations occurs when counting the molecules crossing a fixed plane during some interval of time.

7.2 Effusion

Consider a container of gas with a small hole of area A in the side. Gas will leak (i.e., effuse) out of the hole (see Fig. 7.1). The hole is small, so that the equilibrium of gas in the container is not disturbed. The number of molecules escaping per unit time is just the number of molecules hitting the hole area in the closed box per second, so is given by ΦA, where Φ is the molecular flux. This is the **effusion rate**.

Example 7.4

In the Knudsen method of measuring vapour pressure p from a liquid containing molecules of mass m at temperature T, the liquid is placed in the bottom of a container that has a small hole of area A at the top (see Fig. 7.2). The container is placed on a weighing balance and its weight Mg is measured as a function of time. In equilibrium, the effusion rate is

$$\Phi A = \frac{pA}{\sqrt{2\pi m k_B T}}, \qquad (7.11)$$

so that the rate of change of mass, dM/dt is given by $-m\Phi A$. Hence

$$p = \sqrt{\frac{2\pi k_B T}{m}} \frac{1}{A} \left| \frac{dM}{dt} \right|. \qquad (7.12)$$

Effusion preferentially selects faster molecules. Therefore the speed distribution of molecules effusing through the hole is not Maxwellian. This result seems paradoxical at first glance: aren't the molecules emerging from the box the same ones that were inside beforehand? How can their distribution be different?

The reason is that the faster molecules inside the box travel more quickly and have a greater probability of reaching the hole than their slower cousins.[1] This can be expressed mathematically by noticing that the number of molecules hitting a wall (or a hole) is given by eqn 6.13 and this has an extra factor of v in it. Thus the distribution of molecules effusing through the hole in some interval of time is proportional to

$$v^3 e^{-mv^2/2k_B T}. \qquad (7.13)$$

Note the extra factor of v in this expression compared with the usual Maxwell–Boltzmann distribution in eqn 5.10 (see Fig. 7.3). The molecules

---

[1] An analogy may help here: the foreign tourists who visit your country are not completely representative of the nation from which they have come; this is because they are likely to be at least a little more adventurous than their average countrymen and countrywomen by the very fact that they have actually stepped out of their own borders.

in the Maxwellian gas had an average energy of $\frac{1}{2}m\langle v^2\rangle = \frac{3}{2}k_B T$, but the molecules in the effusing gas have a higher energy, as the following example will demonstrate.

Example 7.5

What is the mean kinetic energy of gas molecules effusing out of a small hole?
Solution:

$$\langle \text{kinetic energy}\rangle = \frac{1}{2}m\langle v^2\rangle \qquad (7.14)$$

$$= \frac{\frac{1}{2}m \int_0^\infty v^2\, v^3\, e^{-\frac{1}{2}mv^2/k_B T}\, dv}{\int_0^\infty v^3\, e^{-\frac{1}{2}mv^2/k_B T}\, dv}$$

$$= \frac{1}{2}m\left(\frac{2k_B T}{m}\right)\frac{\int_0^\infty u^2 e^{-u}\, du}{\int_0^\infty u e^{-u}\, du}$$

where the substitution $u = mv^2/2k_B T$ has been made. Using the standard integral $\int_0^\infty x^n e^{-x}\, dx = n!$ (see Appendix C.1), we have that

$$\langle \text{kinetic energy}\rangle = 2k_B T. \qquad (7.15)$$

This is larger by a factor of $\frac{4}{3}$ than the mean kinetic energy of molecules in the gas. This is because effusion preferentially selects higher energy molecules.

The hole has to be small. How small? The diameter of the hole has to be much less[2] than the mean free path λ, defined in Section 8.3.

[2] This is because, as we shall see in Section 8.3, the mean free path controls the characteristic distance between collisions. If the hole is small on this scale, molecules can effuse out without the rest of the gas "noticing", i.e., without a pressure gradient developing close to the hole.

Example 7.6

Consider a container divided by a partition with a small hole, diameter D, containing the same gas on each side. The gas on the left-hand side has temperature T_1 and pressure p_1. The gas on the right-hand side has temperature T_2 and pressure p_2.

If $D \gg \lambda$, $p_1 = p_2$.

If $D \ll \lambda$, we are in the effusion regime and the system will achieve equilibrium when the molecular fluxes balance, so that

$$\Phi_1 = \Phi_2, \qquad (7.16)$$

so that, using eqn 7.9 we may write

$$\frac{p_1}{\sqrt{T_1}} = \frac{p_2}{\sqrt{T_2}}. \qquad (7.17)$$

This is called the **Knudsen effect**, after Martin Knudsen (1871–1949).

A final example gives an approximate derivation of the flow rate of gas down a pipe at low pressures.

Example 7.7

Estimate the mass flow rate of gas down a long pipe of length L and diameter D at very low pressures in terms of the difference in pressures $p_1 - p_2$ between the two ends of the pipe.

Solution:

This type of flow is known as **Knudsen flow**. At very low pressures, molecules collide with the walls of the tube much more often than they do with each other. Let us define a coordinate x, which measures the distance along the pipe. The net flux $\Phi(x)$ of molecules flowing down the pipe at position x can be estimated by subtracting the molecules effusing down the pipe since their last collision (roughly a distance D upstream) from the molecules effusing up the pipe since their last collision (roughly a distance D downstream). Thus

$$\Phi(x) \approx \frac{1}{4}\langle v \rangle [n(x-D) - n(x+D)], \tag{7.18}$$

where $n(x)$ is the number density of molecules at position x. Using $p = \frac{1}{3}nm\langle v^2 \rangle$ (eqn 6.15), this can be written

$$\Phi(x) \approx \frac{3}{4m}\frac{\langle v \rangle}{\langle v^2 \rangle}[p(x-D) - p(x+D)]. \tag{7.19}$$

We can write

$$p(x-D) - p(x+D) \approx -2D\frac{\mathrm{d}p}{\mathrm{d}x}, \tag{7.20}$$

but also notice that in steady state Φ must be the same along the tube, so that

$$\frac{\mathrm{d}p}{\mathrm{d}x} = \frac{p_2 - p_1}{L}. \tag{7.21}$$

Hence the mass flow rate $\dot{M} = m\Phi(\pi D^2/4)$ (where $\pi D^2/4$ is the cross-sectional area of the pipe) is given by

$$\dot{M} \approx \frac{3}{8}\frac{\langle v \rangle}{\langle v^2 \rangle}\pi D^3 \frac{p_1 - p_2}{L}. \tag{7.22}$$

With eqns 5.13 and 5.14, we have that

$$\frac{\langle v \rangle^2}{\langle v^2 \rangle} = \frac{8}{3\pi}, \tag{7.23}$$

and hence our estimate of the Knudsen flow rate is

$$\dot{M} \approx \frac{D^3}{\langle v \rangle}\frac{p_1 - p_2}{L}. \tag{7.24}$$

Note that the flow rate is proportional to D^3, so it is much more efficient to pump gas through wide pipes to obtain high flow rates.

Chapter summary

- The molecular flux, Φ, is the number of molecules striking unit area per second and is given by

$$\Phi = \frac{1}{4} n \langle v \rangle.$$

- This expression, together with the ideal gas equation, can be used to derive an alternative expression for the particle flux:

$$\Phi = \frac{p}{\sqrt{2\pi m k_B T}}.$$

- These expressions also govern molecular effusion through a small hole.

Exercises

(7.1) In a vacuum chamber designed for surface–science experiments, the pressure of residual gas is kept as low as possible so that surfaces can be kept clean. The coverage of a surface by a single monolayer requires about 10^{19} atoms per m². What pressure would be needed to deposit less than one monolayer per hour from residual gas? You may assume that if a molecule hits the surface, it sticks.

(7.2) A vessel contains a monatomic gas at temperature T. Use the Maxwell–Boltzmann distribution of speeds to calculate the mean kinetic energy of the molecules.
Molecules of the gas stream through a small hole into a vacuum. A box is opened for a short time and catches some of the molecules. Neglecting the thermal capacity of the box, calculate the final temperature of the gas trapped in the box.

(7.3) A closed vessel is partially filled with liquid mercury; there is a hole of area 10^{-7} m² above the liquid level. The vessel is placed in a region of high vacuum at 273 K and after 30 days is found to be lighter by 2.4×10^{-5} kg. Estimate the vapour pressure of mercury at 273 K. (The relative molecular mass of mercury is 200.59.)

(7.4) Calculate the mean speed and most probable speed for a molecule of mass m which has effused out of an enclosure at temperature T. Which of the two speeds is the larger?

(7.5) A gas effuses into a vacuum through a small hole of area A. The particles are then collimated by passing through a very small circular hole of radius a, in a screen a distance d from the first hole. Show that the rate at which particles emerge from the second hole is $\frac{1}{4} n A \langle v \rangle (a^2/d^2)$, where n is the particle density and $\langle v \rangle$ is the average speed. (Assume that no collisions take place after the gas effuses through the second hole, and that $d \gg a$.)

(7.6) Show that if a gas were allowed to leak through a small hole into an evacuated sphere and the particles condensed where they first hit the surface they would form a uniform coating.

(7.7) An astronaut goes for a space walk and her space suit is pressurized to 1 atm. Unfortunately, a tiny piece of space dust punctures her suit and it develops a small hole of radius $1\,\mu\mathrm{m}$. What force does she feel due to the effusing gas?

(7.8) Show that the time dependence of the pressure inside an oven (volume V) containing hot gas (molecular mass m, temperature T) with a small hole of area A is given by

$$p(t) = p(0) e^{-t/\tau}, \tag{7.25}$$

with

$$\tau = \frac{V}{A} \sqrt{\frac{2\pi m}{k_B T}}. \tag{7.26}$$

8 The mean free path and collisions

8.1 The mean collision time 70
8.2 The collision cross-section 71
8.3 The mean free path 73
Chapter summary 74
Exercises 74

[1] It turns out that large-angle scattering dominates transport processes in most gases (described in Chapter 9) and is largely independent of energy and therefore temperature; this allows us to use a rigid-sphere model of collisions, i.e. to model atoms in a gas as billiard balls.

At room temperature, the rms speed of O_2 or N_2 is about 500 ms^{-1}. Processes such as the diffusion of one gas into another would therefore be almost instantaneous, were it not for the occurrence of *collisions* between molecules. Collisions are fundamentally quantum mechanical events, but in a dilute gas, molecules spend most of their time between collisions and so we can consider them as classical billiard balls and ignore the details of what actually happens during a collision. All that we care about is that after collisions the molecules' velocities become essentially randomized.[1] In this chapter we will model the effect of collisions in a gas and develop the concepts of a mean collision time, the collision cross-section and the mean free path.

8.1 The mean collision time

In this section, we aim to calculate the average time between molecular collisions. Let us consider a particular molecule moving in a gas of other similar molecules. To make things simple to start with, we suppose that the molecule under consideration is travelling at speed v and that the other molecules in the gas are stationary. This is clearly a gross over-simplification, but we will relax this assumption later. We will also attribute a collision cross-section σ to each molecule, which is something like the cross-sectional area of our molecule. Again, we will refine this definition later in the chapter.

In a time $\mathrm{d}t$, our molecule will sweep out a volume $\sigma v \mathrm{d}t$. If another molecule happens to lie inside this volume, there will be a collision. With n molecules per unit volume, the probability of a collision in time $\mathrm{d}t$ is therefore $n\sigma v \mathrm{d}t$. Let us define $P(t)$ as follows:

$$P(t) = \text{the probability of a molecule not colliding up to time } t. \quad (8.1)$$

Elementary calculus then implies that

$$P(t + \mathrm{d}t) = P(t) + \frac{\mathrm{d}P}{\mathrm{d}t}\mathrm{d}t, \quad (8.2)$$

but $P(t + \mathrm{d}t)$ is also the probability of a molecule not colliding up to time t *multiplied by* the probability of not colliding in subsequent time $\mathrm{d}t$, i.e.,

$$P(t + \mathrm{d}t) = P(t)(1 - n\sigma v \mathrm{d}t). \quad (8.3)$$

Hence rearranging gives

$$\frac{1}{P}\frac{dP}{dt} = -n\sigma v \tag{8.4}$$

and therefore that (using $P(0) = 1$)

$$P(t) = e^{-n\sigma vt}. \tag{8.5}$$

Now the probability of surviving without collision up to time t but then colliding in the next dt is

$$e^{-n\sigma vt} n\sigma v dt. \tag{8.6}$$

We can check that this is a proper probability by integrating it,

$$\int_0^\infty e^{-n\sigma vt} n\sigma v dt = 1, \tag{8.7}$$

and confirming that it is equal to unity. Here, use has been made of the integral

$$\int_0^\infty e^{-x} dx = 0! = 1 \tag{8.8}$$

(see Appendix C.1). We are now in a position to calculate the **mean scattering time** τ, which is the average time elapsed between collisions for a given molecule. This is given by

$$\begin{aligned}
\tau &= \int_0^\infty t\, e^{-n\sigma vt} n\sigma v\, dt \\
&= \frac{1}{n\sigma v} \int_0^\infty (n\sigma vt) e^{-n\sigma vt} d(n\sigma vt) \\
&= \frac{1}{n\sigma v} \int_0^\infty x e^{-x} dx
\end{aligned} \tag{8.9}$$

where the integral has been simplified by the substitution $x = n\sigma vt$. Hence we find that

$$\boxed{\tau = \frac{1}{n\sigma v},} \tag{8.10}$$

where use has been made of the integral (again, see Appendix C.1)

$$\int_0^\infty x e^{-x} dx = 1! = 1. \tag{8.11}$$

8.2 The collision cross-section

In this section we will consider the factor σ in much more detail. To be as general as possible, we will consider two spherical molecules of radii a_1 and a_2 with a **hard-sphere potential** between them (see Fig. 8.1).

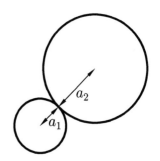

Fig. 8.1 Two spherical molecules of radii a_1 and a_2 with a hard-sphere potential between them.

This implies that there is a potential energy function $V(R)$ that depends on the relative separation R of their centres, and is given by

$$V(R) = \begin{cases} 0 & R > a_1 + a_2 \\ \infty & R \leq a_1 + a_2 \end{cases} \tag{8.12}$$

and this is sketched in Fig. 8.2.

The **impact parameter** b between two moving molecules is defined as the distance of closest approach that would result if the molecular trajectories were undeflected by the collision. Thus for a hard-sphere potential there is only a collision if the impact parameter $b < a_1 + a_2$. Focus on one of these molecules (let's say the one with radius a_1). This is depicted in Fig. 8.3. Now imagine molecules of the other type (with radius a_2) nearby. A collision will only take place if the centre of these other molecules comes inside a tube of radius $a_1 + a_2$ (so that the molecule labelled A would not collide, whereas B and C would). Thus our first molecule can be considered to sweep out an imaginary tube of space of cross-sectional area $\pi(a_1 + a_2)^2$ that defines its "personal space". The area of this tube is called the **collision cross-section** σ and is then given by

$$\sigma = \pi(a_1 + a_2)^2. \tag{8.13}$$

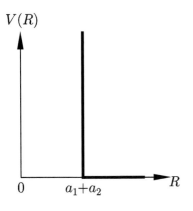

Fig. 8.2 The hard-sphere potential $V(R)$.

If $a_1 = a_2 = a$, then

$$\sigma = \pi d^2 \tag{8.14}$$

where $d = 2a$ is the molecular diameter.

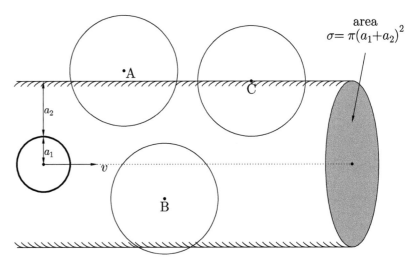

Fig. 8.3 A molecule sweeps out an imaginary tube of space of cross-sectional area $\sigma = \pi(a_1 + a_2)^2$. If the centre of another molecule enters this tube, there will be a collision.

[2] But not too low a temperature, or quantum effects become important.

[3] Cross-sections in nuclear and particle physics can be much larger than the size of the object, expressing the fact that an object (in this case a particle) can react strongly with things a long distance away from it.

Is the hard-sphere potential correct? It is a good approximation at lower temperatures,[2] but progressively worsens as the temperature increases. Molecules are not really hard spheres but slightly squashy objects, and when they move at higher speeds and plough into each other with more momentum, you need more of a direct hit to cause a collision. Thus as the gas is warmed, the molecules may appear to have a smaller cross-sectional area.[3]

8.3 The mean free path

Having derived the mean collision time, it is tempting to derive the **mean free path** as

$$\lambda = \langle v \rangle \tau = \frac{\langle v \rangle}{n\sigma v} \qquad (8.15)$$

but what should we take as v? A first guess is to use $\langle v \rangle$, but that turns out to be not quite right. What has gone wrong?

Our approach to molecular scattering has been to focus on one molecule as the moving one, and think of all of the others as sitting ducks, fixed in space waiting patiently for a collision to occur. The reality is quite different: all molecules are whizzing around. We should therefore take v as the average relative velocity, i.e., $\langle v_r \rangle$, where

$$\boldsymbol{v}_r = \boldsymbol{v}_1 - \boldsymbol{v}_2 \qquad (8.16)$$

and $\boldsymbol{v}_1$ and $\boldsymbol{v}_2$ are the velocities of two molecules labelled 1 and 2. Now,

$$v_r^2 = v_1^2 + v_2^2 - 2\boldsymbol{v}_1 \cdot \boldsymbol{v}_2, \qquad (8.17)$$

so that

$$\langle v_r^2 \rangle = \langle v_1^2 \rangle + \langle v_2^2 \rangle = 2\langle v^2 \rangle, \qquad (8.18)$$

because $\langle \boldsymbol{v}_1 \cdot \boldsymbol{v}_2 \rangle = 0$ (which follows because $\langle \cos\theta \rangle = 0$). The quantity which we want is $\langle v_r \rangle$, but what we have an expression for is $\langle v_r^2 \rangle$. If the probability distribution describing molecular speed is a Maxwell–Boltzmann distribution, then the error in writing $\langle v_r \rangle \approx \sqrt{\langle v_r^2 \rangle}$ is small,[4] so to a reasonable degree of approximation we can write

$$\langle v_r \rangle \approx \sqrt{\langle v_r^2 \rangle} \approx \sqrt{2}\langle v \rangle \qquad (8.19)$$

and hence we obtain an expression for λ as follows:[5]

$$\boxed{\lambda = \frac{1}{\sqrt{2}n\sigma}.} \qquad (8.20)$$

Substitution of $p = nk_BT$ yields the expression

$$\lambda = \frac{k_BT}{\sqrt{2}p\sigma}. \qquad (8.21)$$

To increase the mean free path by a certain factor, the pressure needs to be decreased by the same factor.

Example 8.1

Calculate the mean free path for a gas of N_2 at room temperature and pressure. (For N_2, take the molecular diameter to be $d = 0.37$ nm.)
Solution:
The collision cross-section is $\pi d^2 = 4.3 \times 10^{-19}$ m^2. We have $p \approx 10^5$ Pa and $T \approx 300$ K, so the number density is $n = p/k_BT \approx 10^5/(1.38 \times 10^{-23} \times 300) \approx 2 \times 10^{25}$ m^{-3}. This leads to $\lambda = 1/(\sqrt{2}n\sigma) = 6.8 \times 10^{-8}$ m.

---

[4] Equation 7.23 implies that $\langle v \rangle / \sqrt{\langle v^2 \rangle} = \sqrt{\frac{8}{3\pi}} = 0.92$, so the error is less than 10%.

[5] Although this derivation has used an approximation, it turns out that eqn 8.20 is exact. A brief version of the full derivation is given here. Consider a first class of molecules which move at velocity $\boldsymbol{v}$ and consider only collisions with a second class of molecules which move at velocity $\boldsymbol{u}$. In a frame moving at velocity $\boldsymbol{u}$, this second class of molecules are stationary and offer a total cross-section of $n\sigma f(\boldsymbol{u})\,d\boldsymbol{u}$, where $f(\boldsymbol{u}) = g(u_x)g(u_y)g(u_z)$ is a Maxwell–Boltzmann distribution for the vector $\boldsymbol{u} = (u_x, u_y, u_z)$. In unit time, the total volume swept out by these targets relative to the first class of molecules (which in this frame move at velocity $\boldsymbol{v} - \boldsymbol{u}$) is $|\boldsymbol{v} - \boldsymbol{u}|n\sigma f(\boldsymbol{u})\,d\boldsymbol{u}$. The number of encounters per second is obtained by multiplying this volume by the probability of finding one of the first class of molecules in unit volume, giving $|\boldsymbol{v} - \boldsymbol{u}|n\sigma f(\boldsymbol{u})\,d\boldsymbol{u} f(\boldsymbol{v})\,d\boldsymbol{v}$. The collision rate R is therefore obtained by integrating over all $\boldsymbol{u}$ and $\boldsymbol{v}$ giving

$$R = n\sigma \int\int |\boldsymbol{v} - \boldsymbol{u}| f(\boldsymbol{u})\,d\boldsymbol{u} f(\boldsymbol{v})\,d\boldsymbol{v},$$

which writing $\boldsymbol{x} = (\boldsymbol{v} - \boldsymbol{u})/\sqrt{2}$ and $\boldsymbol{y} = (\boldsymbol{v} + \boldsymbol{u})/\sqrt{2}$ can be transformed into

$$R = n\sigma\sqrt{2}\int |\boldsymbol{x}|f(\boldsymbol{x})\,d\boldsymbol{x}\int f(\boldsymbol{y})\,d\boldsymbol{y},$$

where the first integral yields $\langle v \rangle$ and the second integral is unity. Hence $R = n\sigma\sqrt{2}\langle v \rangle$ and the mean free path is $\lambda = \langle v \rangle/R = 1/(\sqrt{2}n\sigma)$.

Notice that both λ and τ *decrease* with increasing pressure at fixed temperature. Thus the frequency of collisions *increases* with increasing pressure.

Chapter summary

- The mean scattering time is given by
$$\tau = \frac{1}{n\sigma \langle v_r \rangle},$$
where the collision cross-section is $\sigma = \pi d^2$, d is the molecular diameter and $\langle v_r \rangle \approx \sqrt{2} \langle v \rangle$.

- The mean free path is
$$\lambda = \frac{1}{\sqrt{2} n \sigma}.$$

Exercises

(8.1) What is the mean free path of an N_2 molecule in an ultra-high-vacuum chamber at a pressure of 10^{-10} mbar? What is the mean collision time? The chamber has a diameter of 0.5 m. On average, how many collisions will the molecule make with the chamber walls compared with collisions with other molecules? If the pressure is suddenly raised to 10^{-6} mbar, how do these results change?

(8.2) (a) Show that the root mean square free path is given by $\sqrt{2}\lambda$ where λ is the mean free path.

(b) What is the most probable free path length?

(c) What percentage of molecules travel a distance greater than (i) λ, (ii) 2λ, (iii) 5λ?

(8.3) Show that particles hitting a plane boundary have travelled a distance $2\lambda/3$ perpendicular to the plane since their last collision, on average.

(8.4) A diffuse cloud of neutral hydrogen atoms in space has a temperature of 50 K and number density $500 \, \text{cm}^{-3}$. Estimate the mean scattering time (in years) between hydrogen atoms in the cloud. Estimate the mean free path (in astronomical units). (1 astronomical unit is the Earth–Sun distance; see Appendix A for a numerical value.)

Part III

Transport and thermal diffusion

In the third part of this book, we use our results from the kinetic theory of gases to derive various transport properties of gases and then apply this to solving the thermal diffusion equation. This part is structured as follows:

- In Chapter 9, we use the intuition developed from considering molecular collisions and the mean free path to determine various *transport properties*, in particular *viscosity*, *thermal conductivity*, and *diffusion*. These correspond to the transport of momentum, heat, and particles respectively.
- In Chapter 10, we derive the *thermal diffusion equation*, which shows how heat is transported between regions of different temperature. This equation is a differential equation and can be applied to a variety of physical situations, and we show how to solve it in certain cases of high symmetry.

9 Transport properties in gases

9.1 Viscosity 76
9.2 Thermal conductivity 81
9.3 Diffusion 83
9.4 More detailed theory 86
Chapter summary 88
Further reading 88
Exercises 89

In this chapter, we wish to describe how a gas can transport momentum, energy, or particles from one place to another. The model we have used so far has been that of a gas in equilibrium, so that none of its macroscopic parameters are time-dependent. Now we consider *non-equilibrium* situations, but still in the *steady state*, i.e., so that the system parameters are time-independent, but the surroundings will be time-dependent. The phenomena we want to treat are called **transport properties** and we will consider

(1) **Viscosity**, which is the transport of *momentum*,
(2) **Thermal conductivity**, which is the transport of *heat*, and
(3) **Diffusion**, which is the transport of *particles*.

9.1 Viscosity

[1] This proportionality was suggested by Isaac Newton and holds for many liquids and most gases, which are thus termed **Newtonian fluids**. Non-Newtonian fluids have a viscosity that is a function of the applied shear stress.

[2] Also used is the **kinematic viscosity** ν, defined by $\nu = \eta/\rho$, where ρ is the density. This is useful because one often wants to compare the viscous forces with inertial forces. The unit of kinematic viscosity is $\mathrm{m^2\,s^{-1}}$.

Viscosity is the measure of the resistance of a fluid to the deformation produced by a shear stress. For straight, parallel, and uniform flow, the shear stress between the layers is proportional[1] to the velocity gradient in the direction perpendicular to the layers. The constant of proportionality, given the symbol η, is called the **coefficient of viscosity**, the **dynamic viscosity**, or simply the **viscosity**.[2]

Consider the scenario in Fig. 9.1 in which a fluid is sandwiched between two plates of area A, which each lie in the xy plane. A shear stress $\tau_{xz} = F/A$ is applied to the fluid by sliding the top plate over it at speed u while keeping the bottom plate stationary. A shear force F is applied. A velocity gradient $\mathrm{d}\langle u_x \rangle / \mathrm{d}z$ is set up, so that $\langle u_x \rangle = 0$ near the bottom plate and $\langle u_x \rangle = u$ near the top plate. If the fluid is a gas, then this extra motion in the x-direction is superimposed on the Maxwell–Boltzmann motion in the x, y and z directions (and hence the use of the average $\langle u_x \rangle$, rather than u_x).

The viscosity η is then defined by

$$\tau_{xz} = \frac{F}{A} = \eta \frac{\mathrm{d}\langle u_x \rangle}{\mathrm{d}z}. \tag{9.1}$$

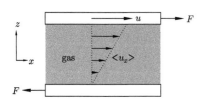

Fig. 9.1 A fluid is sandwiched between two plates of area A which each lie in an xy plane (see text).

The units of viscosity are $\mathrm{Pa\,s}$ ($= \mathrm{N\,m^{-2}\,s}$). Force is rate of change of momentum, and hence transverse momentum is being transported through the fluid. This is achieved because molecules travelling in the $+z$ direction move from a layer in which $\langle u_x \rangle$ is smaller to one in which $\langle u_x \rangle$ is larger, and hence they transfer net momentum to that layer in the $-x$ direction. Molecules travelling parallel to $-z$ have the opposite effect. Hence, the shear stress τ_{xz} is equal to the transverse momentum transported across each square metre per second, and τ_{xz} is equal to a flux of momentum (though note that there must be a minus sign involved, because the momentum flux must be from regions of high transverse velocity to regions of low transverse velocity, which is in the opposite direction to the velocity gradient). The velocity gradient $\partial \langle u_x \rangle / \partial z$ therefore drives a momentum flux Π_z, according to

$$\Pi_z = -\eta \frac{\partial \langle u_x \rangle}{\partial z}. \tag{9.2}$$

The viscosity can be calculated using kinetic theory as follows:

Recall first that we showed before in eqn 6.13 that the number of molecules hitting unit area per second is $v \cos\theta \, n \, f(v) \, dv \, \frac{1}{2} \sin\theta \, d\theta$. Consider molecules travelling at an angle θ to the z-axis (see Fig. 9.2). Then molecules crossing a plane of constant z will have travelled on average a distance λ since their last collision, and so they will have travelled a distance $\lambda \cos\theta$ parallel to the z-axis since their last collision. Over that distance there is an average increase in $\langle u_x \rangle$ given by $(\partial \langle u_x \rangle / \partial z) \lambda \cos\theta$, so these upwards-travelling molecules bring an excess momentum in the x-direction given by[3]

$$-m \left(\frac{\partial \langle u_x \rangle}{\partial z} \right) \lambda \cos\theta. \tag{9.3}$$

Hence the total x-momentum transported across unit area perpendicular to z in unit time is the momentum flux Π_z given by

$$\begin{aligned}
\Pi_z &= \int_0^\infty \int_0^\pi v \cos\theta \, n \, f(v) \, dv \, \frac{1}{2} \sin\theta \, d\theta \cdot m \left(-\frac{\partial \langle u_x \rangle}{\partial z} \right) \lambda \cos\theta \\
&= \frac{1}{2} nm\lambda \int_0^\infty v f(v) \, dv \left(-\frac{\partial \langle u_x \rangle}{\partial z} \right) \int_0^\pi \cos^2\theta \sin\theta \, d\theta \\
&= -\frac{1}{3} nm\lambda \langle v \rangle \left(\frac{\partial \langle u_x \rangle}{\partial z} \right).
\end{aligned} \tag{9.4}$$

Hence the viscosity is given by

$$\boxed{\eta = \frac{1}{3} nm\lambda \langle v \rangle.} \tag{9.5}$$

Equation 9.5 has some important consequences.

- η **is independent of pressure.**
  Because $\lambda \approx 1/(\sqrt{2} n\sigma) \propto n^{-1}$, the viscosity is independent of n and hence (at constant temperature) it is independent of pressure. This is at first sight a weird result: as you increase the pressure, and hence n, you should be better at transmitting momentum

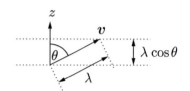

Fig. 9.2 Molecular velocity v for molecules travelling at an angle θ to the z-axis. These will have travelled on average a distance λ since their last collision, and so they will have travelled a distance $\lambda \cos\theta$ parallel to the z-axis since their last collision.

[3]The negative sign is because the molecules moving in the $+z$ direction are moving up the velocity gradient from a slower to a faster region and so bring a deficit in x-momentum if $\left(\frac{\partial \langle u_x \rangle}{\partial z} \right)$ is positive. It is the same reason for the negative sign in eqn 9.2.

For this calculation, the integration over angle θ runs from 0 to π, since we wish to sum over the molecules travelling in all possible directions.

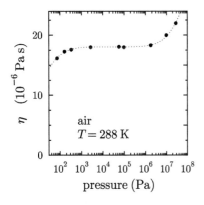

Fig. 9.3 The apparent viscosity of air as a function of pressure at 288 K. It is found to be constant over a wide range of pressure.

because you have more molecules to do it with. However, your mean free path reduces correspondingly, so that each molecule becomes less effective at transmitting momentum in such a way as to precisely cancel out the effect of having more of them.

This result holds impressively well over quite a range of pressures (see Fig. 9.3) although it begins to fail at very low or very high pressures.

- $\eta \propto T^{1/2}$.
  Because η is independent of n, the only temperature dependence is from $\langle v \rangle \propto T^{1/2}$, and hence $\eta \propto T^{1/2}$. Note therefore that the viscosity of gases *increases* with T, which is different for most liquids, which get runnier (i.e., *less* viscous) when you heat them.

- Substituting in $\lambda = (\sqrt{2}n\sigma)^{-1}$, $\sigma = \pi d^2$ and $\langle v \rangle = (8k_\mathrm{B}T/\pi m)^{1/2}$ yields a more useful (though less memorable) expression for the viscosity:

$$\eta = \frac{2}{3\pi d^2}\left(\frac{mk_\mathrm{B}T}{\pi}\right)^{1/2}. \tag{9.6}$$

- Equation 9.6 predicts that the viscosity will be proportional to $\sqrt{m}/d^2$ at constant temperature. This proportionality holds very well, as shown in Fig. 9.4.

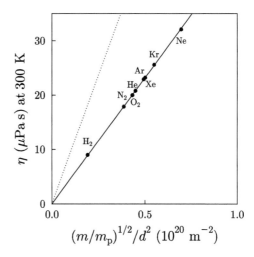

Fig. 9.4 The dependence of the viscosity of various gases on $\sqrt{m}/d^2$. The dotted line is the prediction of eqn 9.6. The solid line is the prediction of eqn 9.45.

- Various *approximations* have gone into this approach, and a condition for their validity is that

$$L \gg \lambda \gg d, \tag{9.7}$$

where L is the size of the container holding the gas and d is the molecular diameter. We need $\lambda \gg d$ (pressure not too high) so that we can neglect collisions involving more than two particles. We need $\lambda \ll L$ (pressure not too low) so that molecules mainly collide with each other and not with the container walls.[4] If λ is of the

[4] A gas in which $\lambda \gg L$, so that molecules rarely collide with each other, is called a **Knudsen gas**.

same order of magnitude or greater than L, most of a molecule's collisions will be with the container walls. Figure 9.3 indeed shows that the pressure-independence of the viscosity begins to break down when the pressure is too low or too high.

- The factor of $\frac{1}{3}$ in eqn 9.5 is not quite right, so that eqn 9.6 leads to the dotted line in Fig. 9.4. To get a precise numerical factor, you need to consider the fact that the velocity distribution is different in different layers (because of the shear stress applied) and then average over the distribution of path lengths. This will be done in Section 9.4 and leads to a prediction that gives the solid line in Fig. 9.4.

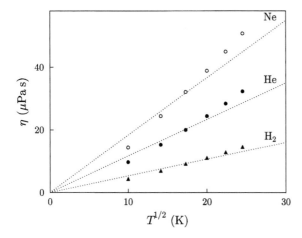

Fig. 9.5 The temperature dependence of the viscosity of various gases. The agreement with the predicted $T^{1/2}$ behaviour is satisfactory as a first approximation, but not very good in detail.

- The measured temperature dependence of the viscosity of various gases broadly agrees with our prediction that $\eta \propto \sqrt{T}$, as shown in Fig. 9.5, but the agreement is not quite perfect. The reason for this is that the collision cross-section, $\sigma = \pi d^2$, is actually temperature-dependent. At high temperatures, molecules move faster and hence have to collide more directly to have a proper momentum-randomizing collision. We have been assuming that molecules behave as perfect hard spheres and that any collision perfectly randomizes the molecular motion, but this is not precisely true. This means that the effective molecular diameter shrinks as you increase the temperature, increasing the viscosity over and above the expected $\sqrt{T}$ dependence. This is evident in the data presented in Fig. 9.5.

- Viscosity can be measured by the damping of torsional oscillations in the apparatus shown in the box.

Measurement of viscosity

Maxwell developed a method for measuring the viscosity of a gas by observing the damping rate of oscillations of a disc suspended from a fixed support by a torsion fibre. It is positioned halfway between two, fixed horizontal discs and oscillates parallel to them in the gas. This is shown in Fig. 9.6(a), with the fixed horizontal discs shaded and the oscillating disc in white. The damping of the torsional oscillations is from the viscous damping due to the gas trapped on each side of the oscillating disc between the fixed discs. The fixed discs are mounted inside a vacuum chamber in which the composition and pressure of the gas to be measured can be varied.

A very accurate method is the *rotating–cylinder method*, in which gas is confined between two vertical coaxial cylinders. It is shown in Fig. 9.6(b). The outer cylinder (inner radius b) is rotated by a motor at a constant angular speed ω_0, while the inner cylinder (outer radius a) is suspended by a torsion fibre from a fixed support. The torque G on the outer cylinder is transmitted via the gas to the inner cylinder and a resulting torque on the torsion fibre. The velocity $u(r)$ is related to the angular velocity $\omega(r)$ by $u(r) = r\omega(r)$, and we expect that ω varies all the way from 0 at $r = a$ to ω_0 at $r = b$. The velocity gradient is thus

$$\frac{du}{dr} = \omega + r\frac{d\omega}{dr}, \qquad (9.8)$$

but the first term on the right-hand side simply corresponds to the velocity gradient due to rigid rotation and does not contribute to the viscous shearing stress, which is thus $\eta r d\omega/dr$. The force F on a cylindrical element of gas (of length l) is then just this viscous stress multiplied by the area of the cylinder $2\pi r l$, i.e.,

$$F = 2\pi r l \eta \times r\frac{d\omega}{dr}, \qquad (9.9)$$

and so the torque $G = rF$ on this cylindrical element is

$$G = 2\pi r^3 l \eta \frac{d\omega}{dr}. \qquad (9.10)$$

In the steady state, there is no change in viscous torque from the outer to the inner cylinder (if there were, angular acceleration would be induced somewhere and the system would change) so this torque is transmitted to the suspended cylinder. Hence rearranging and integrating give

$$G\int_a^b \frac{dr}{r^3} = 2\pi l \eta \int_0^{\omega_0} d\omega = 2\pi l \eta \omega_0, \qquad (9.11)$$

so that

$$\eta = \frac{G}{4\pi\omega_0 l}\left(\frac{1}{a^2} - \frac{1}{b^2}\right). \qquad (9.12)$$

The torque G is related to the angular deflection ϕ of the inner cylinder by $G = \alpha\phi$. The angular deflection can be measured using a light beam reflected from a small mirror attached to the torsion fibre. The coefficient α is known as the **torsion constant**. This can be found by measuring the period T of torsional oscillations of an object of moment of inertia I suspended from the wire, which is

$$T = 2\pi\sqrt{\frac{I}{\alpha}}. \qquad (9.13)$$

Knowledge of I and T yields α which can be used with the measured ϕ to obtain G and hence η.

(a)

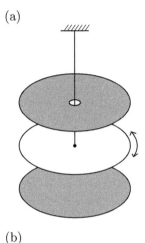

(b)

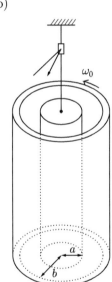

Fig. 9.6 Measuring viscosity by (a) Maxwell's method and (b) the rotating–cylinder method.

9.2 Thermal conductivity

We have defined **heat** as "thermal energy in transit".[5] It quantifies the transfer of energy in response to a temperature gradient. The amount of heat that flows along a temperature gradient depends on the thermal conductivity of the material, which we will now define.

[5] See Chapter 2.

Thermal conductivity can be considered in one dimension using the diagram shown in Fig. 9.7. Heat flows from hot to cold, and so flows against the temperature gradient. The flow of heat can be described by a heat flux vector $\boldsymbol{J}$, whose direction lies along the direction of flow of heat and whose magnitude is equal to the heat energy flowing per unit time per unit area (measured in $\mathrm{J\,s^{-1}\,m^{-2}} = \mathrm{W\,m^{-2}}$). The heat flux J_z in the z-direction is given by

$$J_z = -\kappa \left(\frac{\partial T}{\partial z}\right), \tag{9.14}$$

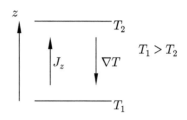

Fig. 9.7 Heat flows in the opposite direction to the temperature gradient.

where the negative sign is because heat flows "downhill". The constant κ is called the **thermal conductivity**[6] of the gas. In general, in three dimensions we can write that the heat flux $\boldsymbol{J}$ is related to temperature using

[6] Thermal conductivity has units $\mathrm{W\,m^{-1}\,K^{-1}}$.

$$\boldsymbol{J} = -\kappa \nabla T. \tag{9.15}$$

How do molecules in a gas "carry" heat? Gas molecules have energy, and as we found in eqn 5.17 their mean translational kinetic energy $\langle \frac{1}{2}mv^2 \rangle = \frac{3}{2}k_\mathrm{B} T$ depends on the temperature. Therefore to increase the temperature of a gas by 1 K, one has to increase the mean kinetic energy by $\frac{3}{2}k_\mathrm{B}$ per molecule. The heat capacity[7] C of the gas is the heat required to increase the temperature of gas by 1 K. The heat capacity C_molecule of a gas molecule is therefore equal to $\frac{3}{2}k_\mathrm{B}$, though we will later see that it can be larger than this if the molecule can store energy in forms other than translational kinetic energy.[8]

[7] See Section 2.2.

[8] Other forms include rotational kinetic energy or vibrational energy, if the gas molecules are polyatomic.

The derivation of the thermal conductivity of a gas is very similar to that for viscosity. Consider molecules travelling along the z-axis. Then molecules crossing a plane of constant z will have travelled on average a distance λ since their last collision, and so they will have travelled a distance $\lambda \cos\theta$ parallel to the z-axis since their last collision. Therefore they bring a deficit of thermal energy given by

$$C_\mathrm{molecule} \times \Delta T = C_\mathrm{molecule} \frac{\partial T}{\partial z} \lambda \cos\theta, \tag{9.16}$$

where C_molecule is the heat capacity of a single molecule. Hence the total thermal energy transported across unit area in unit time, i.e., the heat flux, is given by

$$\begin{aligned}
J_z &= \int_0^\infty dv \int_0^\pi \left(-C_\mathrm{molecule} \frac{\partial T}{\partial z} \lambda \cos\theta\right) v \cos\theta\, n f(v) \frac{1}{2} \sin\theta d\theta \\
&= -\frac{1}{2} n C_\mathrm{molecule} \lambda \int_0^\infty v f(v)\, dv \frac{\partial T}{\partial z} \int_0^\pi \cos^2\theta \sin\theta\, d\theta \\
&= -\frac{1}{3} n C_\mathrm{molecule} \lambda \langle v \rangle \frac{\partial T}{\partial z}.
\end{aligned} \tag{9.17}$$

Hence the thermal conductivity κ is given by

$$\kappa = \frac{1}{3} C_V \lambda \langle v \rangle, \quad (9.18)$$

where $C_V = n C_{\text{molecule}}$ is the heat capacity per unit volume (though the subscript V here refers to a temperature change at constant volume). Equation 9.18 has some important consequences.

- κ is independent of pressure.
  The argument is the same as for η. Because $\lambda \approx 1/(\sqrt{2} n \sigma) \propto n^{-1}$, κ is independent of n and hence (at constant temperature) it is independent of pressure.

- $\kappa \propto T^{1/2}$.
  The argument is also the same as for η. Because κ is independent of n, the only temperature dependence is from $\langle v \rangle \propto \sqrt{T}$, and hence $\kappa \propto T^{1/2}$. This holds quite well for a number of gases (see Fig. 9.8).

- As for viscosity, substituting in $\lambda = (\sqrt{2} n \sigma)^{-1}$, $\sigma = \pi d^2$ and $\langle v \rangle = (8 k_B T / \pi m)^{1/2}$ yields a more useful (though less memorable) expression for the thermal conductivity:

$$\kappa = \frac{2}{3 \pi d^2} C_{\text{molecule}} \left(\frac{k_B T}{\pi m} \right)^{1/2}. \quad (9.19)$$

- $L \gg \lambda \gg d$ is again the relevant condition for our treatment to hold.

- Equation 9.19 predicts that the thermal conductivity will be proportional to $1/(\sqrt{m} d^2)$ at constant temperature. This holds very well, as shown in Fig. 9.9.

- Thermal conductivity can be measured by various techniques; see the box.

The similarity of η and κ would suggest that

$$\frac{\kappa}{\eta} = \frac{C_{\text{molecule}}}{m}. \quad (9.20)$$

The ratio C_{molecule}/m is the specific heat capacity[9] c_V (the subscript V indicating a measurement at constant volume), so equivalently

$$\kappa = c_V \eta. \quad (9.21)$$

However, neither of these relations hold too well. Faster molecules cross a given plane more often than slow ones. These carry more kinetic energy and therefore do carry more heat. However, they don't necessarily carry more average momentum in the x-direction. We will return to this point in Section 9.4.

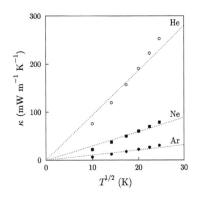

Fig. 9.8 The thermal conductivity of various gases as a function of temperature. The agreement with the predicted $T^{1/2}$ behaviour is satisfactory as a first approximation, but not very good in detail.

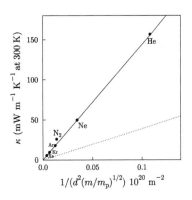

Fig. 9.9 The dependence of the thermal conductivity of various gases on $1/(\sqrt{m} d^2)$. The dotted line is the prediction of eqn 9.19. The solid line is the prediction of eqn 9.46, which works very well for the monatomic noble gases, but a little less well for diatomic N_2.

[9] See Section 2.2.

Measurement of thermal conductivity

The thermal conductivity κ can be measured using the *hot-wire method*. Gas fills the space between two coaxial cylinders (inner cylinder radius a, outer cylinder radius b) as shown in Fig. 9.10. The outer cylinder is connected to a constant-temperature bath of temperature T_b, while heat is generated in the inner cylinder (the hot wire) at rate Q per unit length of the cylinder (measured in units of W m^{-1}). The temperature of the inner cylinder rises to T_a. The rate Q can be connected with the radial heat flux J_r using

$$Q = 2\pi r J_r, \tag{9.22}$$

and J_r itself is given by $-\kappa \partial T/\partial r$, as in eqn 9.14. Hence

$$Q = -2\pi r \kappa \left(\frac{\partial T}{\partial r}\right), \tag{9.23}$$

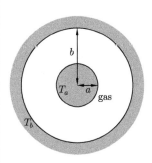

Fig. 9.10 The hot-wire method for measuring thermal conductivity.

and rearranging and integrating yields

$$Q \int_a^b \frac{\mathrm{d}r}{r} = -2\pi\kappa \int_{T_a}^{T_b} \mathrm{d}T, \tag{9.24}$$

and hence

$$\kappa = \frac{Q}{2\pi} \frac{\ln b/a}{T_a - T_b}. \tag{9.25}$$

Since Q is known (it is the power supplied to heat the inner cylinder) and T_a and T_b can be measured, the value of κ can be deduced.

An important application of this technique is in the **Pirani gauge**, which is commonly used in vacuum systems to measure pressure. A sensor wire is heated electrically, and the pressure of the gas is determined by measuring the current needed to keep the wire at a constant temperature. (The resistance of the wire is temperature dependent, so the temperature is estimated by measuring the resistance of the wire.) The Pirani gauge thus relies on the fact that at low pressure the thermal conductivity *is* a function of pressure (since the condition $\lambda \ll L$, where L is a linear dimension in the gauge, is not met). In fact, a typical Pirani gauge will not work to detect pressures much above 1 mbar because, above these pressures, the thermal conductivity of the gases no longer changes with pressure. The thermal conductivity of each gas is different, so the gauge has to be calibrated for the individual gas being measured.

9.3 Diffusion

Consider a distribution of similar molecules, some of which are labelled (e.g., by being radioactive). Let there be $n^*(z)$ of these labelled molecules per unit volume, but note that n^* is allowed to be a function of the z coordinate. The flux Φ_z of labelled molecules parallel to the z-direction (measured in m^{-2}s^{-1}) is[10]

$$\Phi_z = -D\left(\frac{\partial n^*}{\partial z}\right), \tag{9.26}$$

where D is the **coefficient of self-diffusion**.[11] Now consider a thin slab of gas of thickness $\mathrm{d}z$ and area A, as shown in Fig. 9.11. The flux into the slab is

$$A\Phi_z, \tag{9.27}$$

[10] In three dimensions, this equation is written $\Phi = -D\nabla n^*$. This is a statement of **Fick's law**, named after Adolf Fick (1829–1901).

[11] We use the phrase self-diffusion because the molecules that are diffusing are the same (apart from being labelled) as the molecules into which they are diffusing. Later we will consider diffusion of molecules into dissimilar molecules.

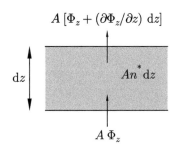

Fig. 9.11 The fluxes into and out of a thin slab of gas of thickness dz and area A.

[12]See also Appendix C.12.

and the flux out of the slab is

$$A\left(\Phi_z + \frac{\partial \Phi_z}{\partial z}dz\right). \tag{9.28}$$

The difference in these two fluxes must be balanced by the time-dependent changes in the number of labelled particles inside the region. Hence

$$\frac{\partial}{\partial t}(n^* A\, dz) = -A\frac{\partial \Phi_z}{\partial z}\, dz, \tag{9.29}$$

so that

$$\frac{\partial n^*}{\partial t} = -\frac{\partial \Phi_z}{\partial z}, \tag{9.30}$$

and hence that

$$\frac{\partial n^*}{\partial t} = D\frac{\partial^2 n^*}{\partial z^2}. \tag{9.31}$$

This is the **diffusion equation**. A derivation of the diffusion equation in three dimensions is shown in the box.[12]

Three-dimensional derivation of the diffusion equation

The total number of labelled particles that flow out of a closed surface S per second is given by the integral

$$\int_S \mathbf{\Phi} \cdot d\mathbf{S}, \tag{9.32}$$

and this must be balanced by the rate of decrease of labelled particles inside the volume V surrounded by S, i.e.,

$$\int_S \mathbf{\Phi} \cdot d\mathbf{S} = -\frac{\partial}{\partial t}\int_V n^*\, dV. \tag{9.33}$$

The divergence theorem implies that

$$\int_S \mathbf{\Phi} \cdot d\mathbf{S} = \int_V \nabla \cdot \mathbf{\Phi}\, dV, \tag{9.34}$$

and hence that

$$\nabla \cdot \mathbf{\Phi} = -\frac{\partial n^*}{\partial t}. \tag{9.35}$$

Substituting in $\mathbf{\Phi} = -D\nabla n^*$ then yields the diffusion equation, which is

$$\boxed{\frac{\partial n^*}{\partial t} = D\nabla^2 n^*.} \tag{9.36}$$

A kinetic theory derivation of D proceeds as follows. The excess labelled molecules hitting unit area per second is

$$\begin{aligned}\Phi_z &= \int_0^\pi d\theta \int_0^\infty dv\, v\cos\theta f(v)\frac{1}{2}\sin\theta \left(-\frac{\partial n^*}{\partial z}\lambda\cos\theta\right)\\ &= -\frac{1}{3}\lambda\langle v\rangle\frac{\partial n^*}{\partial z},\end{aligned} \tag{9.37}$$

and hence
$$D = \frac{1}{3}\lambda \langle v \rangle. \qquad (9.38)$$

This equation has some important implications:

- $D \propto p^{-1}$

  In this case, there is no factor of n, but $\lambda \propto 1/n$ and hence $D \propto n^{-1}$ and at fixed temperature $D \propto p^{-1}$ (this holds quite well experimentally, see Fig. 9.12).

- $D \propto T^{3/2}$

  Because $p = nk_BT$ and $\langle v \rangle \propto T^{1/2}$, we have that $D \propto T^{3/2}$ at fixed pressure.

- $D\rho = \eta$

  The only difference between the formula for D and that for η is a factor of $\rho = nm$, and so
  $$D\rho = \eta. \qquad (9.39)$$

- $D \propto m^{-1/2} d^{-2}$, which is the same dependence as thermal conductivity.

- The less memorable formula for D is, as before, obtained by substituting in the expressions for $\langle v \rangle$ and λ, yielding
$$D = \frac{2}{3\pi n d^2} \left(\frac{k_B T}{\pi m} \right)^{1/2}. \qquad (9.40)$$

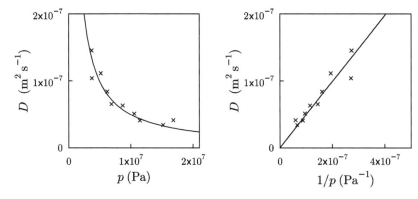

Fig. 9.12 Diffusion as a function of pressure.

This section has been about self-diffusion, where labelled atoms (or molecules) diffuse amongst unlabelled, but otherwise identical, atoms (or molecules). Experimentally, it is easier to measure the diffusion of atoms (or molecules) of one type (call them type 1, mass m_1, diameter d_1) amongst atoms (or molecules) of another type (call them type 2, mass m_2, diameter d_2). In this case the diffusion constant D_{12} is used which is given by eqn 9.40 with d replaced by $(d_1+d_2)/2$ and m replaced by $2m_1m_2/(m_1+m_2)$, so that

$$D_{12} = \frac{2}{3\pi n(\frac{1}{2}[d_1+d_2])^2} \left(\frac{k_B T(m_1+m_2)}{2\pi m_1 m_2} \right)^{1/2}. \qquad (9.41)$$

9.4 More detailed theory

The treatment of the transport properties presented so far in this chapter has the merit that it allows one to get the basic dependences fairly straightforwardly, and gives good insight as to what is going on. However, some of the details of the predictions are not in complete agreement with experiment and it is the purpose of this section to offer a critique of this approach and see how things might be improved. This section contains more advanced material than considered in the rest of this chapter and can be skipped at first reading.

One effect, which we have ignored, is the persistence of velocity after a collision. Our assumption has been that following a collision, a molecule's velocity becomes completely randomized and is completely uncorrelated with its velocity before the collision. However, although that is the simplest approximation to take, it is not correct. After most collisions, a molecule will retain some component of its velocity in the direction of its original motion. Moreover, our treatment has implicitly assumed a Maxwellian distribution of molecular velocities and that the different components of v are uncorrelated with each other, so that they can be considered to be independent random variables.[13] However, these components are actually partially correlated with each other and so are not independent random variables.

[13] See Section 3.6.

A further effect which becomes important at low pressure is the presence of boundaries; the details of the collisions of molecules with walls of a container can be quite important, and such collisions become more important as the pressure is reduced so that the mean free path increases.

Yet another consideration is the interconversion between the internal energy of a molecule and its translational degrees of freedom. As we will see in later chapters, the heat capacity of a molecule contains terms not only due to its translational motion ($C_{\text{molecule}} = \frac{3}{2}k_\text{B}$) but also due to its rotational and vibrational degrees of freedom. Collisions can give rise to processes where a molecule's energy can be redistributed throughout these different degrees of freedom. Thus if the molar heat capacity C_V can be written as the sum of two terms, $C_V = C_V' + C_V''$, where C_V' is due to translational degrees of freedom and C_V'' is due to other degrees of freedom, then it turns out that eqn 9.21 should be amended to give

$$\kappa = \left(\frac{5}{2}C_V' + C_V''\right)\eta. \tag{9.42}$$

The $\frac{5}{2}$ factor reflects the correlations that exist between momentum, energy, and translational motion. The most energetic molecules are the most rapid and therefore possess longer mean free paths. This leads to **Eucken's formula**, which states that

$$\kappa = \frac{1}{4}(9\gamma - 5)\eta C_V. \tag{9.43}$$

For an ideal monatomic gas $\gamma = \frac{5}{3}$ and hence

$$\kappa = \frac{5}{2}\eta C_V, \tag{9.44}$$

which supersedes eqn 9.21.

A more accurate treatment of the effects mentioned in this section has been performed by Chapman and Enskog (in the twentieth century); the methods used go beyond the scope of this text, but we summarize the results.

- The viscosity, which was written as $\eta = (2/3\pi d^2)(mk_\text{B}T/\pi)^{1/2}$ in eqn 9.6, should be replaced by

$$\eta = \frac{5}{16}\frac{1}{d^2}\left(\frac{mk_\text{B}T}{\pi}\right)^{1/2}, \qquad (9.45)$$

i.e., the $2/3\pi$ should be replaced by $5/16$.

- The corrected formula for κ (which we had evaluated in eqn 9.19) can be obtained from this expression of η using Eucken's formula, eqn 9.43, and hence reads

$$\kappa = \frac{25}{32d^2}C_\text{molecule}\left(\frac{k_\text{B}T}{\pi m}\right)^{1/2}, \qquad (9.46)$$

i.e., the $2/3\pi$ should be replaced by $25/32$.

- The formula for D, which appears in eqn 9.40, should now be replaced by

$$D = \frac{3}{8}\frac{1}{nd^2}\left(\frac{k_\text{B}T}{\pi m}\right)^{1/2}, \qquad (9.47)$$

i.e., the $2/3\pi$ should be replaced by $3/8$. Similarly, eqn 9.41 should be replaced by

$$D = \frac{3}{8n(\frac{1}{2}[d_1+d_2])^2}\left(\frac{k_\text{B}T(m_1+m_2)}{2\pi m_1 m_2}\right)^{1/2}. \qquad (9.48)$$

This also alters other conclusions, such as eqn 9.39, which becomes

$$D\rho = \frac{\frac{3}{8}\eta}{\frac{5}{16}} = \frac{6\eta}{5}. \qquad (9.49)$$

Chapter summary

- Viscosity, η, defined by $\Pi_z = -\eta\, \partial \langle u_x \rangle / \partial z$ is (approximately)
$$\eta = \frac{1}{3} nm\lambda \langle v \rangle.$$

- Thermal conductivity, κ, defined by $J_z = -\kappa\, \partial T / \partial z$ is (approximately)
$$\kappa = \frac{1}{3} C_V \lambda \langle v \rangle.$$

- Diffusion, D, defined by $\Phi_z = -D\, \partial n^* / \partial z$ is (approximately)
$$D = \frac{1}{3} \lambda \langle v \rangle.$$

- These relationships assume that
$$L \gg \lambda \gg d.$$

The results of a more detailed theory have been summarized (and serve only to alter the numerical factors at the start of each equation).

- The predicted pressure, temperature, molecular mass and molecular diameter dependences are:

η	κ	D
$\propto p^0$	$\propto p^0$	$\propto p^{-1}$
$\propto T^{1/2}$	$\propto T^{1/2}$	$\propto T^{3/2}$
$\propto m^{1/2} d^{-2}$	$\propto m^{-1/2} d^{-2}$	$\propto m^{-1/2} d^{-2}$

(In this table, $\propto p^0$ means independent of pressure.)

Further reading

Chapman and Cowling (1970) is the classic treatise describing the more advanced treatment of transport properties in gases.

Exercises

(9.1) Is air more viscous than water? Compare the dynamic viscosity η and the kinematic viscosity $\nu = \eta/\rho$ using the following data:

	ρ (kg m^{-3})	η (Pa s)
Air	1.3	17.4×10^{-6}
Water	1000	1.0×10^{-3}

(9.2) Obtain an expression for the thermal conductivity of a gas at ordinary pressures. The thermal conductivity of argon (atomic weight 40) at STP is 1.6×10^{-2} Wm^{-1}K^{-1}. Use this to calculate the mean free path in argon at STP. Express the mean free path in terms of an effective atomic radius for collisions and find the value of this radius. Solid argon has a close–packed cubic structure in which, if the atoms are regarded as hard spheres, 0.74 of the volume of the structure is filled. The density of solid argon is 1.6×10^3 kg m^{-3}. Compare the effective atomic radius obtained from this information with your effective collision radius. Comment on your result.

(9.3) Define the coefficient of viscosity. Use kinetic theory to show that the coefficient of viscosity of a gas is given, with suitable approximations, by

$$\eta = K\rho \langle c \rangle \lambda$$

where ρ is the density of the gas, λ is the mean free path of the gas molecules, $\langle c \rangle$ is their mean speed, and K is a number that depends on the approximations you make.

In 1660 Boyle set up a pendulum inside a vessel that was attached to a pump that could remove air from the vessel. He was surprised to find that there was no observable change in the rate of damping of the swings of the pendulum when the pump was set going. Explain his observation in terms of the above formula.

Make a rough order-of-magnitude estimate of the lower limit to the pressure that Boyle obtained; use reasonable assumptions concerning the apparatus that Boyle might have used. [The viscosity of air at atmospheric pressure and at 293 K is 18.2 μN s m^{-2}.]

Explain why the damping is nearly independent of pressure despite the fact that fewer molecules collide with the pendulum as the pressure is reduced.

(9.4) Two plane discs, each of radius 5 cm, are mounted coaxially with their adjacent surfaces 1 mm apart. They are in a chamber containing Ar gas at STP (viscosity 2.1×10^{-5} N s m^{-2}) and are free to rotate about their common axis. One of them rotates with an angular velocity of 10 rad s^{-1}. Find the torque that must be applied to the other to keep it stationary.

(9.5) Measurements of the viscosity, η, of argon gas (^{40}Ar) over a range of pressures yield the following results at two temperatures:

at 500 K $\eta \approx 3.5 \times 10^{-5}$ kg m^{-1} s^{-1};
at 2000 K $\eta \approx 8.0 \times 10^{-5}$ kg m^{-1} s^{-1}.

The viscosity is found to be approximately independent of pressure. Discuss the extent to which these data are consistent with (i) simple kinetic theory, and (ii) the diameter of the argon atom (0.34 nm) deduced from the density of solid argon at low temperatures.

(9.6) In Section 11.3, we will define the ratio of C_p to C_V as given by the number γ. We will also show that $C_p = C_V + R$, where the heat capacities here are per mole. Show that these definitions lead to

$$C_V = \frac{R}{(\gamma - 1)}. \quad (9.50)$$

Starting with the formulae $C_V = C_V' + C_V''$ and $\kappa = \left(\frac{5}{2}C_V' + C_V''\right)\eta$, show that if $C_V'/R = \frac{3}{2}$, then

$$\kappa = \frac{1}{4}(9\gamma - 5)\eta C_V, \quad (9.51)$$

which is Eucken's formula. Deduce the value of γ for each of the following monatomic gases measured at room temperatures.

Species	$\kappa/(\eta C_V)$
He	2.45
Ne	2.52
Ar	2.48
Kr	2.54
Xe	2.58

Deduce what proportion of the heat capacity of the molecules is associated with the translational degrees of freedom for these gases. (Hint: notice the word "monatomic".)

10 The thermal diffusion equation

10.1 Derivation of the thermal diffusion equation 90
10.2 The one-dimensional thermal diffusion equation 91
10.3 The steady state 94
10.4 The thermal diffusion equation for a sphere 94
10.5 Newton's law of cooling 99
10.6 The Prandtl number 100
10.7 Sources of heat 101
10.8 Particle diffusion 102
Chapter summary 103
Exercises 103

This section assumes familiarity with solving differential equations (see, e.g., Boas (1983), Riley *et al.* (2006)). It can be omitted at first reading.

In the previous chapter, we have seen how the thermal conductivity of a gas can be calculated using kinetic theory. In this chapter, we look at solving problems involving the thermal conductivity of matter using a technique developed by mathematicians in the late eighteenth and early nineteenth centuries. The key equation describes thermal diffusion, i.e., how heat appears to "diffuse" from one place to the other, and most of this chapter introduces techniques for solving this equation.

10.1 Derivation of the thermal diffusion equation

Recall from eqn 9.15 that the heat flux $\boldsymbol{J}$ is given by

$$\boldsymbol{J} = -\kappa \nabla T. \tag{10.1}$$

This equation is very similar mathematically to the equation for particle flux $\boldsymbol{\Phi}$ in eqn 9.26 which is, in three dimensions,

$$\boldsymbol{\Phi} = -D\nabla n, \tag{10.2}$$

where D is the diffusion constant, and also to the flow of electrical current given by the current density $\boldsymbol{J}_e$ defined by

$$\boldsymbol{J}_e = \sigma \boldsymbol{E} = -\sigma \nabla \phi, \tag{10.3}$$

where σ is the conductivity, $\boldsymbol{E}$ is the electric field and ϕ here is the electric potential. Because of this mathematical similarity, an equation that is analogous to the diffusion equation (eqn 9.36) holds in each case. We will derive the thermal diffusion equation in this section.

In fact in all these phenomena, there needs to be some account of the fact that you can't destroy energy, or particles, or charge. (We will only treat the thermal case here.) The total heat flow out of a closed surface S (as in Fig. 10.1) is given by the integral

$$\int_S \boldsymbol{J} \cdot \mathrm{d}\boldsymbol{S}, \tag{10.4}$$

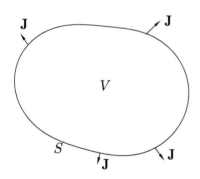

Fig. 10.1 A closed surface S encloses a volume V. The total heat flow out of S is given by $\int_S \boldsymbol{J} \cdot \mathrm{d}\boldsymbol{S}$.

and is a quantity with the dimension of power. It is therefore equal to the rate which the material inside the surface is losing energy. This can be expressed as the rate of change of the total thermal energy inside the volume V surrounded by the closed surface S. The thermal energy can be written as the volume integral $\int_V CT\,dV$, where C here is the heat capacity per unit volume (measured in $\mathrm{J\,K^{-1}\,m^{-3}}$) and is equal to ρc, where ρ is the density and c is the heat capacity per unit mass (the specific heat capacity, see Section 2.2). Hence

$$\int_S \boldsymbol{J}\cdot d\boldsymbol{S} = -\frac{\partial}{\partial t}\int_V CT\,dV. \qquad (10.5)$$

The divergence theorem implies that

$$\int_S \boldsymbol{J}\cdot d\boldsymbol{S} = \int_V \nabla\cdot\boldsymbol{J}\,dV, \qquad (10.6)$$

and hence that

$$\nabla\cdot\boldsymbol{J} = -C\frac{\partial T}{\partial t}. \qquad (10.7)$$

Substituting in eqn 10.1 then yields the **thermal diffusion equation** which is

$$\boxed{\frac{\partial T}{\partial t} = D\nabla^2 T,} \qquad (10.8)$$

where $D = \kappa/C$ is the **thermal diffusivity**. Since κ has units $\mathrm{W\,m^{-1}\,K^{-1}}$ and $C = \rho c$ has units $\mathrm{J\,K^{-1}\,m^{-3}}$, D has units $\mathrm{m^2\,s^{-1}}$.

We haven't worried about what the "zero" of thermal energy is; there could also be an additive, time-independent, constant in the expression for total thermal energy, but since we are going to differentiate this with respect to time to obtain the rate of change of thermal energy, it doesn't matter. We are also assuming C is temperature-independent for the time being.

10.2 The one-dimensional thermal diffusion equation

In one dimension, this equation becomes

$$\frac{\partial T}{\partial t} = D\frac{\partial^2 T}{\partial x^2}, \qquad (10.9)$$

and can be solved using conventional methods.

Example 10.1

Solution of the one-dimensional thermal diffusion equation

The one-dimensional thermal diffusion equation looks a bit like a wave equation. Therefore, one method to solve eqn 10.9 is to look for wave-like solutions of the form

$$T(x,t) \propto \exp(\mathrm{i}(kx - \omega t)), \qquad (10.10)$$

where $k = 2\pi/\lambda$ is the wave vector, $\omega = 2\pi f$ is the angular frequency, λ is the wavelength and f is the frequency. Substitution of this equation into eqn 10.9 yields

$$-\mathrm{i}\omega = -Dk^2 \qquad (10.11)$$

and hence
$$k^2 = \frac{i\omega}{D} \tag{10.12}$$
so that
$$k = \pm(1+i)\sqrt{\frac{\omega}{2D}}. \tag{10.13}$$

The spatial part of the wave, which looks like $\exp(ikx)$, can either be of the form
$$\exp\left((i-1)\sqrt{\frac{\omega}{2D}}x\right), \quad \text{which blows up as } x \to -\infty, \tag{10.14}$$
or
$$\exp\left((-i+1)\sqrt{\frac{\omega}{2D}}x\right), \quad \text{which blows up as } x \to \infty. \tag{10.15}$$

Let us now solve a problem in which a boundary condition is applied at $x = 0$ and a solution is desired in the region $x > 0$. We don't want solutions that blow up as $x \to \infty$ and pick the first type of solution (i.e., eqn 10.14). Hence our general solution for $x \geq 0$ can be written as
$$T(x,t) = \sum_\omega A(\omega) \exp(-i\omega t) \exp\left((i-1)\sqrt{\frac{\omega}{2D}}x\right), \tag{10.16}$$

where we have summed over all possible frequencies. To find which frequencies are needed, we have to be specific about the boundary condition for which we want to solve.

Let us imagine that we want to solve the one-dimensional problem of the propagation of sinusoidal temperature waves into the ground. The waves could be due to the alternation of day and night (for a wave with period 1 day), or winter and summer (for a wave with period 1 year). The boundary condition can be written as
$$T(0,t) = T_0 + \Delta T \cos \Omega t. \tag{10.17}$$

This boundary condition can be rewritten
$$T(0,t) = T_0 + \frac{\Delta T}{2} e^{i\Omega t} + \frac{\Delta T}{2} e^{-i\Omega t}. \tag{10.18}$$

However, at $x = 0$ the general solution (eqn 10.16) becomes
$$T(0,t) = \sum_\omega A(\omega) \exp(-i\omega t). \tag{10.19}$$

Comparison of eqns 10.18 and 10.19 implies that the only non-zero values of $A(\omega)$ are
$$A(0) = T_0, \quad A(-\Omega) = \frac{\Delta T}{2}, \quad \text{and} \quad A(\Omega) = \frac{\Delta T}{2}. \tag{10.20}$$

Hence the solution to our problem for $x \geq 0$ is
$$T(x,t) = T_0 + \Delta T\, e^{-x/\delta} \cos\left(\Omega t - \frac{x}{\delta}\right), \tag{10.21}$$

where
$$\delta = \sqrt{\frac{2D}{\Omega}} = \sqrt{\frac{2\kappa}{\Omega C}} \qquad (10.22)$$
is known as the **skin depth**. The solution in eqn 10.21 is plotted in Fig. 10.2. [Note that the use of the term skin depth brings out the analogy between this effect and the skin depth that arises when electromagnetic waves are incident on a metal surface, see e.g. Griffiths (2003).]

We note the following important features of this solution:

- T falls off exponentially as $e^{-x/\delta}$.
- There is a phase shift of x/δ radians in the oscillations.
- $\delta \propto \Omega^{-1/2}$ so that faster oscillations fall off faster.

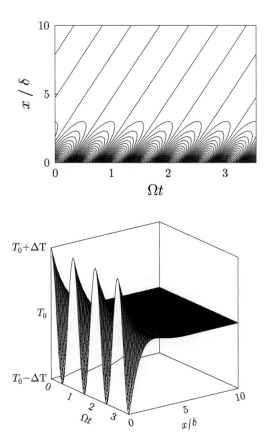

Fig. 10.2 A contour plot and a three-dimensional surface plot of eqn 10.21, showing that the temperature falls off exponentially as $e^{-x/\delta}$. The contour plot shows that there is a phase shift in the oscillations as x increases.

10.3 The steady state

If the system has reached a **steady state**, its properties are not time-dependent. This includes the temperature, so that

$$\frac{\partial T}{\partial t} = 0. \tag{10.23}$$

Hence in this case, the thermal diffusion equation reduces to

$$\nabla^2 T = 0, \tag{10.24}$$

which is Laplace's equation. Note that the thermal diffusivity $D = \kappa/C$ plays no rôle in this equation. However, there is still a heat flux $\boldsymbol{J} = -\kappa \nabla T$ and so the thermal conductivity κ is still relevant.

Example 10.2

The plane $x = 0$ is maintained at a temperature T_1 and the plane $x = L$ is maintained at a temperature $T_2 < T_1$. Find the heat flux.
Solution:
The steady state implies that we must use Laplace's equation in one dimension so $\partial^2 T/\partial x^2 = 0$. Integrating twice and putting in the boundary conditions yields

$$T = \frac{(T_2 - T_1)x}{L} + T_1 \text{ for } 0 \le x \le L, \tag{10.25}$$

and hence the heat flux is

$$J = -\kappa \left(\frac{\partial T}{\partial x}\right) = \frac{\kappa}{L}(T_1 - T_2). \tag{10.26}$$

The quantity $\frac{\kappa}{L}$ is called the **thermal conductance** or sometimes the **U value** and is measured in $\mathrm{W\,m^{-2}\,K^{-1}}$. Its reciprocal $\frac{L}{\kappa}$ is called the **thermal resistance** or sometimes the **R value** and is measured in $\mathrm{m^2\,K\,W^{-1}}$. The thermal resistance of duvets is measured in togs, where 1 tog is equal to $0.1~\mathrm{m^2\,K\,W^{-1}}$.

10.4 The thermal diffusion equation for a sphere

Very often, heat transfer problems have spherical symmetry (e.g., the cooling of the Earth or the Sun). In this section we will show that one can also solve the (rather forbidding looking) problem of the thermal diffusion equation in a system with spherical symmetry. In spherical polar coordinates, we have in general that $\nabla^2 T$ is given by[1]

---

[1] See Appendix B.

$$\nabla^2 T = \frac{1}{r^2}\frac{\partial}{\partial r}\left(r^2 \frac{\partial T}{\partial r}\right) + \frac{1}{r^2 \sin\theta}\frac{\partial}{\partial \theta}\left(\sin\theta \frac{\partial T}{\partial \theta}\right) + \frac{1}{r^2 \sin^2\theta}\frac{\partial^2 T}{\partial \phi^2}, \quad (10.27)$$

so that if T is not a function of θ or ϕ we can write

$$\nabla^2 T = \frac{1}{r^2}\frac{\partial}{\partial r}\left(r^2 \frac{\partial T}{\partial r}\right), \quad (10.28)$$

and hence the diffusion equation becomes

$$\boxed{\frac{\partial T}{\partial t} = \frac{\kappa}{C}\frac{1}{r^2}\frac{\partial}{\partial r}\left(r^2 \frac{\partial T}{\partial r}\right).} \quad (10.29)$$

Example 10.3
The thermal diffusion equation for a sphere in the steady state
In the steady state, $\partial T/\partial t = 0$ and hence we need to solve

$$\frac{1}{r^2}\frac{\partial}{\partial r}\left(r^2 \frac{\partial T}{\partial r}\right) = 0. \quad (10.30)$$

Now if T is independent of r, $\partial T/\partial r = 0$ and this will be a solution. Moreover, if $r^2(\partial T/\partial r)$ is independent of r, this will generate another solution. Now $r^2(\partial T/\partial r) = $ constant implies that $T \propto r^{-1}$. Hence a general solution is

$$T = A + \frac{B}{r}, \quad (10.31)$$

where A and B are constants. This should not surprise us if we know some electromagnetism, as we are solving Laplace's equation in spherical coordinates assuming spherical symmetry, and in electromagnetism the solution for the electric potential in this case is an arbitrary constant plus a Coulomb potential, proportional to $1/r$.

A practical problem one often needs to solve is cooking a quantity of meat. The meat is initially at some cool temperature (the temperature of the kitchen or of the refrigerator) and it is placed into a hot oven. The skill in cooking is getting the inside up to temperature. How long does it take? The next example shows how to calculate this for the (rather artificial) example of a spherical chicken!

Example 10.4
The spherical chicken
A spherical chicken[2] of radius a at initial temperature T_0 is placed into an oven at temperature T_1 at time $t = 0$ (see Fig. 10.3). The boundary conditions are that the oven is at temperature T_1 so that

$$T(a, t) = T_1, \quad (10.32)$$

[2] The methods in this example can also be applied to a spherical nut roast.

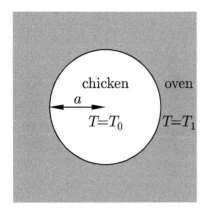

Fig. 10.3 Initial condition of a spherical chicken of radius a at initial temperature T_0, which is placed into an oven at temperature T_1 at time $t = 0$.

and the chicken is originally at temperature T_0, so that for $r < a$

$$T(r, 0) = T_0. \tag{10.33}$$

We want to obtain the temperature as a function of time at the centre of the chicken, i.e., $T(0, t)$.

Solution:

We will show how we can transform this to a one-dimensional diffusion equation. This is accomplished using a substitution

$$T(r, t) = T_1 + \frac{B(r, t)}{r}, \tag{10.34}$$

where $B(r, t)$ is now a function of r and t. This substitution is motivated by the solution to the steady–state problem in eqn 10.31 and of course means that we can write B as $B = r(T - T_1)$.

We now need to work out some partial differentials:

$$\frac{\partial T}{\partial t} = \frac{1}{r}\frac{\partial B}{\partial t}, \tag{10.35}$$

$$\frac{\partial T}{\partial r} = -\frac{B}{r^2} + \frac{1}{r}\frac{\partial B}{\partial r}, \tag{10.36}$$

and hence multiplying eqn 10.36 by r^2 we have that

$$r^2 \frac{\partial T}{\partial r} = -B + r\frac{\partial B}{\partial r}, \tag{10.37}$$

and therefore

$$\frac{\partial}{\partial r}\left[r^2 \frac{\partial T}{\partial r}\right] = r\frac{\partial^2 B}{\partial r^2}, \tag{10.38}$$

which means that eqn 10.29 becomes

$$\frac{\partial B}{\partial t} = D\frac{\partial^2 B}{\partial r^2}, \tag{10.39}$$

where $D = \kappa/C$. This is a one-dimensional diffusion equation and is therefore much easier to solve than the one with which we started.

The new boundary conditions can be rewritten as follows:

(1) Because $B = r(T - T_1)$ we have that $B = 0$ when $r = 0$:

$$B(0, t) = 0; \tag{10.40}$$

(2) Because $T = T_1$ at $r = a$ we have that:

$$B(a, t) = 0; \tag{10.41}$$

(3) Because $T = T_0$ at $t = 0$ we have that (for $r < a$):

$$B(r, 0) = r(T_0 - T_1). \tag{10.42}$$

We look for wave-like solutions with these boundary conditions and hence are led to try
$$B = \sin(kr)e^{-i\omega t}, \qquad (10.43)$$
and substituting this into eqn 10.39 yields
$$i\omega = Dk^2. \qquad (10.44)$$
The relation $ka = n\pi$ where n is an integer fits the first two boundary conditions and hence
$$i\omega = D\left(\frac{n\pi}{a}\right)^2, \qquad (10.45)$$
and hence our general solution is
$$B(r,t) = \sum_{n=1}^{\infty} A_n \sin\left(\frac{n\pi r}{a}\right) e^{-D\left(\frac{n\pi}{a}\right)^2 t}. \qquad (10.46)$$

To find A_n, we need to match this solution at $t = 0$ using our third boundary condition. Hence
$$r(T_0 - T_1) = \sum_{n=1}^{\infty} A_n \sin\left(\frac{n\pi r}{a}\right). \qquad (10.47)$$

We multiply both sides by $\sin\left(\frac{m\pi r}{a}\right)$ and integrate, so that
$$\int_0^a \sin\left(\frac{m\pi r}{a}\right) r(T_0 - T_1)\, dr = \sum_{n=1}^{\infty} A_n \int_0^a \sin\left(\frac{m\pi r}{a}\right) \sin\left(\frac{n\pi r}{a}\right) dr. \qquad (10.48)$$

The right-hand side yields $A_m a/2$ and the left-hand side can be integrated by parts. This yields
$$A_m = \frac{2a}{m\pi}(T_1 - T_0)(-1)^m, \qquad (10.49)$$
and hence by substituting this back into eqn 10.46 we obtain
$$B(r,t) = \frac{2a}{\pi}(T_1 - T_0) \sum_{n=1}^{\infty} \frac{(-1)^n}{n} \sin\left(\frac{n\pi r}{a}\right) e^{-D(n\pi/a)^2 t}. \qquad (10.50)$$

Putting this back into eqn 10.34 shows that the temperature $T(r,t)$ inside the chicken ($r \leq a$) behaves as
$$T(r,t) = T_1 + \frac{2a}{\pi}(T_1 - T_0) \sum_{n=1}^{\infty} \frac{(-1)^n}{n} \frac{\sin(n\pi r/a)}{r} e^{-D(n\pi/a)^2 t}. \qquad (10.51)$$

The centre of the chicken has temperature
$$T(0,t) = T_1 + 2(T_1 - T_0) \sum_{n=1}^{\infty} (-1)^n e^{-D(n\pi/a)^2 t}, \qquad (10.52)$$
which is deduced from eqn 10.51 using the fact that as $r \to 0$,
$$\frac{1}{r} \sin\left(\frac{n\pi r}{a}\right) \to \frac{n\pi}{a}. \qquad (10.53)$$

> This trick of multiplying both sides by $\sin(m\pi r/a)$ and integrating is a useful strategy in problems involving Fourier series (the expression in eqn 10.47 is a Fourier series). The trick relies on the fact that the two functions $\sin(n\pi r/a)$ and $\sin(m\pi r/a)$ are orthogonal unless $m = n$.

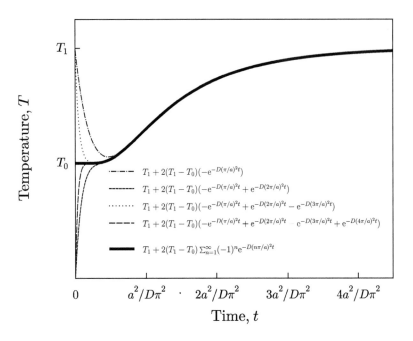

Fig. 10.4 The sum of the first few terms of eqn 10.52, shown together with $T(0, t)$ evaluated from all terms (thick solid line). The sums of only the first few terms fail near $t = 0$ and one needs more and more terms to give an accurate estimate of the temperatures as t gets closer to 0 (although this is the region where one knows what the temperature is anyway!).

The expression in eqn 10.52 (see Fig. 10.4) becomes dominated by the first exponential in the sum as time t increases, so that

$$T(0, t) \approx T_1 - 2(T_1 - T_0)e^{-D(\pi/a)^2 t}, \qquad (10.54)$$

for $t \gg a^2/D\pi^2$. Analogous behaviour is of course found for a warm sphere cooling in a colder environment. A cooling or warming body thus behaves like a low-pass filter, with the smallest exponent dominating at long times. The smaller the sphere, the shorter the time before it warms or cools, according to a simple exponential law.

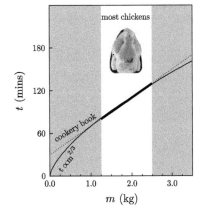

Fig. 10.5 The cooking time for a chicken according to eqn 10.55 (solid line) and the cook's rule of 40 minutes per kg plus 30 minutes "for the pot" given in many cookery books. The two rules agree for most normal-sized chickens. [Note 1 kg is approximately 2.2 lb.]

This example shows that the cooking time t is proportional to a^2. It therefore scales with the surface area $4\pi a^2$ and not the volume $\frac{4}{3}\pi a^3$. The mass m of the chicken is proportional to its volume (assuming the density of chickens is constant) and therefore

$$t \propto m^{2/3}. \qquad (10.55)$$

However, cookery books give a different "law" for cooking chickens: they often quote a rule which is something like 40 minutes per kg plus 30 minutes "for the pot". This is clearly nonsense, since the pot doesn't need cooking, and the rule fails for stir-frying small pieces of chicken (which cook in seconds in a hot pan and clearly don't need 30 minutes added on). However, the two approaches give approximately the same answer for most normal-sized chickens (see Fig. 10.5).

Example 10.5

The surface of a spherical animal of radius a is maintained at a temperature T_0 by its internal metabolism. It sits in a medium of thermal conductivity κ which is at a lower temperature T_1 (as measured at a large distance from the animal). Assuming steady–state conditions, find the rate at which the animal loses heat.

Solution:
In the region outside the animal, $\partial T/\partial t = 0$ and hence by eqns 10.30 and 10.31 we have that $T(r) = A + B/r$ where A and B are constants. Since $T(a) = T_0$ and $T(r) \to T_1$ for $r \to \infty$, we have $A = T_1$ and $B = a(T_0 - T_1)$. The heat flux is radial and is given by $J = -\kappa \partial T/\partial r = \kappa a(T_0 - T_1)/r^2$ and so at the surface $(r = a)$ is given by $J = \kappa(T_0 - T_1)/a$. The total amount of heat lost at the surface per second is therefore obtained by multiplying J by the surface area of the sphere, yielding $4\pi a^2 J = 4\pi \kappa a(T_0 - T_1)$. Note that the heat lost per second is proportional to a, even though the heat generated by the animal presumably scales with its volume and hence with a^3. Therefore heat loss is much more important for small animals than for large ones.

10.5 Newton's law of cooling

Newton's law of cooling states that the temperature of a cooling body falls exponentially towards the temperature of its surroundings with a rate proportional to the area of contact between the body and the environment. The results of the previous section indicate that it is an approximation to reality, as a cooling sphere only cools exponentially at long times.

Newton's law of cooling is often stated as follows: the heat loss of a solid or liquid surface (a hot central–heating pipe or the exposed surface of a cup of tea) to the surrounding gas (usually air, which is free to convect the heat away) is proportional to the area of contact multiplied by the temperature difference between the solid/liquid and the gas. Mathematically, this can be expressed as an equation for the heat flux $\boldsymbol{J}$, which is

$$\boldsymbol{J} = \boldsymbol{h}\Delta T, \tag{10.56}$$

where ΔT is the temperature difference between the body and its environment and $\boldsymbol{h}$ is a vector whose direction is normal to the surface of the body and whose magnitude $h = |\boldsymbol{h}|$ is a heat transfer coefficient. In general, h depends on the temperature of the body and its surroundings and varies over the surface, so that Newton's "law" of cooling is more of an empirical relation.

The steps leading from eqn 10.56 to an exponential decay of temperature are demonstrated in the following example.

Example 10.6

A polystyrene cup containing tea at temperature T_{hot} at $t = 0$ stands for a while in a room with air temperature T_{air}. The heat loss through the surface area A exposed to the air is, according to Newton's law of cooling, proportional to $A(T(t) - T_{\text{air}})$, where $T(t)$ is the temperature of the tea at time t. Ignoring the heat lost by other means, we have that

$$-C\frac{\partial T}{\partial t} = JA = hA(T - T_{\text{air}}), \tag{10.57}$$

where J is the heat flux, C is the heat capacity of the cup of tea and h is a constant, so that

$$T = T_{\text{air}} + (T_{\text{hot}} - T_{\text{air}})e^{-\lambda t} \tag{10.58}$$

where $\lambda = Ah/C$.

What makes these types of calculation of heat transfer so difficult is that heat transfer from bodies into their surrounding gas or liquid is often dominated by **convection**.[3] Convection can be defined as the transfer of heat by the motion of or within a fluid (i.e., within a liquid or a gas). Convection is often driven by the fact that warmer fluid expands and rises, while colder fluid contracts and sinks; this causes currents in the fluid to be set up, which rather efficiently transfer heat. Our analysis of the thermal conductivity in a gas ignores such currents. Convection is a very complicated process and can depend on the precise details of the geometry of the surroundings. A third form of heat transfer is by thermal radiation, and this will be the subject of chapter 23.

[3] One can either have **forced convection**, in which fluid is driven past the cooling body by some external input of work (provided by means of a pump, fan, propulsive motion of an aircraft, etc.), or **free convection**, in which any external fluid motion is driven only by the temperature difference between the cooling body and the surrounding fluid. Newton's law of cooling is actually only correct for forced convection, while for free convection (which one should probably use for the example of the cooling of a cup of tea in air) the heat transfer coefficient is temperature dependent ($h \propto (\Delta T)^{1/4}$ for laminar flow, $h \propto (\Delta T)^{1/3}$ in the turbulent regime). We examine convection in stars in more detail in Section 35.3.2.

10.6 The Prandtl number

How valid is it to ignore convection? It's clearly fine to ignore it in a solid, but for a fluid we need to know the relative strength of the diffusion of momentum and heat. Convection dominates if momentum diffusion dominates (because convection involves transport of the gas itself) but conduction dominates if heat diffusion dominates. We can express these two diffusivities using the kinematic viscosity $\nu = \eta/\rho$ (with units m^{2}s^{-1}) and the thermal diffusivity $D = \kappa/\rho c_p$ (also with units m^{2}s^{-1}), where ρ is the density. To examine their relative magnitudes, we define the **Prandtl number** as the dimensionless ratio σ_p obtained by dividing ν by D, so that

$$\sigma_p = \frac{\nu}{D} = \frac{\eta c_p}{\kappa}. \tag{10.59}$$

For an ideal gas, we can use $c_p/c_V = \gamma = \frac{5}{3}$, and using eqn 9.21 (which states that $\kappa = c_V \eta$) we arrive at $\sigma_p = \frac{5}{3}$. However, eqn 9.21 resulted from an approximate treatment, and the corrected version is eqn 9.44 (which states that $\kappa = \frac{5}{2}\eta c_V$), and hence we arrive at

$$\sigma_p = \frac{2}{3}. \tag{10.60}$$

For many gases, the Prandtl number is found to be around this value. It is between 100 and 40000 for engine oil and around 0.015 for mercury. When $\sigma_p \gg 1$, diffusion of momentum (i.e., viscosity) dominates over diffusion of heat (i.e., thermal conductivity), and convection is the dominant mode of heat transport. When $\sigma_p \ll 1$ the reverse is true, and thermal conduction dominates the heat transport.

10.7 Sources of heat

If heat is generated at a rate H per unit volume (so H is measured in $\mathrm{W\,m^{-3}}$), this will add to the divergence of $\boldsymbol{J}$ so that eqn 10.7 becomes

$$\nabla \cdot \boldsymbol{J} = -C\frac{\partial T}{\partial t} + H, \tag{10.61}$$

and hence the thermal diffusion equation becomes

$$\nabla^2 T = \frac{C}{\kappa}\frac{\partial T}{\partial t} - \frac{H}{\kappa}, \tag{10.62}$$

or equivalently

$$\boxed{\frac{\partial T}{\partial t} = D\nabla^2 T + \frac{H}{C}.} \tag{10.63}$$

Example 10.7

A metallic bar of length L with both ends maintained at $T = T_0$ passes a current, which generates heat H per unit length of the bar per second. Find the temperature at the centre of the bar in steady state.
Solution: In steady state,

$$\frac{\partial T}{\partial t} = 0, \tag{10.64}$$

and so

$$\frac{\partial^2 T}{\partial x^2} = -\frac{H}{\kappa}. \tag{10.65}$$

Integrating this twice yields

$$T = \alpha x + \beta - \frac{H}{2\kappa}x^2, \tag{10.66}$$

where α and β are constants of integration. The boundary conditions imply that

$$T - T_0 = \frac{H}{2\kappa}x(L - x), \tag{10.67}$$

so that at $x = L/2$ we have that the temperature is

$$T = T_0 + \frac{HL^2}{8\kappa}. \tag{10.68}$$

10.8 Particle diffusion

This chapter has been concerned with the diffusion of heat, but as stated at the beginning of the chapter, the same laws apply to diffusion of particles. Just as heat diffuses down a temperature gradient ∇T from hot to cold, so particles diffuse down a concentration gradient ∇n from high concentration to low concentration. The mathematics are analogous since the diffusion equation $\partial n/\partial t = D\nabla^2 n$ (with D as a diffusion constant) is essentially the same as the thermal diffusion equation $\partial T/\partial t = D\nabla^2 T$ (with $D = \kappa/C$ as the thermal diffusivity). The techniques presented in this chapter can be used to solve many problems in diffusion physics.

Example 10.8

A sphere of radius a is placed in an infinite medium containing certain particles with number density n_0. The sphere absorbs these particles with great efficiency so that the number density at distance $r = a$ from the centre of the sphere is zero. Find the rate of absorption of the particles by the sphere.

Solution:

This problem is entirely analogous to that of Example 10.5. Using the same methods, we find

$$n(r) = n_0 \left(1 - \frac{a}{r}\right) \qquad (10.69)$$

outside the sphere, so that the flux Φ at the surface is

$$\Phi = -D\left(\frac{\partial n}{\partial r}\right)_{r=a} = \frac{Dn_0}{a}, \qquad (10.70)$$

and so the total rate of absorption is obtained by multiplying Φ by the surface area of the sphere which gives

$$\text{rate of absorption} = 4\pi a n_0. \qquad (10.71)$$

Notice that this rate is (again) proportional to the radius a and not to the area (or even the volume). This has important implications in biology. Bacteria absorb oxygen from their environment and this is at a maximum rate $4\pi a n_0$ (assuming them to be spherical and maximally efficient absorbers), but their consumption of oxygen scales with their volume and hence with a^3. This sets a maximum limit on the size of a bacterium, because if it is too big the bacterium will not be able to supply its internal oxygen needs. Large organisms are multicellular.

Chapter summary

- The thermal diffusion equation (in the absence of a heat source) is

$$\boxed{\frac{\partial T}{\partial t} = D\nabla^2 T,} \qquad (10.72)$$

where $D = \kappa/C$ is the thermal diffusivity.

- "Steady state" implies that

$$\frac{\partial}{\partial t}(\text{physical quantity}) = 0. \qquad (10.73)$$

- If heat is generated at a rate H per unit volume per unit time, then the thermal diffusion equation becomes

$$\frac{\partial T}{\partial t} = D\nabla^2 T + \frac{H}{C}. \qquad (10.74)$$

- Newton's law of cooling states that the heat loss from a solid or liquid surface is proportional to the area of the surface multiplied by the temperature difference between the solid/liquid and the gas.

- The particle diffusion equation is

$$\frac{\partial n}{\partial t} = D\nabla^2 n, \qquad (10.75)$$

where D is the diffusion constant.

Exercises

(10.1) One face of a thick uniform layer is subject to sinusoidal temperature variations of angular frequency ω. Show that damped sinusoidal temperature oscillations propagate into the layer and give an expression for the decay length of the oscillation amplitude. A cellar is built underground and is covered by a ceiling, which is 3 m thick and made of limestone. The outside temperature is subject to daily fluctuations of amplitude 10°C and annual fluctuations of 20°C. Estimate the magnitude of the daily and annual temperature variations within the cellar. Assuming that January is the coldest month of the year, when will the cellar's temperature be at its lowest?

[The thermal conductivity of limestone is $1.6\,\mathrm{W\,m^{-1}\,K^{-1}}$, and the heat capacity of limestone is $2.5 \times 10^6\,\mathrm{J\,K^{-1}\,m^{-3}}$.]

(10.2) (a) A cylindrical wire of thermal conductivity κ, radius a and resistivity ρ uniformly carries a current I. The temperature of its surface is fixed at T_0 using water cooling. Show that the temperature $T(r)$ inside the wire at radius r is given by

$$T(r) = T_0 + \frac{\rho I^2}{4\pi^2 a^4 \kappa}(a^2 - r^2).$$

(b) The wire is now placed in air at temperature T_{air} and the wire loses heat from its surface according to Newton's law of cooling (so that the heat flux from the surface of the wire is given by $\alpha(T(a) - T_{\mathrm{air}})$, where α is a constant). Find the temperature $T(r)$.

(10.3) Show that for the problem of a spherical chicken being cooked in an oven considered in Example 10.4, the temperature T gets 90% of the way from T_0 to T_1 after a time $\sim a^2 \ln 20/\pi^2 D$.

(10.4) A microprocessor has an array of metal fins attached to it, whose purpose is to remove heat generated within the processor. Each fin may be represented by a long thin cylindrical copper rod with one end attached to the processor; heat received by the rod through this end is lost to the surroundings through its sides.
Show that the temperature $T(x,t)$ at location x along the rod at time t obeys the equation

$$\rho C_p \frac{\partial T}{\partial t} = \kappa \frac{\partial^2 T}{\partial x^2} - \frac{2}{a} R(T),$$

where a is the radius of the rod, and $R(T)$ is the rate of heat loss per unit area of surface at temperature T. The surroundings of the rod are at temperature T_0. Assume that $R(T)$ has the form of Newton's law of cooling, namely

$$R(T) = A(T - T_0).$$

In the steady state:
(a) obtain an expression for T as a function of x for the case of an infinitely long rod whose hot end has temperature T_m;
(b) show that the heat that can be transported away by a long rod (with radius a) is proportional to $a^{3/2}$, provided that A is independent of a.
In practice the rod is not infinitely long. What length does it need to have for the results above to be approximately valid? The radius of the rod, a, is 1.5 mm.
[The thermal conductivity of copper is $380\,\text{W}\,\text{m}^{-1}\,\text{K}^{-1}$. The cooling constant $A = 250\,\text{W}\,\text{m}^{-2}\,\text{K}^{-1}$.]

(10.5) For oscillations at frequency ω, a viscous skin depth δ_v can be defined by

$$\delta_\text{v} = \left(\frac{2\eta}{\rho\omega}\right)^{1/2}, \tag{10.76}$$

analogously to the thermal skin depth

$$\delta = \left(\frac{2\kappa}{\rho c_p \omega}\right)^{1/2} \tag{10.77}$$

defined in this chapter. Show that

$$\left(\frac{\delta_\text{v}}{\delta}\right)^2 = \sigma_\text{p}, \tag{10.78}$$

where σ_p is the Prandtl number (see eqn 10.59).

(10.6) For thermal waves, calculate the magnitude of the group velocity. This shows that the thermal diffusion equation cannot hold exactly since the velocity of propagation can become larger than that of any particles that could carry heat through the material. We now consider a modification of the thermal diffusion equation which fixes this problem. Consider the number density n of thermal carriers in a material. In equilibrium, $n = n_0$, so that

$$\left(\frac{\partial n}{\partial t}\right) = -\boldsymbol{v}\cdot\nabla n + \frac{n - n_0}{\tau}, \tag{10.79}$$

where τ is a relaxation time and $\boldsymbol{v}$ is the carrier velocity. Multiply this equation by $\hbar\omega\tau\boldsymbol{v}$, where $\hbar\omega$ is the energy of a carrier, and sum over all $\boldsymbol{k}$ states. Using the fact that $\sum_{\boldsymbol{k}} n_0 \boldsymbol{v} = 0$ and $\boldsymbol{J} = \sum_{\boldsymbol{k}} \hbar\omega n \boldsymbol{v}$, and that $|n - n_0| \ll n_0$ show that

$$\boldsymbol{J} + \tau \frac{\mathrm{d}\boldsymbol{J}}{\mathrm{d}t} = -\kappa \nabla T, \tag{10.80}$$

and hence the modified thermal diffusion equation becomes

$$\frac{\partial T}{\partial t} + \tau \frac{\partial^2 T}{\partial t^2} = D\nabla^2 T. \tag{10.81}$$

Show that this modified equation gives a group velocity whose magnitude remains finite. Is this modification ever necessary?

(10.7) A series of N large, flat rectangular slabs with thickness Δx_i and thermal conductivity κ_i are placed on top of one another. The top and bottom surfaces are maintained at temperature T_i and T_f respectively. Show that the heat flux J through the slabs is given by $J = (T_i - T_f)/\sum_i R_i$, where $R_i = \Delta x_i / \kappa_i$.

(10.8) The space between two concentric cylinders is filled with material of thermal conductivity κ. The inner (outer) cylinder has radius r_1 (r_2) and is maintained at temperature T_1 (T_2). Derive an expression for the heat flow per unit length between the cylinders.

(10.9) A pipe of radius R is maintained at a uniform temperature T. To reduce heat loss from the pipe, it is lagged by an insulating material of thermal conductivity κ. The lagged pipe has radius $r > R$. Assume that all surfaces lose heat according to Newton's law of cooling $\boldsymbol{J} = \boldsymbol{h}\Delta T$, where $h = |\boldsymbol{h}|$ can be taken to be a constant. Show that the heat loss per unit length of pipe is inversely proportional to

$$\frac{1}{hr} + \frac{1}{\kappa}\ln\left(\frac{r}{R}\right), \tag{10.82}$$

and hence show that thin lagging doesn't reduce heat loss if $R < \kappa/h$.

Jean Baptiste Joseph Fourier (1768–1830)

Fourier was born in Auxerre, France, the son of a tailor. He was schooled there in the École Royale Militaire where he showed early mathematical promise. In 1787 he entered a Benedictine abbey to train for the priesthood, but the pull of science was too great and he never followed that vocation, instead becoming a teacher at his old school in Auxerre. He was also interested in politics, and unfortunately there was a lot of it around at the time; Fourier became embroiled in the Revolutionary ferment and in 1794 came close to being guillotined, but following Robespierre's execution by the same means, the political tide turned in Fourier's favour. He was able to study at the École Normale in Paris under such luminaries as Lagrange and Laplace, and in 1795 took up a chair at the École Polytechnique.

Fig. 10.6 J. B. J. Fourier

Fourier joined Napoleon on his invasion of Egypt in 1798, becoming governor of Lower Egypt in the process. There he carried out archaeological explorations and later wrote a book about Egypt (which Napoleon then edited to make the history sections more favourable to himself). Nelson's defeat of the French fleet in late 1798 rendered Fourier isolated there, but he nevertheless set up political institutions. He managed to slink back to France in 1801 to resume his academic post, but Napoleon (a hard man to refuse) sent him back to an administrative position in Grenoble where he ended up on such highbrow activities as supervising the draining of swamps and organizing the construction of a road between Grenoble and Turin. He nevertheless found enough time to work on experiments on the propagation of heat and published, in 1807, his memoir on this subject. Lagrange and Laplace criticized his mathematics (Fourier had been forced to invent new techniques to solve the problem, which we now call Fourier series, and this was fearsomely unfamiliar stuff at the time), while the notoriously difficult Biot (he of the Biot–Savart law fame) claimed that Fourier had ignored his own crucial work on the subject (Fourier had discounted it, as Biot's work on this subject was wrong). Fourier's work won him a prize, but reservations about its importance or correctness remained.

In 1815, Napoleon was exiled to Elba and Fourier managed to avoid Napoleon who was due to pass through Grenoble en route out of France. When Napoleon escaped, he brought an army to Grenoble and Fourier avoided him again, earning Napoleon's displeasure, but he managed to patch things up and got himself made Prefect of Rhône, a position from which he resigned as soon as he could. Following Napoleon's final defeat at Waterloo, Fourier became somewhat out of favour in political circles and was able to continue working on physics and mathematics back in Paris. In 1822 he published his *Théorie analytique de chaleur* (Analytical Theory of Heat) which included all his work on thermal diffusion and the use of Fourier series, a work that was to prove influential with many later thermodynamicists of the nineteenth century.

In 1824, Fourier wrote an essay that pointed towards what we now call the greenhouse effect; he realised that the insulating effect of the atmosphere might increase the Earth's surface temperature. He understood the way planets lose heat via infrared radiation (though he called it "chaleur obscure"). Since so much of his scientific work had been bound up with the nature of heat (even his work on Fourier series was only performed so he could solve heat problems) he became, in his later years, somewhat obsessed by the imagined healing powers of heat. He kept his house overheated, and wore excessively warm clothes, in order to maximize the effect of the supposedly life-giving heat. He died in 1830 after falling down the stairs.

Part IV

The first law

In this part we are now ready to think about *energy* in some detail and hence introduce the *first law of thermodynamics*. This part is structured as follows:

- In Chapter 11, we present the notion of a *function of state*, of which *internal energy* is one of the most useful. We discuss in detail the *first law of thermodynamics*, which states that energy is conserved and heat is a form of energy. We derive expressions for the *heat capacity* measured at constant volume or pressure for an ideal gas.
- In Chapter 12 we introduce the key concept of *reversibility* and discuss *isothermal* and *adiabatic* processes.

11 Energy

11.1 Some definitions 108
11.2 The first law of thermodynamics 110
11.3 Heat capacity 112
Chapter summary 115
Exercises 115

In this chapter we are going to focus on one of the key concepts in thermal physics, that of *energy*. What happens when energy is changed from one form to another? How much work can you get out of a quantity of heat? These are key questions to be answered. We are now beginning a study of **thermodynamics** proper, and in this chapter we will introduce the first law of thermodynamics. Before the first law, the most important concept in this chapter, we will introduce some additional ideas.

11.1 Some definitions

11.1.1 A system in thermal equilibrium

In thermodynamics, we define a **system** to be whatever part of the Universe we select for study. Near the system are its **surroundings**. We recall from Section 4.1 that a system is in *thermal equilibrium* when its macroscopic observables (such as its pressure or its temperature) have ceased to change with time. If you take a gas in a container, which has been held at a certain stable temperature for a considerable period of time, the gas is likely to be in thermal equilibrium. A system in thermal equilibrium having a particular set of macroscopic observables is said to be in a particular **equilibrium state**. If however, you *suddenly* apply a lot of heat to one side of the box, then *initially at least*, the gas is likely to be in a **non-equilibrium state**.

11.1.2 Functions of state

A system is in an equilibrium state if macroscopic observable properties have fixed, definite values, independent of "how they got there". These properties are **functions of state** (sometimes called **variables of state**). A **function of state** is any physical quantity that has a well-defined value for each equilibrium state of the system. Thus, in thermal equilibrium these variables of state have no time dependence. Examples are volume, pressure, temperature, and internal energy, and we will introduce a lot more in what follows. Examples of quantities that are not functions of state include the position of particle number 4325667, the total work done on a system, and the total heat put into the system. Later, we will show in detail why work and heat are not functions of state. However, the point can be understood as follows: the fact that your hands are warm or cold depends on their current temperature (a function of state), independently of how you got them to that temperature. For example, you can get to the same final thermodynamic state of having warm hands by different combinations of working and heating, e.g., you can end up with warm hands by rubbing them together (using the muscles in your arms to do work on them) or putting them in a toaster[1] (adding heat).

[1] NB: don't try this at home.

We now give a more mathematical treatment of what is meant by a function of state. Let the state of a system be described by parameters $\boldsymbol{x} = (x_1, x_2, \ldots)$ and let $f(\boldsymbol{x})$ be some function of state. [Note that this could be a very trivial function, such as $f(\boldsymbol{x}) = x_1$, since what we've called "parameters" are themselves functions of state. But we want to allow for more complicated functions of state which might be combinations of these "parameters".] Then if the system parameters change from $\boldsymbol{x}_i$ to $\boldsymbol{x}_f$, the change in f is

$$\Delta f = \int_{\boldsymbol{x}_i}^{\boldsymbol{x}_f} \mathrm{d}f = f(\boldsymbol{x}_f) - f(\boldsymbol{x}_i). \tag{11.1}$$

This only depends on the end points $\boldsymbol{x}_i$ and $\boldsymbol{x}_f$. The quantity $\mathrm{d}f$ is an **exact differential** (see Appendix C.7) and functions of state have exact differentials. By contrast, a quantity that is represented by an **inexact differential** is not a function of state. The following example illustrates these kinds of differential.

(a)

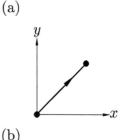

(b)

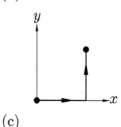

Example 11.1

Let a system be described by two parameters, x and y. Let $f = xy$ so that

$$\mathrm{d}f = \mathrm{d}(xy) = y\,\mathrm{d}x + x\,\mathrm{d}y. \tag{11.2}$$

(c)

Then if (x, y) changes from $(0, 0)$ to $(1, 1)$, the change in f is given by

$$\Delta f = \int_{(0,0)}^{(1,1)} \mathrm{d}f = [xy]_{(0,0)}^{(1,1)} = (1 \times 1) - (0 \times 0) = 1. \tag{11.3}$$

This answer is independent of the exact path taken (it could be any of those shown in Fig. 11.1) because $\mathrm{d}f$ is an exact differential.

Now consider[2] $\bar{\mathrm{d}}g = y\,\mathrm{d}x$. The change in g when (x, y) changes from $(0, 0)$ to $(1, 1)$ along the path shown in Fig. 11.1(a) is given by

Fig. 11.1 Three possible paths between the points $(x, y) = (0, 0)$ and $(x, y) = (1, 1)$.

$$\Delta g = \int_{(0,0)}^{(1,1)} y\,\mathrm{d}x = \int_0^1 x\,\mathrm{d}x = \frac{1}{2}. \tag{11.4}$$

[2] We put a line through quantities such as the d in $\mathrm{d}g$ to signify that it is an inexact differential.

However when the integral is not carried out along the line $y = x$, but along the path shown in Fig. 11.1(b), it is given by

$$\Delta g = \int_{(0,0)}^{(1,0)} y\,\mathrm{d}x + \int_{(1,0)}^{(1,1)} y\,\mathrm{d}x = 0. \tag{11.5}$$

If the integral is taken along the path shown in Fig. 11.1(c), yet another result would be obtained, but we are not going to attempt to calculate that!

Hence we find that the value of Δg depends on the path taken, and this is because đg is an inexact differential.[3]

[3]Note that if x is taken to be volume, V, and y is taken to be pressure, p, then the quantity f is proportional to temperature, while đg is the negative of the work đ$W = -p\,dV$. This demonstrates that temperature is a function of state and work is not.

Recall from Section 1.2 that functions of state can either be:

- **extensive** (proportional to system size), e.g., energy, volume, magnetization, mass, or

- **intensive** (independent of system size), e.g., temperature, pressure, magnetic field, density, energy density.

In general one can find an **equation of state** that connects functions of state: for a gas this takes the form $f(p, V, T) = 0$. An example is the equation of state for an ideal gas, $pV = nRT$, which we met in eqn 1.12.

11.2 The first law of thermodynamics

Though the idea that heat and work are both forms of energy seems obvious to a modern physicist, the idea took some getting used to. Lavoisier had, in 1789, proposed that heat was a weightless, conserved fluid called *caloric*. Caloric was a fundamental element that couldn't be created or destroyed. Lavoisier's notion "explained" a number of phenomena, such as combustion (fuels have stored caloric which is released on burning). Rumford in 1798 realized that something was wrong with the caloric theory: heating could be produced by friction, and if you keep on drilling through a cannon barrel (to take the example that drew the problem to his attention) almost limitless supplies of heat can be extracted. Where does all this caloric come from? Mayer quantified this in 1842 with an elegant experiment in which he frictionally generated heat in paper pulp and measured the temperature rise. Joule[4] independently performed similar experiments, but more accurately, in the period 1840–1845 (and his results became better known so that he was able to claim the credit!) Joule let a mass tied to a string slowly descend a certain height, while the other end of the string turns a paddle wheel immersed in a certain mass of water. The turning of the paddle frictionally heats the water. After a number of descents, Joule measured the temperature rise of the water. In this way he was able to deduce the "mechanical equivalent of heat". He also measured the heat output of a resistor (which, in modern units, is equal to $I^2 R$, where I is the current and R the resistance). He was able to show that the same heat was produced for the same energy used, independent of the method of delivery. This implied that heat is a form of energy. Joule's experiments therefore consigned the caloric theory of heat to a footnote in history.

[4]The SI unit of energy is named after Joule. $1\,\mathrm{J} = 1\,\mathrm{N\,m}$. Older units are still in use in some places: 1 calorie is defined as the energy required to raise 1 g of water by 1°C (actually from 14.5°C to 15.5°C at sea level) and $1\,\mathrm{cal} = 4.184\,\mathrm{J}$. The energy contained in food is usually measured in kilocalories (kcal), where $1\,\mathrm{kcal} = 1000\,\mathrm{cal}$. Older books sometimes used the erg: $1\,\mathrm{erg} = 10^{-7}\,\mathrm{J}$. The British thermal unit (Btu) is an archaic unit, no longer commonly used in Britain: $1\,\mathrm{Btu} = 1055\,\mathrm{J}$. The foot-pound is $1\,\mathrm{ft\,lb} = 1.356\,\mathrm{J}$. Electricity bills often record energy in kilowatt hours ($1\,\mathrm{kWh} = 3.6\,\mathrm{MJ}$). Useful in atomic physics is the electron volt: $1\,\mathrm{eV} = 1.602 \times 10^{-19}\,\mathrm{J}$.

However, it was Mayer and later Helmholtz who elevated the experimental observations into a grand principle, which we can state as follows:

> **The first law of thermodynamics**
> Energy is conserved and heat and work are both forms of energy.

A system has an **internal energy** U, which is the sum of the energy of all the internal degrees of freedom that the system possesses. U is a function of state because it has a well–defined value for each equilibrium state of the system. We can change the internal energy of the system by heating it or by doing work on it. The heat Q and work W are not functions of state since they concern the manner in which energy is delivered to (or extracted from) the system. After the event of delivering energy to the system, you have no way of telling which of Q or W was added to (or subtracted from) the system by examining the system's state.

The following analogy may be helpful: your personal bank balance behaves something like the internal energy U in that it acts like a function of state of your finances; cheques and cash are like heat and work in that they both result in a change in your bank balance, but after they have been paid in, you can't tell by simply looking at the value of your bank balance by which method the money was paid in.

The change in internal energy U of a system can be written

$$\Delta U = \Delta Q + \Delta W, \tag{11.6}$$

where ΔQ is the heat supplied **to** the system and ΔW is the work done **on** the system. Note the convention: ΔQ is positive for heat supplied to the system; if ΔQ is negative, heat is extracted from the system; ΔW is positive for work done on the system; if ΔW is negative, the system does work on its surroundings.

We define a **thermally isolated system** as a system that cannot exchange heat with its surroundings. In this case we find that $\Delta U = \Delta W$, because no heat can pass in or out of a thermally isolated system.

For a differential change, we write eqn 11.6 as

$$\boxed{dU = đQ + đW,} \tag{11.7}$$

where $đW$ and $đQ$ are inexact differentials.

The work done on stretching a wire by a distance dx with a tension F is (see Fig. 11.2(a))

$$đW = F\,dx. \tag{11.8}$$

The work done by compressing a gas (pressure p, volume V) by a piston can be calculated in a similar fashion (see Fig. 11.2(b)). In this case the force is $F = pA$, where A is the area of the piston, and $A\,dx = -dV$, so that

$$\boxed{đW = -p\,dV.} \tag{11.9}$$

In this equation, the negative sign ensures that the work $đW$ done on the system is positive when dV is negative, i.e., when the gas is being compressed.

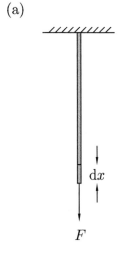

(a)

(b)

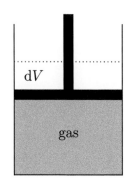

Fig. 11.2 (a) The work done stretching a wire by a distance dx is $F\,dx$. (b) The work done compressing a gas is $-p\,dV$.

It turns out that eqn 11.9 is only strictly true for a reversible change, a point we will explain further in Section 12.1. The idea is that if the piston is not frictionless, or if you move the piston too suddenly and generate shock waves, you will need to do more work to compress the gas because more heat is dissipated in the process.

11.3 Heat capacity

We now want to understand in greater detail how adding heat can change the internal energy of gas. In general, the internal energy will be a function of temperature and volume, so that we can write $U = U(T, V)$. Hence a small change in U can be related to changes in T and V by

$$\mathrm{d}U = \left(\frac{\partial U}{\partial T}\right)_V \mathrm{d}T + \left(\frac{\partial U}{\partial V}\right)_T \mathrm{d}V. \tag{11.10}$$

Rearranging eqn 11.7 with eqn 11.9 yields

$$\mathrm{d}Q = \mathrm{d}U + p\,\mathrm{d}V, \tag{11.11}$$

and now using eqn 11.10 we have that

$$\mathrm{d}Q = \left(\frac{\partial U}{\partial T}\right)_V \mathrm{d}T + \left[\left(\frac{\partial U}{\partial V}\right)_T + p\right]\mathrm{d}V. \tag{11.12}$$

We can divide eqn 11.12 by $\mathrm{d}T$ to obtain

$$\frac{\mathrm{d}Q}{\mathrm{d}T} = \left(\frac{\partial U}{\partial T}\right)_V + \left[\left(\frac{\partial U}{\partial V}\right)_T + p\right]\frac{\mathrm{d}V}{\mathrm{d}T}, \tag{11.13}$$

which is valid for any change in T or V. However, what we want to know is what is the amount of heat we have to add to effect a change of temperature under certain constraints. The first constraint is that of keeping the volume constant. We recall the definition of the heat capacity at constant volume C_V (see Section 2.2, eqn 2.6) as

$$C_V = \left(\frac{\partial Q}{\partial T}\right)_V. \tag{11.14}$$

From eqn 11.13, this constraint knocks out the second term and implies that

$$C_V = \left(\frac{\partial U}{\partial T}\right)_V. \tag{11.15}$$

The heat capacity at constant pressure is then, using eqns 2.7 and 11.13, given by

$$C_p = \left(\frac{\partial Q}{\partial T}\right)_p \tag{11.16}$$

$$= \left(\frac{\partial U}{\partial T}\right)_V + \left[\left(\frac{\partial U}{\partial V}\right)_T + p\right]\left(\frac{\partial V}{\partial T}\right)_p \tag{11.17}$$

so that
$$C_p - C_V = \left[\left(\frac{\partial U}{\partial V}\right)_T + p\right]\left(\frac{\partial V}{\partial T}\right)_p. \qquad (11.18)$$

Recall from Section 2.2 that heat capacities are measured in J K^{-1} and refer to the heat capacity of a certain quantity of gas. We will sometimes wish to talk about the heat capacity per mole of gas, or sometimes the heat capacity per mass of gas. We will use small c for the latter, known as the *specific heat capacities*:

$$c_V = \frac{C_V}{M} \qquad (11.19)$$

$$c_p = \frac{C_p}{M}, \qquad (11.20)$$

where M is the mass of the material. Specific heat capacities are measured in $\text{J K}^{-1}\text{kg}^{-1}$.

Example 11.2

Heat capacity of an ideal monatomic gas

For an ideal monatomic gas, the internal energy U is due to the kinetic energy, and hence $U = \frac{3}{2}RT$ per mole (see eqn 5.17; this result arises from the kinetic theory of gases). This means that U is only a function of temperature. Hence

$$\left(\frac{\partial U}{\partial V}\right)_T = 0. \qquad (11.21)$$

The equation of state for 1 mole of ideal gas is

$$pV = RT, \qquad (11.22)$$

so that

$$V = \frac{RT}{p}, \qquad (11.23)$$

and hence

$$\left(\frac{\partial V}{\partial T}\right)_p = \frac{R}{p}, \qquad (11.24)$$

and hence using eqns 11.18, 11.21 and 11.24 we have that

$$C_p - C_V = \left[\left(\frac{\partial U}{\partial V}\right)_T + p\right]\left(\frac{\partial V}{\partial T}\right)_p = R. \qquad (11.25)$$

Because $U = \frac{3}{2}RT$, we therefore have that

$$C_V = \left(\frac{\partial U}{\partial T}\right)_V = \frac{3}{2}R \text{ per mole}, \qquad (11.26)$$

and

$$C_p = C_V + R = \frac{5}{2}R \text{ per mole}. \qquad (11.27)$$

Example 11.3

Is it always true that $dU = C_V \, dT$?

Solution:

No, in general eqn 11.10 and eqn 11.15 imply that

$$dU = C_V \, dT + \left(\frac{\partial U}{\partial V}\right)_T dV. \tag{11.28}$$

For an ideal gas, $\left(\frac{\partial U}{\partial V}\right)_T = 0$ (eqn 11.21) so it is true that

$$dU = C_V \, dT, \tag{11.29}$$

but for non-ideal gases, $\left(\frac{\partial U}{\partial V}\right)_T \neq 0$ and hence $dU \neq C_v \, dT$.

The ratio of C_p to C_V turns out to be a very useful quantity (we will see why in the following chapter) and therefore we give it a special name. We define the **adiabatic index**[5] γ as the *ratio* of C_p and C_V, so that

[5] γ is sometimes called the **adiabatic exponent**.

$$\gamma = \frac{C_p}{C_V}. \tag{11.30}$$

The reason for the name will become clear in the following chapter.

Example 11.4

What is γ for an ideal monatomic gas?

Solution:

Using the results from the previous example[6]

[6] If the gas is not monatomic, γ can take a different value; see Section 19.2.

$$\gamma = \frac{C_p}{C_V} = \frac{C_V + R}{C_V} = 1 + \frac{R}{C_V} = \frac{5}{3}. \tag{11.31}$$

Example 11.5

Assuming $U = C_V T$ for an ideal gas, find (i) the internal energy per unit mass and (ii) the internal energy per unit volume.

Solution: Using the ideal gas equation $pV = Nk_B T$ and the density $\rho = Nm/V$ (where m is the mass of one molecule), we find that

$$\frac{p}{\rho} = \frac{k_B T}{m}. \tag{11.32}$$

Using eqn 11.31, we have that the heat capacity per mole is given by

$$C_V = \frac{R}{\gamma - 1}. \qquad (11.33)$$

Hence, we can write that the internal energy for one mole of gas is

$$U = C_V T = \frac{RT}{\gamma - 1} = \frac{N_A k_B T}{\gamma - 1}. \qquad (11.34)$$

The molar mass is mN_A, and so dividing eqn 11.34 by the molar mass, yields $\tilde{u}$, the internal energy per unit mass, given by

$$\tilde{u} = \frac{p}{\rho(\gamma - 1)}. \qquad (11.35)$$

Multiplying $\tilde{u}$ by the density ρ gives u, the internal energy per unit volume, as

$$u = \rho \tilde{u} = \frac{p}{\gamma - 1}. \qquad (11.36)$$

Chapter summary

- Functions of state have exact differentials.
- The first law of thermodynamics states that "energy is conserved and heat is a form of energy".
- $dU = đW + đQ$.
- For a reversible change, $đW = -p\,dV$.
- $C_V = \left(\frac{\partial Q}{\partial T}\right)_V = \left(\frac{\partial U}{\partial T}\right)_V$.
- $C_p = \left(\frac{\partial Q}{\partial T}\right)_P$ and $C_p - C_V = R$ for a mole of ideal gas.
- The adiabatic index is $\gamma = C_p/C_V$.

Exercises

(11.1) One mole of ideal monatomic gas is confined in a cylinder by a piston and is maintained at a constant temperature T_0 by thermal contact with a heat reservoir. The gas slowly expands from V_1 to V_2 while being held at the same temperature T_0. Why does the internal energy of the gas not change? Calculate the work done by the gas and the heat flow into the gas.

(11.2) Show that, for an ideal gas,

$$\frac{R}{C_V} = \gamma - 1 \qquad (11.37)$$

and
$$\frac{R}{C_p} = \frac{\gamma - 1}{\gamma}, \quad (11.38)$$

where C_V and C_p are the heat capacities per mole.

(11.3) Consider the differential
$$dz = 2xy\, dx + (x^2 + 2y)\, dy. \quad (11.39)$$

Evaluate the integral $\int_{(x_1,y_1)}^{(x_2,y_2)} dz$ along the paths consisting of straight-line segments
(i) $(x_1, y_1) \to (x_2, y_1)$ and then $(x_2, y_1) \to (x_2, y_2)$.
(ii) $(x_1, y_1) \to (x_1, y_2)$ and then $(x_1, y_2) \to (x_2, y_2)$.
Is dz an exact differential?

(11.4) In polar coordinates, $x = r\cos\theta$ and $y = r\sin\theta$. The definition of x implies that
$$\frac{\partial x}{\partial r} = \cos\theta = \frac{x}{r}. \quad (11.40)$$

But we also have $x^2 + y^2 = r^2$, so differentiating with respect to r gives
$$2x\frac{\partial x}{\partial r} = 2r \quad \Longrightarrow \quad \frac{\partial x}{\partial r} = \frac{r}{x}. \quad (11.41)$$

But eqns 11.40 and 11.41 imply that
$$\frac{\partial x}{\partial r} = \frac{\partial r}{\partial x}. \quad (11.42)$$

What's gone wrong?

(11.5) In the comic song by Flanders and Swann about the laws of thermodynamics, they summarize the first law by the statement:

Heat is work and work is heat

Is that a good summary?

Antoine Lavoisier (1743–1794)

All flammable materials contain the odourless, colourless, tasteless substance *phlogiston*, and the process of burning them releases this phlogiston into the air. The burned material is said to be "dephlogistonated".

Fig. 11.3 Antoine Lavoisier

That this notion is completely untrue was first shown by Antoine Lavoisier, who was born into a wealthy Parisian family. Lavoisier showed that both sulphur and phosphorous increase in weight once burned, but the weight gain was lost from the air. He demonstrated that it was oxygen that was responsible for combustion, not phlogiston, and also that oxygen was responsible for the rusting of metals (his oxygen work was helped by results communicated to him by Joseph Priestley, and Lavoisier was a little lax in giving Priestley credit for this). Lavoisier showed that hydrogen and oxygen combined to make water and also identified the concept of an element as a fundamental substance that could not be broken down into simpler constituents by chemical processes. Lavoisier combined great experimental skill (and in this, he was ably assisted by his wife) and theoretical insight and is considered a founder of modern chemistry. Unfortunately, he added to his list of elemental substances both light and *caloric*, his proposed fluid which carried heat. Thus while ridding science of an unnecessary mythical substance (phlogiston), he introduced another one (caloric).

Lavoisier was a tax collector and thus found himself in the firing line when the French revolution started, the fact that he ploughed his dubiously gotten gains into scientific research cutting no ice with revolutionaries. He had unfortunately made an enemy of Jean-Paul Marat, a journalist with an interest in science who in 1780 had wanted to join the French Academy of Sciences, but was blocked by Lavoisier. In 1792 Marat, now a firebrand revolutionary leader, demanded Lavoisier's death. Although Marat was himself assassinated in 1793 (while lying in his bath), Lavoisier was guillotined the following year.

Benjamin Thompson [Count Rumford] (1753–1814)

Thompson was born in rural Massachusetts and had an early interest in science. In 1772, as a humble doctor's apprentice, he married a rich heiress, moved to Rumford, New Hampshire, and got himself appointed as a major in a local militia. He threw his lot in with the British during the American Revolution, feeding them information about the location of American forces and performing scientific work on the force of gunpowder. His British loyalties made him few friends in the land of his birth and he fled to Britain, abandoning his wife.

Fig. 11.4 Benjamin Thompson

He subsequently fell out with the British and moved, in 1785, to Bavaria where he worked for Elector Karl Theodor who made him a Count, and henceforth he was known as Count Rumford. He organized the poor workhouses, established the cultivation of the potato in Bavaria and invented Rumford soup. He continued to work on science, sometimes erratically (he believed that gases and liquids were perfect insulators of heat) but sometimes brilliantly; he noticed that the drilling of metal cannon barrels produced apparently limitless amounts of heat and his subsequent understanding of the production of heat by friction allowed him to put an end to Lavoisier's caloric theory. Not content with simply destroying Lavoisier's theory, he married Lavoisier's widow in 1804, though they separated four years later (Rumford unkindly remarked that Antoine Lavoisier had been lucky to have been guillotined than to have stayed married to her!). In 1799, Rumford founded the Royal Institution of Great Britain, establishing Davy as the first lecturer (Michael Faraday was appointed there 14 years later). He also endowed a medal for the Royal Society and a chair at Harvard. Rumford was also a prolific inventor and gave the world the Rumford fireplace, the double boiler, a drip coffeepot and, perhaps improbably, baked Alaska (though Rumford's priority on the latter invention is not universally accepted).

12 Isothermal and adiabatic processes

12.1 Reversibility 118
12.2 Isothermal expansion of an ideal gas 120
12.3 Adiabatic expansion of an ideal gas 121
12.4 Adiabatic atmosphere 121
Chapter summary 123
Exercises 123

In this chapter we will apply the results of the previous chapter to illustrate some properties concerning isothermal and adiabatic expansions of gases. These results will assume that the expansions are reversible, and so the first part of this chapter explores the key concept of reversibility. This will be important for our discussion of entropy in subsequent chapters.

12.1 Reversibility

The laws of physics are reversible, so that if any process is allowed, then the time-reversed process can also occur. For example, if you could film the molecules in a gas bouncing off each other and the container walls, then when watching the film it would be hard to tell whether the film was being played forwards or backwards.

However, there are plenty of processes that you see in nature which seem to be irreversible. For example, consider an egg rolling off the edge of a table and smashing on the floor. Potential energy is converted into kinetic energy as the egg falls, and ultimately the energy ends up as a small amount of heat in the broken egg and the floor. The law of conservation of energy does not forbid the conversion of that heat back into kinetic energy of the reassembled egg which would then leap off the ground and back on to the table. However, this is never observed to happen. As another example, consider a battery driving a current I through a resistor with resistance R and dissipating heat at a rate $I^2 R$ into the environment. Again, one never finds heat being absorbed by a resistor from its environment, resulting in the generation of a spontaneous current that can used to recharge the battery.

Lots of processes are like this, in which the final outcome is some potential, chemical, or kinetic energy that gets converted into heat, which is then dissipated into the environment. As we shall see, the reason seems to be that there are lots more ways that the energy can be distributed in heat than in any other way, and this is therefore the most *probable outcome*. To try and understand this statistical nature of reversibility, it is helpful to consider the following example.

Example 12.1

We return to the situation described in Example 4.1. To recap, you are given a large box containing 100 identical coins. With the lid on the box, you give it a really good long and hard shake, so that you can hear the coins flipping, rattling, and being generally tossed around. Now you open the lid and look inside the box. Some of the coins will be lying with heads facing up and some with tails facing up. We assume that each of the 2^{100} possible configurations (the microstates) are equally likely to be found. *Each of these is equally likely* and so each has a probability of occurrence of approximately 10^{-30}. However, the measurement made is counting the number of heads and the number of tails (the macrostates), and the results of this measurement are not equally likely. In Example 4.1 we showed that of the $\approx 10^{30}$ individual microstates, a large number ($\approx 4 \times 10^{27}$) corresponded to 50 heads and 50 tails, but only one microstate corresponded to 100 heads and 0 tails.

Now, imagine that you had in fact carefully prepared the coins so that they were lying heads up. Following a good shake, the coins will most probably be a mixture of heads and tails. If, on the other hand, you carefully prepared a mixed arrangement of heads and tails, a good shake of the box is very unlikely to achieve a state in which all the coins lie with heads facing up. The process of shaking the box seems *almost always* to randomize the number of heads and tails, and this is an irreversible process.

This shows that the statistical behaviour of large systems is such as to make certain outcomes (such as a box of coins with mixed heads and tails) more likely than certain others (such as a box of coins containing coins the same way up). The statistics of large numbers therefore seems to drive many physical changes in an irreversible direction. How can we carry out a process in a reversible fashion?

The early researchers in thermodynamics wrestled with this problem, which was of enormous practical importance in the design of engines, in which you want to waste as little heat as possible to make your engine as efficient as possible. It was realized that when gases are expanded or compressed, it is possible to convert energy irreversibly into heat, and this will generally occur when we perform the expansion or the compression very fast, causing shock waves to be propagated through the gas (we will consider this effect in more detail in Chapter 32). However, it is possible to perform the expansion or compression *reversibly* if we do it sufficiently slowly so that *the gas remains in equilibrium throughout the entire process* and passes seamlessly from one equilibrium state to the next, each equilibrium state differing from the previous one by an infinitesimal change in the system parameters. Such a process is said to be **quasistatic**, since the process is almost in completely unchanging static equilibrium. As we shall see, heat can nevertheless be absorbed or

120 *Isothermal and adiabatic processes*

[1] This is an important point: reversibility does not necessarily exclude the generation of heat. However, reversibility does require the absence of friction; a vehicle braking and coming to a complete stop, converting its kinetic energy into heat through friction in the brakes, is an irreversible process.

emitted in the process, while still maintaining reversibility.[1] In contrast, for an irreversible process, a non-zero change (rather than a sequence of infinitesimal changes) is made to the system, and therefore the system is not in equilibrium throughout the process.

An important (but given the name, perhaps not surprising) property of reversible processes is that you can run them in reverse. This fact we will use a great deal in Chapter 13. Of course, it would take an infinite amount of time for a strictly reversible process to occur, so most processes we term reversible are approximations to the "real thing".

12.2 Isothermal expansion of an ideal gas

In this section, we will calculate the heat change in a reversible isothermal expansion of an ideal gas. The word **isothermal** means "at constant temperature", and hence in an isothermal process

$$\Delta T = 0. \tag{12.1}$$

For an ideal gas, we showed in eqn 11.29 that $dU = C_V \, dT$, and so this means that for an isothermal change

$$\Delta U = 0, \tag{12.2}$$

since U is a function of temperature only. Equation 12.2 implies that $dU = 0$ and hence from eqn 11.7

$$đW = -đQ, \tag{12.3}$$

so that the work done by the gas on its surroundings as it expands is equal to the heat absorbed by the gas. We can use $đW = -p\,dV$ (eqn 11.9), which is the correct expression for the work done in a reversible expansion. Hence the heat absorbed by the gas during an isothermal expansion from volume V_1 to volume V_2 of 1 mole of an ideal gas at temperature T is

$$\Delta Q = \int đQ \tag{12.4}$$

$$= -\int đW \tag{12.5}$$

$$= \int_{V_1}^{V_2} p \, dV \tag{12.6}$$

$$= \int_{V_1}^{V_2} \frac{RT}{V} \, dV \tag{12.7}$$

$$= RT \ln \frac{V_2}{V_1}. \tag{12.8}$$

For an expansion, $V_2 > V_1$, and so $\Delta Q > 0$. The internal energy has stayed the same, but the volume has increased so that the energy density has gone down. The energy density and the pressure are proportional to one another[2], so that pressure will also have decreased.

[2] See eqn 6.25.

12.3 Adiabatic expansion of an ideal gas

The word **adiathermal** means "without flow of heat". A system bounded by adiathermal walls is said to be *thermally isolated*. Any work done on such a system produces an adiathermal change. We define a change to be **adiabatic** if it is both adiathermal and reversible. In an adiabatic expansion, therefore, there is no flow of heat and we have

$$đQ = 0. \tag{12.9}$$

The first law of thermodynamics therefore implies that

$$dU = đW. \tag{12.10}$$

For an ideal gas, $dU = C_V dT$, and using $đW = -p\, dV$ for a reversible change, we find that, for 1 mole of ideal gas,

$$C_V\, dT = -p\, dV = -\frac{RT}{V}\, dV, \tag{12.11}$$

so that

$$\ln\frac{T_2}{T_1} = -\frac{R}{C_V}\ln\frac{V_2}{V_1}. \tag{12.12}$$

C_V here is per mole, since we are dealing with 1 mole of ideal gas.

Now $C_p = C_V + R$, and dividing this by C_V yields

$$\gamma = \frac{C_p}{C_V} = 1 + \frac{R}{C_V}, \tag{12.13}$$

and therefore $-(R/C_V) = 1 - \gamma$, so that eqn 12.12 becomes

$$TV^{\gamma-1} = \text{constant}, \tag{12.14}$$

or equivalently (using $pV \propto T$ for an ideal gas)

$$p^{1-\gamma}T^{\gamma} = \text{constant} \tag{12.15}$$

and

$$\boxed{pV^{\gamma} = \text{constant},} \tag{12.16}$$

the last equation probably being the most memorable.

Figure 12.1 shows **isotherms** (lines of constant temperature, as would be followed in an isothermal expansion) and **adiabats** (lines followed by an adiabatic expansion in which heat cannot enter or leave the system) for an ideal gas on a graph of pressure against volume. At each point, the adiabats have a steeper gradient than the isotherms, a fact we will return to in a later chapter.

12.4 Adiabatic atmosphere

The hydrostatic equation (eqn 4.23) expresses the additional pressure due to a thickness dz of atmosphere with density ρ and is

$$dp = -\rho g\, dz. \tag{12.17}$$

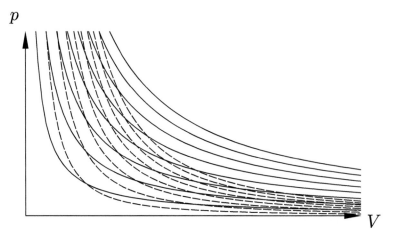

Fig. 12.1 Isotherms (solid lines) and adiabats (dashed lines).

Since $p = nk_B T$ and $\rho = nm$, where m is the mass of one molecule, we can write $\rho = mp/k_B T$ and hence

$$\frac{dp}{dz} = -\frac{mgp}{k_B T}, \qquad (12.18)$$

which implies that

$$T\frac{dp}{p} = -\frac{mg}{k_B}dz. \qquad (12.19)$$

For an isothermal atmosphere, T is a constant, and one obtains the results of Example 4.4. This assumes that the whole atmosphere is at a uniform temperature, which is unrealistic. A much better approximation (although nevertheless still an approximation to reality) is that each parcel of air[3] does not exchange heat with its surroundings. This means that if a parcel of air rises, it expands adiabatically. In this case, eqn 12.19 can be solved by recalling that for an adiabatic expansion $p^{1-\gamma}T^\gamma$ is a constant (see eqn 12.15) and hence that

$$(1-\gamma)\frac{dp}{p} + \gamma\frac{dT}{T} = 0. \qquad (12.20)$$

Substituting this into eqn 12.19 yields

$$\frac{dT}{dz} = -\left(\frac{\gamma-1}{\gamma}\right)\frac{mg}{k_B}, \qquad (12.21)$$

which is an expression relating the rate of decrease of temperature with height, predicting it to be linear. We can rewrite $(\gamma-1)/\gamma = R/C_p$, and using $R = N_A k_B$ and writing the molar mass $M_{\text{molar}} = N_A m$ we can write eqn 12.21 as

$$\frac{dT}{dz} = -\frac{M_{\text{molar}}g}{C_p}. \qquad (12.22)$$

[3] Atmospheric physicists call a "bit" of air a "parcel".

The quantity $M_{\text{molar}}g/C_p$ is known as the **adiabatic lapse rate**. For dry air (mostly nitrogen), it comes out as 9.7 K km^{-1}. Experimental values in the atmosphere are closer to 6–7 K km^{-1} (partly because the atmosphere isn't dry, and latent heat effects, due to the heat needed to evaporate water droplets [and sometimes thaw ice crystals], are also important).

Chapter summary

- In an isothermal expansion $\Delta T = 0$.
- An adiabatic change is both adiathermal (no flow of heat) and reversible. In an adiabatic expansion of an ideal gas, pV^γ is constant.

Exercises

(12.1) In an adiabatic expansion of an ideal gas, pV^γ is constant. Show also that

$$TV^{\gamma-1} = \text{constant}, \quad (12.23)$$
$$T = \text{constant} \times p^{1-1/\gamma}. \quad (12.24)$$

(12.2) Assume that gases behave according to a law given by $pV = f(T)$, where $f(T)$ is a function of temperature. Show that this implies

$$\left(\frac{\partial p}{\partial T}\right)_V = \frac{1}{V}\frac{df}{dT}, \quad (12.25)$$
$$\left(\frac{\partial V}{\partial T}\right)_p = \frac{1}{p}\frac{df}{dT}. \quad (12.26)$$

Show also that

$$\left(\frac{\partial Q}{\partial V}\right)_p = C_p \left(\frac{\partial T}{\partial V}\right)_p, \quad (12.27)$$
$$\left(\frac{\partial Q}{\partial p}\right)_V = C_V \left(\frac{\partial T}{\partial p}\right)_V. \quad (12.28)$$

In an adiabatic change, we have that

$$dQ = \left(\frac{\partial Q}{\partial p}\right)_V dp + \left(\frac{\partial Q}{\partial V}\right)_p dV = 0. \quad (12.29)$$

Hence show that pV^γ is a constant.

(12.3) Explain why we can write

$$dQ = C_p\, dT + A\, dp \quad \text{and} \quad (12.30)$$
$$dQ = C_V\, dT + B\, dV, \quad (12.31)$$

where A and B are constants. Subtract these equations and show that

$$(C_p - C_V)dT = B\, dV - A\, dp, \quad (12.32)$$

and that at constant temperature

$$\left(\frac{\partial p}{\partial V}\right)_T = \frac{B}{A}. \quad (12.33)$$

In an adiabatic change, show that

$$dp = -(C_p/A)dT, \quad (12.34)$$
$$dV = -(C_V/B)dT. \quad (12.35)$$

Hence show that in an adiabatic change, we have that

$$\left(\frac{\partial p}{\partial V}\right)_{\text{adiabatic}} = \gamma \left(\frac{\partial p}{\partial V}\right)_T, \quad (12.36)$$
$$\left(\frac{\partial V}{\partial T}\right)_{\text{adiabatic}} = \frac{1}{1-\gamma}\left(\frac{\partial V}{\partial T}\right)_p, \quad (12.37)$$
$$\left(\frac{\partial p}{\partial T}\right)_{\text{adiabatic}} = \frac{\gamma}{\gamma-1}\left(\frac{\partial p}{\partial T}\right)_V. \quad (12.38)$$

(12.4) Using eqn 12.36, relate the gradients of adiabats and isotherms on a p–V diagram.

(12.5) Two thermally insulated cylinders, A and B, of equal volume, both equipped with pistons, are connected by a valve. Initially A has its piston fully withdrawn and contains a perfect monatomic gas at temperature T, while B has its piston fully inserted, and the valve is closed. Calculate the final temperature of the gas after the following operations, which each start with the same initial arrangement. The thermal capacity of the cylinders is to be ignored.
(a) The valve is fully opened and the gas slowly drawn into B by pulling out the piston B; piston A remains stationary.
(b) Piston B is fully withdrawn and the valve is opened slightly; the gas is then driven as far as it will go into B by pushing home piston A at such a rate that the pressure in A remains constant: the cylinders are in thermal contact.

(12.6) In Rüchhardt's method of measuring γ, illustrated in Fig. 12.2, a ball of mass m is placed snugly inside a tube (cross-sectional area A) connected to a container of gas (volume V). The pressure p of the gas inside the container is slightly greater than atmospheric pressure p_0 because of the downwards force of the ball, so that

$$p = p_0 + \frac{mg}{A}. \qquad (12.39)$$

Show that if the ball is given a slight downwards displacement, it will undergo simple harmonic motion with period τ given by

$$\tau = 2\pi\sqrt{\frac{mV}{\gamma p A^2}}. \qquad (12.40)$$

[You may neglect friction. As the oscillations are fairly rapid, the changes in p and V that occur can be treated as occurring adiabatically.]
In Rinkel's 1929 modification of this experiment, the ball is held in position in the neck where the gas pressure p in the container is exactly equal to air pressure, and then let drop, the distance L that it falls before it starts to go up again is measured. Show that this distance is given by

$$mgL = \frac{\gamma p A^2 L^2}{2V}. \qquad (12.41)$$

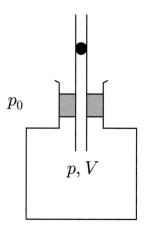

Fig. 12.2 Rüchhardt's apparatus for measuring γ. A ball of mass m oscillates up and down inside a tube.

Part V

The second law

In this part we introduce the *second law of thermodynamics* and follow its consequences. This part is structured as follows:

- In Chapter 13, we consider *heat engines*, which are cyclic processes that convert heat into work. We state various forms of the *second law of thermodynamics* and prove their equivalence, in particular showing that no engine can be more efficient than a *Carnot engine*. We also prove *Clausius' theorem*, which applies to any cyclic process.

- In Chapter 14 we show how the results from the preceding chapter lead to the concept of *entropy*. We derive the important equation $dU = TdS - pdV$, which combines the first and second laws of thermodynamics. We also introduce the *Joule expansion* and use it to discuss the statistical interpretation of entropy and *Maxwell's demon*.

- There is a very deep connection between *entropy* and *information*, and we explore this in Chapter 15, briefly touching on *data compression* and *quantum information*.

13 Heat engines and the second law

13.1 The second law of thermodynamics   126
13.2 The Carnot engine   127
13.3 Carnot's theorem   130
13.4 Equivalence of Clausius' and Kelvin's statements   131
13.5 Examples of heat engines   131
13.6 Heat engines running backwards   133
13.7 Clausius' theorem   134
Chapter summary   137
Further reading   137
Exercises   137

[1] A **reservoir** in this context is a body, which is sufficiently large that we can consider it to have essentially infinite heat capacity. This means that you can keep sucking heat out of it, or dumping heat into it, without its temperature changing. See Section 4.6.

[2] The "in isolation" phrase is very important here. In a refrigerator, heat is sucked out of cold food and squirted out of the back into your warm kitchen, so that it flows in the "wrong" direction: from cold to hot. However, this process is not happening in isolation. Work is being done by the refrigerator motor and electrical power is being consumed, adding to your electricity bill.

In this chapter, we introduce the second law of thermodynamics, probably the most important and far-reaching of all concepts in thermal physics. We are going to illustrate it with an application to the theory of "heat engines", which are machines that produce work from a temperature difference between two reservoirs.[1] It was by considering these engines that such nineteenth century physicists as Carnot, Clausius and Kelvin came to develop their different statements of the second law of thermodynamics. However, as we will see in subsequent chapters, the second law of thermodynamics has a wider applicability, affecting all types of processes in large systems and bringing insights in information theory and cosmology. In this chapter, we will begin by stating two alternative forms of the second law of thermodynamics and then discuss how these statements impact on the efficiency of heat engines.

13.1 The second law of thermodynamics

The second law of thermodynamics can be formulated as a statement about the direction of heat flow that occurs as a system approaches equilibrium (and hence there is a connection with the direction of the "arrow of time"). Heat is always observed to flow from a hot body to a cold body, and the reverse process, in isolation,[2] never occurs. Therefore, following Clausius, we can state the second law of thermodynamics as follows:

> **Clausius' statement of the second law of thermodynamics**
> "No process is possible whose sole result is the transfer of heat from a colder to a hotter body."

It turns out that an equivalent statement of the second law of thermodynamics can be made, concerning how easy it is to change energy between different forms, in particular between work and heat. It is very easy to convert work into heat. For example, pick up a brick of mass m and carry it up to the top of a building of height h (thus doing work on it equal to mgh) and then let it fall back to ground level by dropping it off the top (being careful not to hit passing pedestrians). All the work that you've done in carrying the brick to the top of the building will be dissipated in heat (and a small amount of sound energy) as the brick hits the ground. However, conversion of heat into work is much harder, and in fact the complete conversion of heat into work is impossible. This point is expressed in Kelvin's statement of the second law of thermodynamics:

> **Kelvin's statement of the second law of thermodynamics**:
> "No process is possible whose sole result is the complete conversion of heat into work."

These two statements of the second law of thermodynamics do not seem to be obviously connected, but the equivalence of these two statements will be shown in Section 13.4.

13.2 The Carnot engine

Kelvin's statement of the second law of thermodynamics says that you can't *completely* convert heat into work. However, it does not forbid *some* conversion of heat into work. How good a conversion from heat to work is possible? To answer this question, we have to introduce the concept of an engine. We define an **engine** as a system operating a cyclic process that converts heat into work. It has to be cyclic so that it can be continuously operated, producing a steady power.

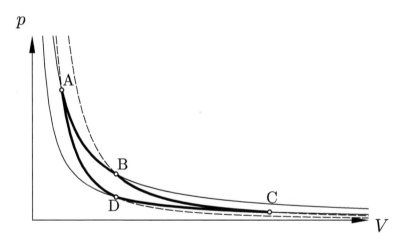

Fig. 13.1 A Carnot cycle consists of two reversible adiabats (BC and DA) and two reversible isotherms (AB and CD). The Carnot cycle is here shown on a p–V plot. It is operated in the direction A→B→C→D→A, i.e., clockwise around the solid curve. Heat Q_h enters in the isotherm A→B and heat Q_ℓ leaves in the isotherm C→D.

One such engine is the **Carnot engine**, which is based on a process called a **Carnot cycle** and which is illustrated in Fig. 13.1. An equivalent plot which is easier to sketch is shown in Fig. 13.2. The Carnot cycle consists of two reversible adiabats and two reversible isotherms for an ideal gas. The engine operates between two heat reservoirs, one at a higher temperature T_h and one at a lower temperature T_ℓ. Heat enters and leaves only during the reversible isotherms (because no heat can

Fig. 13.2 A Carnot cycle can be drawn on replotted axes where the isotherms are shown as horizontal lines (T is constant for an isotherm) and the adiabats are shown as vertical lines (where the quantity S, which must be some function of pV^γ, is constant in an adiabatic expansion; in Chapter 14 we will give a physical interpretation of S).

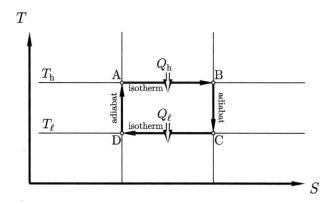

enter or leave during an adiabat). Heat Q_h enters during the expansion A→B and heat Q_ℓ leaves during the compression C→D. Because the process is cyclic, the change of internal energy (a state function) in going round the cycle is zero. Hence the work output by the engine, W, is given by

$$W = Q_h - Q_\ell. \tag{13.1}$$

Example 13.1

Find an expression for Q_h/Q_ℓ for an ideal gas undergoing a Carnot cycle in terms of the temperatures T_h and T_ℓ.

Solution:

Using the results of Section 12.2, we can write down

$$\text{A} \rightarrow \text{B}: \qquad Q_h = RT_h \ln \frac{V_B}{V_A}, \tag{13.2}$$

$$\text{B} \rightarrow \text{C}: \qquad \left(\frac{T_h}{T_\ell}\right) = \left(\frac{V_C}{V_B}\right)^{\gamma-1}, \tag{13.3}$$

$$\text{C} \rightarrow \text{D}: \qquad Q_\ell = -RT_\ell \ln \frac{V_D}{V_C}, \tag{13.4}$$

$$\text{D} \rightarrow \text{A}: \qquad \left(\frac{T_\ell}{T_h}\right) = \left(\frac{V_A}{V_D}\right)^{\gamma-1}. \tag{13.5}$$

Equations 13.3 and 13.5 lead to

$$\frac{V_B}{V_A} = \frac{V_C}{V_D}, \tag{13.6}$$

and dividing eqn 13.2 by eqn 13.4 and substituting in eqn 13.6 leads to

$$\boxed{\frac{Q_h}{Q_\ell} = \frac{T_h}{T_\ell}.} \tag{13.7}$$

This is a key result.[3]

[3] In fact, when we later prove in Section 13.3 that all reversible engines have this efficiency, one can use eqn 13.7 as a thermodynamic *definition* of temperature. In this book, we have preferred to define temperature using a statistical argument via eqn 4.7.

The Carnot engine is shown schematically in Fig. 13.3. It is drawn as a machine with heat input Q_h from a reservoir at temperature T_h, drawn as a horizontal line, and two outputs, one of work W and the other of heat Q_ℓ, which passes into the reservoir at temperature T_ℓ.

The concept of **efficiency** is important to characterize engines. It is the ratio of "what you want to achieve' to "what you have to do to achieve it". For an engine, what you want to achieve is work (to pull a train up a hill for example) and what you have to do to achieve it is to put heat in (by shovelling coal into the furnace), keeping the hot reservoir at T_h and providing heat Q_h for the engine. We therefore define the efficiency η of an engine as the ratio of the work out to the heat in. Thus

$$\eta = \frac{W}{Q_h}. \tag{13.8}$$

Note that since the work out cannot be greater than the heat in (i.e., $W < Q_h$) we must have that $\eta < 1$. The efficiency must be below 100%.

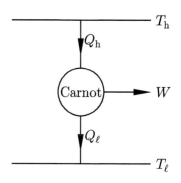

Fig. 13.3 A Carnot engine shown schematically. In diagrams such as this one, the arrows are labelled with the heat and work flowing in one cycle of the engine.

Example 13.2

For the Carnot engine, the efficiency can be calculated using eqns 13.1, 13.7, and 13.8 as follows: substituting eqn 13.1 into 13.8 yields

$$\eta_{\text{Carnot}} = \frac{Q_h - Q_\ell}{Q_h}, \tag{13.9}$$

and eqn 13.7 then implies that

$$\eta_{\text{Carnot}} = \frac{T_h - T_\ell}{T_h} = 1 - \frac{T_\ell}{T_h}. \tag{13.10}$$

How does this efficiency compare to that of a real engine? It turns out that real engines are much less efficient than Carnot engines.

Example 13.3

A power station steam turbine operates between $T_h \sim 800$ K and $T_\ell = 300$ K. If it were a Carnot engine, it could achieve an efficiency of $\eta_{\text{Carnot}} = (T_h - T_\ell)/T_h \approx 60\%$, but in fact real power stations do not achieve the maximum efficiency and figures closer to 40% are typical.

13.3 Carnot's theorem

The Carnot engine is in fact the most efficient engine possible! This is stated in Carnot's theorem, as follows:

> **Carnot's theorem**
> Of all the heat engines working between two given temperatures, none is more efficient than a Carnot engine.

Remarkably, one can *prove* Carnot's theorem on the basis of Clausius' statement of the second law of thermodynamics.[4] The proof follows a *reductio ad absurdum* argument.

[4]This means that Carnot's theorem is, in itself, a statement of the second law of thermodynamics.

Proof: Imagine that E is an engine that is more efficient than a Carnot engine (i.e., $\eta_E > \eta_{\text{Carnot}}$). The Carnot engine is reversible so one can run it in reverse. Engine E, and a Carnot engine run in reverse, are connected together as shown in Fig. 13.4. Now since $\eta_E > \eta_{\text{Carnot}}$, we have that

$$\frac{W}{Q'_h} > \frac{W}{Q_h}, \tag{13.11}$$

and so

$$Q_h > Q'_h. \tag{13.12}$$

The first law of thermodynamics implies that

$$W = Q'_h - Q'_\ell = Q_h - Q_\ell, \tag{13.13}$$

so that

$$Q_h - Q'_h = Q_\ell - Q'_\ell. \tag{13.14}$$

Now $Q_h - Q'_h$ is positive because of eqn 13.12, and therefore so is $Q_\ell - Q'_\ell$. The expression $Q_h - Q'_h$ is the net amount of heat dumped into the reservoir at temperature T_h. The expression $Q_\ell - Q'_\ell$ is the net amount of heat extracted from the reservoir at temperature T_ℓ. Because both these expressions are positive, the combined system shown in Fig. 13.4 simply extracts heat from the reservoir at T_ℓ and dumps it into the reservoir at T_h. This violates Clausius' statement of the second law of thermodynamics, and therefore engine E cannot exist.

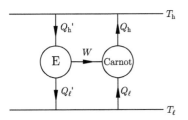

Fig. 13.4 A hypothetical engine E, which is more efficient than a Carnot engine, is connected to a Carnot engine.

> **Corollary:** All reversible engines working between two temperatures have the same efficiency η_{Carnot}.

Proof: Imagine another reversible engine R. Its efficiency $\eta_R \leq \eta_{\text{Carnot}}$ by Carnot's theorem. We run it in reverse and connect it to a Carnot engine going forwards, as shown in Fig. 13.5. This arrangement will simply transfer heat from the cold reservoir to the hot reservoir and violates Clausius' statement of the second law of thermodynamics *unless* $\eta_R = \eta_{\text{Carnot}}$. Therefore all reversible engines have the same efficiency

$$\eta_{\text{Carnot}} = \frac{T_h - T_\ell}{T_h}. \tag{13.15}$$

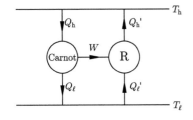

Fig. 13.5 A hypothetical reversible engine R is connected to a Carnot engine.

13.4 Equivalence of Clausius' and Kelvin's statements

We first prove the proposition that if a system violates Kelvin's statement of the second law of thermodynamics, it violates Clausius' statement of the second law of thermodynamics.

Proof: If a system violates Kelvin's statement of the second law of thermodynamics, one could connect it to a Carnot engine as shown in Fig. 13.6. The first law implies that

$$Q'_h = W \tag{13.16}$$

and that

$$Q_h = W + Q_\ell. \tag{13.17}$$

The heat dumped in the reservoir at temperature T_h is

$$Q_h - Q'_h = Q_\ell. \tag{13.18}$$

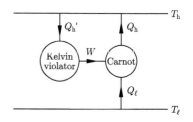

Fig. 13.6 A Kelvin violator is connected to a Carnot engine.

This is also equal to the heat extracted from the reservoir at temperature T_ℓ. The combined process therefore has the net result of transferring heat Q_ℓ from the reservoir at T_ℓ to the reservoir at T_h as its sole effect and thus violates Clausius' statement of the second law of thermodynamics. Therefore the Kelvin violator does not exist.

We now prove the opposite proposition, that if a system violates Clausius' statement of the second law of thermodynamics, it violates Kelvin's statement of the second law of thermodynamics.

Proof: If a system violates Clausius' statement of the second law of thermodynamics, one could connect it to a Carnot engine as shown in Fig. 13.7. The first law implies that

$$Q_h - Q_\ell = W. \tag{13.19}$$

The sole effect of this process is thus to convert heat $Q_h - Q_\ell$ into work and thus violates Kelvin's statement.

We have thus shown the equivalence of Clausius' and Kelvin's statements of the second law of thermodynamics.

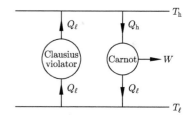

Fig. 13.7 A Clausius violator is connected to a Carnot engine.

13.5 Examples of heat engines

One of the first engines to be constructed was made in the first century by Hero of Alexandria, and is sketched in Fig. 13.8(a). It consists of a hollow sphere with a pair of bent pipes projecting from it. Steam is fed via another pair of pipes and once expelled through the bent pipes causes rotational motion. Though Hero's engine convincingly converts heat into work, and thus qualifies as a *bona fide* heat engine, it was little more than an entertaining toy. More practical was the engine sketched in Fig. 13.8(b), which was designed by Thomas Newcomen (1664–1729).

This was one of the first practical steam engines and was used for pumping water out of mines. Steam is used to push the piston upwards. Then cold water is injected from the tank and condenses the steam, reducing the pressure in the piston. Atmospheric pressure then pushes the piston down and raises the beam on the other side of the fulcrum. The problem with Newcomen's engine was that one had then to heat up the steam chamber again before steam could be readmitted and so it was extremely inefficient. James Watt (1736–1819) famously improved the design so that condensation took place in a separate chamber, which was connected to the steam cylinder by a pipe. This work led the foundation of the industrial revolution.

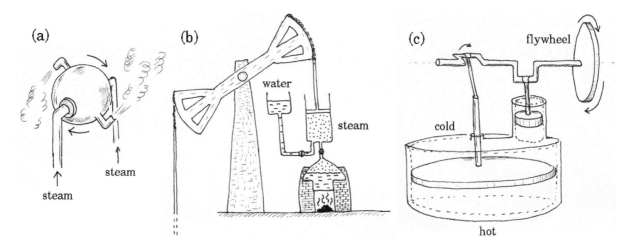

Fig. 13.8 Sketches of (a) Hero's engine, (b) Newcomen's engine, and (c) Stirling's engine.

Another design of an engine is **Stirling's engine**, the brainchild of the Rev. Robert Stirling (1790–1878), which is sketched in Fig. 13.8(c). It works purely by the repeated heating and cooling of a sealed amount of gas. In the particular engine shown in Fig. 13.8(c), the crankshaft is driven by the two pistons in an oscillatory fashion, but the 90° bend ensures that the two pistons move out of phase. The motion is driven by a temperature differential between the top and bottom surfaces of the engine. The design is very simple and contains no valves and operates at relatively low pressures. However, such an engine literally has to "warm up" to establish the temperature differential and so it is harder to regulate power output.

One of the most popular engines is the **internal combustion engine** used in most automobile applications. Rather than externally heating water to produce steam (as with Newcomen's and Watt's engines) or to produce a temperature differential (as with Stirling's engine), here the burning of fuel inside the engine's combustion chamber generates the high temperature and pressure necessary to produce useful work. Different fuels can be used to drive these engines, including diesel, gasoline, natural gas, and even biofuels, such as ethanol. These engines all produce carbon dioxide, and this has important consequences for Earth's atmosphere, as we shall discuss in Chapter 37. There are many different types of internal combustion engines, including piston engines (in which pressure is converted into rotating motion using a set of pistons), combustion turbines (in which gas flow is used to spin a turbine's blades), and jet engines (in which a fast moving jet of gas is used to generate thrust).[5]

[5] In Exercise 13.5 we consider the Otto cycle, which models the diesel engine, a type of internal combustion engine.

13.6 Heat engines running backwards

In this section we discuss two applications of heat engines in which the engine is run in reverse, putting in work to move heat around.

Example 13.4

(a) **The refrigerator**
The refrigerator is a heat engine that is run backwards so that you put work in and cause a heat flow from a cold reservoir to a hot reservoir (see Fig. 13.9). In this case, the cold reservoir is the food inside the refrigerator that you wish to keep cold and the hot reservoir is usually your kitchen. For a refrigerator, we must define the efficiency in a different way from the efficiency of a heat engine. This is because what you want to achieve is "heat sucked out of the contents of the refrigerator" and what you have to do to achieve it is "electrical work" from the mains electricity supply. Thus we define the efficiency of a refrigerator as

$$\eta = \frac{Q_\ell}{W}. \tag{13.20}$$

For a refrigerator fitted with a Carnot engine, it is then easy to show that

$$\eta_{\text{Carnot}} = \frac{T_\ell}{T_\text{h} - T_\ell}, \tag{13.21}$$

which can yield an efficiency above 100%.

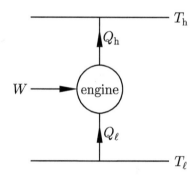

Fig. 13.9 A refrigerator or a heat pump. Both devices are heat engines run in reverse (i.e., reversing the arrows on the cycle shown in Fig. 13.3).

(b) **The heat pump**
A heat pump is essentially a refrigerator (Fig. 13.9 applies also for a heat pump), but it is utilized in a different way. It is used to pump heat from a reservoir, to a place where it is desired to add heat. For example, the reservoir could be the soil/rock several metres underground and heat could be pumped out of the reservoir into a house which needs heating. In one cycle of the engine, we want to add heat Q_h to the house, and now W is the work we must apply (in the form of electrical work) to accomplish this. The efficiency of a heat pump is therefore defined as

$$\eta = \frac{Q_\text{h}}{W}. \tag{13.22}$$

Note that $Q_h > W$ and so $\eta > 1$. The efficiency is always above 100%! (See Exercise 13.1.) This shows why heat pumps are attractive[6] for heating. It is always possible to turn work into heat with 100% efficiency (an electric fire turns electrical work into heat in this way), but a heat pump can allow you to get even more heat into your house for the same electrical work (and hence for the same electricity bill!).

[6] However, the capital cost means that heat pumps have not become popular until recently.

For a heat pump fitted with a Carnot engine, it is easy to show that

$$\eta_{\text{Carnot}} = \frac{T_h}{T_h - T_\ell}. \tag{13.23}$$

13.7 Clausius' theorem

Consider a Carnot cycle. In one cycle, heat Q_h enters and heat Q_ℓ leaves. Heat is therefore not a conserved quantity of the cycle. However, we found in eqn 13.7 that for a Carnot cycle

$$\frac{Q_h}{Q_\ell} = \frac{T_h}{T_\ell}, \tag{13.24}$$

and so if we define[7] ΔQ_{rev} as the heat entering the system at each point, we have that

[7] The subscript "rev" on ΔQ_{rev} is there to remind us that we are dealing with a reversible engine.

$$\sum_{\text{cycle}} \frac{\Delta Q_{\text{rev}}}{T} = \frac{Q_h}{T_h} + \frac{(-Q_\ell)}{T_\ell} = 0, \tag{13.25}$$

and so $\Delta Q_{\text{rev}}/T$ sums to zero around the cycle. Replacing the sum by an integral, we could write

$$\oint \frac{dQ_{\text{rev}}}{T} = 0 \tag{13.26}$$

for this Carnot cycle.

Our argument so far has been in terms of a Carnot cycle operating between two distinct heat reservoirs. Real engine cycles can be much more complicated than this in that their "working substance" changes temperature in a much more complicated way and, moreover, real engines do not behave perfectly reversibly.[8] Therefore we would like to generalize our treatment so that it can be applied to a general cycle operating between a whole series of reservoirs and we would like the cycle to be either reversible or irreversible. Our general cycle is illustrated in Fig. 13.10(a). For this cycle, heat dQ_i enters at a particular part of the cycle. At this point the system is connected to a reservoir, which is at temperature T_i. The total work extracted from the cycle is ΔW, given by

[8] You need to get the energy out of a real engine quickly, so you do not have time to do everything quasistatically!

$$\Delta W = \sum_{\text{cycle}} dQ_i, \tag{13.27}$$

from the first law of thermodynamics. The sum here is taken around the whole cycle, indicated schematically by the dotted circle in Fig. 13.10(a).

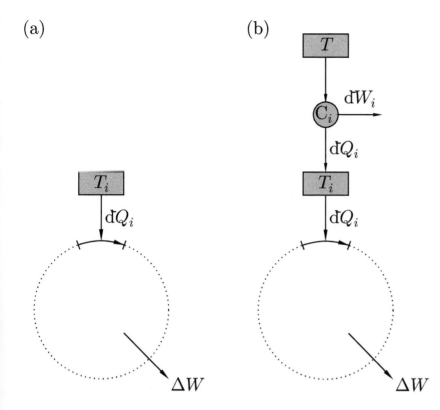

Fig. 13.10 (a) A general cycle in which heat dQ_i enters in part of the cycle from a reservoir at temperature T_i. Work ΔW is extracted from each cycle. (b) The same cycle, but showing the heat dQ_i entering the reservoir at T_i from a reservoir at temperature T via a Carnot engine (labelled C_i).

Next we imagine that the heat at each point is supplied via a Carnot engine, which is connected between a reservoir at temperature T and the reservoir at temperature T_i (see Fig. 13.10(b)). The reservoir at T is common for all the Carnot engines connected at all points of the cycle. Each Carnot engine produces work dW_i, and for a Carnot engine we know that

$$\frac{\text{heat to reservoir at } T_i}{T_i} = \frac{\text{heat from reservoir at } T}{T}, \qquad (13.28)$$

and hence

$$\frac{dQ_i}{T_i} = \frac{dQ_i + dW_i}{T}. \qquad (13.29)$$

Rearranging, we have that

$$dW_i = dQ_i\left(\frac{T}{T_i} - 1\right). \qquad (13.30)$$

The thermodynamic system in Fig. 13.10(b) looks at first sight to do nothing other than convert heat to work, which is not allowed according to Kelvin's statement of the second law of thermodynamics, and hence we must insist that this is not the case. Hence

$$\text{total work produced per cycle} = \Delta W + \sum_{\text{cycle}} dW_i \leq 0. \qquad (13.31)$$

Using eqns 13.27, 13.30, and 13.31, we therefore have that

$$T \sum_{\text{cycle}} \frac{dQ_i}{T_i} \leq 0. \tag{13.32}$$

Since $T > 0$, we have that

$$\sum_{\text{cycle}} \frac{dQ_i}{T_i} \leq 0, \tag{13.33}$$

and replacing the sum by an integral, we can write this as

$$\oint \frac{dQ}{T} \leq 0, \tag{13.34}$$

which is known as the **Clausius inequality**, embodied in the expression of Clausius' theorem:

Clausius' theorem
For any closed cycle, $\oint \frac{dQ}{T} \leq 0$, where equality necessarily holds for a reversible cycle.

Example 13.5

Two bodies with temperature-independent heat capacities C_h and C_ℓ are used as reservoirs for a Carnot heat engine (see Fig. 13.11). Derive an expression for the total work obtainable.
Solution: In an infinitesimal change we have that

$$dQ_h = -C_h \, dT_h \tag{13.35}$$
$$dQ_\ell = C_\ell \, dT_\ell, \tag{13.36}$$

and for a Carnot engine we have that

$$\frac{dQ_\ell}{T_\ell} = \frac{dQ_h}{T_h}, \tag{13.37}$$

and integrating gives $\int_{T_\ell}^{T_f} \frac{dQ_\ell}{T_\ell} = \int_{T_h}^{T_f} \frac{dQ_h}{T_h}$ and hence

$$C_\ell \ln \frac{T_f}{T_\ell} = -C_h \ln \frac{T_f}{T_h}, \tag{13.38}$$

where T_f is the final temperature of each reservoir. Thus

$$T_f^{C_h + C_\ell} = T_h^{C_h} T_\ell^{C_\ell}. \tag{13.39}$$

The total heat extracted from each reservoir is $\Delta Q_h = C_h(T_h - T_f)$ and $\Delta Q_\ell = C_\ell(T_f - T_\ell)$ respectively and so the total work is

$$\Delta W = \Delta Q_h - \Delta Q_\ell = C_h T_h + C_\ell T_\ell - (C_h + C_\ell) T_f. \tag{13.40}$$

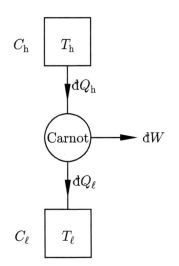

Fig. 13.11 A Carnot engine shown schematically. In diagrams such as this one, the arrows are labelled with the heat and work flowing in one cycle of the engine.

Chapter summary

- No process is possible whose sole result is the transfer of heat from a colder to a hotter body. (Clausius' statement of the second law of thermodynamics)
- No process is possible whose sole result is the complete conversion of heat into work. (Kelvin's statement of the second law of thermodynamics)
- Of all the heat engines working between two given temperatures, none is more efficient than a Carnot engine. (Carnot's theorem)
- All the above are equivalent statements of the second law of thermodynamics.
- All reversible engines operating between temperatures T_h and T_ℓ have the efficiency of a Carnot engine: $\eta_{\text{Carnot}} = (T_h - T_\ell)/T_h$.
- For a Carnot engine:
$$\frac{Q_h}{Q_\ell} = \frac{T_h}{T_\ell}.$$
- Clausius' theorem states that for any closed cycle, $\oint \frac{\mathrm{d}Q}{T} \leq 0$ where equality necessarily holds for a reversible cycle.

Further reading

An entertaining account of how steam engines really work may be found in Semmens and Goldfinch (2000). A short account of Watt's development of his engine is in Marsden (2002).

Exercises

(13.1) A heat pump has an efficiency greater than 100%. Does this violate the laws of thermodynamics?

(13.2) What is the maximum possible efficiency of an engine operating between two thermal reservoirs, one at 100°C and the other at 0°C?

(13.3) The history of science is littered with various schemes for producing **perpetual motion**. A machine that does this is sometimes referred to as a *perpetuum mobile*, which is the Latin term for a perpetual motion machine.

- A perpetual motion machine of the first kind produces more energy than it uses.
- A perpetual motion machine of the second kind produces exactly the same amount of energy as it uses, but it continues running forever indefinitely by converting all its waste heat back into mechanical work.

Give a critique of these two types of machine and state which laws of thermodynamics they each break, if any.

(13.4) A possible ideal-gas cycle operates as follows:
(i) from an initial state (p_1, V_1) the gas is cooled at constant pressure to (p_1, V_2);
(ii) the gas is heated at constant volume to (p_2, V_2);
(iii) the gas expands adiabatically back to (p_1, V_1).
Assuming constant heat capacities, show that the thermal efficiency is

$$1 - \gamma \frac{(V_1/V_2) - 1}{(p_2/p_1) - 1}. \tag{13.41}$$

(You may quote the fact that in an adiabatic change of an ideal gas, pV^γ stays constant, where $\gamma = c_p/c_V$.)

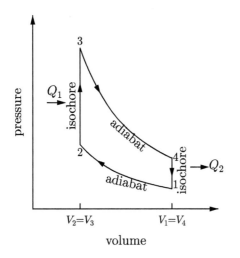

Fig. 13.12 The Otto cycle. (An isochore is a line of constant volume.)

(13.5) Show that the efficiency of the standard Otto cycle (shown in Fig. 13.12) is $1 - r^{1-\gamma}$, where $r = V_1/V_2$ is the compression ratio. The **Otto cycle** is the four-stroke cycle in internal combustion engines in cars, lorries, and electrical generators.

(13.6) An ideal air conditioner operating on a Carnot cycle absorbs heat Q_2 from a house at temperature T_2 and discharges Q_1 to the outside at temperature T_1, consuming electrical energy E. Heat leakage into the house follows Newton's law,

$$Q = A[T_1 - T_2], \tag{13.42}$$

where A is a constant. Derive an expression for T_2 in terms of T_1, E, and A for continuous operation when the steady state has been reached.

The air conditioner is controlled by a thermostat. The system is designed so that with the thermostat set at $20°C$ and outside temperature $30°C$ the system operates at 30% of the maximum electrical energy input. Find the highest outside temperature for which the house may be maintained inside at $20°C$.

(13.7) Two identical bodies of constant heat capacity C_p at temperatures T_1 and T_2 respectively are used as reservoirs for a heat engine. If the bodies remain at constant pressure, show that the amount of work obtainable is

$$W = C_p (T_1 + T_2 - 2T_f), \tag{13.43}$$

where T_f is the final temperature attained by both bodies. Show that if the most efficient engine is used, then $T_f^2 = T_1 T_2$.

(13.8) A building is maintained at a temperature T by means of an ideal heat pump, which uses a river at temperature T_0 as a source of heat. The heat pump consumes power W, and the building loses heat to its surroundings at a rate $\alpha(T - T_0)$, where α is a positive constant. Show that T is given by

$$T = T_0 + \frac{W}{2\alpha} \left(1 + \sqrt{1 + 4\alpha T_0/W}\right). \tag{13.44}$$

(13.9) Three identical bodies of constant thermal capacity are at temperatures 300 K, 300 K, and 100 K. If no work or heat is supplied from outside, what is the highest temperature to which any one of these bodies can be raised by the operation of heat engines? If you set this problem up correctly you may have to solve a cubic equation. This looks hard to solve but in fact you can deduce one of the roots [Hint: what is the highest temperature of the bodies if you do nothing to connect them?].

(13.10) In a heat engine, heat can diffuse between the hot reservoir and the cold reservoir and in Chapter 10 we showed that this takes place on a timescale which scales with the square of the linear size of the system (see Example 10.4). The mechanical timescale of an engine typically scales simply with the linear size of the engine. Explain why this means that heat engines don't work on very small scales. [This is one reason why the "engines" powering biological systems, which have to be extremely small, are not heat engines. Instead, useful energy is extracted directly from chemical bonds. Heat engines also often run on chemical fuel but use the fuel to heat one of the reservoirs and then extract work from the temperature difference thereby generated.]

Sadi Carnot (1796–1832)

Sadi Carnot's father, Lazare Carnot (1753–1823), was an engineer and mathematician who founded the École Polytechnique in Paris, was briefly Napoleon Bonaparte's minister of war and served as his military governor of Antwerp. After Napoleon's defeat, Lazare Carnot was forced into exile. He fled to Warsaw in 1815 and then moved to Magdeburg in Germany in 1816.

Fig. 13.13 Sadi Carnot

It was there in 1818 that he saw a steam engine, and both he and his son Sadi Carnot, who visited him there in 1821, became hooked on the problem of understanding how it worked.

Sadi Carnot had been educated as a child by his father. In 1812 he entered the École Polytechnique and studied with Poisson and Ampère. He then moved to Metz and studied military engineering, worked for a while as a military engineer, and then moved back to Paris in 1819. There he became interested in a variety of industrial problems as well as the theory of gases. He had now become skilled in tackling various problems, but it was his visit to Magdeburg that proved crucial in bringing him the problem that was to be his life's most important work. In this, his father's influence was a significant factor in the solution to the problem. Lazare Carnot had been obsessed by the operation of machines all his life and had been particularly interested in thinking about the operation of water wheels. In a water wheel, falling water can be made to produce useful mechanical work. The water falls from a reservoir of high potential energy to a reservoir of low potential energy, and on the way down, the water turns a wheel which then drives some useful machine such as a flour mill. Lazare Carnot had thought a great deal about how you could make such systems as efficient as possible and convert as much of the potential energy of the water as possible into useful work.

Sadi Carnot was struck by the analogy between such a water wheel and a steam engine, in which heat (rather than water) flows from a reservoir at high temperature to a reservoir at low temperature. Carnot's genius was that rather than focus on the details of the steam engine he decided to consider an engine in abstracted form, focusing purely on the flow of heat between two thermal reservoirs. He idealized the workings of an engine as consisting of simple gas cycles (in what we now know as a Carnot cycle) and worked out its efficiency. He realized that to be as efficient as possible, the engine had to pass slowly through a series of equilibrium states and that it therefore had to be reversible. At any stage, you could reverse its operation and send it the other way around the cycle. He was then able to use this fact to prove that all reversible heat engines operating between two temperatures had the same efficiency.

This work was summarized in his paper on the subject, *Réflexions sur la puissance motrice du feu et sur les machines propres à développer cette puissance* (Reflections on the motive power of fire and machines fitted to develop that power), which was published in 1824. Carnot's paper was favourably reviewed, but had little immediate impact. Few could see the relevance of his work, or at least see past the abstract argument and the unfamiliar notions of idealized engine cycles; his introduction, in which he praised the technical superiority of English engine designers, may not have helped win his French audience. Carnot died in 1832 during a cholera epidemic, and most of his papers were destroyed (the standard precaution following a cholera fatality). The French physicist Émile Clapeyron later noticed his work and published his own paper on it in 1834. However, it was yet another decade before the work simultaneously came to the notice of a young German student, Rudolf Clausius, and a recent graduate of Cambridge University, William Thomson (later Lord Kelvin), who would each individually make much of Carnot's ideas. In particular, Clausius patched up and modernized Carnot's arguments (which had assumed the validity of the prevailing, but subsequently discredited, caloric theory of heat) and was motivated by Carnot's ideas to introduce the concept of entropy.

14 Entropy

14.1 Definition of entropy 140
14.2 Irreversible change 140
14.3 The first law revisited 142
14.4 The Joule expansion 144
14.5 The statistical basis for entropy 146
14.6 The entropy of mixing 147
14.7 Maxwell's demon 149
14.8 Entropy and probability 150
Chapter summary 153
Exercises 153

In this chapter we will use the results from Chapter 13 to define a quantity called entropy and to understand how entropy changes in reversible and irreversible processes. We will also consider the statistical basis for entropy, and use this to understand the entropy of mixing, the apparent conundrum of Maxwell's demon and the connection between entropy and probability.

14.1 Definition of entropy

In this section, we introduce a thermodynamic definition of entropy. We begin by recalling from eqn 13.26 that $\oint đQ_{\text{rev}}/T = 0$. This means that the integral

$$\int_A^B \frac{đQ_{\text{rev}}}{T}$$

is path independent (see Appendix C.7). Therefore the quantity $đQ_{\text{rev}}/T$ is an exact differential and we can write down a new state function which we call entropy. We therefore *define* the **entropy** S by

$$\boxed{dS = \frac{đQ_{\text{rev}}}{T},} \qquad (14.1)$$

so that

$$S(B) - S(A) = \int_A^B \frac{đQ_{\text{rev}}}{T}, \qquad (14.2)$$

and S is a function of state. For an adiabatic process (a reversible adiathermal process) we have that

$$đQ_{\text{rev}} = 0. \qquad (14.3)$$

Hence an adiabatic process involves no change in entropy (the process is also called **isentropic**).

14.2 Irreversible change

Entropy S is defined in terms of reversible changes of heat. Since S is a state function, then the integral of S around a closed loop is zero, so that

$$\oint \frac{đQ_{\text{rev}}}{T} = 0. \qquad (14.4)$$

Let us now consider a loop which contains an irreversible section (A→B) and a reversible section (B→A), as shown in Fig. 14.1. The Clausius inequality (eqn 13.34) implies that, integrating around this loop, we have that

$$\oint \frac{\mathrm{d}Q}{T} \leq 0. \tag{14.5}$$

Writing out the left-hand side in detail, we have that

$$\int_A^B \frac{\mathrm{d}Q}{T} + \int_B^A \frac{\mathrm{d}Q_{\mathrm{rev}}}{T} \leq 0, \tag{14.6}$$

and hence rearranging gives

$$\int_A^B \frac{\mathrm{d}Q}{T} \leq \int_A^B \frac{\mathrm{d}Q_{\mathrm{rev}}}{T}. \tag{14.7}$$

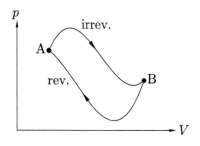

Fig. 14.1 An irreversible and a reversible change between two points A and B in p–V parameter space.

This is true however close A and B get to each other, so in general we can write that the change in entropy $\mathrm{d}S$ is given by

$$\mathrm{d}S = \frac{\mathrm{d}Q_{\mathrm{rev}}}{T} \geq \frac{\mathrm{d}Q}{T}. \tag{14.8}$$

The equality in this expression is only obtained (somewhat trivially) if the process on the right-hand side is actually reversible. Note that because S is a state function, the entropy change in going from A to B is independent of the route.

Consider a thermally isolated system. In such a system $\mathrm{d}Q = 0$ for any process, so that the above inequality becomes

$$\boxed{\mathrm{d}S \geq 0.} \tag{14.9}$$

This is a very important equation and is, in fact, another statement of the second law of thermodynamics. It shows that any change for this thermally isolated system always results in the entropy either staying the same (for a reversible change)[1] or increasing (for an irreversible change). This gives us yet another statement of the second law, namely that: "the entropy of an isolated system tends to a maximum." We can tentatively apply these ideas to the Universe as a whole, under the assumption that the Universe itself is a thermally isolated system:

[1] For a reversible process in a thermally isolated system, $T\,\mathrm{d}S \equiv \mathrm{d}Q_{\mathrm{rev}} = 0$ because no heat can flow in or out.

Application to the Universe
Assuming that the Universe can be treated as an isolated system, the first two laws of thermodynamics become:

(1) $U_{\mathrm{Universe}} = $ constant.
(2) S_{Universe} can only increase.

The following example illustrates how the entropy of a particular system and a reservoir, as well as that of the Universe (taken to be the system plus reservoir), changes in an irreversible process.

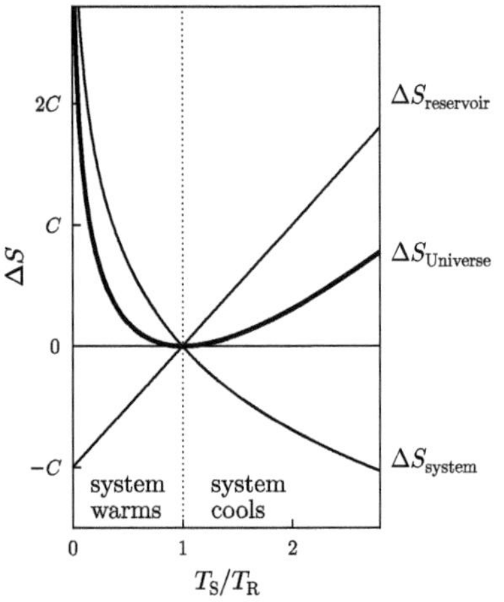

Example 14.1

A large reservoir at temperature T_R is placed in thermal contact with a small system at temperature T_S. They both end up at the temperature of the reservoir, T_R. The heat transferred from the reservoir to the system is $\Delta Q = C(T_R - T_S)$, where C is the heat capacity of the system.

- If $T_R > T_S$, heat is transferred from reservoir to system, the system warms and its entropy increases; the entropy of the reservoir decreases, because heat flows out of it.
- If $T_R < T_S$, heat is transferred from system to reservoir, the system cools and its entropy decreases; the entropy of the reservoir increases, because heat flows into it.

Let us calculate these entropy changes in detail: The entropy change in the reservoir, which has constant temperature T_R, is

$$\Delta S_{\text{reservoir}} = \int \frac{\dbar Q}{T_R} = \frac{1}{T_R} \int \dbar Q = -\frac{\Delta Q}{T_R} = \frac{C(T_S - T_R)}{T_R}, \quad (14.10)$$

while the entropy change in the system is

$$\Delta S_{\text{system}} = \int \frac{\dbar Q}{T} = \int_{T_S}^{T_R} \frac{C \, dT}{T} = C \ln \frac{T_R}{T_S}. \quad (14.11)$$

Hence, the total entropy change in the Universe is

$$\Delta S_{\text{Universe}} = \Delta S_{\text{system}} + \Delta S_{\text{reservoir}} = C \left[\ln \frac{T_R}{T_S} + \frac{T_S}{T_R} - 1 \right]. \quad (14.12)$$

These expressions are plotted in Fig. 14.2 and demonstrate that even though $\Delta S_{\text{reservoir}}$ and ΔS_{system} can each be positive or negative, we always have that

$$\Delta S_{\text{Universe}} \geq 0. \quad (14.13)$$

Fig. 14.2 The entropy change in the simple process in which a small system is placed in contact with a large reservoir.

14.3 The first law revisited

Using our new notion of entropy, it is possible to obtain a much more elegant and useful statement of the first law of thermodynamics. We recall from eqn 11.7 that the first law is given by

$$dU = \dbar Q + \dbar W. \quad (14.14)$$

Now, for a reversible change only, we have that

$$\dbar Q = T dS \quad (14.15)$$

and
$$đW = -pdV. \qquad (14.16)$$

Combining these, we find that
$$dU = TdS - pdV. \qquad (14.17)$$

Constructing this equation, we stress, has assumed that the change is reversible. However, since all the quantities in eqn 14.17 are functions of state, and are therefore path independent, this equation holds for irreversible processes as well! For an irreversible change, $đQ \leq TdS$ and also $đW \geq -pdV$, but with $đQ$ being smaller than for the reversible case and $đW$ being larger than for the reversible case so that dU is the same whether the change is reversible or irreversible.

Therefore, we *always* have that:
$$\boxed{dU = TdS - pdV.} \qquad (14.18)$$

This equation implies that the internal energy U changes when either S or V changes. Thus, the function U can be written in terms of the variables S and V, which are its so-called **natural variables**. These variables are both *extensive* (i.e., they scale with the size of the system).[2] The variables p and T are both *intensive* (i.e., they do not scale with the size of the system) and behave a bit like *forces*, since they show how the internal energy changes with respect to some parameter. In fact, since mathematically we can write dU as

[2] See Section 11.1.2.

$$dU = \left(\frac{\partial U}{\partial S}\right)_V dS + \left(\frac{\partial U}{\partial V}\right)_S dV, \qquad (14.19)$$

we can make the identification of T and p using

$$T = \left(\frac{\partial U}{\partial S}\right)_V \quad \text{and} \qquad (14.20)$$

$$p = -\left(\frac{\partial U}{\partial V}\right)_S. \qquad (14.21)$$

The ratio of p and T can also be written in terms of the variables U, S and V, as follows:

$$\frac{p}{T} = -\left(\frac{\partial U}{\partial V}\right)_S \left(\frac{\partial S}{\partial U}\right)_V, \qquad (14.22)$$

using the reciprocal theorem (see eqn C.41). Hence

$$\frac{p}{T} = \left(\frac{\partial S}{\partial V}\right)_U, \qquad (14.23)$$

using the reciprocity theorem (see eqn C.42). These equations are used in the following example.

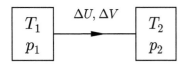

Fig. 14.3 Two systems, 1 and 2, which are able to exchange volume and internal energy.

Example 14.2

Consider two systems, with pressures p_1 and p_2 and temperatures T_1 and T_2. If internal energy ΔU is transferred from system 1 to system 2, and volume ΔV is transferred from system 1 to system 2 (see Fig. 14.3), find the change of entropy. Show that equilibrium results when $T_1 = T_2$ and $p_1 = p_2$.

Solution:

Equation 14.18 can be rewritten as

$$dS = \frac{1}{T}dU + \frac{p}{T}dV. \tag{14.24}$$

If we now apply this to our problem, the change in total entropy $\Delta S = \Delta S_1 + \Delta S_2$ is then straightforwardly

$$\Delta S = \left(\frac{1}{T_1} - \frac{1}{T_2}\right)\Delta U + \left(\frac{p_1}{T_1} - \frac{p_2}{T_2}\right)\Delta V. \tag{14.25}$$

Equation 14.9 shows that the entropy always increases in any physical process. Thus, when equilibrium is achieved, the entropy will have achieved a maximum, so that $\Delta S = 0$. This means that the joint system cannot increase its entropy by further exchanging volume or internal energy between system 1 and system 2. $\Delta S = 0$ can only be achieved when $T_1 = T_2$ and $p_1 = p_2$.

Eqn 14.18 is an important equation that will be used a great deal in subsequent chapters. Before proceeding, we pause to summarize the most important equations in this section and state their applicability.

Summary
$dU = đQ + đW$ always true
$đQ = T\,dS$ only true for reversible changes
$đW = -p\,dV$ only true for reversible changes
$dU = T\,dS - p\,dV$ always true
For irreversible changes: $đQ \leq T\,dS$, $đW \geq -p\,dV$

14.4 The Joule expansion

In this section, we describe in detail an irreversible process known as the **Joule expansion** (see Fig. 14.4). One mole of ideal gas (pressure p_i, temperature T_i) is confined to the left-hand side of a thermally isolated container and occupies a volume V_0. The right-hand side of the container (also volume V_0) is evacuated. The tap between the two parts of the container is then suddenly opened and the gas fills the entire container of volume $2V_0$ (and has new temperature T_f and pressure p_f). Both containers are assumed to be thermally isolated from their surroundings. For the initial state, the ideal gas law implies that

$$p_i V_0 = RT_i, \qquad (14.26)$$

and for the final state that

$$p_f(2V_0) = RT_f. \qquad (14.27)$$

Since the system is thermally isolated from its surroundings, $\Delta U = 0$. Also, since U is only a function of T for an ideal gas, $\Delta T = 0$ and hence $T_i = T_f$. This implies that $p_i V_0 = p_f(2V_0)$, so that the pressure halves, i.e.,

$$p_f = \frac{p_i}{2}. \qquad (14.28)$$

It is hard to calculate directly the change of entropy of a gas in a Joule expansion along the route that it takes from its initial state to the final state. The pressure and volume of the system are undefined during the process immediately after the partition is removed since the gas is in a non-equilibrium state. However, entropy is a function of state and therefore for the purposes of the calculation, we can take another route from the initial state to the final state since changes of functions of state are independent of the route taken. Let us calculate the change in entropy for a reversible isothermal expansion of the gas from volume V_0 to volume $2V_0$ (as indicated in Fig. 14.5). Since the internal energy is constant in the isothermal expansion of an ideal gas, $dU = 0$, and hence the new form of the first law in eqn 14.18 gives us $T\,dS = p\,dV$, so that

$$\Delta S = \int_i^f dS = \int_{V_0}^{2V_0} \frac{p\,dV}{T} = \int_{V_0}^{2V_0} \frac{R\,dV}{V} = R\ln 2. \qquad (14.29)$$

Since S is a function of state, this increase in entropy $R\ln 2$ is also the change of entropy for the Joule expansion.

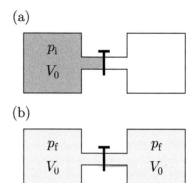

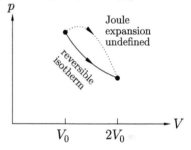

Fig. 14.4 The Joule expansion between volume V_0 and volume $2V_0$. One mole of ideal gas (pressure p_i, temperature T_i) is confined to the left-hand side of a container in a volume V_0. The container is thermally isolated from its surroundings. The tap between the two parts of the container is then suddenly opened and the gas fills the entire container of volume $2V_0$ (and has new temperature T_f and pressure p_f).

Fig. 14.5 The Joule expansion between volume V_0 and volume $2V_0$ and a reversible isothermal expansion of a gas between the same volumes. The path in the p–V plane for the Joule expansion is undefined, whereas it is well defined for the reversible isothermal expansion. In each case however, the start and end points are well defined. Since entropy is a function of state, the change in entropy for the two processes is the same, regardless of route.

Example 14.3

What is the change of entropy in the gas, surroundings, and Universe during a Joule expansion?

Solution:

Above, we have worked out ΔS_{gas} for the reversible isothermal expansion and the Joule expansion: they have to be the same. What about the surroundings and the Universe in each case?

For the reversible isothermal expansion of the gas, we deduce the change of entropy in the surroundings so that the entropy in the Universe does not increase (because we are dealing with a reversible situation).

$$\begin{aligned}
\Delta S_{\text{gas}} &= R\ln 2, \\
\Delta S_{\text{surroundings}} &= -R\ln 2, \\
\Delta S_{\text{Universe}} &= \Delta S_{\text{gas}} + \Delta S_{\text{surroundings}} = 0. \qquad (14.30)
\end{aligned}$$

Notice that the entropy of the surroundings goes down. This does not contradict the second law of thermodynamics. The entropy of something can decrease if that something is not isolated. Here the surroundings are not isolated because they are able to exchange heat with the system.

For the Joule expansion, the system is thermally isolated so that the entropy of the surroundings does not change. Hence

$$\begin{aligned} \Delta S_{\text{gas}} &= R \ln 2, \\ \Delta S_{\text{surroundings}} &= 0, \\ \Delta S_{\text{Universe}} &= \Delta S_{\text{gas}} + \Delta S_{\text{surroundings}} = R \ln 2. \end{aligned} \quad (14.31)$$

Once the Joule expansion has occurred, you can only put the gas back in the left-hand side by compressing it. The best[3] you can do is to do this reversibly, by a reversible isothermal compression, which takes work ΔW given (for 1 mole of gas) by

[3] In other words, the method involving the least work.

$$\Delta W = -\int_{2V_0}^{V_0} p \, dV = -\int_{2V_0}^{V_0} \frac{RT}{V} \, dV = RT \ln 2 = T \Delta S_{\text{gas}}. \quad (14.32)$$

The increase of entropy in a Joule expansion is thus $\Delta W / T$.

A paradox?

- In the Joule expansion, the system is thermally isolated so no heat can be exchanged: $\Delta Q = 0$.
- No work is done: $\Delta W = 0$.
- Hence $\Delta U = 0$ (so for an ideal gas, $\Delta T = 0$).
- But if $\Delta Q = 0$, doesn't that imply that $\Delta S = \Delta Q / T = 0$?

The above reasoning is correct, until the very end: the answer to the question in the last point is NO! The equation $đQ = T dS$ is only true for reversible changes. In general $đQ \leq T dS$, and here we have $\Delta Q = 0$ and $\Delta S = R \ln 2$, so we have that $\Delta Q \leq T \Delta S$.

14.5 The statistical basis for entropy

We now want to show that in addition to defining entropy via thermodynamics, i.e., using $dS = đQ_{\text{rev}} / T$, it is also possible to define entropy via statistics. We will motivate this as follows:

As we showed in eqn 14.20, the first law $dU = T dS - p dV$ implies that

$$T = \left(\frac{\partial U}{\partial S} \right)_V, \quad (14.33)$$

or equivalently
$$\frac{1}{T} = \left(\frac{\partial S}{\partial U}\right)_V. \tag{14.34}$$

Now, recall from eqn 4.7 that
$$\frac{1}{k_B T} = \frac{d \ln \Omega}{dE}. \tag{14.35}$$

Comparing these last two equations motivates the identification of S with $k_B \ln \Omega$, i.e.,
$$\boxed{S = k_B \ln \Omega.} \tag{14.36}$$

This is the expression for the entropy of a system that is in a particular macrostate in terms of Ω, the number of microstates associated with that macrostate. We are assuming that the system is in a particular macrostate with fixed energy, and this situation is known as the *microcanonical ensemble* (see Section 4.5). Later in this chapter (see Section 14.8), and also later in the book, we will generalize this result to express the entropy for more complicated situations. Nevertheless, this expression is sufficiently important that it was inscribed on Boltzmann's tombstone, although on the tombstone the symbol Ω is written as a "W".[4] In the following example, we will apply this expression to understanding the Joule expansion, which we introduced in Section 14.4.

[4]See page 31.

Example 14.4
Joule expansion
Following a Joule expansion, each molecule can be either on the left-hand side or the right-hand side of the container. For each molecule there are therefore two ways of placing it. For one mole (N_A molecules) there are 2^{N_A} ways of placing them. The number of microstates associated with the gas being in a container twice as big as the initial volume is larger by a multiplicative factor
$$2^{N_A}, \tag{14.37}$$
so that the additional entropy is
$$\Delta S = k_B \ln 2^{N_A} = k_B N_A \ln 2 = R \ln 2, \tag{14.38}$$
which is the same expression as written in eqn 14.29.

14.6 The entropy of mixing

Consider two different ideal gases (call them 1 and 2) which are in separate vessels with volumes xV and $(1-x)V$ respectively at the same pressures p and temperatures T (see Fig. 14.6). Since the pressures and

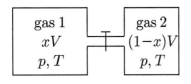

Fig. 14.6 Gas 1 is confined in a vessel of volume xV, while gas 2 is confined in a vessel of volume $(1-x)V$. Both gases are at pressure p and temperature T. Mixing occurs once the tap on the pipe connecting the two vessels is opened.

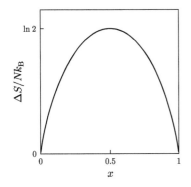

Fig. 14.7 The entropy of mixing according to eqn 14.40.

temperatures are the same on each side, and since $p = (N/V)k_B T$, the number of molecules of gas 1 is xN and of gas 2 is $(1-x)N$, where N is the total number of molecules.

If the tap on the pipe connecting the two vessels is opened, the gases will spontaneously mix, resulting in an increase in entropy, known as the **entropy of mixing**. As for the Joule expansion, we can imagine going from the starting state (gas 1 in the first vessel, gas 2 in the second vessel) to the final state (a homogeneous mixture of gas 1 and gas 2 distributed throughout both vessels) via a reversible route, so that we *imagine* a reversible expansion of gas 1 from xV into the combined volume V and a reversible expansion of gas 2 from $(1-x)V$ into the combined volume V. For an isothermal expansion of an ideal gas, the internal energy doesn't change and hence $T\,dS = p\,dV$ so that $dS = (p/T)\,dV = Nk_B\,dV/V$, using the ideal gas law. This means that the entropy of mixing for our problem is

$$\Delta S = xNk_B \int_{xV}^{V} \frac{dV_1}{V_1} + (1-x)Nk_B \int_{(1-x)V}^{V} \frac{dV_2}{V_2} \quad (14.39)$$

and hence

$$\Delta S = -Nk_B(x\ln x + (1-x)\ln(1-x)). \quad (14.40)$$

This equation is plotted in Fig. 14.7. As expected, there is no entropy increase when $x = 0$ or $x = 1$. The maximum entropy change occurs when $x = \frac{1}{2}$, in which case $\Delta S = Nk_B \ln 2$. This of course corresponds to the equilibrium state in which no further increase of entropy is possible.

This expression for $x = \frac{1}{2}$ also admits to a very simple statistical interpretation. Before the mixing of the gases takes place, we know that gas 1 is only in the first vessel and gas 2 is only in the second vessel. After mixing, each molecule can exist in additional "microstates"; for every microstate with a molecule of gas 1 on the left there is now an additional one with a molecule of gas 1 now on the right. Therefore Ω must be multiplied by 2^N and hence S must increase by $k_B \ln 2^N$, which is $Nk_B \ln 2$.

This treatment has a profound consequence: distinguishability is an important concept! We have assumed that there is some tangible *difference* between gas 1 and gas 2, so that there is some way to label whether a particular molecule is gas 1 or gas 2. For example, if the two gases were nitrogen and oxygen, one could measure the mass of the molecules to determine which was which. But what if the two gases were actually the same? Physically, we would expect that mixing them would have no observable consequences, so there should be no increase in entropy. Thus mixing should only increase entropy if the gases really are distinguishable. We will return to this issue of distinguishability in Chapter 21.

14.7 Maxwell's demon

In 1867, James Clerk Maxwell came up with an intriguing puzzle via a thought experiment. This has turned out to be much more illuminating and hard to solve than he might ever have imagined. The thought experiment can be stated as follows: imagine performing a Joule expansion on a gas. A gas is initially in one chamber, which is connected via a closed tap to a second chamber containing only a vacuum (see Fig. 14.4). The tap is opened and the gas in the first chamber expands to fill both chambers. Equilibrium is established and the pressure in each chamber is now half of what it was in the first chamber at the start. The Joule expansion is formally irreversible as there is no way to get the gas back into the initial chamber without doing work. Or is there? Maxwell imagined that the tap was operated by a microscopic intelligent creature, now called **Maxwell's demon**, who was able to watch the individual molecules bouncing around close to the tap (see Fig. 14.8). If the demon sees a gas molecule heading from the second chamber back into the first, it quickly opens the tap and then shuts it straight away, just letting the molecule through. If it spots a gas molecule heading from the first chamber back into the second chamber, it keeps the tap closed. The demon does no work[5] and yet it can make sure that the gas molecules in the second chamber all go back into the first chamber. Thus it creates a pressure difference between the two chambers where none existed before the demon started its mischief.

[5] It does no work in the pdV sense, though it does do some in the brain sense.

Fig. 14.8 Maxwell's demon watches the gas molecules in chambers A and B and *intelligently* opens and shuts the trap door connecting the chambers. The demon is therefore able to reverse the Joule expansion and only let molecules travel from B to A, thus *apparently* contravening the second law of thermodynamics.

Now, a similar demon could be employed to make hot molecules go the wrong way (i.e., so that heat flows the wrong way, from cold to hot – this in fact was Maxwell's original implementation of the demon), or even to sort out molecules of different types (and thus subvert the "entropy of mixing", see Section 14.6). It looks as if the demon could therefore cause entropy to decrease in a system with no consequent increase in entropy anywhere else. In short, Maxwell's demon appears to make a mockery out of the second law of thermodynamics. How on earth does it get away with it?

Many very good minds have addressed this problem. One early idea was that the demon needs to make measurements of where all the gas molecules are, and to do this would need to shine light on the molecules; thus the process of observation of the molecules might be thought to rescue us from Maxwell's demon. However, this idea turned out not to be correct as it was found to be possible, even in principle, to detect a molecule with arbitrarily little work and dissipation. Remarkably, it turns out that because a demon needs to have a memory to operate (so that it can remember where it has observed a molecule and any other results of its measurement process), this act of storing *information* (actually it is the act of erasing information, as we will discuss below) is associated with an increase of entropy, and this increase cancels out any decrease in entropy that the demon might be able to effect in the system. This connection between information and entropy is an extremely important insight and will be explored in Chapter 15.

The demon is in fact a type of computational device that processes and stores information about the world. It is possible to design a computational process that proceeds entirely reversibly, and therefore has no increase in entropy associated with it. However, the act of erasing information is irreversible (as anyone who has ever failed to backup their data and then had their computer crash will testify). Erasing information always has an associated increase in entropy (of $k_\mathrm{B} \ln 2$ per bit, as we shall see in Chapter 15); Maxwell's demon can operate reversibly therefore, but only if it has a large enough hard disc that it doesn't ever need to clear space to continue operating. The Maxwell demon therefore beautifully illustrates the connection between entropy and information.

14.8 Entropy and probability

The entropy that you measure is due to the number of different states in which the system can exist, according to $S = k_\mathrm{B} \ln \Omega$ (eqn 14.36). However, each state may consist of a large number of microstates that we can't directly measure. Since the system could exist in any one of those microstates, there is extra entropy associated with them. An example should make this idea clear.

14.8 Entropy and probability

Example 14.5

A system has five possible equally likely states in which it can exist, and which of those states it occupies can be distinguished by some easy physical measurement. The entropy is therefore, using eqn 14.36,

$$S = k_B \ln 5. \tag{14.41}$$

However, each of those five states is made up of three equally likely microstates and it is not possible to measure easily which of those microstates it is in. The extra entropy associated with these microstates is $S_{\text{micro}} = k_B \ln 3$. The system therefore *really* has $3 \times 5 = 15$ states and the total entropy is therefore $S_{\text{tot}} = k_B \ln 15$. This can be decomposed into

$$S_{\text{tot}} = S + S_{\text{micro}}. \tag{14.42}$$

Now let us suppose that a system can have N different, equally likely microstates. As usual, it is hard to measure the details of these microstates directly, but let us assume that they are there. These microstates are divided into various groups (we will call these groups *macrostates*) with n_i microstates contained in the ith macrostate. The macrostates are easier to distinguish using experiment because they correspond to some macroscopic, measurable property. We must have that the sum of all the microstates in each macrostate is equal to the total number of microstates, so that

$$\sum_i n_i = N. \tag{14.43}$$

The probability P_i of finding the system in the ith macrostate is then given by

$$P_i = \frac{n_i}{N}. \tag{14.44}$$

Equation 14.43 then implies that $\sum P_i = 1$ as required. The total entropy is of course $S_{\text{tot}} = k_B \ln N$, though we can't measure that directly (having no information about the microstates which is easily accessible). Nevertheless, S_{tot} is equal to the sum of the entropy associated with the freedom of being able to be in different macrostates, which is our measured entropy S, and the entropy S_{micro} associated with it being able to be in different microstates within a macrostate. Putting this statement in an equation, we have

$$S_{\text{tot}} = S + S_{\text{micro}}, \tag{14.45}$$

which is identical to eqn 14.42. The entropy associated with being able to be in different microstates (the aspect we can't measure) is given by

$$S_{\text{micro}} = \langle S_i \rangle = \sum_i P_i S_i, \tag{14.46}$$

where $S_i = k_B \ln n_i$ is the entropy of the microstates in the ith macrostate and, to recap, P_i is the probability of a particular macrostate being occupied. Hence

$$\begin{aligned} S &= S_{\text{tot}} - S_{\text{micro}} \\ &= k_B \left(\ln N - \sum_i P_i \ln n_i \right) \\ &= k_B \sum_i P_i (\ln N - \ln n_i), \end{aligned} \qquad (14.47)$$

and using $\ln N - \ln n_i = -\ln(n_i/N) = -\ln P_i$ (from eqn 14.44) yields **Gibbs' expression for the entropy**:

$$\boxed{S = -k_B \sum_i P_i \ln P_i.} \qquad (14.48)$$

Example 14.6
Find the entropy for a system with Ω macrostates, each with probability $P_i = 1/\Omega$ (i.e., assuming the microcanonical ensemble).
Solution:
Using eqn 14.48, substitution of $P_i = 1/\Omega$ yields

$$S = -k_B \sum_i P_i \ln P_i = -k_B \sum_{i=1}^{\Omega} \frac{1}{\Omega} \ln \frac{1}{\Omega} = -k_B \ln \frac{1}{\Omega} = k_B \ln \Omega, \quad (14.49)$$

which is the same as eqn 14.36.

A connection between the Boltzmann probability and the expression for entropy in eqn 14.48 is demonstrated in the following example.

Example 14.7
Maximize $S = -k_B \sum_i P_i \ln P_i$ (eqn 14.48) subject to the constraints that $\sum P_i = 1$ and $\sum_i P_i E_i = U$.
Solution:
Use the method of Lagrange multipliers,[6] in which we maximize

$$\frac{S}{k_B} - \alpha \times (\text{constraint 1}) - \beta \times (\text{constraint 2}) \qquad (14.50)$$

where α and β are Lagrange multipliers. Thus we vary this expression with respect to one of the probabilities P_j and get

$$\frac{\partial}{\partial P_j} \left(\sum_i -P_i \ln P_i - \alpha P_i - \beta P_i E_i \right) = 0, \qquad (14.51)$$

[6]See Appendix C.13.

so that
$$-\ln P_j - 1 - \alpha - \beta E_j = 0. \tag{14.52}$$

This can be rearranged to give
$$P_j = \frac{e^{-\beta E_j}}{e^{1+\alpha}}, \tag{14.53}$$

so that with $Z = e^{1+\alpha}$ we have
$$P_j = \frac{e^{-\beta E_j}}{Z}, \tag{14.54}$$

which is our familiar expression for the Boltzmann probability (eqn 4.13).

Chapter summary

- Entropy is defined by $dS = đQ_{\text{rev}}/T$.
- The entropy of an isolated system tends to a maximum.
- The entropy of an isolated system attains this maximum at equilibrium.
- The laws of thermodynamics can be stated as follows:
  (1) $U_{\text{Universe}} = \text{constant}$.
  (2) S_{Universe} can only increase.
- These can be combined to give $dU = T\,dS - p\,dV$, which always holds.
- The statistical definition of entropy is $S = k_B \ln \Omega$.
- The general definition of entropy, due to Gibbs, is
  $S = -k_B \sum_i P_i \ln P_i$.

Exercises

(14.1) A mug of tea has been left to cool from 90 °C to 18 °C. If there is 0.2 kg of tea in the mug, and the tea has specific heat capacity 4200 J K^{-1} kg^{-1}, show that the entropy of the tea has decreased by 185.7 J K^{-1}. Comment on the sign of this result.

(14.2) In a free expansion of a perfect gas (also called Joule expansion), we know U does not change, and no work is done. However, the entropy must increase because the process is irreversible. Are these statements compatible with the first law $dU = T\,dS - p\,dV$?

(14.3) A 10 Ω resistor is held at a temperature of 300 K. A current of 5 A is passed through the resistor for 2 minutes. Ignoring changes in the source of the current, what is the change of entropy in (a) the resistor and (b) the Universe?

(14.4) Calculate the change of entropy

(a) of a bath containing water, initially at 20 °C, when it is placed in thermal contact with a very large heat reservoir at 80 °C,

(b) of the reservoir when process (a) occurs,

(c) of the bath and of the reservoir if the bath is brought to 80 °C through the operation of a Carnot engine between them.

The bath and its contents have total heat capacity 10^4 J K^{-1}.

[Hint for (c): which of the heat transfers considered in parts (a) and (b) change when you use a Carnot engine, and by how much? Where does the difference in heat energy go?]

(14.5) A block of lead of heat capacity 1 kJ K^{-1} is cooled from 200 K to 100 K in two ways.
(a) It is plunged into a large liquid bath at 100 K.
(b) The block is first cooled to 150 K in one liquid bath and then to 100 K in another bath.
Calculate the entropy changes in the system comprising block plus baths in cooling from 200 K to 100 K in these two cases. Prove that in the limit of an infinite number of intermediate baths the total entropy change is zero.

(14.6) Calculate the changes in entropy of the Universe as a result of the following processes:
(a) A capacitor of capacitance 1 µF is connected to a battery of emf. 100 V at 0 °C. (NB think carefully about what happens when a capacitor is charged from a battery.)
(b) The same capacitor, after being charged to 100 V, is discharged through a resistor at 0 °C.
(c) One mole of gas at 0 °C is expanded reversibly and isothermally to twice its initial volume.
(d) One mole of gas at 0 °C is expanded reversibly and adiabatically to twice its initial volume.
(e) The same expansion as in (d) is carried out by opening a valve to an evacuated container of equal volume.

(14.7) Consider n moles of a gas, initially confined within a volume V and held at temperature T. The gas is expanded to a total volume αV, where α is a constant, by (a) a reversible isothermal expansion and (b) removing a partition and allowing a free expansion into the vacuum. Both cases are illustrated in Fig. 14.9. Assuming the gas is ideal, derive an expression for the change of entropy of the gas in each case.

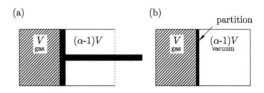

Fig. 14.9 Diagram showing n moles of gas, initially confined within a volume V.

Repeat this calculation for case (a), assuming that the gas obeys the van der Waals equation of state

$$\left(p + \frac{n^2 a}{V^2}\right)(V - nb) = nRT. \quad (14.55)$$

Show further that for case (b) the temperature of the van der Waals gas falls by an amount proportional to $(\alpha - 1)/\alpha$.

(14.8) The probability of a system being in the ith microstate is

$$P_i = e^{-\beta E_i}/Z, \quad (14.56)$$

where E_i is the energy of the ith microstate and β and Z are constants. Show that the entropy is given by

$$S/k_B = \ln Z + \beta U, \quad (14.57)$$

where $U = \sum_i P_i E_i$ is the internal energy.

(14.9) Use the Gibbs expression for entropy (eqn 14.48) to derive the formula for the entropy of mixing (eqn 14.40).

Julius Robert von Mayer (1814–1878)

Robert Mayer studied medicine in Tübingen and took the somewhat unusual career route of signing up as a ship's doctor with a Dutch vessel bound for the East Indies. While letting blood from sailors in the tropics, he noticed that their venous blood was redder than observed back home and concluded that the metabolic oxidation rate in hotter climates was slower. Since a constant body temperature was required for life, the body must reduce its oxidation rate because oxidation of material from food produces internal heat. Though there was some questionable physiological reasoning in his logic, Mayer was on to something. He had realized that energy was something that needed to be *conserved* in any physical process. Back in Heilbronn, Germany, Mayer set to work on a measurement of the mechanical equivalent of heat and wrote a paper in 1841, which was the first statement of the conservation of energy (though he used the word "force"). Mayer's work predated the ideas of Joule and Helmholtz (though his experiment was not as accurate as Joule's) and his notion of the conservation of energy had a wider scope than that of Helmholtz; not only were mechanical energy and heat convertible, but his principle could be applied to tides, meteorites, solar energy, and living things. His paper was eventually published in 1842, but received little acclaim. A later more detailed paper in 1845 was rejected and he published it privately.

Fig. 14.10 Robert Mayer

Mayer then went through a bit of a bad patch, to put it mildly: others began to get the credit for ideas he thought he had pioneered, three of his children died in the late 1840's and he attempted suicide in 1850, jumping out of a third-storey window, but only succeeding in permanently laming himself. In 1851 he checked into a mental institution where he received sometimes brutal treatment and was discharged in 1853, with the doctors unable to offer him any hope of a cure. In 1858, he was even referred to as being dead in a lecture by Liebig (famous for his condenser, and editor of the journal that had accepted Mayer's 1842 paper). Mayer's scientific reputation began to recover in the 1860's and he was awarded the Copley Medal of the Royal Society of London in 1871, the year after it was awarded to Joule.

James Prescott Joule (1818–1889)

James Joule was the son of a wealthy brewer in Salford, near Manchester, England. Joule was educated at home, and his tutors included John Dalton, the father of modern atomic theory. In 1833, illness forced his father to retire, and Joule was left in charge of the family brewery. He had a passion for scientific research and set up a laboratory, working there in the early morning and late evening so that he could continue his day job. In 1840, he showed that the heat dissipated by an electric current I in a resistor R was proportional to $I^2 R$ (what we now call **Joule heating**). In 1846, Joule discovered the phenomenon of magnetostriction (by which a magnet changes its length when magnetized). However Joule's work did not impress the Royal Society and he was dismissed as a mere provincial dilettante. However, Joule was undeterred and he decided to work on the convertibility of energy and to try to measure the mechanical equivalent of heat.

Fig. 14.11 James Joule

In his most famous experiment he measured the increase in temperature of a thermally insulated barrel of water, stirred by a paddle wheel, which was driven by a falling weight. But this was just one of an exhaustive series of meticulously performed experiments that aimed to determine the mechanical equivalent of heat, using electrical circuits, chemical reactions, viscous heating, mechanical contraptions, and gas compression. He even attempted to measure the temperature difference between water at the top and bottom of a waterfall, an opportunity afforded to him by being in Switzerland on his honeymoon!

Joule's obsessive industry paid off: his completely different experimental methods gave consistent results.

Part of Joule's success was in designing thermometers with unprecedented accuracy; they could measure temperature changes as small as 1/200 degrees Fahrenheit. This was necessary as the effects he was looking for tended to be small. His methods proved to be accurate and even his early measurements were within several percent of the modern accepted value of the mechanical equivalent of heat, and his 1850 experiment was within 1 percent. However, the smallness of the effect led to scepticism, particularly from the scientific establishment, who had all had proper educations, didn't spend their days making beer and knew that you couldn't measure temperature differences as tiny as Joule claimed to have observed.

However the tide began to turn in Joule's favour in the late 1840's. Helmholtz recognized Joule's contribution to the conservation of energy in his paper of 1847. In the same year, Joule gave a talk at a British Association meeting in Oxford where Stokes, Faraday, and Thomson were in attendance. Thomson was intrigued and the two struck up a correspondence, resulting in a fruitful collaboration between the two between 1852 and 1856. They measured the temperature fall in the expansion of a gas, and discovered the Joule–Thomson effect.

Joule refused all academic appointments, preferring to work independently. Though without advanced education, Joule had excellent instincts and was an early defender of the kinetic theory of gases, and felt his way towards a kinetic theory of heat, perhaps because of his youthful exposure to Dalton's teachings. On Joule's gravestone is inscribed the number "772.55", the number of foot-pounds required to heat a pound of water by one degree Fahrenheit. It is fitting that today, mechanical and thermal energy are measured in the same unit: the Joule.

Rudolf Clausius (1822–1888)

Rudolf Clausius studied mathematics and physics in Berlin, and was awarded his doctorate in Halle University for work on the colour of the sky. Clausius turned his attention to the theory of heat and, in 1850, he published a paper that essentially saw him picking up the baton left by Sadi Carnot (via an 1834 paper by Emile Clapeyron) and running with it. He defined the internal energy, U, of a system and wrote that the change of heat was given by $dQ = dU + (1/J)p\,dV$, where the factor J (the mechanical equivalent of heat) was necessary to convert mechanical energy $p\,dV$ into the same units as thermal energy (a conversion which in today's units is, of course, unnecessary). He also showed that in a Carnot process, the integral round a closed loop of $f(T)\,dQ$ was zero, where $f(T)$ was some function of temperature.

Fig. 14.12 Rudolf Clausius

His work brought him a professorship in Berlin, though he subsequently moved to chairs in Zürich (1855), Würzburg (1867), and Bonn (1869). In 1854, he wrote a paper in which he stated that heat cannot of itself pass from a colder to a warmer body, a statement of the second law of thermodynamics. He also showed that his function $f(T)$ could be written (in modern notation) as $f(T) = 1/T$. In 1865 he was ready to give $f(T)\,dQ$ a name, defining the entropy (a word he made up to sound like "energy" but contain "trope" meaning "turning", as in the word "heliotrope", a plant which turns towards the Sun) using $dS = dQ/T$ for a reversible process. He also summarized the first and second laws of thermodynamics by stating that the energy of the world is constant and its entropy tends to a maximum.

When Bismarck started the Franco-Prussian war, Clausius patriotically ran a volunteer ambulance corps of Bonn students in 1870–1871, carrying off the wounded from battles in Vionville and Gravelotte. He was wounded in the knee, but received the Iron Cross for his efforts in 1871. He was no less zealous in defending Germany's pre-eminence in thermal physics in various priority disputes, being provoked into siding with Mayer's claim over Joule's, and in various debates with Tait, Thomson, and Maxwell. Clausius however showed little interest in the work of Boltzmann and Gibbs that aimed to understand the molecular origin of the irreversibility that he had discovered and named.

Information theory

15

In this chapter we are going to examine the concept of information and relate it to thermodynamic entropy. At first sight, this seems a slightly crazy thing to do. What on earth do something to do with heat engines and something to do with bits and bytes have in common? It turns out that there is a very deep connection between these two concepts. To understand why, we begin our account by trying to formulate one definition of information.

15.1 Information and Shannon entropy	157
15.2 Information and thermodynamics	159
15.3 Data compression	160
15.4 Quantum information	162
15.5 Conditional and joint probabilities	165
15.6 Bayes' theorem	165
Chapter summary	168
Further reading	168
Exercises	169

15.1 Information and Shannon entropy

Consider the following three true statements about Isaac Newton and his birthday.[1]

(1) Isaac Newton's birthday falls on a particular day of the year.
(2) Isaac Newton's birthday falls in the second half of the year.
(3) Isaac Newton's birthday falls on the 25th of a month.

The first statement has, by any sensible measure, no information content. *All* birthdays fall on a particular day of the year. The second statement has more information content: at least we now know which half of the year his birthday is. The third statement is much more specific and has the greatest information content.[2]

How do we quantify information content? Well, one property we could notice is that the greater the probability of the statement being true *in the absence of any prior information*, the less the information content of the statement. Thus if you knew no prior information about Newton's birthday, then you would say that statement 1 has probability $P_1 = 1$, statement 2 has probability $P_2 = \frac{1}{2}$, and statement 3 has probability[3] $P_3 = \frac{12}{365}$; so as the probability decreases, the information content increases. Moreover, since the useful statements 2 and 3 are independent, then if you are given statements 2 and 3 together, their information contents should *add*. Moreover, the probability of statements 2 and 3 *both* being true, in the absence of prior information, is $P_2 \times P_3 = \frac{6}{365}$. Since the probability of two independent statements being true is the *product* of their individual probabilities, and since it is natural to assume that information content is *additive*, one is motivated to adopt the definition of information which was proposed by Claude Shannon (1916–2001) as follows:

[1] The statements assume that dates are expressed according to the calendar which was used in Newton's day. The Gregorian calendar was not adopted in England until 1742.

[2] In fact, Newton was born on December 25th, 1642. Converting this Julian calendar date to the (currently used) Gregorian calendar gives January 4th, 1643, so Newton's dates are usually given as 1643–1727.

[3] We are using the fact that 1642 was not a leap year!

The **information** content Q of a statement is defined by

$$Q = -k \log P, \tag{15.1}$$

where P is the probability of the statement and k is a positive constant.[4] If we use $\log_2$ (log to the base 2) for the logarithm in this expression and also $k = 1$, then the information Q is measured in **bits**. If instead we use $\ln \equiv \log_e$ and choose $k = k_B$, then we have a definition that, as we shall see, will match what we have found in thermodynamics. In this chapter, we will stick with the former convention since bits are a useful quantity with which to think about information.

Thus, if we have a set of statements with probability P_i, with corresponding information $Q_i = -k \log P_i$, then the average information content S is given by

$$S = \langle Q \rangle = \sum_i Q_i P_i = -k \sum_i P_i \log P_i. \tag{15.2}$$

The average information is called the **Shannon entropy**.

[4] We need k to be a positive constant so that as P goes up, Q goes down.

Example 15.1

- A fair die produces outcomes 1, 2, 3, 4, 5, and 6 with probabilities $\frac{1}{6}, \frac{1}{6}, \frac{1}{6}, \frac{1}{6}, \frac{1}{6}, \frac{1}{6}$. The information associated with each outcome is $Q = -k \log \frac{1}{6} = k \log 6$ and the average information content is then $S = k \log 6$. Taking $k = 1$ and using log to the base 2 gives a Shannon entropy of 2.58 bits.
- A biased die produces outcomes 1, 2, 3, 4, 5, and 6 with probabilities $\frac{1}{10}, \frac{1}{10}, \frac{1}{10}, \frac{1}{10}, \frac{1}{10}, \frac{1}{2}$. The information contents associated with the outcomes are $k \log 10$, $k \log 10$, $k \log 10$, $k \log 10$, $k \log 10$, and $k \log 2$. (These are 3.32, 3.32, 3.32, 3.32, 3.32, and 1 bit respectively.) If we take $k = 1$ again, the Shannon entropy is then $S = k(5 \times \frac{1}{10} \log 10 + \frac{1}{2} \log 2) = k(\log \sqrt{20})$ (this is 2.16 bits). This Shannon entropy is smaller than in the case of the fair die.

The Shannon entropy quantifies how much information we gain, on average, following a measurement of a particular quantity. (Another way of looking at it is to say the Shannon entropy quantifies the amount of *uncertainty* we have about a quantity *before* we measure it.) To make these ideas more concrete, let us study a simple example in which there are only two possible outcomes of a particular random process (such as the tossing of a coin, or asking the question "will it rain tomorrow?").

Example 15.2

What is the Shannon entropy for a Bernoulli trial (a two-outcome random variable[5]) with probabilities P and $1-P$ of the two outcomes?

[5]See Section 3.7

Solution:

$$S = -\sum_i P_i \log P_i = -P \log P - (1-P) \log(1-P), \qquad (15.3)$$

where we have set $k = 1$. This behaviour is sketched in Fig. 15.1. The Shannon entropy has a maximum when $P = \frac{1}{2}$ (greatest uncertainty about the outcome, or greatest information gained, 1 bit, following a trial) and a minimum when $P = 0$ or 1 (least uncertainty about the outcome, or least information gained, 0 bit, following a trial).

The information associated with each of the two possible outcomes is also shown in Fig. 15.1 as dotted lines. The information associated with the outcome having probability P is given by $Q_1 = -\log_2 P$ and decreases as P increases. Clearly when this outcome is very unlikely (P small) the information associated with getting that outcome is very large (Q_1 is many bits of information). However, such an outcome doesn't happen very often so it doesn't contribute much to the average information (i.e., to the Shannon entropy, the solid line in Fig. 15.1). When this outcome is almost certain (P almost 1) it contributes a lot to the average information but has very little information content. For the other outcome, with probability $1 - P$, $Q_2 = -\log_2(1 - P)$ and the behaviour is simply a mirror image of this. The maximum average information is when $P = 1 - P = \frac{1}{2}$ and both outcomes have 1 bit of information associated with them.

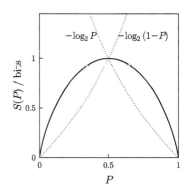

Fig. 15.1 The Shannon entropy of a Bernoulli trial (a two-outcome random variable) with probabilities of the two outcomes given by P and $1 - P$. The units are chosen so that the Shannon entropy is in bits. Also shown is the information associated with each outcome (dotted lines).

15.2 Information and thermodynamics

Remarkably, the formula for Shannon entropy in eqn 15.2 is identical (apart from whether you take your constant as k or k_B) to Gibbs' expression for thermodynamic entropy in eqn 14.48. This gives us a useful perspective on what thermodynamic entropy is. It is a measure of our uncertainty of a system, based on our limited knowledge of its properties and ignorance about which of its microstates it is in. In making inferences on the basis of partial information, we can assign probabilities on the basis that we maximize entropy subject to the constraints provided by what is known about the system. This is exactly what we did in Example 14.7, when we maximized the Gibbs' entropy of an isolated system subject to the constraint that the total energy U was constant; hey presto, we found that we recovered the Boltzmann probability distribution. With this viewpoint, one can begin to understand thermodynamics from an information theory viewpoint.

However, not only does information theory apply to physical systems, but as pointed out by Rolf Landauer (1927–1999), information itself is a physical quantity. Imagine a physical computing device which has stored N bits of information and is connected to a thermal reservoir of temperature T. The bits can be either one or zero. Now we decide to physically erase that information. Erasure must be irreversible. There must be no vestige of the original stored information left in the erased state of the system. Let us erase the information by resetting all the bits to zero.[6] Then this irreversible process reduces the number of states of the system by a factor 2^N and hence the entropy of the system goes down by $Nk_B \ln 2$, or $k_B \ln 2$ per bit. For the total entropy of the Universe not to decrease, the entropy of the surroundings must go up by $k_B \ln 2$ per bit and so we must dissipate heat in the surroundings equal to $k_B T \ln 2$ per bit erased.

[6]We could equally well reset the bits to one.

This connection between entropy and information helps us in our understanding of Maxwell's demon discussed in Section 14.7. By performing computations about molecules and their velocities, the demon has to store information. Each bit of information is associated with entropy, as becomes clear when the demon has to free up some space on its hard disk to continue computing. The process of erasing one bit of information gives rise to an increase of entropy of $k_B \ln 2$. If Maxwell's demon reverses the Joule expansion of 1 mole of gas, it might therefore seem that it has decreased the entropy of the Universe by $N_A k_B \ln 2 = R \ln 2$, but it will have had to store at least N_A bits of information to do this. Assuming that Maxwell's demons only have on-board a storage capacity of a few hundred gigabytes, which is much less than N_A bits, the demon will have had to erase its disk many, many times in the process of its operation, thus leading to an increase in entropy of the Universe which at least equals, and probably outweighs, the decrease of entropy of the Universe it was aiming to achieve.

If the demon is somehow fitted with a vast on-board memory so that it doesn't have to erase its memory to do the computation, then the increase in entropy of the Universe can be delayed until the demon needs to free up some memory space. Eventually, one supposes, as the demon begins to age and becomes forgetful, the Universe will reclaim all that entropy!

15.3 Data compression

Information must be stored, or sometimes transmitted from one place to another. It is therefore useful if it can be compressed down to its minimum possible size. This really begs the question what the actual irreducible amount of *real* information in a particular block of data really is; many messages, political speeches, and even sometimes book chapters, contain large amounts of extraneous padding that is not really needed. Of course, when we compress a file on a computer we often get something that is unreadable to human beings. The English language has various quirks, such as when you see a letter "q" it is almost always followed by a "u", so is that second "u" really needed when you know it is coming? A good data compression algorithm will get rid of extra things like that, plus much more besides. Hence, the question of how many bits are in a given source of data seems like a useful question for computer scientists to attempt to answer; in fact we will see it has implications for physics!

We will not prove **Shannon's noiseless channel coding theorem** here, but motivate it and then state it.

Example 15.3

Let us consider the simplest case in which our data are stored in the form of the binary digits "0" and "1". Let us further suppose that the data contain "0" with probability P and "1" with probability $1 - P$. If $P = \frac{1}{2}$ then our data cannot really be compressed, as each bit of data contains real information. Let us now suppose that $P = 0.9$ so that the data contain more "0"s than "1"s. In this case, the data contain less information, and it is not hard to find a way of taking advantage of this. For example, let us read the data into our compression algorithm in pairs of bits, rather than one bit at a time, and make the following transformations:

$$\begin{aligned} 00 &\rightarrow 0 \\ 10 &\rightarrow 10 \\ 01 &\rightarrow 110 \\ 11 &\rightarrow 1110 \end{aligned}$$

In each of the transformations, we end on a single "0", which lets the decompression algorithm know that it can start reading the next sequence. Now, of course, although the pair of symbols "00" has been compressed to "0", saving a bit, the pair of symbols "01" has been enlarged to "110" and "11" has been even more enlarged to "1110", costing one extra or two extra bits respectively. However, "00" is very likely to occur (probability 0.81) while "01" and "11" are much less likely to occur (probabilities 0.09 and 0.01 respectively), so overall we save bits using this compression scheme.

This example gives us a clue as to how to compress data more generally. The aim is to identify in a sequence of data what the typical sequences are and then efficiently code only those. When the amount of data becomes very large, then anything other than these typical sequences is very unlikely to occur. Because there are fewer typical sequences than there are sequences in general, a saving can be made. Hence, let us divide up some data into sequences of length n. Assuming the elements in the data do not depend on each other, then the probability of finding a sequence $x_1, x_2, \ldots, x_n$ is

$$P(x_1, x_2, \ldots, x_n) = P(x_1)P(x_2)\ldots P(x_n) \approx P^{nP}(1-P)^{n(1-P)}, \quad (15.4)$$

for typical sequences. Taking logarithms to base 2 of both sides gives

$$-\log_2 P(x_1, x_2, \ldots, x_n) \approx -nP\log_2 P - n(1-P)\log_2(1-P) = nS, \quad (15.5)$$

where S is the entropy for a Bernoulli trial with probability P. Hence

$$P(x_1, x_2, \ldots, x_n) \approx \frac{1}{2^{nS}}. \quad (15.6)$$

This shows that there are at most only 2^{nS} typical sequences and hence it only requires nS bits to code them. As n becomes larger, and the typical sequences become longer, the possibility of this scheme failing becomes smaller and smaller.

A compression algorithm will take a typical sequence of n terms $x_1, x_2, \ldots, x_n$ and turn them into a string of length nR. Hence, the smaller R is, the greater the compression. Shannon's noiseless channel coding theorem states that if we have a source of information with entropy S, and if $R > S$, then there exists a reliable compression scheme of compression factor R. Conversely, if $R < S$ then any compression scheme will not be reliable. Thus the entropy S sets the ultimate compression limit on a set of data.

15.4 Quantum information

This section shows how the concept of information can be extended to quantum systems and assumes familiarity with the main results of quantum mechanics.

In this chapter we have seen that in classical systems the information content is connected with the probability. In quantum systems, these probabilities are replaced by **density matrices**. A density matrix is used to describe the statistical state of a quantum system, as can arise for a quantum system in thermal equilibrium at finite temperature. A summary of the main results concerning density matrices is given in the box on page 163.

For quantum systems, the information is represented by the operator $-k\log\rho$, where ρ is the density matrix; as before we take $k = 1$. Hence the average information, or entropy, would be $\langle -\log\rho\rangle$. This leads to the definition of the **von Neumann entropy** S as[7]

$$S(\rho) = -\text{Tr}(\rho\log\rho). \quad (15.7)$$

If the eigenvalues of ρ are $\lambda_1, \lambda_2 \ldots$, then the von Neumann entropy becomes

$$S(\rho) = -\sum_i \lambda_i \log \lambda_i, \quad (15.8)$$

which looks like the Shannon entropy.

[7]The operator Tr means the **trace** of the following matrix, i.e., the sum of the diagonal elements.

The density matrix

- If a quantum system is in one of a number of states $|\psi_i\rangle$ with probability P_i, then the density matrix ρ for the system is defined by
$$\rho = \sum_i P_i |\psi_i\rangle\langle\psi_i|. \tag{15.9}$$

- As an example, think of a three-state system and think of $|\psi_1\rangle$ as a column vector $\begin{pmatrix} 1 \\ 0 \\ 0 \end{pmatrix}$, and hence $\langle\psi_1|$ as a row vector $(1,0,0)$, and similarly for $|\psi_2\rangle$, $\langle\psi_2|$, $|\psi_3\rangle$ and $\langle\psi_3|$. Then

$$\begin{aligned}
\rho &= P_1 \begin{pmatrix} 1 & 0 & 0 \\ 0 & 0 & 0 \\ 0 & 0 & 0 \end{pmatrix} + P_2 \begin{pmatrix} 0 & 0 & 0 \\ 0 & 1 & 0 \\ 0 & 0 & 0 \end{pmatrix} + P_3 \begin{pmatrix} 0 & 0 & 0 \\ 0 & 0 & 0 \\ 0 & 0 & 1 \end{pmatrix} \\
&= \begin{pmatrix} P_1 & 0 & 0 \\ 0 & P_2 & 0 \\ 0 & 0 & P_3 \end{pmatrix}.
\end{aligned} \tag{15.10}$$

This form of the density matrix looks very simple, but this is only because we have expressed it in a very simple basis.

- If $P_j \neq 0$ and $P_{i\neq j} = 0$, then the system is said to be in a **pure state** and ρ can be written in the simple form
$$\rho = |\psi_j\rangle\langle\psi_j|. \tag{15.11}$$
Otherwise, it is said to be in a **mixed state**.

- One can show that the expectation value $\langle \hat{A} \rangle$ of a quantum mechanical operator $\hat{A}$ is equal to
$$\langle \hat{A} \rangle = \text{Tr}(\hat{A}\rho). \tag{15.12}$$

- One can also prove that
$$\text{Tr}\rho = 1, \tag{15.13}$$
where $\text{Tr}\rho$ means the trace of the density matrix. This expresses the fact that the sum of the probabilities must equal unity, and is in fact a special case of eqn 15.12, setting $\hat{A} = 1$.

- One can also show that $\text{Tr}\rho^2 \leq 1$ with equality if and only if the state is pure.

- For a system in thermal equilibrium at temperature T, P_i is given by the Boltzmann factor $e^{-\beta E_i}$ where E_i is an eigenvalue of the Hamiltonian $\hat{H}$. The **thermal density matrix** ρ_th is
$$\rho_\text{th} = \sum_i e^{-\beta E_i} |\psi_i\rangle\langle\psi_i| = \exp(-\beta\hat{H}). \tag{15.14}$$

[8] A pure state is defined in the box on page 163.

[9] Note that we take $0 \ln 0 = 0$.

[10] To show this, use Lagrange multipliers.

Example 15.4

Show that the entropy of a pure state[8] is zero. How can you maximize the entropy?
Solution:
(i) As shown in the box on page 163, the trace of the density matrix is equal to one ($\text{Tr}\rho = 1$), and hence the sum of the eigenvalues of the density matrix is
$$\sum \lambda_i = 1. \qquad (15.15)$$
For a pure state only one eigenvalue will be one and all the other eigenvalues will be zero, and hence[9] $S(\rho) = 0$, i.e., the entropy of a pure state is zero. This is not surprising, since for a pure state there is no "uncertainty" about the state of the system.
(ii) The entropy $S(\rho) = -\sum_i \lambda_i \log \lambda_i$ is maximized[10] when $\lambda_i = 1/n$ for all i, where n is the dimension of the density matrix. In this case, the entropy is $S(\rho) = n \times (-\frac{1}{n} \log \frac{1}{n}) = \log n$. This corresponds to there being maximal uncertainty in its precise state.

[11] An arbitary qubit can be written as $|\psi\rangle = \alpha|0\rangle + \beta|1\rangle$ where $|\alpha|^2 + |\beta|^2 = 1$.

[12] Einstein called entanglement "spooky action at a distance", and used it to argue against the Copenhagen interpretation of quantum mechanics and show that quantum mechanics is incomplete.

[13] It turns out that a unitary operator, such as the time-evolution operator, acting on a state leaves the entropy unchanged. This is akin to our results in thermodynamics that reversibility is connected with the preservation of entropy.

Classical information is made up only of sequences of "0"s and "1"s (in a sense, all information can be broken down into a series of "yes/no" questions). Quantum information is composed of quantum bits (known as **qubits**), that are two-level quantum systems which can be represented by linear combinations[11] of the states $|0\rangle$ and $|1\rangle$. Quantum mechanical states can also be *entangled* with each other. The phenomenon of **entanglement**[12] has no classical counterpart. Quantum information therefore also contains entangled superpositions such as $(|01\rangle+|10\rangle)/\sqrt{2}$. Here the quantum states of two objects must be described with reference to each other; measurement of the first bit in the sequence to be a 0 forces the second bit to be 1; if the measurement of the first bit gives a 1, the second bit has to be 0; these correlations persist in an entangled quantum system even if the individual objects encoding each bit are spatially separated. Entangled systems cannot be described by pure states of the individual subsystems, and this is where entropy plays a rôle, as a quantifier of the degree of mixing of states. If the overall system is pure, the entropy of its subsystems can be used to measure its degree of entanglement with the other subsystems.[13]

In this text we do not have space to provide many details about the subject of quantum information, which is a rapidly developing area of current research. Suffice it to say that the processing of information in quantum mechanical systems has some intriguing facets, which are not present in the study of classical information. Entanglement of bits is just one example. As another example, the **no-cloning theorem** states that it is impossible to make a copy of non-orthogonal quantum mechanical states (for classical systems, there is no physical mechanism to stop you copying information, only copyright laws). All of these features lead to the very rich structure of quantum information theory.

15.5 Conditional and joint probabilities

To explore some implications of information theory in more depth we need to introduce some more ideas from probability theory. Now the probability of something often depends on information about what has happened before. Whether it rains tomorrow may depend on whether it has actually rained today. This means that having the information about whether it has rained today may affect how you assign the probability of it raining tomorrow. Not having that information may lead to a different result. This allows us to define the **conditional probability** $P(A|B)$ as the probability that event A occurs *given* that event B has happened. We can also define the **joint probability** $P(A \cap B)$ as the probability that event A and event B both occur. The joint probability $P(A \cap B)$ is equal to the probability that event B occurred multiplied by the probability that A occurred, given that B did, i.e.,

$$P(A \cap B) = P(A|B)P(B), \quad (15.16)$$

and, equally well,

$$P(A \cap B) = P(B|A)P(A). \quad (15.17)$$

If A and B are independent events, then $P(A|B) = P(A)$ (because the probability that A occurs is independent of whether B has occurred or not) and hence

$$P(A \cap B) = P(A)P(B). \quad (15.18)$$

Now consider the case where there are a number of mutually exclusive events A_i such that

$$\sum_i P(A_i) = 1. \quad (15.19)$$

Then we can write the probability of some other event X as

$$P(X) = \sum_i P(X|A_i)P(A_i). \quad (15.20)$$

In the following section, these ideas will be used to prove a very important theorem.

15.6 Bayes' theorem

Very often, you know that if you are given some hypothesis H you can use it to compute the probability of some outcome O assuming that hypothesis (i.e., you can compute $P(O|H)$). But what you often want to do is the reverse: you know the outcome because it has actually occurred and you want to choose an explanation out of the possible hypotheses. In other words, given the outcome you want to know the probability that the hypothesis is true. This transformation of $P(O|H)$ into $P(H|O)$ can be accomplished using **Bayes' theorem**.[14] This can be stated as follows:

$$P(A|B) = \frac{P(B|A)P(A)}{P(B)}. \quad (15.21)$$

[14] Named after Thomas Bayes (1702–1761), although the modern form is due to Pierre-Simon Laplace (1749–1827).

Here $P(A)$ is called the **prior probability**, since it is the probability of A occurring without any knowledge as to the outcome of B. The quantity which you derive is $P(A|B)$, the **posterior probability**. The proof of Bayes' theorem is very simple: one simply equates eqns 15.16 and 15.17 and rearranges.

Example 15.5

It is known that one per cent of a group of athletes are using illegal drugs to boost their performance. The drug test is 95% accurate (and so will give a correct diagnosis 95% of the time). A particular athlete is tested and gets a positive result. Is he guilty?

Solution:

The prior probabilities are

$$P(\mathsf{D}) = 0.01 \qquad (15.22)$$
$$P(\bar{\mathsf{D}}) = 0.99,$$

where D means "taking drugs" and $\bar{\mathsf{D}}$ means "not taking drugs". We will also define Y to mean "test positive" and $\bar{\mathsf{Y}}$ to mean "test negative". Since he tested positive, what we want to know is the probability of his guilt, which is $P(\mathsf{D}|\mathsf{Y})$. Because the drug test is 95% accurate, we have

$$\begin{aligned}
P(\mathsf{Y}|\mathsf{D}) &= 0.95 \quad \text{(true positive)} \\
P(\mathsf{Y}|\bar{\mathsf{D}}) &= 0.05 \quad \text{(false positive)} \\
P(\bar{\mathsf{Y}}|\bar{\mathsf{D}}) &= 0.95 \quad \text{(true negative)} \\
P(\bar{\mathsf{Y}}|\mathsf{D}) &= 0.05 \quad \text{(false negative)}.
\end{aligned} \qquad (15.23)$$

The probability $P(\mathsf{Y})$ of a positive test is given by eqn 15.20 as

$$P(\mathsf{Y}) = P(\mathsf{Y}|\mathsf{D})P(\mathsf{D}) + P(\mathsf{Y}|\bar{\mathsf{D}})P(\bar{\mathsf{D}}) = 0.95 \times 0.01 + 0.05 \times 0.99 \approx 0.06. \qquad (15.24)$$

Bayes' theorem then gives

$$P(\mathsf{D}|\mathsf{Y}) = \frac{P(\mathsf{Y}|\mathsf{D})P(\mathsf{D})}{P(\mathsf{Y})} = 0.16. \qquad (15.25)$$

Hence there is only a 16% probability that he took the drug. This surprising result occurs because although the test is very accurate, the case of illegal drug use in athletes is actually very rare (at least under the assumptions given in this example) and so most positive results are false positives.

The next example demonstrates very powerfully that the probabilities you assign depend very strongly on the information you are given, and sometimes in a surprising way.

Example 15.6
Mrs Trellis (from North Wales) has two children, born three years apart. One of them is a boy. What is the probability that Mrs Trellis has a daughter? [Not *all* of the information given to you here is relevant!] If, instead, you had been told that "Mrs Trellis has two children and the taller of her children was a boy", would that have changed your answer?
Solution:
This is another question that emphasizes the fact that probability all depends on the information you know. Some of the information you are given here is indeed irrelevant (the three years apart and the North Wales are irrelevant). The information you have is that one of the children is a boy. There are now three possibilities for the sexes of Mrs Trellis' children (in order of seniority):

(1) boy; boy,
(2) boy; girl,
(3) girl; boy.

The fourth possibility you might think of, "girl; girl", is discounted by the information that one of the children is a boy. Thus the probability that Mrs Trellis has a daughter is $\frac{2}{3}$ [assuming of course that Mrs Trellis has a 50:50 chance of producing a male or female baby at every birth]. The reason that the answer to this question is not $\frac{1}{2}$ is that we don't know which of Mrs Trellis' two children our initial bit of information refers to (i.e., that the child is a boy), whether it refers to the older or the younger one.

Forget older versus younger, we could distinguish between the two children in many different ways: in order of height, weight, number of freckles, etc. Thus the table of possibilities listed above could be written, not in order of seniority, but in order of height, darkness of hair, blueness of eyes, etc. So, if instead we were told that it was the taller of the children that was a boy, then amazingly that additional information changes the probabilities. All our attention is now focused on the other child, the shorter one, who can either be male or female. It's now a probability of $\frac{1}{2}$ that the shorter child is a daughter.

Astonishingly, knowledge of the height of one of the children alters the probability of sex, even though we have assumed that height and sex are uncorrelated. If you like, we could have replaced the statement "the taller of the children was a boy" with "the child with a first name earlier in the alphabet was a boy" and that would also have the same effect! This demonstrates the important rôle of *distinguishability* in statistics, a concept that will return!

In physics, we try to make inferences about the world based on what we can measure. Those inferences are made on the basis of probability and information theory and this feeds into the Shannon entropy. When we cover the indistinguishability of particles in a gas in Chapter 21 we will find that this has real thermodynamic implications and the above example prepares us not to be surprised by this.

Furthermore, information theory provides a rationale for setting up probability distributions on the basis of partial knowledge; one simply maximizes the entropy of the distribution subject to the constraints provided by the data. This so-called **maximum entropy estimate** is the least biased estimate consistent with the given data.[15] Thermodynamics also gives the best description of the properties of a system that has so many ($\approx 10^{23}$) particles that one cannot follow it precisely; the Boltzmann probability obtained by maximizing the Gibbs entropy[16] is the least-biased estimate of the probability consistent with the constraint that a system has fixed internal energy U.

[15] This approach was used in Example 14.7; see also Exercises 15.3 and 22.1.

[16] Example 14.7.

Chapter summary

- The information Q is given by $Q = -\ln P$ where P is the probability.
- The entropy is the average information $S = \langle Q \rangle = -\sum_i P_i \log P_i$.
- The quantum mechanical generalization of this is the von Neumann entropy given by $S(\rho) = -\text{Tr}(\rho \log \rho)$ where ρ is the density matrix.
- Bayes' theorem relates the posterior probability (which is a conditional probability) to the prior probability.

Further reading

The results that we have stated in this chapter concerning Shannon's coding theorems, and which we considered only for the case of Bernoulli trials, i.e., for binary outputs, can be proved for the general case. Shannon also studied communication over noisy channels, in which the presence of noise randomly flips bits with a certain probability. In this case it is also possible to show how much information can be reliably transmitted using such a channel (essentially how many times you have to "repeat" the message to get yourself "heard", though actually this is done using error-correcting codes). Further information may be found in Feynman (1996) and Mackay (2003). An excellent account of the problem of Maxwell's demon may be found in Leff and Rex (2003). Quantum information theory has become a very hot research topic in the last few years and an excellent introduction is Nielsen and Chuang (2000).

Exercises

(15.1) In a typical microchip, a bit is stored by a 5 fF capacitor using a voltage of 3 V. Calculate the energy stored in eV per bit and compare this with the minimum heat dissipation by erasure, which is $k_B T \ln 2$ per bit, at room temperature.

(15.2) A particular logic gate takes two binary inputs A and B and has two binary outputs A' and B'. Its truth table is

A	B	A'	B'
0	0	1	1
0	1	1	0
1	0	0	1
1	1	0	0

and the operations producing these outputs are $A' = \text{NOT } A$ and $B' = \text{NOT } B$. The input has a Shannon entropy of 2 bits. Show that the output has a Shannon entropy of 2 bits.
A second logic gate has a truth table given by

A	B	A'	B'
0	0	0	0
0	1	1	0
1	0	1	0
1	1	1	1

This can be achieved using $A' = A \text{ OR } B$ and $B' = A \text{ AND } B$. Show that the output now has an entropy of $\frac{3}{2}$ bits. What is the crucial difference between the two logic gates?

(15.3) Maximize the Shannon entropy $S = -k \sum_i P_i \log P_i$ subject to the constraints that $\sum P_i = 1$ and $\langle f(x) \rangle = \sum P_i f(x_i)$ and show that

$$P_i = \frac{1}{Z(\beta)} e^{-\beta f(x_i)}, \quad (15.26)$$

$$Z(\beta) = \sum e^{-\beta f(x_i)}, \quad (15.27)$$

$$\langle f(x) \rangle = -\frac{\text{d}}{\text{d}\beta} \ln Z(\beta). \quad (15.28)$$

(15.4) Noise in a communication channel flips bits at random with probability P. Argue that the entropy associated with this process is

$$S = -P \log P - (1-P) \log(1-P). \quad (15.29)$$

It turns out that the rate R at which we can pass information along this noisy channel is $1 - S$. (This is an application of Shannon's noisy channel coding theorem, and a nice proof of this theorem is given on page 548 of Nielsen and Chuang (2000).)

(15.5) (a) The **relative entropy** measures the closeness of two probability distributions P and Q and is defined by

$$S(P||Q) = \sum P_i \log \left(\frac{P_i}{Q_i}\right) = -S_P - \sum P_i \log Q_i, \quad (15.30)$$

where $S_P = -\sum P_i \log P_i$. Show that $S(P||Q) \geq 0$ with equality if and only if $P_i = Q_i$ for all i.
(b) If i takes N values with probability P_i, then show that

$$S(P||Q) = -S_P + \log N, \quad (15.31)$$

where $Q_i = 1/N$ for all i. Hence show that

$$S_P \leq \log N, \quad (15.32)$$

with equality if and only if P_i is uniformly distributed between all N outcomes.

(15.6) In a TV game show, a contestant is shown three closed doors. Behind one of the doors is a shiny expensive sports car, but behind the other two are goats. The contestant chooses one of the doors at random (she has, after all, a one-in-three chance of winning the car). The game show host (who knows where the car is really located) flings open one of the other two doors to reveal a goat. He grins at the contestant and says: "Well done, you didn't pick the goat behind this door." (Audience applauds sycophantically.) He then adds, still grinning: "But do you want to swap and choose the other closed door or stick with your original choice?" What should she do?

Part VI

Thermodynamics in action

In this part we use the laws of thermodynamics developed in Part V to solve real problems in thermodynamics. Part VI is structured as follows:

- In Chapter 16 we derive various functions of state, called *thermodynamic potentials*, in particular the *enthalpy*, *Helmholtz function* and *Gibbs function*, and show how they can be used to investigate thermodynamic systems under various *constraints*. We introduce the *Maxwell relations*, which allow us to relate various partial differentials in thermal physics.
- In Chapter 17 we show that the results derived so far can be extended straightforwardly to a variety of different thermodynamic systems other than the ideal gas.
- In Chapter 18 we introduce the *third law of thermodynamics*, which is really an addendum to the second law, and explain some of its consequences.

16 Thermodynamic potentials

16.1 Internal energy, U 172
16.2 Enthalpy, H 173
16.3 Helmholtz function, F 174
16.4 Gibbs function, G 175
16.5 Constraints   176
16.6 Maxwell's relations   179
Chapter summary   187
Exercises   187

The internal energy U of a system is a function of state, which means that a system undergoes the same change in U when we move it from one equilibrium state to another, irrespective of which route we take through parameter space. This makes U a very useful quantity, though not a uniquely useful quantity. In fact, we can make a number of other functions of state, simply by adding to U various other combinations of the functions of state p, V, T, and S in such a way as to give the resulting quantity the dimensions of energy. These new functions of state are called **thermodynamic potentials**, and examples include $U + TS$, $U - pV$, $U + 2pV - 3TS$. However, most thermodynamic potentials that one could pick are really not very useful (including the ones we've just quoted as examples!) but three of them are extremely useful and are given special symbols: $H = U + pV$, $F = U - TS$ and $G = U + pV - TS$. In this chapter, we will explore why these three quantities are so useful. First, however, we will review some properties concerning the internal energy U.

16.1 Internal energy, U

Let us review the results concerning the internal energy that were derived in Section 14.3. Changes in the internal energy U of a system are given by the first law of thermodynamics written in the form (eqn 14.17):

$$dU = TdS - pdV. \quad (16.1)$$

[1]See Section 14.3.

This equation shows that the *natural variables*[1] to describe U are S and V, since changes in U are due to changes in S or V. Hence we write $U = U(S, V)$ to show that U is a function of S and V. Moreover, if S and V are held constant for the system, then

$$dU = 0, \quad (16.2)$$

which is the same as saying that U is a constant. Equation 16.1 implies that the temperature T can be expressed as a differential of U using

$$T = \left(\frac{\partial U}{\partial S}\right)_V, \quad (16.3)$$

and similarly the pressure p can be expressed as

$$p = -\left(\frac{\partial U}{\partial V}\right)_S. \quad (16.4)$$

We also have that for **isochoric** processes (where isochoric means that V is constant),
$$dU = TdS, \tag{16.5}$$
and for reversible[2] isochoric processes
$$dU = đQ_{\text{rev}} = C_V\,dT, \tag{16.6}$$
and hence
$$\Delta U = \int_{T_1}^{T_2} C_V\,dT. \tag{16.7}$$
This is only true for systems held at constant volume; we would like to be able to extend this to systems held at constant pressure (an easier constraint to apply experimentally), and this can be achieved using the thermodynamic potential called enthalpy, which we describe next.

[2] For a reversible process, $dQ = TdS$, see Section 14.3.

16.2 Enthalpy, H

We define the **enthalpy** H by
$$\boxed{H = U + pV.} \tag{16.8}$$
This definition together with eqn 16.1 implies that
$$\begin{aligned} dH &= TdS - pdV + pdV + Vdp \\ &= TdS + Vdp. \end{aligned} \tag{16.9}$$
The natural variables for H are thus S and p, and we have that $H = H(S,p)$. We can therefore immediately write down that for a **isobaric** (i.e., constant pressure) process,
$$dH = TdS, \tag{16.10}$$
and for a reversible isobaric process
$$dH = đQ_{\text{rev}} = C_p\,dT, \tag{16.11}$$
so that
$$\Delta H = \int_{T_1}^{T_2} C_p\,dT. \tag{16.12}$$
This shows the importance of H, that for reversible isobaric processes the enthalpy represents the heat absorbed by the system.[3] Isobaric conditions are relatively easy to obtain: an experiment that is open to the air in a laboratory is usually at constant pressure since pressure is provided by the atmosphere.[4] We also conclude from eqn 16.9 that if both S and p are constant, we have that $dH = 0$.

Equation 16.9 also implies that
$$T = \left(\frac{\partial H}{\partial S}\right)_p, \tag{16.13}$$

[3] If you add heat to the system at constant pressure, the enthalpy H of the system goes up. If heat is provided by the system to its surroundings H goes down.

[4] At a given latitude, the atmosphere provides a constant pressure, small changes due to weather fronts notwithstanding.

and
$$V = \left(\frac{\partial H}{\partial p}\right)_S. \tag{16.14}$$

Both U and H suffer from the drawback that one of their natural variables is the entropy S, which is not a very easy parameter to vary in a lab. It would be more convenient if we could substitute that for the temperature T, which is, of course, a much easier quantity to control and to vary. This is accomplished for both of our next two functions of state, the Helmholtz and Gibbs functions.

16.3 Helmholtz function, F

We define the **Helmholtz function** using

$$\boxed{F = U - TS.} \tag{16.15}$$

Hence we find that

$$\begin{aligned} \mathrm{d}F &= T\mathrm{d}S - p\mathrm{d}V - T\mathrm{d}S - S\mathrm{d}T \\ &= -S\mathrm{d}T - p\mathrm{d}V. \end{aligned} \tag{16.16}$$

This implies that the natural variables for F are V and T, and we can therefore write $F = F(T, V)$. For an *isothermal* process (constant T), we can simplify eqn 16.16 further and write that

$$\mathrm{d}F = -p\mathrm{d}V, \tag{16.17}$$

and hence

$$\Delta F = -\int_{V_1}^{V_2} p\,\mathrm{d}V. \tag{16.18}$$

Hence a positive change in F represents reversible work done on the system by the surroundings, while a negative change in F represents reversible work done on the surroundings by the system. As we shall see in Section 16.5, F represents the maximum amount of work you can get out of a system at constant temperature, since the system will do work on its surroundings until its Helmholtz function reaches a minimum. Equation 16.16 implies that the entropy S can be written as

$$S = -\left(\frac{\partial F}{\partial T}\right)_V, \tag{16.19}$$

and the pressure p as

$$p = -\left(\frac{\partial F}{\partial V}\right)_T. \tag{16.20}$$

If T and V are constant, we have that $\mathrm{d}F = 0$ and F is a constant.

16.4 Gibbs function, G

We define the **Gibbs function** using

$$\boxed{G = H - TS.} \qquad (16.21)$$

Hence we find that

$$\begin{aligned} dG &= TdS + Vdp - TdS - SdT \\ &= -SdT + Vdp, \end{aligned} \qquad (16.22)$$

and the natural variables of G are T and p. [Hence we can write $G = G(T,p)$.]

Having T and p as natural variables is particularly convenient as T and p are the easiest quantities to manipulate and control for most experimental systems. In particular, note that if T and p are constant, $dG = 0$. Hence G is conserved in any isothermal isobaric process.[5]

The expression in eqn 16.22 allows us to write down expressions for entropy and volume as follows:

$$S = -\left(\frac{\partial G}{\partial T}\right)_p \qquad (16.23)$$

and

$$V = \left(\frac{\partial G}{\partial p}\right)_T. \qquad (16.24)$$

We have now defined the four main thermodynamic potentials, which are useful in much of thermal physics: the internal energy U, the enthalpy H, the Helmholtz function F, and the Gibbs function G. Before proceeding further, we summarize the main equations which we have used so far.

[5] For example, at a phase transition between two different phases (call them phase 1 and phase 2), there is phase coexistence between the two phases at the same pressure at the transition temperature. Hence the specific Gibbs functions (the Gibbs functions per unit mass) for phase 1 and phase 2 must be equal at the phase transition. This will be particularly useful for us in Chapter 28.

Function of state			Differential	Natural variables	First derivatives	
Internal energy	U		$dU = TdS - pdV$	$U = U(S,V)$	$T = \left(\frac{\partial U}{\partial S}\right)_V,$	$p = -\left(\frac{\partial U}{\partial V}\right)_S$
Enthalpy	$H = U + pV$		$dH = TdS + Vdp$	$H = H(S,p)$	$T = \left(\frac{\partial H}{\partial S}\right)_p,$	$V = \left(\frac{\partial H}{\partial p}\right)_S$
Helmholtz function	$F = U - TS$		$dF = -SdT - pdV$	$F = F(T,V)$	$S = -\left(\frac{\partial F}{\partial T}\right)_V,$	$p = -\left(\frac{\partial F}{\partial V}\right)_T$
Gibbs function	$G = H - TS$		$dG = -SdT + Vdp$	$G = G(T,p)$	$S = -\left(\frac{\partial G}{\partial T}\right)_p,$	$V = \left(\frac{\partial G}{\partial p}\right)_T$

Note that to derive these equations quickly, all you need to do is memorize the definitions of H, F and G and the first law in the form $dU = TdS - pdV$ and the rest can be written down straightforwardly.

Example 16.1

Show that $U = -T^2 \left(\frac{\partial}{\partial T}\right)_V \frac{F}{T}$ and $H = -T^2 \left(\frac{\partial}{\partial T}\right)_p \frac{G}{T}$.

Solution:

Using the expressions

$$S = -\left(\frac{\partial F}{\partial T}\right)_V \quad \text{and} \quad S = -\left(\frac{\partial G}{\partial T}\right)_p,$$

we can write down

$$U = F + TS = F - T\left(\frac{\partial F}{\partial T}\right)_V = -T^2 \left(\frac{\partial (F/T)}{\partial T}\right)_V \quad (16.25)$$

and

$$H = G + TS = G - T\left(\frac{\partial G}{\partial T}\right)_p = -T^2 \left(\frac{\partial (G/T)}{\partial T}\right)_p. \quad (16.26)$$

These equations are known as the **Gibbs–Helmholtz equations** and are useful in chemical thermodynamics.

16.5 Constraints

We have seen that the thermodynamic potentials are valid functions of state and have particular properties. But we have not yet seen how they might be useful, and there might be a suspicion lurking that H, F, and G are rather artificial objects whereas U, the internal energy, is the only natural one. This is not the case, as we shall now show.[6] However, which of these functions of state is the most useful one depends on the *context* of the problem, and in particular on the type of constraint that is applied to the system.

Consider a large mass sitting on the top of a cliff, near the edge. This system has the potential to provide useful work, since one could connect the mass to a pulley system, lower the mass down the cliff edge and extract mechanical work. When the mass lies at the bottom of the cliff, no more useful work can be obtained. It would be very useful to have a quantity that depends on the amount of available useful work a system can provide, and we call such a quantity the **free energy**. In working out what the free energy is in any particular situation, we have to remember that a system can exchange energy with its surroundings, and how it does that rather depends on what sort of *constraint* the surroundings apply to the system. We shall first demonstrate this using a particular case, and then proceed to the general case.

Consider first a system with fixed volume, held at a temperature T by its contact with the surroundings. If heat $đQ$ enters the system,

---

[6] A further weakness with the "internal energy", which will become apparent later, is that it is only for a box of gas that it is obvious what "internal" means. For a box of gas, internal energy clearly means that energy which is inside the gas, associated with the molecules in the gas. However, if the thermodynamic system is a magnetic material in a magnetic field, should "internal energy" only mean energy inside the magnetic material, or should it also include the field energy in the surroundings or associated with the coil causing the magnetic field? We return to this issue in Chapter 17.

the entropy S_0 of the surroundings changes by $dS_0 = -đQ/T$ and the change in entropy of the system, dS, must be such that the total change in entropy of the Universe must be greater than, or equal to, zero (i.e., $dS + dS_0 \geq 0$). Hence $dS - đQ/T \geq 0$ and so $T\,dS \geq đQ$. Now by the first law, $đQ = dU - đW$ and so the work added to the system must satisfy

$$đW \geq dU - T\,dS. \tag{16.27}$$

Now since T is fixed, $dF = d(U - TS) = dU - T\,dS$, and hence eqn 16.27 can be written

$$đW \geq dF. \tag{16.28}$$

What we have shown is that adding work to the system increases the system's Helmholtz function (which we may now call a Helmholtz free energy[7]). In a reversible process, $đW = dF$ and the work added to the system goes directly into an increase of Helmholtz free energy. If we extract a certain amount of work from the system ($đW < 0$), then this will be associated with at least as big a drop in the sample's Helmholtz free energy (equality only being obtained in a reversible process). Returning to our analogy, adding work to the system hauls the mass up to the top of the cliff and gives it the potential to do work in the future (adding free energy to the system), extracting work from the system occurs by letting the mass drop down the cliff and reduces its potential to provide work in the future (subtracting free energy from the system).

Another example is a quantity of oil, which stores free energy that can be released when the oil is burned. However, how that free energy is defined depends on how the oil is burned. If it burns inside a sealed drum containing only oil and air, then the combustion will take place in a fixed volume. In this case, the relevant free energy is the Helmholtz function, as above. However, if the oil is burned in the open air, then the combustion products will need to push against the atmosphere and the free energy will be the Gibbs function,[8] as we shall show.

Note that if the system is mechanically isolated from its surroundings, so that no work can be applied or extracted, then $đW = 0$ and eqn 16.28 becomes

$$dF \leq 0. \tag{16.29}$$

Thus any change in F will be negative. As the system settles down towards equilibrium, all processes will tend to force F downwards. Once the system has reached equilibrium, F will be constant at this minimum level. Hence equilibrium can only be achieved by minimizing F.

We now need to repeat the argument we used to justify eqns 16.28 and 16.29 for more general constraints. In general, a system is able to exchange heat with its surroundings and also, if the system's volume changes, it may do work on its surroundings. Let us now consider a system in contact with surroundings at temperature T_0 and pressure p_0 (see Fig. 16.1). As described above, if heat $đQ$ enters the system, the entropy change of the system satisfies $T_0\,dS \geq đQ$. In the general case, we write the first law as

$$đQ = dU - đW - (-p_0\,dV), \tag{16.30}$$

[7] This is because useful work could be extracted back out again, and hence it is a free energy in the sense we have defined.

[8] For this example, the constraint applied by the atmosphere is the fixing of pressure.

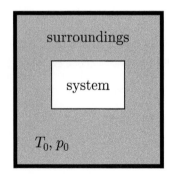

Fig. 16.1 A system in contact with surroundings at temperature T_0 and pressure p_0.

where we have explicitly separated the mechanical work đW added to the system from the work $-p_0\,dV$ done by the surroundings due to the volume change of the system. Putting this all together gives

$$\text{đ}W \geq dU + p_0 dV - T_0 dS. \tag{16.31}$$

We now define the **availability** A by

$$A = U + p_0 V - T_0 S, \tag{16.32}$$

and because p_0 and T_0 are constants, we have

$$dA = dU + p_0 dV - T_0 dS. \tag{16.33}$$

Hence eqn 16.31 becomes

$$\text{đ}W \geq dA, \tag{16.34}$$

which generalizes eqn 16.28. Changes in availability provide free energy "available" for doing work. A will change its form depending on the type of constraint, as shown below. First, note that just as we found eqn 16.29 for the specific case of fixed V and T, in the general case the availability can be used to express a general minimization principle. If the system is mechanically isolated, then

$$\boxed{dA \leq 0,} \tag{16.35}$$

which generalizes eqn 16.29. We have derived this inequality from the second law of thermodynamics. It demonstrates that changes in A are always negative. All processes will tend to force A downwards towards a minimum value. Once the system has reached equilibrium, A will be constant at this minimum level. Hence equilibrium can only be achieved by minimizing A. However, the type of equilibrium achieved depends on the nature of the constraints, as we will now show.

- **System thermally isolated and with fixed volume:**
  Since no heat can enter the system and the system can do no work on its surroundings, $dU = 0$. Hence eqn 16.33 becomes $dA = -T_0\,dS$ and therefore $dA \leq 0$ implies that $dS \geq 0$. Thus we must *maximize S* to find the equilibrium state.

- **System with fixed volume at constant temperature:**
  $dA = dU - T_0 dS \leq 0$, but the temperature is fixed, $dT = 0$, and so $dF = dU - T_0 dS - S dT = dU - T_0 dS$, leading to

  $$dA = dF \leq 0, \tag{16.36}$$

  so we must *minimize F* to find the equilibrium state.[9]

- **System at constant pressure and temperature:**
  Eqn 16.33 gives $dA = dU - T_0 dS + p_0 dV \leq 0$. We can write dG (from the definition $G = H - TS$) as

  $$dG = dU + p_0\,dV + V\,dp - T_0\,dS - S\,dT = dU - T_0 dS + p_0 dV, \tag{16.37}$$

  since $dp = dT = 0$, and hence

  $$dA = dG \leq 0, \tag{16.38}$$

  so we must *minimize G* to find the equilibrium state.[10]

[9] For the constraint of fixed volume and constant temperature, F can be interpreted as the Helmholtz *free energy*.

[10] For the constraint of fixed pressure and constant temperature, G can be interpreted as the Gibbs *free energy*.

Example 16.2

Chemistry laboratories are usually at constant pressure. If a chemical reaction is carried out at constant pressure, then by eqn 16.11 we have

$$\Delta H = \Delta Q, \qquad (16.39)$$

and hence ΔH is the reversible heat added to the system, i.e., the heat absorbed by the reaction. (Recall that our convention is that ΔQ is the heat *entering* the system, and in this case the system is the reacting chemicals.)

- If $\Delta H < 0$, the reaction is called **exothermic** and heat will be emitted.
- If $\Delta H > 0$, the reaction is called **endothermic** and heat will be absorbed.

However, this does not tell you whether or not a chemical reaction will actually proceed. Usually reactions occur[11] at constant T and p, so if the system is trying to minimize its availability, then we need to consider ΔG. The second law of thermodynamics (via eqn 16.35 and hence eqn 16.38) therefore implies that a chemical system will minimize G, so that if $\Delta G < 0$, the reaction may spontaneously occur.[12]

[11] The temperature may rise *during* a reaction, but if the final products cool to the original temperature, one only needs to think about the beginning and end points, since G is a function of state.

[12] However, one may also need to consider the kinetics of the reaction. Often a reaction has to pass via a metastable intermediate state, which may have a higher Gibbs function, so the system cannot spontaneously lower its Gibbs function without having it slightly raised first. This gives a reaction an **activation energy** that must be added before the reaction can proceed, even though the completion of the reaction gives you all that energy back and more.

16.6 Maxwell's relations

In this section, we are going to derive four equations, which are known as Maxwell's relations. These equations are very useful in solving problems in thermodynamics, since each one relates a partial differential between quantities that can be hard to measure to a partial differential between quantities that can be much easier to measure. The derivation proceeds along the following lines: a state functon f is a function of variables x and y. A change in f can be written as

$$df = \left(\frac{\partial f}{\partial x}\right)_y dx + \left(\frac{\partial f}{\partial y}\right)_x dy. \qquad (16.40)$$

Because df is an exact differential (see Appendix C.7), we have

$$\left(\frac{\partial^2 f}{\partial x \partial y}\right) = \left(\frac{\partial^2 f}{\partial y \partial x}\right). \qquad (16.41)$$

Hence writing

$$F_x = \left(\frac{\partial f}{\partial x}\right)_y \text{ and } F_y = \left(\frac{\partial f}{\partial y}\right)_x, \qquad (16.42)$$

we have

$$\left(\frac{\partial F_y}{\partial x}\right) = \left(\frac{\partial F_x}{\partial y}\right). \qquad (16.43)$$

We can now apply this idea to each of the state functions U, H, F, and G in turn.

Example 16.3

The Maxwell relation based on G can be derived as follows. We write down an expression for $\mathrm{d}G$:

$$\mathrm{d}G = -S\mathrm{d}T + V\mathrm{d}p. \tag{16.44}$$

We can also write

$$\mathrm{d}G = \left(\frac{\partial G}{\partial T}\right)_p \mathrm{d}T + \left(\frac{\partial G}{\partial p}\right)_T \mathrm{d}p, \tag{16.45}$$

and hence we can write $S = -(\partial G/\partial T)_p$ and $V = (\partial G/\partial p)_T$. Because $\mathrm{d}G$ is an exact differential, we have

$$\left(\frac{\partial^2 G}{\partial T \partial p}\right) = \left(\frac{\partial^2 G}{\partial p \partial T}\right), \tag{16.46}$$

and hence we have the following Maxwell relation:

$$-\left(\frac{\partial S}{\partial p}\right)_T = \left(\frac{\partial V}{\partial T}\right)_p. \tag{16.47}$$

This reasoning can be applied to each of the thermodynamic potentials U, H, F, and G to yield the four **Maxwell's relations**:

Maxwell's relations:

$$\left(\frac{\partial T}{\partial V}\right)_S = -\left(\frac{\partial p}{\partial S}\right)_V \tag{16.48}$$

$$\left(\frac{\partial T}{\partial p}\right)_S = \left(\frac{\partial V}{\partial S}\right)_p \tag{16.49}$$

$$\left(\frac{\partial S}{\partial V}\right)_T = \left(\frac{\partial p}{\partial T}\right)_V \tag{16.50}$$

$$\left(\frac{\partial S}{\partial p}\right)_T = -\left(\frac{\partial V}{\partial T}\right)_p \tag{16.51}$$

We have said that Maxwell's relations relate a partial differential that corresponds to something that can be easily measured to a partial differential that cannot. For example, in eqn 16.51 the term $(\partial V/\partial T)_p$ on the right-hand side tells you how the volume changes as you increase the temperature while keeping the pressure fixed. This is related to a quantity called the isobaric expansivity[13] and is a quantity you can easily imagine being something one could measure in a laboratory. However, the term on the left-hand side of eqn 16.51, $(\partial S/\partial p)_T$, is much more

[13] See eqn 16.66.

mysterious and it is not obvious how a change of entropy with pressure at constant temperature could actually be measured. Fortunately, a Maxwell relation relates it to something which can be.

Maxwell's relations should not be memorized;[14] rather it is better to remember how to derive them!

A more sophisticated way of deriving these equations based on Jacobians (which may not to be everybody's taste) is outlined in the box below. It has the attractive virtue of producing all four relations in one go by directly relating the work done and heat absorbed in a cyclic process, but the unfortunate vice of requiring easy familiarity with the use of Jacobian transformations.

[14] If you do, however, insist on memorizing them, then lots of mnemonics exist. One useful way of remembering them is as follows. Each Maxwell relation is of the form

$$\left(\frac{\partial *}{\partial \ddagger}\right)_\star = \pm \left(\frac{\partial \dagger}{\partial \star}\right)_\ddagger$$

where the pairs of symbols which are similar to each other ($\star$ and $*$, or $\dagger$ and $\ddagger$) signify **conjugate variables**, so that their product has the dimensions of energy: e.g. T and S, and p and V. Thus you can notice that, for each Maxwell relation, terms diagonally opposite each other are conjugate variables. The variable held constant is conjugate to the one on the top of the partial differential. Another point is that you always have a minus sign when V and T are on the same side of equation.

An alternative derivation of Maxwell's relations

The following derivation is more elegant, but requires a knowledge of Jacobians (see Appendix C.9). Consider a cyclic process that can be described in both the T–S and p–V planes. The internal energy U is a state function and therefore doesn't change in a cycle, so $\oint dU = 0$, which implies that $\oint p\,dV = \oint T\,dS$, and hence

$$\int\int dp\,dV = \int\int dT\,dS. \qquad (16.52)$$

This says that the work done (the area enclosed by the cycle in the p–V plane) is equal to the heat absorbed (the area enclosed by the cycle in the T–S plane). However, one can also write

$$\int\int dp\,dV \frac{\partial(T,S)}{\partial(p,V)} = \int\int dT\,dS, \qquad (16.53)$$

where $\partial(T,S)/\partial(p,V)$ is the Jacobian of the transformation from the p–V plane to the T–S plane, and so these two equations imply that

$$\boxed{\frac{\partial(T,S)}{\partial(p,V)} = 1.} \qquad (16.54)$$

This equation is sufficient to generate all four Maxwell relations via

$$\frac{\partial(T,S)}{\partial(x,y)} = \frac{\partial(p,V)}{\partial(x,y)}, \qquad (16.55)$$

where (x,y) are taken as (i) (T,p), (ii) (T,V), (iii) (p,S), and (iv) (S,V), and using the identities in Appendix C.9.

We will now give several examples of how Maxwell's relations can be used to solve problems in thermodynamics.

Example 16.4

Find expressions for $(\partial C_p/\partial p)_T$ and $(\partial C_V/\partial V)_T$ in terms of p, V, and T.

Solution:
By the definitions of C_V and C_p we have that

$$C_V = \left(\frac{\partial Q}{\partial T}\right)_V = T\left(\frac{\partial S}{\partial T}\right)_V \tag{16.56}$$

and

$$C_p = \left(\frac{\partial Q}{\partial T}\right)_p = T\left(\frac{\partial S}{\partial T}\right)_p. \tag{16.57}$$

Now

$$\left(\frac{\partial C_p}{\partial p}\right)_T = \left(\frac{\partial}{\partial p} T\left(\frac{\partial S}{\partial T}\right)_p\right)_T$$

$$= T\left(\frac{\partial}{\partial p}\left(\frac{\partial S}{\partial T}\right)_p\right)_T$$

$$= T\left(\frac{\partial}{\partial T}\left(\frac{\partial S}{\partial p}\right)_T\right)_p \tag{16.58}$$

and therefore, using one of the Maxwell's relations,

$$\left(\frac{\partial C_p}{\partial p}\right)_T = -T\left(\frac{\partial}{\partial T}\left(\frac{\partial V}{\partial T}\right)_p\right)_p = -T\left(\frac{\partial^2 V}{\partial T^2}\right)_p. \tag{16.59}$$

Similarly

$$\left(\frac{\partial C_V}{\partial V}\right)_T = T\left(\frac{\partial^2 p}{\partial T^2}\right)_V. \tag{16.60}$$

Both the expressions in eqns 16.59 and 16.60 are zero for a perfect gas.

Before proceeding further with the examples, we will pause to list the tools which you have at your disposal to solve these sorts of problems. Any given problem may not require you to use all of these, but you may have to use more than one of these "techniques".

(1) **Write down a thermodynamic potential in terms of particular variables.**
If f is a function of x and y, so that $f = f(x, y)$, you then have immediately that

$$df = \left(\frac{\partial f}{\partial x}\right)_y dx + \left(\frac{\partial f}{\partial y}\right)_x dy. \tag{16.61}$$

(2) **Use Maxwell's relations to transform the partial differential you start with into a more convenient one.**
Use the Maxwell's relations in eqns 16.48–16.51.

(3) **Invert a Maxwell's relation using the reciprocal theorem.**
The reciprocal theorem states that
$$\left(\frac{\partial x}{\partial z}\right)_y = \frac{1}{\left(\frac{\partial z}{\partial x}\right)_y}, \qquad (16.62)$$
and this is proved in Appendix C.6 (see eqn C.41).

(4) **Combine partial differentials using the reciprocity theorem.**
The reciprocity theorem states that
$$\left(\frac{\partial x}{\partial y}\right)_z \left(\frac{\partial y}{\partial z}\right)_x \left(\frac{\partial z}{\partial x}\right)_y = -1, \qquad (16.63)$$
which is proved in Appendix C.6 (see eqn C.42). This can be combined with the reciprocal theorem to write that
$$\left(\frac{\partial x}{\partial y}\right)_z = -\left(\frac{\partial x}{\partial z}\right)_y \left(\frac{\partial z}{\partial y}\right)_x, \qquad (16.64)$$
which is a very useful identity.

(5) **Identify a heat capacity.**
Some of the partial differentials appearing in Maxwell's relations relate to real, measurable properties. As we have seen in Example 16.4, both $\left(\frac{\partial S}{\partial T}\right)_V$ and $\left(\frac{\partial S}{\partial T}\right)_p$ can be related to heat capacities:
$$\frac{C_V}{T} = \left(\frac{\partial S}{\partial T}\right)_V \quad \text{and} \quad \frac{C_p}{T} = \left(\frac{\partial S}{\partial T}\right)_p. \qquad (16.65)$$

(6) **Identify a "generalized susceptibility".**
A **generalized susceptibility** quantifies how much a particular variable changes when a generalized force is applied. A **generalized force** is a variable such as T or p which is a differential of the internal energy with respect to some other parameter.[15] An example of a generalized susceptibility is $\left(\frac{\partial V}{\partial T}\right)_x$ which, you will recall, answers the question "keeping x constant, how much does the volume change when you change the temperature?" It is related to the thermal expansivity at constant x, where x is pressure or entropy. Thus the **isobaric expansivity** β_p is defined as
$$\beta_p = \frac{1}{V}\left(\frac{\partial V}{\partial T}\right)_p, \qquad (16.66)$$
while the **adiabatic expansivity** β_S is defined as
$$\beta_S = \frac{1}{V}\left(\frac{\partial V}{\partial T}\right)_S. \qquad (16.67)$$

Expansivities measure the *fractional* change in volume with a change in temperature.

[15] Recall that $T = (\partial U/\partial S)_V$ and $p = -(\partial U/\partial V)_S$.

Another useful generalized susceptibility is the compressibility. This quantifies how large a fractional volume change you achieve when you apply pressure. The **isothermal compressibility** κ_T is defined as

$$\kappa_T = -\frac{1}{V}\left(\frac{\partial V}{\partial p}\right)_T, \tag{16.68}$$

while the **adiabatic compressibility** κ_S is defined as

$$\kappa_S = -\frac{1}{V}\left(\frac{\partial V}{\partial p}\right)_S. \tag{16.69}$$

Both quantities have a minus sign so that the compressibilities are positive (this is because things get smaller when you press them, so fractional volume changes are negative when positive pressure is applied). None of these expansivities or compressibilities appears directly in a Maxwell relation, but each can easily be related to those that do using the reciprocal and reciprocity theorems.

Example 16.5
By considering $S = S(T, V)$, show that $C_p - C_V = VT\beta_p^2/\kappa_T$.
Solution:
Considering $S = S(T, V)$ allows us to write down immediately that

$$dS = \left(\frac{\partial S}{\partial T}\right)_V dT + \left(\frac{\partial S}{\partial V}\right)_T dV. \tag{16.70}$$

Differentiating this equation with respect to T at constant p yields

$$\left(\frac{\partial S}{\partial T}\right)_p = \left(\frac{\partial S}{\partial T}\right)_V + \left(\frac{\partial S}{\partial V}\right)_T \left(\frac{\partial V}{\partial T}\right)_p. \tag{16.71}$$

Now the first two terms can be replaced by C_p/T and C_V/T respectively, while use of a Maxwell's relation and a partial differential identity (see eqn 16.64) yields

$$\left(\frac{\partial S}{\partial V}\right)_T = \left(\frac{\partial p}{\partial T}\right)_V = -\left(\frac{\partial p}{\partial V}\right)_T \left(\frac{\partial V}{\partial T}\right)_p \tag{16.72}$$

and hence using eqns 16.66 and 16.68 we have that

$$C_p - C_V = \frac{VT\beta_p^2}{\kappa_T}. \tag{16.73}$$

The next example shows how to calculate the entropy of an ideal gas.

Example 16.6

Find the entropy of 1 mole of ideal gas.
Solution:
For one mole of ideal gas $pV = RT$. Consider the entropy S as a function of volume and temperature, i.e.,

$$S = S(T, V), \qquad (16.74)$$

so that

$$dS = \left(\frac{\partial S}{\partial T}\right)_V dT + \left(\frac{\partial S}{\partial V}\right)_T dV \qquad (16.75)$$

$$= \frac{C_V}{T} dT + \left(\frac{\partial p}{\partial T}\right)_V dV, \qquad (16.76)$$

using eqn 16.50 and eqn 16.65. The ideal gas law for 1 mole, $p = RT/V$, implies that

$$\left(\frac{\partial p}{\partial T}\right)_V = R/V, \qquad (16.77)$$

and hence, if we integrate eqn 16.76,

$$S = \int \frac{C_V}{T} dT + \int \frac{R dV}{V}. \qquad (16.78)$$

If C_V is not a function of temperature (which is true for an ideal gas) simple integration yields

$$S = C_V \ln T + R \ln V + \text{constant}. \qquad (16.79)$$

The entropy of an ideal gas increases with increasing temperature and increasing volume.

The final example in this chapter shows how to prove that the ratio of the isothermal and adiabatic compressibilities, κ_T/κ_S, is equal to γ.

Example 16.7

Find the ratio of the isothermal and adiabatic compressibilities.
Solution:
This follows using straightforward manipulations of partial differentials. To begin with, we write

$$\frac{\kappa_T}{\kappa_S} = \frac{\frac{1}{V}\left(\frac{\partial V}{\partial p}\right)_T}{\frac{1}{V}\left(\frac{\partial V}{\partial p}\right)_S}, \qquad (16.80)$$

which follows from the definitions of κ_T and κ_S (eqns 16.68 and 16.69). Then we proceed as follows:

$$\frac{\kappa_T}{\kappa_S} = \frac{-\left(\frac{\partial V}{\partial T}\right)_p \left(\frac{\partial T}{\partial p}\right)_V}{-\left(\frac{\partial V}{\partial S}\right)_p \left(\frac{\partial S}{\partial p}\right)_V} \qquad \text{reciprocity theorem (eqn 16.64)}$$

$$= \frac{\left(\frac{\partial V}{\partial T}\right)_p \left(\frac{\partial S}{\partial V}\right)_p}{\left(\frac{\partial p}{\partial T}\right)_V \left(\frac{\partial S}{\partial p}\right)_V} \qquad \text{reciprocal theorem (eqn 16.62)}$$

$$= \frac{\left(\frac{\partial S}{\partial T}\right)_p}{\left(\frac{\partial S}{\partial T}\right)_V} \qquad \text{simplify numerator and denominator}$$

$$= \frac{C_p/T}{C_V/T}$$

$$= \gamma. \qquad (16.81)$$

We can show that this equation is correct for the case of an ideal gas as follows. Assuming the ideal gas equation $pV \propto T$, we have for constant temperature that

$$\frac{\mathrm{d}p}{p} = -\frac{\mathrm{d}V}{V}, \qquad (16.82)$$

and hence using eqn 16.68 we have

$$\kappa_T = \frac{1}{p}. \qquad (16.83)$$

For an adiabatic change $p \propto V^{-\gamma}$ and hence

$$\frac{\mathrm{d}p}{p} = -\gamma \frac{\mathrm{d}V}{V}, \qquad (16.84)$$

and hence using eqn 16.69 we have

$$\kappa_S = \frac{1}{\gamma p}. \qquad (16.85)$$

This agrees with eqn 16.81 above. We note that because κ_T is larger than κ_S (because $\gamma > 1$), the isotherms always have a smaller gradient than the adiabats on a p–V plot (see Fig. 12.1).

Chapter summary

- We define the following thermodynamic potentials:

  $$U, \quad H = U + pV, \quad F = U - TS, \quad G = H - TS,$$

  which are then related by the following differentials:

  $$\begin{aligned} dU &= TdS - pdV \\ dH &= TdS + Vdp \\ dF &= -SdT - pdV \\ dG &= -SdT + Vdp \end{aligned}$$

- The availability A is given by $A = U + p_0 V - T_0 S$, and for any spontaneous change we have that $dA \leq 0$. This means that a system in contact with a reservoir (temperature T_0, pressure p_0) will minimize A which means

  - minimizing U when S and V are fixed;
  - minimizing H when S and p are fixed;
  - minimizing F when T and V are fixed;
  - minimizing G when T and p are fixed.

- Four Maxwell's relations can be derived from the boxed equations above, and used to solve many problems in thermodynamics.

Exercises

(16.1) (a) Using the first law $dU = TdS - pdV$ to provide a reminder, write down the definitions of the four thermodynamic potentials U, H, F, G (in terms of U, S, T, p, V), and give dU, dH, dF, dG in terms of T, S, p, V and their derivatives.

(b) Derive all the Maxwell's relations.

(16.2) (a) Derive the following general relations

(i) $\left(\dfrac{\partial T}{\partial V}\right)_U = -\dfrac{1}{C_V}\left[T\left(\dfrac{\partial p}{\partial T}\right)_V - p\right]$,

(ii) $\left(\dfrac{\partial T}{\partial V}\right)_S = -\dfrac{1}{C_V} T\left(\dfrac{\partial p}{\partial T}\right)_V$,

(iii) $\left(\dfrac{\partial T}{\partial p}\right)_H = \dfrac{1}{C_p}\left[T\left(\dfrac{\partial V}{\partial T}\right)_p - V\right]$.

In each case the quantity on the left-hand side is the appropriate thing to consider for a particular type of expansion. State what type of expansion each refers to.

(b) Using these relations, verify that for an ideal gas $(\partial T/\partial V)_U = 0$ and $(\partial T/\partial p)_H = 0$, and that $(\partial T/\partial V)_S$ leads to the familiar relation $pV^\gamma =$ constant along an isentrope (a curve of constant entropy).

(16.3) Use the first law of thermodynamics to show that

$$\left(\dfrac{\partial U}{\partial V}\right)_T = \dfrac{C_p - C_V}{V \beta_p} - p, \qquad (16.86)$$

where β_p is the coefficient of volume expansivity and the other symbols have their usual meanings.

(16.4) (a) The natural variables for U are S and V. This means that if you know S and V, you can find $U(S,V)$. Show that this also gives you simple expressions for T and p.

(b) Suppose instead that you know V, T and the function $U(T,V)$ (i.e., you have expressed U in terms of variables that are not all the natural variables of U). Show that this leads to a (much more complicated) expression for p, namely

$$\frac{p}{T} = \int \left(\frac{\partial U}{\partial V}\right)_T \frac{\mathrm{d}T}{T^2} + f(V), \qquad (16.87)$$

where $f(V)$ is some (unknown) function of V.

(16.5) Use thermodynamic arguments to obtain the general result that, for any gas at temperature T, the pressure is given by

$$p = T\left(\frac{\partial p}{\partial T}\right)_V - \left(\frac{\partial U}{\partial V}\right)_T, \qquad (16.88)$$

where U is the total energy of the gas.

(16.6) Show that another expression for the entropy per mole of an ideal gas is

$$S = C_p \ln T - R \ln p + \text{constant}. \qquad (16.89)$$

(16.7) Show that the entropy of an ideal gas can be expressed as

$$S = C_V \ln\left(\frac{p}{\rho^\gamma}\right) + \text{constant}. \qquad (16.90)$$

Hermann von Helmholtz (1821–1894)

Since his family couldn't afford to give him an academic education in physics, the seventeen-year old Helmholtz found himself at a Berlin medical school getting a free four-year medical education, the catch being that he then had to serve as a surgeon in the Prussian army.

Fig. 16.2 H. von Helmholtz

It was during his army service that he submitted a paper "On the conservation of force" (his use of the word "force" is more akin to what we call "energy", the two concepts being poorly differentiated at the time). It was a blow against the notion of a "vital force", an indwelling "life source" which was widely proposed by physiologists to explain biological systems. Helmholtz intuited that such a vital force was mere metaphysical speculation and instead all physical and chemical processes involved the exchange of energy from one form to another, and that "all organic processes are reducible to physics". Thus he began a remarkable career based on his remarkable physical insight into physiology.

In 1849 he was appointed professor of physiology at Königsberg, and six years later took up a professorship in anatomy in Bonn, moving to Heidelberg three years later. During this period he pioneered the application of physical and mathematical techniques to physiology: he invented the opthalmoscope (for looking into the eye), the opthalmometer (for measuring the curvature and refractive errors in the eye) and worked on the problem of three-colour vision; he did pioneering research in physiological acoustics, explaining the operation of the inner ear; he also measured the speed of nerve impulses in a frog.

He even found time to make important contributions to understanding vortices in fluids. In 1871, he was appointed to a chair in Berlin, but this time it was in physics; here he pursued work in electrodynamics, non-Euclidean geometry and physical chemistry. Helmholtz mentored and influenced many highly talented students in Berlin, including Planck, Wien, and Hertz.

Helmholtz's scientific life was characterized by a search for unity and clarity. He once said that "whoever in the pursuit of science, seeks after immediate practical utility may rest assured that he seeks in vain", but there can be only few scientists in history whose work has had the result of greater practical utility.

William Thomson [Lord Kelvin] (1824–1907)

Fig. 16.3 William Thomson

William Thomson was something of a prodigy: born in Belfast, the son of a mathematician, he studied in Glasgow University and then moved to Peterhouse, Cambridge. By the time he had graduated, he had written 12 research papers, the first of the 661 of his career. He became Professor of Natural Philosophy in the University of Glasgow at 22, a Fellow of the Royal Society at 27, was knighted at 42, and in 1892 became Baron Kelvin of Largs (taking his new title from the River Kelvin in Glasgow), an appointment that occurred during his presidency of the Royal Society. When he died, he was buried next to Isaac Newton in Westminster Abbey.

Thomson made pioneering contributions in fundamental electromagnetism and fluid dynamics, but also involved himself in large engineering projects. After working out how to solve the problem of sending signals down very long cables, he was involved in laying the first transatlantic telegraph cables in 1866. In 1893, he headed an international commission to plan the design of the Niagara Falls power station and was convinced by Nikola Tesla, somewhat against his better judgement, to use three-phase AC power rather than his preferred DC power transmission. On this point he was unable to forecast the future (which was of course AC, not DC); in similar vein, he pronounced heavier-than-air flying machines "impossible", thought that radio had "no future", and that war, as a "relic of barbarism" would "become as obsolete as duelling". If only.

It is his progress in thermodynamics that interests us here. Inspired by the meticulous thermometric measurements of Henri Regnault, which he had observed during a postgraduate stay in Paris, Thomson proposed an absolute temperature scale in 1848. Thomson was also profoundly influenced by Fourier's theory of heat (which he had read in his teens) and Carnot's work via the paper of Clapeyron. These had assumed a caloric theory of heat, which Thomson had initially adopted, but his encounter with Joule at the 1847 British Association meeting in Oxford had sown some seeds of doubt in caloric. After much thought, Thomson groped his way towards his "dynamical theory of heat", which he published in 1851, a synthesis of Joule and Carnot, containing a description of the degradation of energy and speculations about the heat death of the Universe. He just missed a full articulation of the concept of entropy, but grasped the essential details of the first and second laws of thermodynamics. His subsequent fruitful collaboration with Joule led to the Joule–Thomson (or Joule–Kelvin) effect.

Thomson also discovered many key results concerning thermoelectricity. His most controversial result was however his estimate of the age of the Earth, based on Fourier's thermal diffusion equation. He concluded that if the Earth had originally been a red-hot globe, and had cooled to its present temperature, its age must be about 10^8 years. This pleased nobody: the Earth was too old for those who believed in a six-thousand year old planet but too young for Darwin's evolution to produce the present biological diversity. Thomson could not have known that radioactivity (undiscovered until the very end of the nineteenth century) acts as an additional heat source in the Earth, allowing the Earth to be nearly two orders of magnitude older than he estimated. His lasting legacy however has been his new temperature scale, so that his "absolute zero", the lowest possible temperature obtainable, is zero *kelvin*.

Josiah Willard Gibbs (1839–1903)

Willard Gibbs was born in New Haven and died in New Haven, living his entire life (a brief postdoctoral period in France and Germany excepted) at Yale, where he remained unmarried. His father was also called Josiah Willard Gibbs and had also been a professor at Yale, though in Sacred Literature rather than in Mathematical Physics. Willard Gibbs' life was quiet and secluded, well away from the centres of intense scientific activity at the time, which were all in Europe. This gave this gentle and scholarly man the opportunity to perform clear-thinking, profound and independent work in chemical thermodynamics, work which turned out to be completely revolutionary, though this took time to be appreciated. Willard Gibbs' key papers were published in a series of installments in the *Transactions of the Connecticut Academy of Sciences*, which was hardly required reading at the time; moreover his mathematical style did not make his papers easily accessible. Maxwell was one of the few who were very impressed.

Fig. 16.4 J. W. Gibbs

Gibbs established the key principles of chemical thermodynamics, defined the free energy and chemical potential, completely described phase equilibria with more than one component and championed a geometric view of thermodynamics. Not only did he substantially formulate thermodynamics and statistical mechanics in the form we know it today, but he also championed the use of vector calculus, in its modern form, to describe electromagnetism (in the face of spirited opposition from various prominent Europeans who maintained that the only way to describe electromagnetism was using quaternions).

Gibbs didn't interact a great deal with scientific colleagues in other institutions; he was privately secure in himself and in his ideas. One contemporary wrote of him: "Unassuming in manner, genial and kindly in his intercourse with his fellow men, never showing impatience or irritation, devoid of personal ambition of the baser sort or of the slightest desire to exalt himself, he went far toward realizing the ideal of the unselfish, Christian gentleman".

Rods, bubbles, and magnets

17

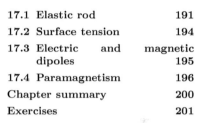

In this book, we have been illustrating the development of thermodynamics using the ideal gas as our chief example. We have written the first law of thermodynamics as

$$dU = T\,dS - p\,dV, \qquad (17.1)$$

and everything has followed from this. However, in this chapter we want to show that thermodynamics can be applied to other types of system. In general we will write the work $đW$ as

$$đW = X\,dx, \qquad (17.2)$$

where X is some (intensive[1]) generalized force and x is some (extensive) generalized displacement. Examples of these are given in Table 17.1. In this chapter we will examine only three of these examples in detail: the elastic rod, the surface tension in a liquid and the assembly of magnetic moments in a paramagnet.

	X	x	$đW$
fluid	$-p$	V	$-p\,dV$
elastic rod	f	L	$f\,dL$
liquid film	γ	A	$\gamma\,dA$
dielectric	$\boldsymbol{E}$	$\boldsymbol{p}_\mathrm{E}$	$-\boldsymbol{p}_\mathrm{E}\cdot d\boldsymbol{E}$
magnetic	$\boldsymbol{B}$	$\boldsymbol{m}$	$-\boldsymbol{m}\cdot d\boldsymbol{B}$

Table 17.1 Generalized force X and generalized displacement x for various different systems. In this table, p = pressure, V = volume, f = tension, L = length, γ = surface tension, A = area, $\boldsymbol{E}$ = electric field, $\boldsymbol{p}_\mathrm{E}$ = electric dipole moment, $\boldsymbol{B}$ = magnetic field, $\boldsymbol{m}$ = magnetic dipole moment.

17.1 Elastic rod	191
17.2 Surface tension	194
17.3 Electric and magnetic dipoles	195
17.4 Paramagnetism	196
Chapter summary	200
Exercises	201

[1]Recall from Section 11.1.2 that intensive variables are independent of the size of the system whereas extensive variables are proportional to the size of the system.

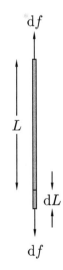

Fig. 17.1 An elastic material of length L and cross-sectional area A is extended a length dL by a tension df.

17.1 Elastic rod

Consider a rod with cross-sectional area A and length L, held at temperature T. The rod is made from any elastic material (such as a metal or rubber) and is placed under an infinitesimal tension df, which leads to the rod extending by an infinitesimal length dL (see Fig. 17.1). We define the **isothermal Young's modulus** E_T as the ratio of **stress** $\sigma = df/A$ to **strain** $\epsilon = dL/L$, so that

$$E_T = \frac{\sigma}{\epsilon} = \frac{L}{A}\left(\frac{\partial f}{\partial L}\right)_T. \tag{17.3}$$

The Young's modulus E_T is always a positive quantity.

There is another useful quantity that characterizes an elastic rod. We can also define the **linear expansivity at constant tension**, α_f, by

$$\alpha_f = \frac{1}{L}\left(\frac{\partial L}{\partial T}\right)_f, \tag{17.4}$$

which is the fractional change in length with temperature. This quantity is positive in most elastic systems (though not rubber). If you hang a weight onto the end of a metal wire (thus keeping the tension f in the wire constant) and heat the wire, it will extend. This implies that $\alpha_f > 0$ for a metal wire. However, if you hang a weight by a piece of rubber and supply heat, you will find that the rubber will contract, which implies that $\alpha_f < 0$ for rubber.

Example 17.1

How does the tension of a wire held at constant length change with temperature?

Solution: Our definitions of E_T and α_f allow us to calculate this. Using eqn C.42, we have that

$$\left(\frac{\partial f}{\partial T}\right)_L = -\left(\frac{\partial f}{\partial L}\right)_T\left(\frac{\partial L}{\partial T}\right)_f = -AE_T\alpha_f, \tag{17.5}$$

where the last step is obtained using eqns 17.3 and 17.4.

This result is familiar to anyone who plays a metal-stringed instrument where $\alpha_f > 0$ and hence $(\partial f/\partial T)_L < 0$ from eqn 17.5; hot weather causes the strings (held at constant length) to slacken (reduce their tension).

We are now in a position to do some thermodynamics on our elastic system. We will rewrite the first law of thermodynamics for this case as

$$dU = T\,dS + f\,dL. \tag{17.6}$$

We can also obtain other thermodynamic potentials, such as the Helmholtz function $F = U - TS$, so that $dF = dU - T\,dS - S\,dT$, and hence

$$dF = -S\,dT + f\,dL. \tag{17.7}$$

Equation 17.7 implies that the entropy S is

$$S = -\left(\frac{\partial F}{\partial T}\right)_L, \tag{17.8}$$

and similarly the tension f is

$$f = \left(\frac{\partial F}{\partial L}\right)_T. \tag{17.9}$$

A Maxwell's-relation–type-step[2] then leads to an expression for the isothermal change in entropy on extension as

$$\left(\frac{\partial S}{\partial L}\right)_T = -\left(\frac{\partial f}{\partial T}\right)_L. \qquad (17.10)$$

The right-hand side of this equation was worked out in eqn 17.5, so that

$$\left(\frac{\partial S}{\partial L}\right)_T = AE_T\alpha_f, \qquad (17.11)$$

where A is the area (presumed not to change), and so stretching the rod (increasing L) results in an *increase* in entropy if $\alpha_f > 0$. This is like the case of an ideal gas for which

$$\left(\frac{\partial S}{\partial V}\right)_T = \left(\frac{\partial p}{\partial T}\right)_V > 0, \qquad (17.12)$$

so that expanding the gas (increasing V) results in an *increase* in entropy. If the entropy of the system goes up as it is expanded isothermally, then heat must be absorbed. For the case of the elastic rod (assuming it is not made of rubber), extending it isothermally (and reversibly) by ΔL would then lead to an absorption of heat ΔQ given by

$$\Delta Q = T\Delta S = AE_T T\alpha_f \Delta L. \qquad (17.13)$$

Why does stretching a wire increase its entropy? Let us consider the case of a metallic wire. This contains many small crystallites, which have low entropy. The action of stretching the wire distorts those small crystallites, and that increases their entropy and so heat is absorbed.[3]

However, for rubber $\alpha_f < 0$, and hence an isothermal extension means that heat is emitted. The action of stretching a piece of rubber at constant temperature results in the alignment of the long rubber molecules, reducing their entropy (see Fig. 17.2) and causing heat to be released.

Example 17.2

The internal energy U for an ideal gas does not change when it is expanded isothermally. How does U change for an elastic rod when it is extended isothermally?

Solution: The change in internal energy on isothermal extension can be worked out from eqn 17.6 and eqn 17.11 by writing

$$\left(\frac{\partial U}{\partial L}\right)_T = T\left(\frac{\partial S}{\partial L}\right)_T + f = f + ATE_T\alpha_f. \qquad (17.14)$$

This is the sum of a positive term expressing the energy going into the rod by work and a term expressing the heat flow into the rod due to an isothermal change of length. (For an ideal gas, a similar analysis applies, but the work done by the gas and the heat that flows into it balance perfectly, so that U does not change.)

---

[2] As in the case of a gas, the Maxwell's relation allows us to relate some differential of entropy (which is hard to measure experimentally, but is telling us something fundamental about the system) to a differential that we can measure in an experiment, here the change in tension with temperature of a rod held at constant length.

[3] For example, the crystallites might distort from cubic to tetragonal symmetry, thus increasing the entropy. In addition, the stretching of the wire may increase the volume per atom in the wire and this also increases the entropy.

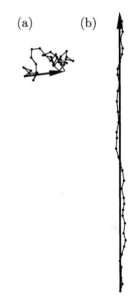

Fig. 17.2 Rubber consists of long-chain molecules. (a) With no force applied, the rubber molecule is quite coiled up and the average end-to-end distance is short, and the entropy is large. This picture has been drawn by taking each segment of the chain to point randomly. (b) With a force applied (along a vertical axis in this diagram), the molecule becomes more aligned with the direction of the applied force, and the end-to-end distance is large, reducing the entropy (see Exercise 17.3).

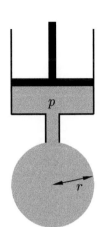

Fig. 17.3 A spherical droplet of liquid of radius r is suspended from a thin pipe connected to a piston, which maintains the pressure p of the liquid.

17.2 Surface tension

We now consider the case of a liquid surface with surface area A. Liquid surfaces cost energy, which is why a liquid will tend to form droplets (or, even better, a single droplet) to minimize this surface energy. The work needed to change the area of a liquid surface is given by

$$\mathrm{d}W = \gamma\, \mathrm{d}A, \tag{17.15}$$

where γ is a parameter known as the **surface tension**.

Consider the arrangement shown in Fig. 17.3. If the piston moves down, work $\mathrm{d}W = F\,\mathrm{d}x = +p\,\mathrm{d}V$ is done on the liquid (which is assumed to be incompressible). The droplet radius will therefore increase by an amount $\mathrm{d}r$ such that $\mathrm{d}V = 4\pi r^2\,\mathrm{d}r$, and the surface area of the droplet will change by an amount

$$\mathrm{d}A = 4\pi(r + \mathrm{d}r)^2 - 4\pi r^2 \approx 8\pi r\,\mathrm{d}r, \tag{17.16}$$

so that

$$\mathrm{d}W = \gamma\,\mathrm{d}A = 8\pi\gamma r\,\mathrm{d}r. \tag{17.17}$$

Equating this to $\mathrm{d}W = F\,\mathrm{d}x = +p\,\mathrm{d}V = p\cdot 4\pi r^2\,\mathrm{d}r$ yields

$$p = \frac{2\gamma}{r}. \tag{17.18}$$

The pressure p in this expression is, of course, really the pressure difference between the pressure in the liquid and the atmospheric pressure against which the surface of the drop pushes.

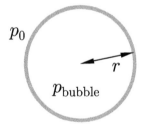

Fig. 17.4 A bubble of radius r has an inner and an outer surface.

Example 17.3

What is the pressure of gas inside a spherical bubble of radius r?
Solution: The bubble (see Fig. 17.4) has *two surfaces*, and so the pressure p_{bubble} of gas inside the bubble, minus the pressure p_0 outside the bubble, has to support two lots of surface tension. Hence, assuming the liquid wall of the bubble is thin (so that the radii of inner and outer walls are both $\approx r$),

$$p_{\text{bubble}} - p_0 = \frac{4\gamma}{r}. \tag{17.19}$$

Notice that surface tension has a microscopic explanation. A molecule in the bulk of the liquid is attracted to its nearest neighbours by intermolecular forces (which is what holds a liquid together), and these forces are applied to a given molecule by its neighbours from all directions. One can think of these forces almost as weak chemical bonds. The molecules at the surface are only attracted by their neighbouring molecules in one direction, back towards the bulk of the liquid, but there is no corresponding attractive force out into the "wild blue yonder". The surface has a higher energy than the bulk because bonds have to be broken in order to make a surface, and γ tells you how much energy you need to form a unit area of surface (which gives an estimate of the size of the intermolecular forces).

We can write the first law of thermodynamics for our surface of area A as

$$\mathrm{d}U = T\,\mathrm{d}S + \gamma\,\mathrm{d}A \tag{17.20}$$

and similarly changes in the Helmholtz function can be written

$$\mathrm{d}F = -S\,\mathrm{d}T + \gamma\,\mathrm{d}A, \tag{17.21}$$

which yields the Maxwell's relation

$$\left(\frac{\partial S}{\partial A}\right)_T = -\left(\frac{\partial \gamma}{\partial T}\right)_A. \tag{17.22}$$

Equation 17.20 implies that

$$\left(\frac{\partial U}{\partial A}\right)_T = T\left(\frac{\partial S}{\partial A}\right)_T + \gamma, \tag{17.23}$$

and hence using eqn 17.22, we have

$$\left(\frac{\partial U}{\partial A}\right)_T = \gamma - T\left(\frac{\partial \gamma}{\partial T}\right)_A, \tag{17.24}$$

the sum of a positive term expressing the energy going into a surface by work and a negative term expressing the heat flow into the surface due to an isothermal change of area. Usually, the surface tension has a temperature dependence as shown in Fig. 17.5, and hence $(\partial \gamma/\partial T)_A < 0$, so in fact both terms contribute a positive amount.

Heat ΔQ is given by

$$\Delta Q = T\left(\frac{\partial S}{\partial A}\right)_T \Delta A = -T\Delta A\left(\frac{\partial \gamma}{\partial T}\right)_A > 0, \tag{17.25}$$

and this is absorbed on isothermally stretching a surface to increase its area by ΔA. This quantity is positive and so heat really is absorbed. Since $\left(\frac{\partial S}{\partial A}\right)_T$ is positive, this shows that the surface has an additional entropy compared with the bulk, in addition to costing extra energy.

17.3 Electric and magnetic dipoles

An electric dipole moment $\boldsymbol{p}_\mathrm{E}$ can interact with an electric field $\boldsymbol{E}$. The potential energy of the dipole in the electric field is $-\boldsymbol{p}_\mathrm{E} \cdot \boldsymbol{E}$. If the electric field changes, the interaction energy can change by

$$\mathrm{d}(-\boldsymbol{p}_\mathrm{E} \cdot \boldsymbol{E}) = -\boldsymbol{p}_\mathrm{E} \cdot \mathrm{d}\boldsymbol{E} - \boldsymbol{E} \cdot \mathrm{d}\boldsymbol{p}_\mathrm{E}. \tag{17.26}$$

There is also some stored energy in the dipole itself. An electric dipole consists of charges $+q$ and $-q$ separated by a distance a, so that the

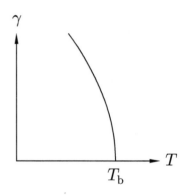

Fig. 17.5 Schematic diagram of the surface tension γ of a liquid as a function of temperature. Since γ must vanish at the boiling temperature T_b, we expect that $(\partial \gamma/\partial T)_A < 0$.

dipole moment has magnitude $p_E = qa$. The force on each charge due to the electric field has magnitude qE. A small change da in the length a means that the dipole moment changes by $dp_E = q\,da$. Modelling the bond between the charges as a spring, the work done on this spring because of the change of length is given by the force qE times the distance da which equals $E(q\,da) = E\,dp_E$. In the case in which the electric field is at an angle to the dipole moment, only the component of the electric field parallel to the dipole moment acts to stretch the spring, so in general we can write this contribution as $+\boldsymbol{E} \cdot d\boldsymbol{p}_E$. Adding this stored energy to the interaction energy from eqn 17.26 gives the work supplied to the system as[4]

$$đW = -\boldsymbol{p}_E \cdot d\boldsymbol{E}. \tag{17.27}$$

Analogous arguments can be used to show that the work supplied to a magnetic dipole is given by

$$đW = -\boldsymbol{m} \cdot d\boldsymbol{B}. \tag{17.28}$$

We consider assemblies of magnetic moments in more detail in the next section.

[4]This is the work added to the system (where system here means the electric dipole and its interaction with the field) and hence is the free energy of the electric dipole. Because the energy of the dipole in the electric field is shared between the dipole and the field (it is an interaction energy, belonging to both parties) it does not make sense to think of energy that is internal to the dipole itself, and the "system" means the interacting dipole and field.

17.4 Paramagnetism

Consider a system of magnetic moments arranged in a lattice at temperature T. We assume that the magnetic moments cannot interact with each other. If the application of a magnetic field causes the magnetic moments to line up, the system is said to exhibit **paramagnetism**. The equivalent formulation of the first law of thermodynamics for a paramagnet is

$$dU = T\,dS - m\,dB, \tag{17.29}$$

where m is the **magnetic moment** and B is the **magnetic field**.[5] The magnetic moment $m = MV$, where M is the **magnetization** and V is the volume. The **magnetic susceptibility** χ is given by

$$\chi = \lim_{H\to 0} \frac{M}{H}. \tag{17.30}$$

For most paramagnets $\chi \ll 1$, so that $M \ll H$ and hence $B = \mu_0(H + M) \approx \mu_0 H$. This implies that we can write the magnetic susceptibility χ as

$$\chi \approx \frac{\mu_0 M}{B}. \tag{17.31}$$

Paramagnetic systems obey **Curie's law**, which states that

$$\chi \propto \frac{1}{T}, \tag{17.32}$$

as shown in Fig. 17.6, and hence

$$\left(\frac{\partial \chi}{\partial T}\right)_B < 0, \tag{17.33}$$

a result that we shall use later.[6]

[5]B is often known as the **magnetic flux density** or the **magnetic induction**, but following common usage, we refer to B as the magnetic field; see Blundell (2001). The magnetic field H (often called the **magnetic field strength**) is related to B and the magnetization M by
$$B = \mu_0(H + M).$$

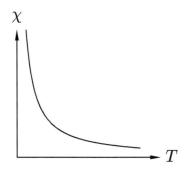

Fig. 17.6 The magnetic susceptibility for a paramagnet follows Curie's law which states that $\chi \propto 1/T$.

[6]Curie's law itself is derived in Example 20.5.

Example 17.4

Show that heat is emitted in an isothermal increase in B (a process known as **isothermal magnetization**) but that temperature is reduced for an adiabatic reduction in B (a process known as **adiabatic demagnetization**).

Solution: Eqn 17.29 implies that changes in the Helmholtz function $F = U - TS$ follows

$$dF = -S\,dT - m\,dB, \qquad (17.34)$$

which yields the Maxwell relation

$$\left(\frac{\partial S}{\partial B}\right)_T = \left(\frac{\partial m}{\partial T}\right)_B \approx \frac{VB}{\mu_0}\left(\frac{\partial \chi}{\partial T}\right)_B, \qquad (17.35)$$

which relates the isothermal change of entropy with field at constant temperature to a differential of the susceptibility χ.

The heat absorbed in an isothermal change of B is

$$\Delta Q = T\left(\frac{\partial S}{\partial B}\right)_T \Delta B = \frac{TVB}{\mu_0}\left(\frac{\partial \chi}{\partial T}\right)_B \Delta B < 0, \qquad (17.36)$$

and since it is negative it implies that heat is actually emitted. The change in temperature in an adiabatic change of B is

$$\left(\frac{\partial T}{\partial B}\right)_S = -\left(\frac{\partial T}{\partial S}\right)_B \left(\frac{\partial S}{\partial B}\right)_T. \qquad (17.37)$$

If we define $C_B = T\left(\frac{\partial S}{\partial T}\right)_B$, the heat capacity at constant B, then substitution of this and eqn 17.35 into eqn 17.37 yields

$$\left(\frac{\partial T}{\partial B}\right)_S = -\frac{TVB}{\mu_0 C_B}\left(\frac{\partial \chi}{\partial T}\right)_B. \qquad (17.38)$$

Equation 17.33 implies that $\left(\frac{\partial T}{\partial B}\right)_S > 0$, and hence we can cool a material using an adiabatic demagnetization, i.e., by reducing the magnetic field on a sample while keeping it at constant entropy. This can yield temperatures as low as a few millikelvin for electronic systems and a few microkelvin for nuclear systems.

This coupling between thermal and magnetic properties is known as the **magnetocaloric effect**.

Let us now consider why adiabatic demagnetization results in the cooling of a material from a microscopic point of view. Consider a sample of a paramagnetic salt, which contains N independent magnetic moments. Without a magnetic field applied, the magnetic moments will point in random directions (because we are assuming that they do not interact with each other) and the system will have no net magnetization. An applied field B will, however, tend to line up the magnetic moments and produce a magnetization. Increasing temperature reduces the magnetization, and increasing magnetic field increases the magnetization. At

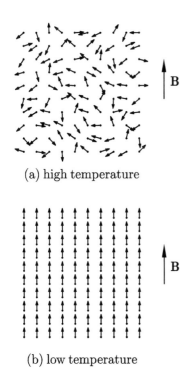

Fig. 17.7 (a) At high temperature, the spins in a paramagnet are in random directions because the thermal energy $k_\mathrm{B} T$ is much larger than the magnetic energy mB. This state has high entropy. (b) At low temperature, the spins become aligned with the field because the thermal energy $k_\mathrm{B} T$ is much smaller than the magnetic energy mB. This state has low entropy.

very high temperature, the magnetic moments all point in random directions and the net magnetization is zero (see Fig. 17.7(a)). The thermal energy $k_\mathrm{B} T$ is so large that all states are equally populated, irrespective of whether or not the state is energetically favourable. If the magnetic moments have angular momentum quantum number $J = \frac{1}{2}$ they can only point parallel or antiparallel to the magnetic field: hence there are $\Omega = 2^N$ ways of arranging up and down magnetic moments. Hence the magnetic contribution to the entropy, S, is

$$S = k_\mathrm{B} \ln \Omega = N k_\mathrm{B} \ln 2. \tag{17.39}$$

In the general case of $J > \frac{1}{2}$, $\Omega = (2J+1)^N$ and the entropy is

$$S = N k_\mathrm{B} \ln(2J+1). \tag{17.40}$$

At lower temperature, the entropy of the paramagnetic salt must reduce as only the lowest energy levels are occupied, corresponding to the average alignment of the magnetic moments with the applied field increasing. At very low temperature, all the magnetic moments will align with the magnetic field to minimize their energy (see Fig. 17.7(b)). In this case there is only one way of arranging the system (with all spins aligned) so $\Omega = 1$ and $S = 0$.

The procedure for magnetically cooling a sample is as follows. The paramagnet is first cooled to a low starting temperature using liquid helium. The magnetic cooling then proceeds via two steps (see also Fig. 17.8).

The first step is **isothermal magnetization**. The energy of a paramagnet is reduced by alignment of the moments parallel to a magnetic field. At a given temperature the alignment of the moments may therefore be enhanced by increasing the strength of an applied magnetic field. This is performed isothermally (see Fig. 17.8, step $a \to b$) by having the sample thermally connected to a bath of liquid helium (the boiling point of helium at atmospheric pressure is 4.2 K), or perhaps with the liquid helium bath at reduced pressure so that the temperature can be less than 4.2 K. The temperature of the sample does not change and the helium bath absorbs the heat liberated by the sample as its energy and entropy decrease. The thermal connection is usually provided by low-pressure helium gas in the sample chamber, which conducts heat between the sample and the chamber walls, the chamber itself sitting inside the helium bath. (The gas is often called "exchange" gas because it allows the sample and the bath to exchange heat.)

The second step is to thermally isolate the sample from the helium bath (by pumping away the exchange gas). The magnetic field is then slowly reduced to zero, slowly so that the process is quasistatic and the entropy is constant. This step is called **adiabatic demagnetization** (see Fig. 17.8, step $b \to c$) and it reduces the temperature of the system. During adiabatic demagnetization the entropy of the sample remains constant; the entropy of the magnetic moments increases (because the moments randomize as the field is turned down) and this is precisely balanced by the decrease in the entropy of the phonons (the lattice vibrations) as the sample cools. Entropy is thus exchanged between the phonons and the spins.

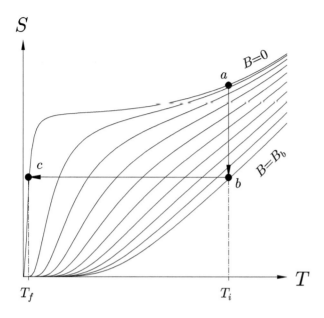

Fig. 17.8 The entropy of a paramagnetic salt as a function of temperature for several different applied magnetic fields between zero and some maximum value, which we will call B_b. Magnetic cooling of a paramagnetic salt from temperature T_i to T_f is accomplished as indicated in two steps: first, isothermal magnetization from a to b by increasing the magnetic field from 0 to B_b at constant temperature T_i; second, adiabatic demagnetization from b to c. The $S(T)$ curves have been calculated assuming $J = \frac{1}{2}$. A term $\propto T^3$ has been added to these curves to simulate the entropy of the lattice vibrations. The curve for $B = 0$ is actually for small, but non-zero, B to simulate the effect of a small residual field.

There is another way of looking at adiabatic demagnetization. Consider the energy levels of magnetic ions in a paramagnetic salt subjected to an applied magnetic field. The population of magnetic ions in each energy level is given by the Boltzmann distribution, as indicated schematically in Fig. 17.9(a). The rate at which the levels decrease in population as the energy increases is determined by the temperature T. When we perform an isothermal magnetization (increasing the applied magnetic field while keeping the temperature constant) we are increasing the spacing between the energy levels of the paramagnetic salt [see Fig. 17.9(b)], but the occupation of each level is determined by the same Boltzmann distribution because the temperature T is constant. Thus the higher–energy levels become depopulated. This depopulation is the result of transitions between energy levels caused by interaction with the surroundings, which are keeping the system at constant temperature. In an adiabatic demagnetization, the external magnetic field is reduced to its original value, closing up the energy levels again. However, because the salt is now thermally isolated, no transitions between energy levels are possible and the populations of each level remain the same [see Fig. 17.9(c)]. Another way of saying this is that in an adiabatic process the entropy $S = -k_B \sum_i P_i \ln P_i$ (eqn 14.48) of the system is constant, and this expression only involves the probability P_i of occupying the ith level, not the energy. Thus the temperature of the paramagnetic salt following the adiabatic demagnetization is lower because the occupancies now correspond to a Boltzmann distribution with a lower temperature.

Does adiabatic demagnetization as a method of cooling have a limit? At first sight it looks as though the entropy would be $S = Nk_B \ln(2J+1)$ at $B = 0$ for all $T > 0$, and therefore with $B \neq 0$, $S \to 0$ only at absolute zero, implying that adiabatic demagnetization might be used to cool all the way to absolute zero. However, in real paramagnetic salts there is always some small residual internal field due to interactions between the moments which ensures that the entropy falls prematurely towards zero when the temperature is a little above absolute zero (see Fig. 17.8). The size of this field puts a limit on the lowest temperature to which the paramagnetic salt can be cooled. In certain paramagnetic salts, which have a very small residual internal field, temperatures of a few millikelvin can be achieved. The failure of Curie's law as we approach $T = 0$ is just one of the consequences of the third law of thermodynamics, which we shall treat in the following chapter.

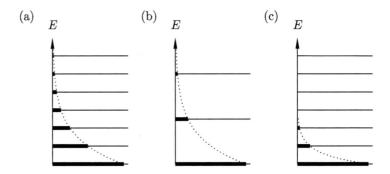

Fig. 17.9 Schematic diagram showing the energy levels in a magnetic system (a) initially, (b) following isothermal magnetization, and (c) following adiabatic demagnetization.

Chapter summary

- The first law for a gas is $dU = T\,dS - p\,dV$. An isothermal expansion results in S increasing (see Fig. 17.10(a)). An adiabatic compression results in T increasing.
- The first law for an elastic rod is $dU = T\,dS + f\,dL$. An isothermal extension of a metal wire results in S increasing (see Fig. 17.10(b)) but for rubber S decreases (see Fig. 17.10(c)). An adiabatic contraction of a metal wire results in T increasing (but for rubber T decreases).
- The first law for a liquid surface is $dU = T\,dS + \gamma\,dA$. An isothermal stretching results in S increasing. An adiabatic contraction results in T increasing.
- The first law is $dU = T\,dS - m\,dB$ for a magnetic system. An isothermal magnetization results in S decreasing (see Fig. 17.10(d)). An adiabatic demagnetization results in T decreasing.

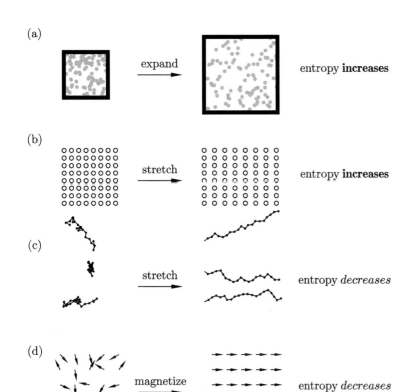

Fig. 17.10 Entropy increases when (a) a gas is expanded isothermally, (b) a metallic rod is stretched isothermally. Entropy decreases when (c) rubber is stretched isothermally and (d) a paramagnet is magnetized isothermally.

Exercises

(17.1) For an elastic rod, show that

$$\left(\frac{\partial C_L}{\partial L}\right)_T = -T\left(\frac{\partial^2 f}{\partial T^2}\right)_L, \qquad (17.41)$$

where C_L is the heat capacity at constant length L.

(17.2) For an elastic rod, show that

$$\left(\frac{\partial T}{\partial L}\right)_S = -\frac{TAE_T\alpha_f}{C_L}. \qquad (17.42)$$

For rubber, explain why this quantity is positive. Hence explain why, if you take a rubber band that has been under tension for some time and suddenly release the tension to zero, the rubber band appears to have cooled.

(17.3) A rubber molecule can be modelled in one dimension as a chain consisting of a series of $N = N_+ + N_-$ links, where N_+ links point in the $+x$ direction, while N_- links point in the $-x$ direction. If the length of one link in the chain is a, show that the length L of the chain is

$$L = a(N_+ - N_-). \qquad (17.43)$$

Show further that the number of ways $\Omega(L)$ of arranging the links to achieve a length L can be written as

$$\Omega(L) = \frac{N!}{N_+!N_-!}, \qquad (17.44)$$

and also that the entropy $S = k_B \ln \Omega(L)$ can be written approximately as

$$S = Nk_B\left[\ln 2 - \frac{L^2}{2N^2a^2}\right] \qquad (17.45)$$

when $L \ll Na$, and hence that S decreases as L increases.

(17.4) The entropy S of a surface can be written as a function of its area A and temperature T. Hence show that

$$\mathrm{d}U = T\,\mathrm{d}S + \gamma\,\mathrm{d}A \qquad (17.46)$$
$$= C_A\,\mathrm{d}T + \left[\gamma - T\left(\frac{\partial \gamma}{\partial T}\right)_A\right]\mathrm{d}A.$$

(17.5) Consider a liquid of density ρ with molar mass M. Explain why the number of molecules per unit area in the surface is approximately

$$(\rho N_A/M)^{2/3}. \qquad (17.47)$$

Hence, the energy contribution per molecule to the surface tension γ is approximately

$$\gamma/(\rho N_A/M)^{2/3}. \qquad (17.48)$$

Evaluate this quantity for water (surface tension at $20\,^\circ$C is approximately $72\,\mathrm{mJ\,m^{-2}}$) and express your answer in eV. Compare your result with the latent heat per molecule (the molar latent heat of water is $4.4 \times 10^4\,\mathrm{J\,mol^{-1}}$).

(17.6) For a stretched rubber band, it is observed experimentally that the tension f is proportional to the temperature T if the length L is held constant. Show that:
(a) the internal energy U is a function of temperature only;
(b) adiabatic stretching of the band results in an increase in temperature;
(c) the band will contract if warmed while kept under constant tension.

(17.7) A soap bubble of radius R_1 and surface tension γ is expanded at constant temperature by forcing in air by driving in a piston containing volume V_{piston} fully home. Show that the work ΔW needed to increase the bubble's radius to R_2 is

$$\Delta W = p_2 V_2 \ln \frac{p_2}{p_1} + 8\pi\gamma(R_2^2 - R_1^2)$$
$$+ p_0(V_2 - V_1 - V_{\mathrm{piston}}), \qquad (17.49)$$

where p_1 and p_2 are the initial and final pressures in the bubble, p_0 is the pressure of the atmosphere and $V_1 = \frac{4}{3}\pi R_1^3$ and $V_2 = \frac{4}{3}\pi R_2^3$.

The third law

18

In Chapter 13, we presented the second law of thermodynamics in various different forms. In Chapter 14, we related this to the concept of entropy and showed that the entropy of an isolated system always either stays the same or increases with time. But what value does the entropy of a system take, and how can you measure it?

One way of measuring the entropy of a system is to measure its heat capacity. For example, if measurements of C_p, the heat capacity at constant pressure, are made as a function of temperature, then using

$$C_p = T\left(\frac{\partial S}{\partial T}\right)_p, \tag{18.1}$$

we can obtain entropy S by integration, so that

$$S = \int \frac{C_p}{T} \mathrm{d}T. \tag{18.2}$$

This is all very well, but when you integrate, you have to worry about constants of integration. Writing eqn 18.2 as a definite integral, we have that the entropy $S(T)$, measured at temperature T, is

$$S(T) = S(T_0) + \int_{T_0}^{T} \frac{C_p}{T} \mathrm{d}T, \tag{18.3}$$

where T_0 is some different temperature (see Fig. 18.1). Thus it seems that we are only able to learn about *changes* in entropy, for example as a system is warmed from T_0 to T, and we are not able to obtain an absolute measurement of entropy itself. The third law of thermodynamics, presented in this chapter, gives us additional information because it provides a value for the entropy at one particular temperature, namely absolute zero.

18.1 Different statements of the third law	203
18.2 Consequences of the third law	205
Chapter summary	208
Exercises	208

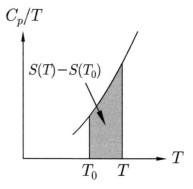

Fig. 18.1 A graphical representation of eqn 18.3.

18.1 Different statements of the third law

Walter H. Nernst (1864–1941) (Fig. 18.2) came up with the first statement of the third law of thermodynamics after examining data on chemical thermodynamics and doing experiments with electrochemical cells. The essential conclusion he came to concerned the change in enthalpy ΔH in a reaction (the heat of the reaction, positive if endothermic, negative if exothermic; see Section 16.5), and the change in Gibbs' function ΔG (that determines in which direction the reaction goes). Since

$G = H - TS$, we expect that

$$\Delta G = \Delta H - T\Delta S, \qquad (18.4)$$

so that as $T \to 0$, $\Delta G \to \Delta H$. Experimental data showed that this was true, but ΔG and ΔH not only came closer together on cooling, but they approached each other asymptotically. On the basis of the data, Nernst also postulated that $\Delta S \to 0$ as $T \to 0$. His statement of the third law, dating from 1906, can be written as

> **Nernst's statement of the third law**
> Near absolute zero, all reactions in a system in internal equilibrium take place with no change in entropy.

Fig. 18.2 W. Nernst

Max Planck (1858–1947) (Fig. 18.3) added more meat to the bones of the statement by making a further hypothesis in 1911, namely that:

> **Planck's statement of the third law**
> The entropy of all systems in internal equilibrium is the same at absolute zero, and may be taken to be zero.

Fig. 18.3 M. Planck

Planck actually made his statement only about perfect crystals. However, it is believed to be true about any system, as long as it is in internal equilibrium (i.e., that all parts of a system are in equilibrium with each other). There are a number of systems, such as ^{4}He and ^{3}He, which are liquids even at very low temperature. Electrons in a metal can be treated as a gas all the way down to $T = 0$. The third law applies to all of these systems. However, note that the systems have to be in *internal equilibrium* for the third law to apply. An example of a system not in equilibrium is a glass, which has frozen-in disorder. For a solid, the lowest-energy phase is the perfect crystal, but the glass phase is higher in energy and is unstable. The glass phase will eventually relax back to the perfect crystalline phase but it may take many centuries to do this, and possibly a time greater than the age of the Universe.[1]

[1] The idea that the glass in the windows of very old cathedrals has flowed over the centuries is popularly believed, but has been debunked, see "Do Cathedral Glasses Flow?", E. D. Zanotto, *Am. J. Phys.* **66**, 392 (1998), see also E. D. Zanotto and P. K. Gupta, *Am. J. Phys.* **67**, 260 (1999).

Planck's choice of zero for the entropy was further motivated by the development of statistical mechanics, a subject we will tackle later in this book. It suffices to say here that the statistical definition of entropy, presented in eqn 14.36 ($S = k_B \ln \Omega$), implies that zero entropy is equivalent to $\Omega = 1$. Thus at absolute zero, when a system finds its ground state, the entropy being equal to zero implies that this ground state is non-degenerate.

At this point, we can raise a potential objection to the third law in Planck's form. Consider a perfect crystal composed of N spinless atoms. We are told by the third law that its entropy is zero. However, let us further suppose that each atom has at its centre a nucleus with angular momentum quantum number I. If no magnetic field is applied to this system, then we appear to have a contradiction. The degeneracy of the nuclear spin is $2I+1$ and if $I > 0$, this will not be equal to one. How can we reconcile this with zero entropy since the non-zero nuclear spin implies that the entropy S of this system should be $S = Nk_B \ln(2I+1)$, to however low a temperature we cool it?

The answer to this apparent contradiction is as follows: in a real system in internal equilibrium, the individual components of the system must be able to exchange energy with each other, i.e., to *interact* with each other. Nuclear spins actually feel a tiny, but non-zero, magnetic field due to the dipolar fields produced by surrounding nuclear spins, and this lifts the degeneracy. Another way of looking at this is to say that the interactions give rise to collective excitations of the nuclear spins. These collective excitations are nuclear spin waves, and the lowest-energy nuclear spin wave, corresponding to the longest-wavelength mode, will be non-degenerate. At sufficiently low temperatures (and this will be extremely low!) only that long-wavelength mode will be thermally occupied and the entropy of the nuclear spin system will be zero.

However, this example raises an important point. If we cool a crystal, we will extract energy from the lattice and its entropy will drop towards zero. However, the nuclear spins will still retain their entropy until cooled to a much lower temperature (reflecting the weaker interactions between nuclear spins compared with the bonds between atoms in the lattice). If we find a method of cooling the nuclei, there might still be some residual entropy associated with the individual nucleons. All these thermodynamic subsystems (the electrons, the nuclear spins, and the nucleons) are very weakly coupled to each other, but their entropies are additive. Francis Simon (1893–1956) (Fig. 18.4) in 1937 called these different subsystems "aspects" and formulated the third law as follows:

Fig. 18.4 F. E. Simon

Simon's statement of the third law
The contribution to the entropy of a system by each aspect of the system which is in internal thermodynamic equilibrium tends to zero as $T \to 0$.

Simon's statement is convenient because it allows us to focus on a particular aspect of interest, knowing that its entropy will tend to zero as T approaches 0, while ignoring the aspects that we don't care about and which might not lose their entropy until much closer to $T = 0$.

18.2 Consequences of the third law

Having provided various statements of the third law, it is time to examine some of its consequences.

- *Heat capacities tend to zero as $T \to 0$*
  This consequence is easy to prove. Any heat capacity C given by

$$C = T\left(\frac{\partial S}{\partial T}\right) = \left(\frac{\partial S}{\partial \ln T}\right) \to 0, \qquad (18.5)$$

because as $T \to 0$, $\ln T \to -\infty$ and $S \to 0$. Hence $C \to 0$.
Note that this result disagrees with the classical prediction of $C = R/2$ per mole per degree of freedom. (We note for future reference that this observation emphasizes the fact that the equipartition theorem, to be presented in Chapter 19, is a high temperature theory and fails at low temperature.)

- *Thermal expansion stops*
  Since $S \to 0$ as $T \to 0$, we have for example that

$$\left(\frac{\partial S}{\partial p}\right)_T \to 0 \tag{18.6}$$

as $T \to 0$, but by a Maxwell relation, this implies that

$$\frac{1}{V}\left(\frac{\partial V}{\partial T}\right)_p \to 0 \tag{18.7}$$

and hence the isobaric expansivity $\beta_p \to 0$.

- *No gases remain ideal as $T \to 0$*
  The ideal monatomic gas has served us well in this book as a simple model that allows us to obtain tractable results. One of these results is eqn 11.25, which states that for an ideal gas, $C_p - C_V = R$ per mole. However, as $T \to 0$, both C_p and C_V tend to zero, and this equation cannot be satisfied. Moreover, we expect that $C_V = 3R/2$ per mole, and as we have seen, this also does not work down to absolute zero. Yet another nail in the coffin of the ideal gas is the expression for its entropy given in eqn 16.79 ($S = C_V \ln T + R \ln V + \text{constant}$). As $T \to 0$, this equation yields $S \to -\infty$, which is as far from zero as you can get!
  Thus we see that the third law forces us to abandon the ideal gas model when thinking about gases at low temperature. Of course, it is at low temperature that the weak interactions between gas molecules (blissfully neglected so far since we have modelled gas molecules as independent entities) become more important. More sophisticated models of gases will be considered in Chapter 26.

- *Curie's law breaks down*
  Curie's law states that the susceptibility χ is proportional to $1/T$ and hence $\chi \to \infty$ as $T \to 0$. However, the third law implies that $(\partial S/\partial B)_T \to 0$ and hence

$$\left(\frac{\partial S}{\partial B}\right)_T = \left(\frac{\partial m}{\partial T}\right)_B = \frac{VB}{\mu_0}\left(\frac{\partial \chi}{\partial T}\right)_B \tag{18.8}$$

must tend to zero. Thus $\left(\frac{\partial \chi}{\partial T}\right) \to 0$, in disagreement with Curie's law. Why does it break down? You may begin to see a theme developing: it is interactions again! Curie's law is derived by considering magnetic moments to be entirely independent, in which case their properties can be determined by considering only the balance between the applied field (driving the moments to align) and temperature (driving the moments to randomize). The susceptibility measures their infinitesimal response to an infinitesimal applied field; this becomes infinite when the thermal fluctuations are removed at $T = 0$. However, if interactions between the magnetic moments are switched on, then an applied field will have much less of an effect because the magnetic moments will already be driven into some partially ordered state by each other.

There is a basic underlying message here: the microscopic parts of a system can behave independently at high temperature, where the thermal energy $k_\mathrm{B}T$ is much larger than any interaction energy. At low temperature, these interactions become important and all notions of independence break down. To paraphrase (badly) the poet John Donne:

> No man is an island, and especially not as $T \to 0$.

- *Unattainability of absolute zero*
  The final point can almost be elevated to the status of another statement of the third law:

  > It is impossible to cool to $T = 0$ in a finite number of steps.

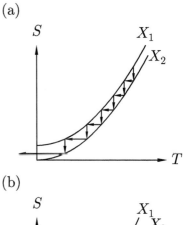

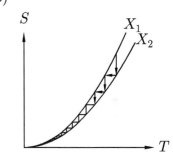

Fig. 18.5 The entropy as a function of temperature for two different values of a parameter X. Cooling is produced by isothermal increases in X (i.e., $X_1 \to X_2$) and adiabatic decreases in X (i.e., $X_2 \to X_1$). (a) If S does not go to 0 as $T \to 0$ it is possible to cool to absolute zero in a finite number of steps. (b) If the third law is obeyed, then it is impossible to cool to absolute zero in a finite number of steps.

This is messy to prove rigorously, but we can justify the argument by reference to Fig. 18.5, which shows plots of S against T for different values of a parameter X (which might be magnetic field, for example). Cooling is produced by isothermal increases in X and adiabatic decreases in X. If the third law did not hold, it would be possible to proceed according to Fig. 18.5(a) and cool all the way to absolute zero. However, because of the third law, the situation is as in Fig. 18.5(b) and the number of steps needed to get to absolute zero becomes infinite.

Before concluding this chapter, we make one remark concerning Carnot engines. Consider a Carnot engine, operating between reservoirs with temperatures T_ℓ and T_h, having an efficiency $\eta = 1 - (T_\ell/T_\mathrm{h})$ (eqn 13.10). If $T_\ell \to 0$, the efficiency η tends to 1. If you operated this Carnot engine, you would then get perfect conversion of heat into work, in violation of Kelvin's statement of the second law of thermodynamics. It seems at first sight that the unattainability of absolute zero (a version of the third law) is a simple consequence of the second law. However, there are difficulties in considering a Carnot engine operating between two reservoirs, one of which is at absolute zero. It is not clear how you can perform an isothermal process at absolute zero, because once a system is at absolute zero it is not possible to get it to change its thermodynamical state without warming it. Thus it is generally believed that the third law is indeed a separate postulate which is independent of the second law. The third law points to the fact that many of our "simple" thermodynamic models, such as the ideal gas equation and Curie's law of paramagnets, need substantial modification if they are to give correct predictions as

$T \to 0$. It is therefore opportune to consider more sophisticated models based on the microscopic properties of real systems, and that brings us to *statistical mechanics*, the subject of the next part of this book.

Chapter summary

- The third law of thermodynamics can be stated in various ways:
- *Nernst*: Near absolute zero, all reactions in a system in internal equilibrium take place with no change in entropy.
- *Planck*: The entropy of all systems in internal equilibrium is the same at absolute zero, and may be taken to be zero.
- *Simon*: The contribution to the entropy of a system by each aspect of the system which is in internal thermodynamic equilibrium tends to zero as $T \to 0$.
- *Unattainability of $T = 0$*: it is impossible to cool to $T = 0$ in a finite number of steps.
- The third law implies that heat capacities and thermal expansivities tend to zero as $T \to 0$.
- Interactions between the constituents of a system become important as $T \to 0$, and this leads to the breakdown of the concept of an ideal gas and also the breakdown of Curie's law.

Exercises

(18.1) Summarize the main consequences of the third law of thermodynamics. Explain how it casts a shadow of doubt on some of the conclusions from various thermodynamic models.

(18.2) Recall from eqn 16.26 that

$$H = G - T \left(\frac{\partial G}{\partial T} \right)_p. \qquad (18.9)$$

Hence show that

$$\Delta G - \Delta H = T \left(\frac{\partial \Delta G}{\partial T} \right)_p, \qquad (18.10)$$

and explain what happens to these terms as the temperature $T \to 0$.

Part VII

Statistical mechanics

In this part we introduce the subject of *statistical mechanics*. This is a thermodynamic theory in which account is taken of the microscopic properties of individual atoms or molecules analysed in a statistical fashion. Statistical mechanics allows macroscopic properties to be calculated from the statistical distribution of the microscopic behaviour of individual atoms and molecules. This part is structured as follows:

- In Chapter 19, we present the *equipartition theorem*, a principle that states that the internal energy of a classical system composed of a large number of particles in thermal equilibrium will distribute itself evenly among each of the quadratic degrees of freedom accessible to the particles of the system.
- In Chapter 20 we introduce the *partition function*, which encodes all the information concerning the states of a system and their thermal occupation. Having the partition function allows you to calculate all the thermodynamic properties of the system.
- In Chapter 21 we calculate the partition function for an ideal gas and use this to define the *quantum concentration*. We show how the *indistinguishability* of molecules affects the statistical properties and has thermodynamic consequences.
- In Chapter 22 we extend our results on partition functions to systems in which the number of particles can vary. This allows us to define the *chemical potential* and introduce the *grand partition function*.
- In Chapter 23, we consider the statistical mechanics of light, which is quantized as *photons*, introducing *black-body radiation*, *radiation pressure*, and the *cosmic microwave background*.
- In Chapter 24, we discuss the analogous behaviour of lattice vibrations, quantized as *phonons*, and introduce the *Einstein model* and *Debye model* of the thermal properties of solids.

19 Equipartition of energy

- 19.1 Equipartition theorem 210
- 19.2 Applications 213
- 19.3 Assumptions made 215
- 19.4 Brownian motion 217
- Chapter summary 218
- Exercises 218

Before introducing the partition function in Chapter 20, which will allow us to calculate many different properties of thermodynamic systems on the basis of their microscopic energy levels (which can be deduced using quantum mechanics), we devote this chapter to the equipartition theorem. This theorem provides a simple, classical theory of thermal systems. It gives remarkably good answers, but only at high temperature, where the details of quantized energy levels can be safely ignored. We will motivate and prove this theorem in the following section, and then apply it to various physical situations in Section 19.2, demonstrating that it provides a rapid and straightforward method for deriving heat capacities. Finally, in Section 19.3, we will critically examine the assumptions that we have made in the derivation of the equipartition theorem.

19.1 Equipartition theorem

Very often in physics one is faced with an energy dependence that is quadratic in some variable.[1] An example would be the kinetic energy E_{KE} of a particle with mass m and velocity v, which is given by

$$E_{\mathrm{KE}} = \frac{1}{2}mv^2. \tag{19.1}$$

Another example would be the potential energy E_{PE} of a mass suspended at one end of a spring with spring constant k and displaced by a distance x from its equilibrium point (see Fig. 19.1). This is given by

$$E_{\mathrm{PE}} = \frac{1}{2}kx^2. \tag{19.2}$$

In fact, the total energy E of a moving mass on the end of a spring is given by the sum of these two terms, so that

$$E = E_{\mathrm{KE}} + E_{\mathrm{PE}} = \frac{1}{2}mv^2 + \frac{1}{2}kx^2, \tag{19.3}$$

and, as the mass undergoes simple harmonic motion, energy is exchanged between E_{KE} and E_{PE}, while the total energy remains fixed.

Let us suppose that a system whose energy has a quadratic dependence on some variable is allowed to interact with a heat bath. It is then able to borrow energy occasionally from its environment, or even give it back into the environment. What mean thermal energy would it have? The

[1] We will show later in Section 19.3 that this quadratic dependence is very common; most potential wells are approximately quadratic near the bottom of the well.

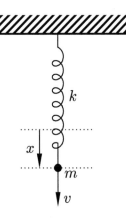

Fig. 19.1 A mass m suspended on a spring with spring constant k. The mass is displaced by a distance x from its equilibrium or "rest" position.

thermal energy would be stored as kinetic or potential energy, so if a mass on a spring is allowed to come into thermal equilibrium with its environment, one could in principle take a very big magnifying glass and see the mass on a spring jiggling *around all by itself* owing to such thermal vibrations. How big would such vibrations be? The calculation is quite straightforward.

Let the energy E of a particular system be given by

$$E = \alpha x^2, \tag{19.4}$$

where α is some positive constant and x is some variable (see Fig. 19.2). Let us also assume that x could in principle take any value with equal probability. The probability $P(x)$ of the system having a particular energy αx^2 is proportional to the Boltzmann factor $e^{-\beta \alpha x^2}$ (see eqn 4.13), so that after normalizing, we have

$$P(x) = \frac{e^{-\beta \alpha x^2}}{\int_{-\infty}^{\infty} e^{-\beta \alpha x^2}\, dx}, \tag{19.5}$$

and the mean energy is

$$\begin{aligned}
\langle E \rangle &= \int_{-\infty}^{\infty} E\, P(x)\, dx \\
&= \frac{\int_{-\infty}^{\infty} \alpha x^2 e^{-\beta \alpha x^2}\, dx}{\int_{-\infty}^{\infty} e^{-\beta \alpha x^2}\, dx} \\
&= \frac{1}{2\beta} \\
&= \frac{1}{2} k_B T. \tag{19.6}
\end{aligned}$$

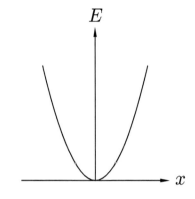

Fig. 19.2 The energy E of a system is $E = \alpha x^2$.

This is a really remarkable result. It is independent of the constant α and gives a mean energy that is proportional to temperature. The theorem can be extended straightforwardly to the energy being the sum of n quadratic terms, as shown in the following example.

Example 19.1

Assume that the energy E of a system can be given by the sum of n independent quadratic terms, so that

$$E = \sum_{i=1}^{n} \alpha_i x_i^2, \tag{19.7}$$

where α_i are constants and x_i are some variables. Assume also that each x_i could in principle take any value with equal probability. Calculate the mean energy.

Solution:

The mean energy $\langle E \rangle$ is given by

$$\langle E \rangle = \int_{-\infty}^{\infty} \cdots \int_{-\infty}^{\infty} E\, P(x_1, x_2, \ldots x_n)\, dx_1\, dx_2 \cdots dx_n. \tag{19.8}$$

This now looks quite complicated when we substitute in the probability as follows

$$\langle E \rangle = \frac{\int_{-\infty}^{\infty} \cdots \int_{-\infty}^{\infty} \left(\sum_{i=1}^{n} \alpha_i x_i^2\right) \exp\left(-\beta \sum_{j=1}^{n} \alpha_j x_j^2\right) \mathrm{d}x_1 \mathrm{d}x_2 \cdots \mathrm{d}x_n}{\int_{-\infty}^{\infty} \cdots \int_{-\infty}^{\infty} \exp\left(-\beta \sum_{j=1}^{n} \alpha_j x_j^2\right) \mathrm{d}x_1 \mathrm{d}x_2 \cdots \mathrm{d}x_n}, \tag{19.9}$$

where i and j have been used to distinguish different sums. This expression can be simplified by recognizing that it is the sum of n similar terms (write out the sums to convince yourself):

$$\langle E \rangle = \sum_{i=1}^{n} \frac{\int_{-\infty}^{\infty} \cdots \int_{-\infty}^{\infty} \alpha_i x_i^2 \exp\left(-\beta \sum_{j=1}^{n} \alpha_j x_j^2\right) \mathrm{d}x_1 \mathrm{d}x_2 \cdots \mathrm{d}x_n}{\int_{-\infty}^{\infty} \cdots \int_{-\infty}^{\infty} \exp\left(-\beta \sum_{j=1}^{n} \alpha_j x_j^2\right) \mathrm{d}x_1 \mathrm{d}x_2 \cdots \mathrm{d}x_n}, \tag{19.10}$$

and then all but one integral cancels between the numerator and denominator of each term, so that

$$\langle E \rangle = \sum_{i=1}^{n} \frac{\int_{-\infty}^{\infty} \alpha_i x_i^2 \exp\left(-\beta \alpha_i x_i^2\right) \mathrm{d}x_i}{\int_{-\infty}^{\infty} \exp\left(-\beta \alpha_i x_i^2\right) \mathrm{d}x_i}. \tag{19.11}$$

Now each term in this sum is the same as the one treated above in eqn 19.6. Hence

$$\langle E \rangle = \sum_{i=1}^{n} \alpha_i \langle x_i^2 \rangle = \sum_{i=1}^{n} \frac{1}{2} k_\mathrm{B} T$$
$$= \frac{n}{2} k_\mathrm{B} T. \tag{19.12}$$

What is often called a vibrational normal mode of a system contains *two* quadratic modes, one for the potential energy and the other for the kinetic energy. Thus to remove any ambiguity we will refer in this chapter to *quadratic modes*.

Each quadratic energy dependence of the system is called a **mode** of the system (or sometimes a **degree of freedom** of the system). The spring, our example at the beginning of this chapter, has two such modes. The result of the example above shows that each mode of the system contributes an amount of energy equal to $\frac{1}{2}k_\mathrm{B}T$ to the total mean energy of the system. This result is the basis of the **equipartition theorem**, which we state as follows:

> **Equipartition theorem**
> If the energy of a classical system is the sum of n quadratic modes, and that system is in contact with a heat reservoir at temperature T, the mean energy of the system is given by $n \times \frac{1}{2}k_\mathrm{B}T$.

The equipartition theorem expresses the fact that energy is "equally partitioned" between all the separate modes of the system, each mode having a mean energy of precisely $\frac{1}{2}k_\mathrm{B}T$.

Example 19.2

We return to our example of a mass on a spring, whose energy is given by the sum of two quadratic energy modes (see eqn 19.3). The equipartition theorem then implies that the mean energy is given by

$$2 \times \frac{1}{2} k_B T = k_B T. \tag{19.13}$$

How big is this energy? At room temperature, $k_B T \approx 4 \times 10^{-21}$ J $\approx$ 0.025 eV, which is a tiny energy. This energy isn't going to set a 10 kg mass on a stiff spring vibrating very much! However, the extraordinary thing about the equipartition theorem is that the result holds *independently of the size of the system*, so that $k_B T = 0.025$ eV is also the mean energy of an atom on the end of a chemical bond (which can be modelled as a spring) at room temperature. For an atom, $k_B T = 0.025$ eV goes a very long way and this explains why atoms in molecules jiggle around a lot at room temperature. We will explore this in more detail below.

19.2 Applications

We now consider four applications of the equipartition theorem.

19.2.1 Translational motion in a monatomic gas

The energy of each atom in a monatomic gas is given by

$$E = \frac{1}{2} m v_x^2 + \frac{1}{2} m v_y^2 + \frac{1}{2} m v_z^2, \tag{19.14}$$

where $\boldsymbol{v} = (v_x, v_y, v_z)$ is the velocity of the atom (see Fig. 19.3). This energy is the sum of three independent quadratic modes, and thus the equipartition theorem gives the mean energy as

$$\langle E \rangle = 3 \times \frac{1}{2} k_B T = \frac{3}{2} k_B T. \tag{19.15}$$

This is in agreement with our earlier derivation of the mean kinetic energy of a gas (see eqn 5.17).

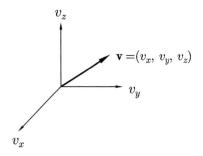

Fig. 19.3 The velocity of a molecule in a gas.

19.2.2 Rotational motion in a diatomic gas

In a diatomic gas, there is an additional possible energy source to consider, namely that of rotational kinetic energy. This adds two terms to the energy

$$\frac{L_1^2}{2I_1} + \frac{L_2^2}{2I_2}, \tag{19.16}$$

where L_1 and L_2 are the angular momenta along the two principal directions shown in Fig. 19.4 and I_1 and I_2 are the corresponding moments of inertia. We do not need to worry about the direction along the diatomic molecule's bond, the axis labelled "3" in Fig. 19.4. (This is because the moment of inertia in this direction is very small (so that the corresponding rotational kinetic energy is very large), so rotational modes in this direction cannot be excited at ordinary temperature; such rotational modes are connected with the individual molecular electronic levels and we will therefore ignore them.)

The total energy is thus the sum of five terms, three due to translational kinetic energy and two due to rotational kinetic energy

$$E = \frac{1}{2}mv_x^2 + \frac{1}{2}mv_y^2 + \frac{1}{2}mv_z^2 + \frac{L_1^2}{2I_1} + \frac{L_2^2}{2I_2}, \quad (19.17)$$

and all of these energy modes are independent of one another. Using the equipartition theorem, we can immediately write down the mean energy as

$$\langle E \rangle = 5 \times \frac{1}{2}k_\mathrm{B}T = \frac{5}{2}k_\mathrm{B}T. \quad (19.18)$$

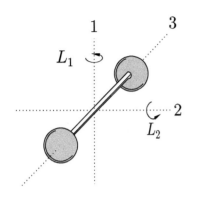

Fig. 19.4 Rotational motion in a diatomic gas.

19.2.3 Vibrational motion in a diatomic gas

If we also include the vibrational motion of the bond linking the two atoms in our diatomic molecule, there are two additional modes to include. The intramolecular bond can be modelled as a spring (see Fig. 19.5), so that the two extra energy terms are the kinetic energy due to *relative* motion of the two atoms and the potential energy in the bond (let us suppose it has spring constant k). Writing the positions of the two atoms as $\boldsymbol{r}_1$ and $\boldsymbol{r}_2$ with respect to some fixed origin, the energy of the molecule can be written

$$E = \frac{1}{2}m(v_x^2 + v_y^2 + v_z^2) + \frac{L_1^2}{2I_1} + \frac{L_2^2}{2I_2} + \frac{1}{2}\mu(\dot{\boldsymbol{r}}_1 - \dot{\boldsymbol{r}}_2)^2 + \frac{1}{2}k(|\boldsymbol{r}_1 - \boldsymbol{r}_2| - \ell_0)^2, \quad (19.19)$$

where $\mu = m_1 m_2/(m_1 + m_2)$ is the reduced mass[2] of the system and ℓ_0 the equilibrium molecular bond length. The equipartition theorem just cares about the number of modes in the system, so the mean energy is simply

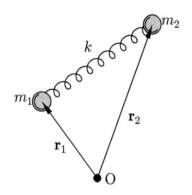

Fig. 19.5 A diatomic molecule can be modelled as two masses connected by a spring.

[2] See Appendix G.

$$\langle E \rangle = 7 \times \frac{1}{2}k_\mathrm{B}T = \frac{7}{2}k_\mathrm{B}T. \quad (19.20)$$

The heat capacity of the systems described above can be obtained by differentiating the energy with respect to temperature. The mean energy is given by

$$\langle E \rangle = \frac{f}{2}k_\mathrm{B}T, \quad (19.21)$$

where f is the number of quadratic modes. This equation implies that

$$C_V \text{ per mole} = \frac{f}{2}R, \quad (19.22)$$

and using eqn 11.27 we have

$$C_p \text{ per mole} = \left(\frac{f}{2} + 1\right) R, \tag{19.23}$$

from which we may derive

$$\gamma = \frac{C_p}{C_V} = \frac{(\frac{f}{2}+1)R}{\frac{f}{2}R} = 1 + \frac{2}{f}. \tag{19.24}$$

We can summarize our results for the heat capacity of gases, per atom or molecule, as follows:

Gas	Modes	f	$\langle E \rangle$	γ
Monatomic	Translational only	3	$\frac{3}{2}k_\text{B}T$	$\frac{5}{3}$
Diatomic	Translational and rotational	5	$\frac{5}{2}k_\text{B}T$	$\frac{7}{5}$
Diatomic	Translational, rotational, and vibrational	7	$\frac{7}{2}k_\text{B}T$	$\frac{9}{7}$

19.2.4 The heat capacity of a solid

In a solid, the atoms are held rigidly in the lattice and there is no possibility of translational motion. However, the atoms can vibrate about their mean positions. Consider a cubic solid (Fig. 19.6) in which each atom is connected by springs (chemical bonds) to six neighbours (one above, one below, one in front, one behind, one to the right, one to the left). Since each spring joins two atoms, then if there are N atoms in the solid, there are $3N$ springs (neglecting the surface of the solid, a reasonable approximation if N is large). Each spring has two quadratic modes of energy (one kinetic, one potential) and hence a mean thermal energy equal to $2 \times \frac{1}{2}k_\text{B}T = k_\text{B}T$. Hence the mean energy of the solid is

$$\langle E \rangle = 3Nk_\text{B}T, \tag{19.25}$$

and the heat capacity is $\partial \langle E \rangle / \partial T = 3Nk_\text{B}$. Because $R = N_A k_\text{B}$, the molar heat capacity of a solid is then expected to be $3N_A k_\text{B} = 3R$. This result agrees quite well with experiment and is known as the Dulong–Petit rule (see Section 24.1).

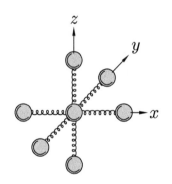

Fig. 19.6 In a cubic solid, each atom is connected by chemical bonds, modelled as springs, to six nearest neighbours, two along each of the three Cartesian axes. Each spring is shared between two atoms.

19.3 Assumptions made

The equipartition theorem seems to be an extremely powerful tool for evaluating thermal energies of systems. However, it does have some limitations, and to discover what these are, it is worth thinking about the assumptions we have made in deriving it.

- We have assumed that the parameter for which we have taken the energy to be quadratic can take any possible value. In the derivation, the variables x_i could be integrated continuously from $-\infty$ to ∞. However, quantum mechanics insists that certain quantities can only take particular "quantized" values. For example, the problem of a mass on a spring is shown by quantum mechanics to have an energy spectrum that is quantized into levels given by $(n + \frac{1}{2})\hbar\omega$. When the thermal energy $k_B T$ is of the same order, or lower than, $\hbar\omega$, the approximation made by ignoring the quantized nature of this energy spectrum is going to be a very bad one. However, when $k_B T \gg \hbar\omega$, the quantized nature of the energy spectrum is going to be largely irrelevant, in much the same way that you don't notice that the different shades of grey in a newspaper photograph are actually made up of lots of little dots if you don't look closely. Thus we come to an important conclusion:

> The equipartition theorem is generally valid only at *high temperature*, so that the thermal energy is larger than the energy gap between quantized energy levels. Results based on the equipartition theorem should emerge as the high-temperature limit of more detailed theories.

- We have assumed throughout that modes are quadratic. Is that always valid? To give a concrete example, imagine that an atom moves with coordinate x in a potential well given by $V(x)$, which is a function that might be more complicated than a quadratic (see, for example, Fig. 19.7). At absolute zero, the atom finds a potential minimum at say x_0 (so that, for the usual reasons, $\partial V/\partial x = 0$ and $\partial^2 V/\partial x^2 > 0$ at $x = x_0$). At temperature $T > 0$, the atom can explore regions away from x_0 by borrowing energy of order $k_B T$ from its environment. Near x_0, the potential $V(x)$ can be expanded[3] as

$$V(x) = V(x_0) + \left(\frac{\partial V}{\partial x}\right)_{x_0}(x - x_0) + \frac{1}{2}\left(\frac{\partial^2 V}{\partial x^2}\right)_{x_0}(x - x_0)^2 + \cdots, \tag{19.26}$$

so that using $\left(\frac{\partial V}{\partial x}\right)_{x_0} = 0$, we find that the potential energy is

$$V(x) = \text{constant} + \frac{1}{2}\left(\frac{\partial^2 V}{\partial x^2}\right)_{x_0}(x - x_0)^2 + \cdots, \tag{19.27}$$

which is a quadratic again. This demonstrates that the bottom of almost all potential wells tends to be approximately quadratic (this is known as the **harmonic approximation**).[4] If the temperature gets too high, the system will be able to access positions far away from x_0 and the approximation of ignoring the higher order (cubic, quartic, etc.) terms (known as the **anharmonic terms**) in the Taylor expansion may become important.

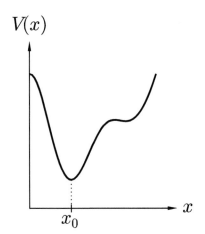

Fig. 19.7 $V(x)$ is a function that is more complicated than a quadratic but has a minimum at $x = x_0$.

[3]Using a Taylor expansion; see Appendix B.

[4]The argument that the bottom of almost all potential wells tends to be approximately quadratic could fail if $(\partial^2 V/\partial x^2)_{x_0}$ turned out to be zero. This would happen if, for example, $V(x) = \alpha(x - x_0)^4$.

However, we have just said that the equipartition theorem is only valid at high temperature. Thus we see that the temperature must be high enough that we can safely ignore the quantum nature of the energy spectrum, but not so high that we invalidate the approximation of treating the relevant potential wells as perfectly quadratic. Fortunately there is plenty of room between these two extremes.

19.4 Brownian motion

We close this chapter with one example in which the effect of the equipartition of energy is encountered.

Example 19.3

Brownian motion

In 1827, Robert Brown[5] used a microscope to observe pollen grains jiggling about in water. He was not the first to make such an observation (any small particles suspended in a fluid will do the same, and are very apparent when looking down a microscope), but this effect has come to be known as **Brownian motion**.

[5]Robert Brown (1773–1858).

The motion is very irregular, consisting of translations and rotations, with grains moving independently, even when moving close to each other. The motion is found to be more active the smaller the particles. The motion is also found to be more active the less viscous the fluid. Brown was able to discount a "vital" explanation of the effect, i.e., that the pollen grains were somehow "alive", but he was not able to give a correct explanation. Something resembling a modern theory of Brownian motion was proposed by Christian Wiener[6] in 1863, though the major breakthrough was made by Einstein in 1905.

[6]Christian Wiener (1826–1896).

We will postpone a full discussion of Brownian motion until Chapter 33, but using the equipartition theorem, the origin of the effect can be understood in outline. Each pollen grain (of mass m) is free to move translationally and so has mean kinetic energy $\frac{1}{2}m\langle v^2 \rangle = \frac{3}{2}k_\mathrm{B}T$. This energy is very small, as we have seen, but leads to a measurable amplitude of vibration for a small pollen grain. The amplitude of vibration is greater for smaller pollen grains because a mean kinetic energy of $\frac{3}{2}k_\mathrm{B}T$ gives more mean square velocity $\langle v^2 \rangle$ to less massive grains. The thermally excited vibrations are resisted by viscous damping, so the motion is expected to be more pronounced in less viscous fluids.

Chapter summary

- The equipartition theorem states that if the energy of a system is the sum of n quadratic modes, and that the system is in contact with a heat reservoir of temperature T, the mean energy of the system is given by $n \times \tfrac{1}{2}k_\text{B}T$.
- The equipartition theorem is a high-temperature result and gives incorrect predictions at low temperature, where the discrete nature of the energy spectrum cannot be ignored.

Exercises

(19.1) What is the mean kinetic energy in eV at room temperature of a gaseous (a) He atom, (b) Xe atom, (c) Ar atom, and (d) Kr atom. [Hint: do you have to make four separate calculations?]

(19.2) Comment on the following values of molar heat capacity in $\text{J K}^{-1}\,\text{mol}^{-1}$, all measured at constant pressure at 298 K.

Al	24.35	Pb	26.44
Ar	20.79	Ne	20.79
Au	25.42	N_2	29.13
Cu	24.44	O_2	29.36
He	20.79	Ag	25.53
H_2	28.82	Xe	20.79
Fe	25.10	Zn	25.40

[Hint: express them in terms of R; which of the substances is a solid and which is gaseous?]

(19.3) A particle at position r is in a potential well $V(r)$ given by
$$V(r) = \frac{A}{r^n} - \frac{B}{r}, \qquad (19.28)$$
where A and B are positive constants and $n > 2$. Show that the bottom of the well is approximately quadratic in r. Hence find the particle's mean thermal energy at temperature T above the bottom of the well assuming the validity of the equipartition theorem in this situation.

(19.4) In Example 19.1, show that
$$\langle x_i^2 \rangle = \frac{k_\text{B}T}{2\alpha_i}. \qquad (19.29)$$

(19.5) If the energy E of a system is not quadratic, but behaves like $E = \alpha |x|$ where $\alpha > 0$, show that the average energy is $\langle E \rangle = k_\text{B}T$.

(19.6) If the energy E of a system behaves like $E = \alpha|x|^n$, where $n = 1, 2, 3 \ldots$ and $\alpha > 0$, show that the average energy is $\langle E \rangle = \xi k_\text{B}T$, where ξ is a numerical constant.

(19.7) A simple pendulum with length ℓ makes an angle θ with the vertical, where $\theta \ll 1$. Show that it oscillates with a period given by $2\pi\sqrt{\ell/g}$. The pendulum is now placed at rest and allowed to come into equilibrium with its surroundings at temperature T. Derive an expression for $\langle \theta^2 \rangle$.

The partition function

20

The probability that a system is in some particular state α is *proportional* to the Boltzmann factor $e^{-\beta E_\alpha}$. We define the **partition function**[1] Z by a sum over all the states of the Boltzmann factors, so that

$$Z = \sum_\alpha e^{-\beta E_\alpha} \qquad (20.1)$$

where the sum is over all states of the system (each one labelled by α). The partition function Z contains all the information about the energies of the states of the system, and the fantastic thing about the partition function is that all thermodynamical quantities can be obtained from it. It behaves like a zipped-up and compressed version of all the properties of the system; once you have Z, you only have to know how to uncompress and unzip it to get functions of state like energy, entropy, Helmholtz function, or heat capacity to simply drop out. We can therefore reduce problem solving in statistical mechanics to two steps:

20.1 Writing down the partition function	220
20.2 Obtaining the functions of state	221
20.3 The big idea	228
20.4 Combining partition functions	228
Chapter summary	231
Exercises	232

[1]The partition function is given the symbol Z because the concept was first coined in German. *Zustandssumme* means "sum over states", which is exactly what Z is. The English name "partition function" reflects the way in which Z measures how energy is "partitioned" between states of the system.

Steps to solving statistical mechanics problems

(1) Write down the partition function Z.
(see Section 20.1)

(2) Go through some standard procedures to obtain the functions of state you want from Z.
(see Section 20.2)

We will outline these two steps in the sections that follow. Before we do that, let us pause to notice an important feature about the partition function.

- The zero of energy is always somewhat arbitrary: one can always choose to measure energy with respect to a different zero, since it is only energy differences that are important. Hence the partition function is defined up to an arbitary multiplicative constant. This seems somewhat strange, but it turns out that many physical quantities are related to the logarithm of the partition function and therefore these quantities are defined up to an additive constant (which might reflect, for example, the rest mass of particles). Other physical quantities, however, are determined by a differential of the logarithm of the partition function and therefore these quantities can be determined precisely.

220 The partition function

This point needs to be remembered whenever the partition function is obtained.

Everything in this chapter refers to what is known as the **single–particle partition function**. We are working out Z for *one* particle of matter, which may well be coupled to a reservoir of other particles, but our attention is only on that single particle of matter. We will defer discussion of how to treat aggregates of particles until the next two chapters. With that in mind, we are now ready to write down some partition functions.

20.1 Writing down the partition function

The partition function contains all the information we need to work out the thermodynamical properties of a system. In this section, we show how you can write down the partition function in the first place.

This procedure is not complicated! Writing down the partition function is nothing more than evaluating eqn 20.1 for different situations. We demonstrate this for a couple of commonly encountered and important examples.

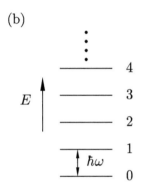

Fig. 20.1 Energy levels of (a) a two-level system and (b) a simple harmonic oscillator.

An alternative form of this result is found by multiplying top and bottom by $e^{\beta \frac{1}{2}\hbar\omega}$ to obtain the result

$$Z = \frac{1}{2\sinh(\beta\hbar\omega/2)}.$$

Example 20.1

(a) The two-level system (see Fig. 20.1(a))
Let the energy of a system be either $-\Delta/2$ or $\Delta/2$. Then

$$Z = \sum_\alpha e^{-\beta E_\alpha} = e^{\beta\Delta/2} + e^{-\beta\Delta/2} = 2\cosh\left(\frac{\beta\Delta}{2}\right), \quad (20.2)$$

where the final result follows from the definition of $\cosh x \equiv \frac{1}{2}(e^x + e^{-x})$ (see Appendix B).

(b) The simple harmonic oscillator (see Fig. 20.1(b))
The energy of the system is $(n+\frac{1}{2})\hbar\omega$ where $n = 0, 1, 2, \ldots$, and hence

$$Z = \sum_\alpha e^{-\beta E_\alpha} = \sum_{n=0}^\infty e^{-\beta(n+\frac{1}{2})\hbar\omega} = e^{-\beta\frac{1}{2}\hbar\omega} \sum_{n=0}^\infty e^{-n\beta\hbar\omega} = \frac{e^{-\frac{1}{2}\beta\hbar\omega}}{1 - e^{-\beta\hbar\omega}}, \quad (20.3)$$

where the sum is evaluated using the standard result for the sum of an infinite geometric progression, see Appendix B.

Two further, slightly more complicated, examples are the set of N equally spaced energy levels and the energy levels appropriate for the rotational states of a diatomic molecule.

Example 20.2

(c) The N-level system (see Fig. 20.2(c))
Let the energy levels of a system be $0, \hbar\omega, 2\hbar\omega, \ldots, (N-1)\hbar\omega$. Then

$$Z = \sum_\alpha e^{-\beta E_\alpha} = \sum_{j=0}^{N-1} e^{-j\beta\hbar\omega} = \frac{1 - e^{-N\beta\hbar\omega}}{1 - e^{-\beta\hbar\omega}}, \qquad (20.4)$$

where the sum is evaluated using the standard result for the sum of a finite geometric progress, see Appendix B.

(d) Rotational energy levels (see Fig. 20.2(d))
The rotational kinetic energy of a molecule with moment of inertia I is given by $\hat{J}^2/2I$, where $\hat{J}$ is the total angular momentum operator. The eigenvalues of $\hat{J}^2$ are given by $\hbar^2 J(J+1)$, where the angular momentum quantum number, J, takes the values $J = 0, 1, 2, \ldots$ The energy levels of this system are given by

$$E_J = \frac{\hbar^2}{2I} J(J+1), \qquad (20.5)$$

and have degeneracy $2J+1$. Hence the partition function is

$$Z = \sum_\alpha e^{-\beta E_\alpha} = \sum_{J=0}^{\infty} (2J+1) e^{-\beta\hbar^2 J(J+1)/2I}, \qquad (20.6)$$

where the factor $(2J+1)$ takes into account the degeneracy of each level.

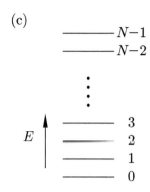

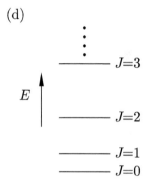

Fig. 20.2 Energy levels of (c) an N-level system and (d) a rotational system.

20.2 Obtaining the functions of state

Once Z has been written down, we can place it in our mathematical sausage machine (see Fig. 20.3), which processes it and spits out fully-fledged thermodynamical functions of state. We now outline the derivations of the components of our sausage machine so that you can derive all these functions of state for any given Z.

- **Internal energy U**
  The internal energy U is given by

$$U = \frac{\sum_i E_i e^{-\beta E_i}}{\sum_i e^{-\beta E_i}}. \qquad (20.7)$$

Now the denominator of this expression is the partition function $Z = \sum_i e^{-\beta E_i}$, but the numerator is simply

$$-\frac{dZ}{d\beta} = \sum_i E_i e^{-\beta E_i}. \qquad (20.8)$$

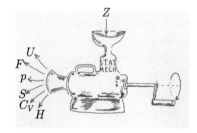

Fig. 20.3 Given Z, it takes only a turn of the handle on our "sausage machine" to produce other functions of state.

Thus $U = -(1/Z)(\mathrm{d}Z/\mathrm{d}\beta)$, or more simply,

$$U = -\frac{\mathrm{d}\ln Z}{\mathrm{d}\beta}. \tag{20.9}$$

This is a useful form since Z is normally expressed in terms of β. If you prefer things in terms of temperature T, then using $\beta = 1/k_\mathrm{B}T$ (and hence $\mathrm{d}/\mathrm{d}\beta = -k_\mathrm{B}T^2(\mathrm{d}/\mathrm{d}T)$) one obtains

$$U = k_\mathrm{B}T^2 \frac{\mathrm{d}\ln Z}{\mathrm{d}T}. \tag{20.10}$$

- *Entropy S*
Since the probability P_j is given by a Boltzmann factor divided by the partition function (so that the sum of the probabilities is one, as can be shown using eqn 20.1), we have $P_j = \mathrm{e}^{-\beta E_j}/Z$ and hence

$$\ln P_j = -\beta E_j - \ln Z. \tag{20.11}$$

Equation 14.48 therefore gives us an expression for the entropy as follows:

$$\begin{aligned} S &= -k_\mathrm{B} \sum_i P_i \ln P_i \\ &= k_\mathrm{B} \sum_i P_i (\beta E_i + \ln Z) \\ &= k_\mathrm{B}(\beta U + \ln Z), \end{aligned} \tag{20.12}$$

where we have used $U = \sum_i P_i E_i$ and $\sum_i P_i = 1$. Substituting the definition of β, namely $\beta = 1/k_\mathrm{B}T$, into this equation gives

$$S = \frac{U}{T} + k_\mathrm{B} \ln Z. \tag{20.13}$$

- *Helmholtz function F*
The Helmholtz function is defined via $F = U - TS$, so using eqn 20.13 we have that

$$F = -k_\mathrm{B} T \ln Z. \tag{20.14}$$

This can also be cast into the memorable form

$$\boxed{Z = \mathrm{e}^{-\beta F}.} \tag{20.15}$$

Once we have an expression for the Helmholtz function, a lot of things come out in the wash. For example, using eqn 16.19 we have that

$$S = -\left(\frac{\partial F}{\partial T}\right)_V = k_\mathrm{B} \ln Z + k_\mathrm{B} T \left(\frac{\partial \ln Z}{\partial T}\right)_V, \tag{20.16}$$

which, using eqn 20.10, is equivalent to eqn 20.13 above. This expression then leads to the heat capacity, via (recall eqn 16.65)

$$C_V = T\left(\frac{\partial S}{\partial T}\right)_V, \tag{20.17}$$

or one can use
$$C_V = \left(\frac{\partial U}{\partial T}\right)_V. \tag{20.18}$$

Either way,
$$C_V = k_B T \left[2\left(\frac{\partial \ln Z}{\partial T}\right)_V + T\left(\frac{\partial^2 \ln Z}{\partial T^2}\right)_V\right]. \tag{20.19}$$

- **Pressure p**
  The pressure can be obtained from F using eqn 16.20, so that
$$p = -\left(\frac{\partial F}{\partial V}\right)_T = k_B T \left(\frac{\partial \ln Z}{\partial V}\right)_T. \tag{20.20}$$

  Having got the pressure we can then write down the enthalpy and the Gibbs function.

- **Enthalpy H**
$$H = U + pV = k_B T \left[T\left(\frac{\partial \ln Z}{\partial T}\right)_V + V\left(\frac{\partial \ln Z}{\partial V}\right)_T\right]. \tag{20.21}$$

- **Gibbs function G**
$$G = F + pV = k_B T \left[-\ln Z + V\left(\frac{\partial \ln Z}{\partial V}\right)_T\right]. \tag{20.22}$$

Function of state		Statistical mechanical expression
U		$-\dfrac{d \ln Z}{d\beta}$
F		$-k_B T \ln Z$
S	$= -\left(\frac{\partial F}{\partial T}\right)_V = \frac{U-F}{T}$	$k_B \ln Z + k_B T \left(\frac{\partial \ln Z}{\partial T}\right)_V$
p	$= -\left(\frac{\partial F}{\partial V}\right)_T$	$k_B T \left(\frac{\partial \ln Z}{\partial V}\right)_T$
H	$= U + pV$	$k_B T \left[T\left(\frac{\partial \ln Z}{\partial T}\right)_V + V\left(\frac{\partial \ln Z}{\partial V}\right)_T\right]$
G	$= F + pV = H - TS$	$k_B T \left[-\ln Z + V\left(\frac{\partial \ln Z}{\partial V}\right)_T\right]$
C_V	$= \left(\frac{\partial U}{\partial T}\right)_V$	$k_B T \left[2\left(\frac{\partial \ln Z}{\partial T}\right)_V + T\left(\frac{\partial^2 \ln Z}{\partial T^2}\right)_V\right]$

Table 20.1 Thermodynamic quantities derived from the partition function Z.

These relations are summarized in Table 20.1. In practice, it is easiest to remember only the relations for U and F, since the others can be derived (using the relations shown in the left column of the table). Now that we have described how the process works, we can set about practising this for different partition functions.

Example 20.3

(a) Two-level system

The partition function for a two-level system (whose energy is either $-\Delta/2$ or $\Delta/2$) is given by eqn 20.2, which states that

$$Z = 2\cosh\left(\frac{\beta\Delta}{2}\right). \tag{20.23}$$

Having obtained Z, we can immediately compute the internal energy U and find that

$$U = -\frac{\mathrm{d}\ln Z}{\mathrm{d}\beta} = -\frac{\Delta}{2}\tanh\left(\frac{\beta\Delta}{2}\right). \tag{20.24}$$

Hence the heat capacity C_V is

$$C_V = \left(\frac{\partial U}{\partial T}\right)_V = k_\mathrm{B}\left(\frac{\beta\Delta}{2}\right)^2 \mathrm{sech}^2\left(\frac{\beta\Delta}{2}\right). \tag{20.25}$$

The Helmholtz function is

$$F = -k_\mathrm{B}T\ln Z = -k_\mathrm{B}T\ln\left[2\cosh\left(\frac{\beta\Delta}{2}\right)\right], \tag{20.26}$$

and hence the entropy is

$$S = \frac{U-F}{T} = -\frac{\Delta}{2T}\tanh\left(\frac{\beta\Delta}{2}\right) + k_\mathrm{B}\ln\left[2\cosh\left(\frac{\beta\Delta}{2}\right)\right]. \tag{20.27}$$

These results are plotted in Fig. 20.4(a). At low temperature, the system is in the lower level and the internal energy U is $-\Delta/2$. The entropy S is $k_\mathrm{B}\ln\Omega$, where Ω is the degeneracy and hence $\Omega = 1$ and so $S = k_\mathrm{B}\ln 1 = 0$. At high temperature, the two levels are each occupied with probability $\frac{1}{2}$, U therefore tends to 0 (which is halfway between $-\Delta/2$ and $\Delta/2$), and the entropy tends to $k_\mathrm{B}\ln 2$ as expected. The entropy rises as the temperature increases because it reflects the freedom of the system to exist in different states, and at high temperature the system has more freedom (in that it can exist in either of the two states). Conversely, cooling corresponds to a kind of "ordering" in which the system can only exist in one state (the lower), and this gives rise to a reduction in the entropy.

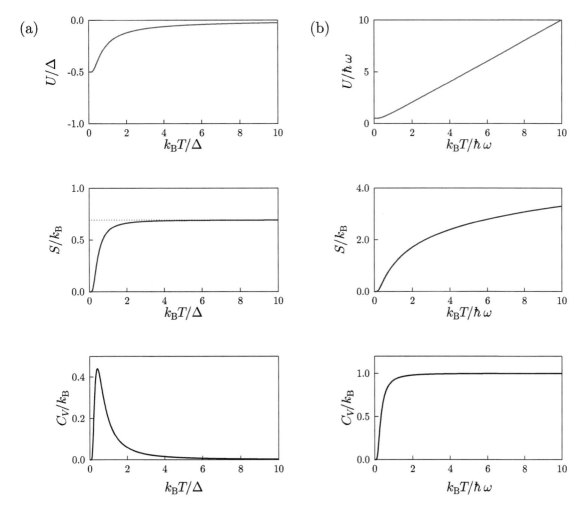

Fig. 20.4 The internal energy U, the entropy S, and the heat capacity C_V for (a) the two-state system (with energy levels $\pm\Delta/2$) and (b) the simple harmonic oscillator.

The heat capacity is very small both (i) at low temperature ($k_B T \ll \Delta$) and (ii) at very high temperature ($k_B T \gg \Delta$), because changes in temperature have no effect on the internal energy when (i) the temperature is so low that only the lower level is occupied and even a small change in temperature won't alter that, and (ii) the temperature is so high that both levels are occupied equally and a small change in temperature won't alter this. At very low temperature, it is hard to change the energy of the system because there is not enough energy to excite transitions from the ground state and therefore the system is "stuck". At very high temperature, it is hard to change the energy of the system because both states are equally occupied. In between, roughly around a temperature $T \approx \Delta/k_B$, the heat capacity rises to a maximum, known as a **Schottky anomaly**,[2] as shown in the lowest panel of Fig. 20.4(a).

[2] Walter Schottky (1886–1976).

This arises because at this temperature, it is possible to thermally excite transitions between the two states of the system. Note, however, that the Schottky anomaly is not a sharp peak, cusp, or spike, as might be associated with a phase transition (see Section 28.7), but is a smooth, fairly broad maximum.

(b) Simple harmonic oscillator

The partition function for the simple harmonic oscillator (from eqn 20.3) is

$$Z = \frac{e^{-\frac{1}{2}\beta\hbar\omega}}{1 - e^{-\beta\hbar\omega}}. \tag{20.28}$$

Hence (referring to Table 20.1), we find that U is given by

$$U = -\frac{d \ln Z}{d\beta} = \hbar\omega \left(\frac{1}{2} + \frac{1}{e^{\beta\hbar\omega} - 1} \right) \tag{20.29}$$

and hence that C_V is

$$C_V = \left(\frac{\partial U}{\partial T} \right)_V = k_B (\beta\hbar\omega)^2 \frac{e^{\beta\hbar\omega}}{(e^{\beta\hbar\omega} - 1)^2}. \tag{20.30}$$

At high temperature, $\beta\hbar\omega \ll 1$ and so $(e^{\beta\hbar\omega} - 1) \approx \beta\hbar\omega$ and $C_V \to k_B$ (the equipartition result). Similarly, $U \to \frac{\hbar\omega}{2} + k_B T \approx k_B T$. The Helmholtz function is (referring to Table 20.1)

$$F = -k_B T \ln Z = \frac{\hbar\omega}{2} + k_B T \ln(1 - e^{-\beta\hbar\omega}), \tag{20.31}$$

and hence the entropy is (referring again to Table 20.1)

$$S = \frac{U - F}{T} = k_B \left(\frac{\beta\hbar\omega}{e^{\beta\hbar\omega} - 1} - \ln(1 - e^{-\beta\hbar\omega}) \right). \tag{20.32}$$

These results are plotted in Fig. 20.4(b). At absolute zero, only the lowest level is occupied, so the internal energy is $\frac{1}{2}\hbar\omega$ and the entropy is $k_B \ln 1 = 0$. The heat capacity is also zero. As the temperature rises, more and more energy levels in the ladder can be occupied, and U rises without limit. The entropy also rises (and follows a dependence which is approximately $k_B \ln(k_B T/\hbar\omega)$ where $k_B T/\hbar\omega$ is approximately the number of occupied levels). Both functions carry on rising because the ladder of energy levels increases without limit. The heat capacity rises to a plateau at $C_V = k_B$, which is the equipartition result (see eqn 19.13).

The results for two further examples are plotted in Fig. 20.5 and are shown without derivation. The first is an N-level system and is shown in Fig. 20.5(a). At low temperature, the behaviour of the thermodynamic functions resembles that of the simple harmonic oscillator, but at higher temperature, U and S begin to saturate and C_V falls, because the system has a limited number of energy levels.

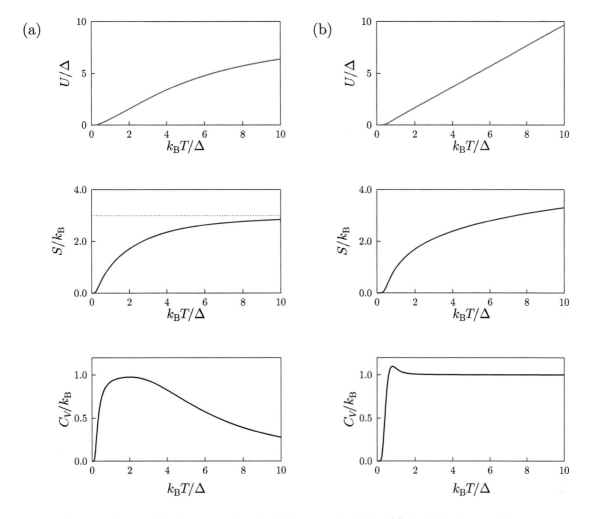

Fig. 20.5 The internal energy U, the entropy S, and the heat capacity C_V for (a) the N-level system (the simulation is shown for $N = 20$) and (b) the rotating diatomic molecule (in this case $\Delta = \hbar^2/2I$ where I is the moment of inertia).

Fig. 20.5(b) shows calculations for the rotating diatomic molecule. This resembles the simple harmonic oscillator at higher temperature (the heat capacity saturates at $C_V = k_B$) but differs at low temperature owing to the detailed difference in the structure of the energy levels. At high temperature, the heat capacity is given by the equipartition result (see eqn 19.13). This can be verified directly using the partition function, which, at high temperature, can be represented by the following integral:

$$Z = \sum_{J=0}^{\infty}(2J+1)e^{-\beta\Delta J(J+1)} \approx \int_0^{\infty}(2J+1)e^{-\beta\Delta J(J+1)}\,dJ, \quad (20.33)$$

where $\Delta = \hbar^2/2I$. Using

$$\frac{d}{dJ}e^{-\beta\Delta J(J+1)} = -(2J+1)\beta\Delta\, e^{-\beta\Delta J(J+1)}, \quad (20.34)$$

we have that

$$Z = -\left[\frac{1}{\beta\Delta}e^{-\beta\Delta J(J+1)}\right]_0^\infty = \frac{1}{\beta\Delta}. \quad (20.35)$$

This implies that $U = -\mathrm{d}\ln Z/\mathrm{d}\beta = 1/\beta = k_\mathrm{B}T$ and hence $C_V = (\mathrm{d}U/\mathrm{d}T)_V = k_\mathrm{B}$.

20.3 The big idea

The examples above illustrate the "big idea" of statistical mechanics: you *describe* a system by its energy levels E_α and *evaluate* its properties by following the prescription given by the two steps:

(1) Write down $Z = \sum_\alpha e^{-\beta E_\alpha}$.
(2) Evaluate various functions of state using the expressions given in Table 20.1.

And that's really all there is to it![3]

You can *understand* the results by comparing the energy $k_\mathrm{B}T$ with the spacings between energy levels.

- If $k_\mathrm{B}T$ is much less than the spacing between the lowest energy level and the first excited level then the system will sit in the lowest level.
- If there are a finite set of levels and $k_\mathrm{B}T$ is much larger than the energy spacing between the lowest and highest levels, then each energy level will be occupied with equal probability.
- If there are an infinite ladder of levels and $k_\mathrm{B}T$ is much larger than the energy spacing between adjacent levels, then the mean energy rises linearly with T and one obtains a result consistent with the equipartition theorem.

[3]Well, almost. The Schrödinger equation can only be solved for a few systems, and if you don't know the energy levels of your system, you can't write down Z. Fortunately, there are quite a number of systems for which you can solve the Schrödinger equation, some of which we are considering in this chapter, and they describe lots and lots of important physical systems, enough to keep us going in this book!

20.4 Combining partition functions

Consider the case when the energy E of a particular system depends on various independent contributions. For example, suppose it is a sum of two contributions a and b, so that the energy levels are given by $E_{i,j}$ where

$$E_{i,j} = E_i^{(a)} + E_j^{(b)}, \quad (20.36)$$

and where $E_i^{(a)}$ is the ith level due to contribution a and $E_j^{(b)}$ is the jth level due to contribution b, so the partition function Z is

$$Z = \sum_i \sum_j e^{-\beta(E_i^{(a)}+E_j^{(b)})} = \sum_i e^{-\beta E_i^{(a)}} \sum_j e^{-\beta E_j^{(b)}} = Z_a Z_b, \quad (20.37)$$

so that the partition functions of the independent contributions *multiply*. Hence also $\ln Z = \ln Z_a + \ln Z_b$, and the effect on functions of state which depend on $\ln Z$ is that the independent contributions *add*.

Example 20.4

(i) The partition function Z for N independent simple harmonic oscillators is given by
$$Z = Z_{\text{SHO}}^N, \tag{20.38}$$
where $Z_{\text{SHO}} = \mathrm{e}^{-\frac{1}{2}\beta\hbar\omega}/(1-\mathrm{e}^{-\beta\hbar\omega})$, from eqn 20.3, is the partition function for a single simple harmonic oscillator.

(ii) A diatomic molecule with *both* vibrational and rotational degrees of freedom has a partition function Z given by
$$Z = Z_{\text{vib}} Z_{\text{rot}}, \tag{20.39}$$
where Z_{vib} is the vibrational partition function $Z_{\text{vib}} = \mathrm{e}^{-\frac{1}{2}\beta\hbar\omega}/(1-\mathrm{e}^{-\beta\hbar\omega})$, from eqn 20.3, and Z_{rot} is the rotational partition function
$$Z_{\text{rot}} = \sum_\alpha \mathrm{e}^{-\beta E_\alpha} = \sum_{J=0}^\infty (2J+1)\mathrm{e}^{-\beta\hbar^2 J(J+1)/2I}. \tag{20.40}$$
from eqn 20.6. For a gas of diatomic molecules, we would also need a factor in the partition function corresponding to translational motion. We will derive this in the following chapter.

The final example of the chapter applies this to a simple magnetic system and allows us to derive Curie's law.[4]

[4] Curie's law was encountered in eqn 17.32.

Example 20.5

The spin-$\frac{1}{2}$ paramagnet

In quantum mechanics, a particle with spin angular momentum equal to $\frac{1}{2}$, placed in a magnetic field B along the z direction, can exist in one of two eigenstates:

- $|\uparrow\rangle$, with angular momentum parallel to the B field, and hence magnetic moment along z equal to $-\mu_\text{B}$ (costing an energy $+\mu_\text{B} B$).
- $|\downarrow\rangle$, with angular momentum antiparallel to the B field, and hence magnetic moment along z equal to $+\mu_\text{B}$ (costing an energy $-\mu_\text{B} B$).

Here $\mu_\text{B} = e\hbar/2m$ is the **Bohr magneton** and we have used the fact that energy $= -\boldsymbol{m} \cdot \boldsymbol{B}$, and also that for a negatively charged particle (the electron) the angular momentum is antiparallel to the magnetic moment.

One spin-$\frac{1}{2}$ particle thus behaves like a two-state system, with the two states having energy E given by: $E = \mu_\text{B} B$ and $E = -\mu_\text{B} B$. Therefore, the single-particle partition function (which we will call Z_1) is simply
$$Z_1 = \mathrm{e}^{\beta\mu_\text{B} B} + \mathrm{e}^{-\beta\mu_\text{B} B} = 2\cosh(\beta\mu_\text{B} B). \tag{20.41}$$

A spin-$\frac{1}{2}$ **paramagnet** is an assembly of N such particles, which are assumed to be *non-interacting*: thus each particle is independent and "does its own thing". Note that although it might be energetically favourable for all the spins to line up along the magnetic field, producing a state like

$$\cdots \uparrow \cdots$$

such a state is not very likely: there is only one microstate associated with it. However, even though it is less energetically favourable, there are lots of microstates associated with having half of the states up and half of them down, e.g.

$$\cdots \uparrow\uparrow\downarrow\uparrow\downarrow\downarrow\downarrow\downarrow\uparrow\uparrow\uparrow\uparrow\downarrow\uparrow\downarrow\downarrow\uparrow\downarrow\uparrow\uparrow\uparrow\downarrow\uparrow\downarrow\uparrow\uparrow\downarrow\downarrow\downarrow\uparrow \cdots$$

The balance between energy U and entropy S is encoded in the Helmholtz function $F = U - TS$ which shows that entropy becomes more important as T gets larger, whereas U is more relevant at low temperature.

Because the spins do not interact with each other the N-particle partition function Z_N can be obtained by multiplying N single-particle partition functions (using the result in eqn 20.37 for combining partition functions of independent systems). Therefore

$$Z_N = Z_1^N, \tag{20.42}$$

and hence F is given by

$$F = -k_\mathrm{B} T \ln Z_N = -N k_\mathrm{B} T \ln \left[2 \cosh(\beta \mu_\mathrm{B} B) \right]. \tag{20.43}$$

We can work out the magnetic moment m of the paramagnet by computing

$$m = -\left(\frac{\partial F}{\partial B} \right)_T = N \mu_\mathrm{B} \tanh(\beta \mu_\mathrm{B} B), \tag{20.44}$$

(see Fig. 20.6) and it is worth considering this equation for a moment. Note that when B gets very big (or T gets very small), the magnetic moment tends to $N\mu_\mathrm{B}$, corresponding to all the magnetic moments pointing up, i.e., to a state like

$$\cdots \uparrow \cdots$$

On the other hand, if B is very small (or T gets very large), the magnetic moment tends to zero, corresponding to a state in which half of the magnetic moments are up and half are down, i.e., to a state like

$$\cdots \uparrow\uparrow\downarrow\uparrow\downarrow\downarrow\downarrow\downarrow\uparrow\uparrow\uparrow\uparrow\downarrow\uparrow\downarrow\downarrow\uparrow\downarrow\uparrow\uparrow\uparrow\downarrow\uparrow\downarrow\uparrow\uparrow\downarrow\downarrow\downarrow\uparrow \cdots$$

We now want to calculate the magnetic susceptibility and show that it leads to what is known as Curie's law. Here is how we do it: the magnetization M is the magnetic moment per unit volume, so writing the volume of the paramagnet as V we have

$$M = \frac{m}{V} = \frac{N \mu_\mathrm{B}}{V} \tanh(\beta \mu_\mathrm{B} B). \tag{20.45}$$

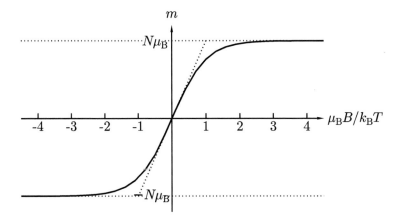

Fig. 20.6 The behaviour of the magnetic moment m as a function of field B and temperature T for a spin-$\frac{1}{2}$ paramagnet, as given by eqn 20.44.

Magnetic susceptibility is measured in a very weak field, so we can look in the limit when B is small so that $\beta\mu_B B \ll 1$ and use $\tanh x \approx x$ for $x \ll 1$, and hence have that

$$M \approx \frac{N\mu_B^2 B}{V k_B T}. \tag{20.46}$$

Recall that $\mathbf{B} = \mu_0(\mathbf{H} + \mathbf{M})$, but for a weakly magnetic material (like a paramagnet), $M \approx \chi H$ and $\chi \ll 1$ is the magnetic susceptibility.[5] Thus we can write that

[5]See eqn 17.30.

$$B \approx \mu_0(1+\chi)H \approx \frac{\mu_0 M}{\chi} \quad \text{and hence} \quad \chi \approx \frac{\mu_0 M}{B}. \tag{20.47}$$

This implies that

$$\chi \approx \frac{N\mu_0 \mu_B^2}{V k_B T}. \tag{20.48}$$

This result obeys **Curie's law:** $\chi \propto 1/T$.

Chapter summary

- The partition function $Z = \sum_\alpha e^{-\beta E_\alpha}$ contains the information needed to find many thermodynamic properties.
- The equations $U = -\mathrm{d}\ln Z/\mathrm{d}\beta$, $F = -k_B T \ln Z$, $S = (U-F)/T$, $p = -\left(\frac{\partial F}{\partial V}\right)_T$, $H = U + pV$, $G = H - TS$ can be used to generate the relevant thermodynamic properties from Z.

Exercises

(20.1) Show that at high temperature, such that $k_B T \gg \hbar\omega$, the partition function of the simple harmonic oscillator is approximately $Z \approx (\beta\hbar\omega)^{-1}$. Hence find U, C, F, and S at high temperature. Repeat the problem for the high temperature limit of the rotational energy levels of the diatomic molecule for which $Z \approx (\beta\hbar^2/2I)^{-1}$ (see eqn 20.35).

(20.2) Show that
$$\ln P_j = \beta(F - E_j). \tag{20.49}$$

(20.3) Show that eqn 20.29 can be rewritten as
$$U = \frac{\hbar\omega}{2}\coth\frac{\beta\hbar\omega}{2}, \tag{20.50}$$
and eqn 20.32 can be rewritten as
$$S = k_B\left[\frac{\hbar\omega\beta}{2}\coth\frac{\beta\hbar\omega}{2} - \ln\left(2\sinh\frac{\beta\hbar\omega}{2}\right)\right]. \tag{20.51}$$

(20.4) Show that the zero-point energy of a simple harmonic oscillator does not contribute to its entropy or heat capacity, but does contribute to its internal energy and Helmholtz function.

(20.5) Show that for N non-interacting spin-$\frac{1}{2}$ particles in a magnetic field B the energy U is given by
$$U = -N\mu_B B \tanh\left(\frac{\mu_B B}{k_B T}\right), \tag{20.52}$$
the heat capacity is given by
$$\frac{C}{Nk_B} = \left(\frac{\mu_B B}{k_B T}\right)^2 \text{sech}^2\left(\frac{\mu_B B}{k_B T}\right), \tag{20.53}$$
and the entropy is given by
$$\frac{S}{Nk_B} = \ln\left[2\cosh\left(\frac{\mu_B B}{k_B T}\right)\right] - \frac{\mu_B B}{k_B T}\tanh\left(\frac{\mu_B B}{k_B T}\right). \tag{20.54}$$

(20.6) A certain magnetic system contains n independent molecules per unit volume, each of which has four energy levels given by 0, $\Delta - g\mu_B B$, Δ, $\Delta + g\mu_B B$ (g is a constant). Write down the partition function, compute the Helmholtz function and hence compute the magnetization M. Hence show that the magnetic susceptibility χ is given by
$$\chi = \lim_{B\to 0}\frac{\mu_0 M}{B} = \frac{2n\mu_0 g^2 \mu_B^2}{k_B T(3 + e^{\Delta/k_B T})}. \tag{20.55}$$

(20.7) The energy E of a system of three independent harmonic oscillators is given by
$$E = (n_x + \tfrac{1}{2})\hbar\omega + (n_y + \tfrac{1}{2})\hbar\omega + (n_z + \tfrac{1}{2})\hbar\omega. \tag{20.56}$$
Show that the partition function Z is given by
$$Z = Z_{SHO}^3, \tag{20.57}$$
where Z_{SHO} is the partition function of a simple harmonic oscillator given in eqn 20.3. Hence show that the Helmholtz function is given by
$$F = \frac{3}{2}\hbar\omega + 3k_B T\ln(1 - e^{-\beta\hbar\omega}), \tag{20.58}$$
and that the heat capacity tends to $3k_B$ at high temperature.

(20.8) The internal levels of an isolated hydrogen atom are given by $E = -R/n^2$ where $R = 13.6$ eV. The degeneracy of each level is given by $2n^2$.
(a) Sketch the energy levels.
(b) Show that
$$Z = \sum_{n=1}^{\infty} 2n^2 \exp\left(\frac{R}{n^2 k_B T}\right). \tag{20.59}$$
Note that when $T \neq 0$, this expression for Z diverges. This is because of the large degeneracy of the hydrogen atom's highly excited states. If the hydrogen atom were to be confined in a box of finite size, this would cut off the highly excited states and Z would not then diverge. By approximating Z as follows:
$$Z \approx \sum_{n=1}^{2} 2n^2 \exp\left(\frac{R}{n^2 k_B T}\right), \tag{20.60}$$
i.e., by ignoring all but the $n = 1$ and $n = 2$ states, estimate the mean energy of a hydrogen atom at 300 K.

(20.9) The energy of a paramagnet can be written as $U = -\boldsymbol{m}\cdot\boldsymbol{B}$. Writing $TS = U - F$, show that if $\boldsymbol{B}$ is varied isothermally then
$$T\,\delta S = -\boldsymbol{B}\cdot\delta\boldsymbol{m}. \tag{20.61}$$
[Hint: use $\boldsymbol{m} = -(\partial F/\partial \boldsymbol{B})_T$.] Show that this is consistent with $\delta U = T\,\delta S - \boldsymbol{m}\cdot\delta\boldsymbol{B}$ (as in eqn 17.29).

Statistical mechanics of an ideal gas

21

The partition function is a sum over all the states of a system of the relevant Boltzmann factors. As we saw in Chapter 20, constructing the partition function is the first step to deriving all the thermodynamic properties of a system. A very important example of this technique is the ideal gas. To determine the partition function of an ideal gas, we have to know what the relevant energy levels are so that we can label the states of the system. Our first step, outlined in the following section, is to work out how many states lie in a certain energy or momentum interval, and this leads us to the *density of states* to be defined below.

21.1 Density of states	233
21.2 Quantum concentration	235
21.3 Distinguishability	236
21.4 Functions of state of the ideal gas	237
21.5 Gibbs paradox	240
21.6 Heat capacity of a diatomic gas	241
Chapter summary	242
Exercises	243

21.1 Density of states

Consider a cubical box of dimensions $L \times L \times L$ and volume $V = L^3$. The box is filled with gas molecules, and we want to consider the momentum states of these gas molecules. It is convenient to label each molecule (we assume that each has mass m) in the gas by its momentum $\boldsymbol{p}$ divided by $\hbar$, i.e., by its **wave vector** $\boldsymbol{k} = \boldsymbol{p}/\hbar$. We assume that the molecules behave like free particles inside the box, but that they are completely confined within the walls of the box. Their wave functions are thus the solution to the Schrödinger equation for the three-dimensional particle-in-a-box problem.[1] We can hence write the wave function of a molecule with wave vector $\boldsymbol{k}$ as[2]

$$\psi(x, y, z) = \left(\frac{2}{L}\right)^{3/2} \sin(k_x x)\, \sin(k_y y)\, \sin(k_z z). \qquad (21.1)$$

The factor $(2/L)^{3/2}$ is simply to ensure that the wave function is normalized over the volume of the box, so that $\int |\psi(x,y,z)|^2 \, dV = 1$. Since the molecules are confined inside the box, we want this wave function to go to zero at the boundaries of the box (the six planes $x = 0$, $x = L$, $y = 0$, $y = L$, $z = 0$, and $z = L$) and this will occur if

$$k_x = \frac{n_x \pi}{L}, \qquad k_y = \frac{n_y \pi}{L}, \qquad k_z = \frac{n_z \pi}{L}, \qquad (21.2)$$

where n_x, n_y, and n_z are integers. We can thus label each state by this triplet of integers.

An allowed state can be represented by a point in three-dimensional $\boldsymbol{k}$-space, and these points are uniformly distributed [in each direction,

[1] We here assume familiarity with basic quantum mechanics.

[2] This wave function is a sum of plane waves travelling in opposite directions. Thus, in this treatment, k_x, k_y, and k_z can only be positive since negating any of them results in the same probability density $|\psi(x, y, z)|^2$.

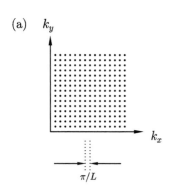

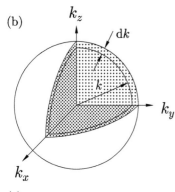

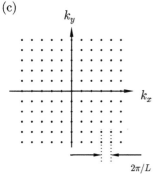

Fig. 21.1 (a) States in k-space are separated by π/L. Each state occupies a volume $(\pi/L)^3$. (b) The density of states can be calculated by considering the volume in k-space between states with wave vector k and states with wave vector $k+\mathrm{d}k$, namely $4\pi k^2 \,\mathrm{d}k$. One octant of the sphere is shown. (c) In Example 21.1, our alternative formulation allows states in k-space to have positive or negative wave vectors and these states are separated by $2\pi/L$. Each state now occupies a volume $(2\pi/L)^3$.

points are separated by a distance π/L, see Fig. 21.1(a)]. A single point in k-space occupies a volume

$$\frac{\pi}{L} \times \frac{\pi}{L} \times \frac{\pi}{L} = \left(\frac{\pi}{L}\right)^3. \qquad (21.3)$$

Let us now focus on the magnitude of the wave vector given by $k = |\mathbf{k}|$. Allowed states with a wave vector whose magnitude lies between k and $k + \mathrm{d}k$ lie on one octant of a spherical shell of radius k and thickness $\mathrm{d}k$ (see Fig. 21.1(b)). It is just one octant since we only allow positive wave vectors in this approach. The volume of this shell is therefore

$$\frac{1}{8} \times 4\pi k^2 \,\mathrm{d}k. \qquad (21.4)$$

The number of allowed states with a wave vector whose magnitude lies between k and $k + \mathrm{d}k$ is described by the function $g(k)\,\mathrm{d}k$, where $g(k)$ is the **density of states**. This number is then given by

$$g(k)\,\mathrm{d}k = \frac{\text{volume in } k\text{-space of one octant of a spherical shell}}{\text{volume in } k\text{-space occupied per allowed state}}. \qquad (21.5)$$

This implies that

$$g(k)\,\mathrm{d}k = \frac{\frac{1}{8} \times 4\pi k^2 \,\mathrm{d}k}{(\pi/L)^3} = \frac{V k^2 \,\mathrm{d}k}{2\pi^2}. \qquad (21.6)$$

Example 21.1

An alternative method of calculating eqn 21.6 is to centre the box of gas at the origin, so that it is bounded by the planes $x = \pm L/2$, $y = \pm L/2$ and $z = \pm L/2$, and to apply periodic boundary conditions.

In this case, the wave function is given by

$$\psi(x,y,z) = \frac{1}{V^{1/2}} e^{\mathrm{i}\mathbf{k}\cdot\mathbf{r}} = \frac{1}{V^{1/2}} e^{\mathrm{i}k_x x} e^{\mathrm{i}k_y y} e^{\mathrm{i}k_z z}. \qquad (21.7)$$

The periodic boundary conditions can now be applied:

$$\psi\left(\frac{L}{2},y,z\right) = \psi\left(-\frac{L}{2},y,z\right), \qquad (21.8)$$

implies that

$$e^{\mathrm{i}k_x L/2} = e^{-\mathrm{i}k_x L/2}, \qquad (21.9)$$

and hence

$$k_x = \frac{2\pi n_x}{L}, \qquad (21.10)$$

where n_x is an integer. Similarly we have that

$$k_y = \frac{2n_y \pi}{L}, \text{ and } k_z = \frac{2n_z \pi}{L}. \qquad (21.11)$$

The points in k-space are now spaced twice as far apart than in our earlier treatment (see Fig. 21.1(c)), but n_x, n_y, and n_z can now be positive or negative, meaning that a complete sphere of values in k-space is used in this formalism. Thus the density of states is now

$$g(k)\,dk = \frac{\text{volume in } k\text{-space of a complete spherical shell}}{\text{volume in } k\text{-space occupied per allowed state}}. \tag{21.12}$$

This implies that

$$g(k)\,dk = \frac{4\pi k^2\,dk}{(2\pi/L)^3} = \frac{Vk^2\,dk}{2\pi^2}, \tag{21.13}$$

as before in eqn 21.6.

Having calculated the density of states in eqn 21.6 (and identically in eqn 21.13), we are now in a position to calculate the partition function of an ideal gas.

21.2 Quantum concentration

The single-particle partition function[3] for the ideal gas is given by a generalization of eqn 20.1 in which we replace the sum by an integral. Hence we have

$$Z_1 = \int_0^\infty e^{-\beta E(k)}\, g(k)\,dk, \tag{21.14}$$

where the energy of a single molecule with wave vector k is given by

$$E(k) = \frac{\hbar^2 k^2}{2m}. \tag{21.15}$$

Hence,

$$Z_1 = \int_0^\infty e^{-\beta \hbar^2 k^2 / 2m} \frac{Vk^2\,dk}{2\pi^2} = \frac{V}{\hbar^3}\left(\frac{mk_BT}{2\pi}\right)^{3/2}, \tag{21.16}$$

which can be written in the appealingly simple form

$$\boxed{Z_1 = V n_Q, \quad \text{where} \quad n_Q = \frac{1}{\hbar^3}\left(\frac{mk_BT}{2\pi}\right)^{3/2},} \tag{21.17}$$

where n_Q is known as the **quantum concentration**. We can define λ_{th}, the **thermal wavelength**, as follows:

$$\lambda_{\text{th}} = n_Q^{-1/3} = \frac{h}{\sqrt{2\pi m k_B T}}, \tag{21.18}$$

and hence we can also write

$$\boxed{Z_1 = \frac{V}{\lambda_{\text{th}}^3}.} \tag{21.19}$$

Equation 21.17 (and 21.19) brings out the important fact that the partition function is proportional to the volume of the system (and also proportional to temperature to the power of 3/2). The importance of this will be seen in the following section.

[3] There is a distinction between the partition function associated with "single-particle states" (where we focus our attention only on a single particle in our system, assuming it has freedom to exist in any state without having to worry about not occupying a state that has already been taken by another particle) and the partition function associated with the whole system. This point will be made clear in the following section. However, we will introduce the subscript 1 at this point to remind ourselves that we are thinking about single-particle states.

21.3 Distinguishability

In this section, we want to attempt to understand what happens for our gas of N molecules, moving on from considering only single–particle states to considering the N-particle state. This is a surprisingly subtle point and to see why, we study the following, much simpler, example.

Example 21.2

Consider a particle which can exist in two states. We model this particle as a thermodynamic system in which the energy can be either 0 or ϵ. The two states of the system are shown in Fig. 21.2(a) and the single-partition function is

$$Z_1 = e^0 + e^{-\beta\epsilon} = 1 + e^{-\beta\epsilon}. \tag{21.20}$$

Now consider two such particles, which behave in the same way, and let us suppose that they are *distinguishable* (for example, they might have different physical locations, or they might have some different attribute, like colour). The possible states of the combined system are shown in Fig. 21.2(b), and we have made them distinguishable in the diagram by depicting them with different symbols. In this case we can write down the two-particle partition function Z_2 as a sum over those four possible states, and hence

$$Z_2 = e^0 + e^{-\beta\epsilon} + e^{-\beta\epsilon} + e^{-2\beta\epsilon}, \tag{21.21}$$

and in this case we see that

$$Z_2 = (Z_1)^2. \tag{21.22}$$

In much the same way, we could work out the N-particle partition function for N distinguishable particles and show that it is given by

$$Z_N = (Z_1)^N. \tag{21.23}$$

However, what happens if the particles are *indistinguishable*? Returning to the combination of two systems, there are now only three possible states of the combined system, as shown in Fig. 21.2(c). The partition function is now

$$Z_2 = e^0 + e^{-\beta\epsilon} + e^{-2\beta\epsilon} \neq (Z_1)^2. \tag{21.24}$$

What has happened is that $(Z_1)^2$ correctly accounts for those states in which the particles are in the same energy level, but has overcounted (by a factor of two) those states in which the particles are in different energy levels. Similarly, for N indistinguishable particles, the N-particle partition function $Z_N \neq (Z_1)^N$ because $(Z_1)^N$ overcounts states in which all N particles are in different states by a factor of $N!$.

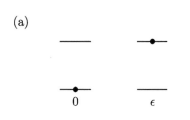
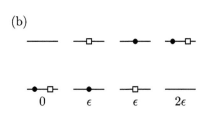
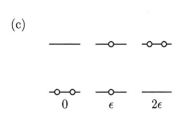

Fig. 21.2 (a) A particle is described by a two-state system with energy 0 or ϵ. (b) The possible states for two such particles if they are distinguishable. (c) The possible states for two such particles if they are indistinguishable.

Let us summarize the results of this example. If the N particles are distinguishable, then we can write the N-particle partition function Z_N as

$$Z_N = (Z_1)^N. \tag{21.25}$$

If they are indistinguishable, then it is much more complicated.[4] However, we can make a rather crafty approximation, as follows. If it is possible to ignore those configurations in which two or more particles are occupying the same energy level, then we can assume exactly the same answer as the distinguishable case and so we only have to worry about the single overcounting factor, which we make when we ignore indistinguishability. If we have N particles all in different states, then that overcounting factor is $N!$ (the number of different arrangements of N distinguishable particles on N distinct sites). Hence we can write the N-particle partition function Z_N for indistinguishable particles as

$$Z_N = \frac{(Z_1)^N}{N!}. \tag{21.26}$$

[4]Note that identical (and hence indistinguishable) particles can be made to be distinguishable if they are **localized**. The particles can then be distinguished by their physical location. Electrons in a gas are indistinguishable if there is no means of labelling which is which, but the electrons sitting in a particular magnetic orbital, one per atom of a magnetic solid, are distinguishable.

This result has assumed that it is possible to ignore those states in which two or more particles occupy the same energy level. When is this approximation possible? We will have only one particle occupying any given state if the system is in a regime when the number of available states is much larger than the number of particles. So for the ideal gas, we require that the number of thermally accessible energy levels must be much larger than the number of molecules in the gas. This occurs when n, the number density of molecules, is much less than the quantum concentration n_Q. Thus the condition for validity of eqn 21.26 for an ideal gas is

$$n \ll n_Q. \tag{21.27}$$

If this condition holds, the N-particle partition function for an ideal gas can be written as

$$Z_N = \frac{1}{N!} \left(\frac{V}{\lambda_{\text{th}}^3} \right)^N. \tag{21.28}$$

The quantum concentration n_Q is plotted in Fig. 21.3 for electrons, protons, N_2 molecules, and C_{60} molecules (known as buckyballs). At room temperature, the quantum concentration of N_2 molecules is much higher than the actual number density of molecules in air ($\approx 10^{25}$ m^{-3}) and so the approximation in eqn 21.26 is a good one. Electrons in a metal have a concentration $\approx 10^{29}$ m^{-3}, which is larger than the quantum concentration for electrons at room temperature, so the approximation in eqn 21.26 will not work for electrons and their quantum properties have to be considered in more detail.

21.4 Functions of state of the ideal gas

Having obtained the partition function of an ideal gas, we are now in a position to use the machinery of statistical mechanics, developed in

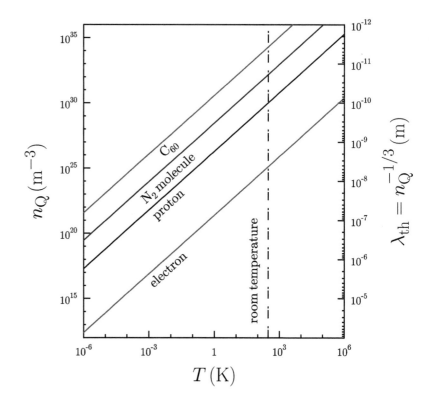

Fig. 21.3 The quantum concentration n_Q and thermal wavelength λ_{th} for electrons, protons, N$_2$ molecules, and buckyballs.

Chapter 20, to derive all the relevant thermodynamic properties. This we do in the following example.

Example 21.3

The partition function for N molecules in a gas is given in eqn 21.28 by

$$Z_N = \frac{1}{N!}\left(\frac{V}{\lambda_{th}^3}\right)^N \propto (VT^{3/2})^N, \tag{21.29}$$

since $\lambda_{th} \propto T^{-1/2}$. Hence we can write

$$\ln Z_N = N \ln V + \frac{3N}{2}\ln T + \text{constants}. \tag{21.30}$$

The internal energy U is given by

$$U = -\frac{\mathrm{d}\ln Z_N}{\mathrm{d}\beta} = \frac{3}{2}Nk_\mathrm{B}T, \tag{21.31}$$

so that the heat capacity is $C_V = \frac{3}{2}Nk_\mathrm{B}$, in agreement with previous results.

The Helmholtz function is

$$F = -k_B T \ln Z_N = -k_B T N \ln V - k_B \frac{3N}{2} T \ln T - k_B T \times \text{constants}, \tag{21.32}$$

so that

$$p = -\left(\frac{\partial F}{\partial V}\right)_T = \frac{N k_B T}{V} = n k_B T, \tag{21.33}$$

which is, reassuringly, the ideal gas equation. This also gives the enthalpy H via

$$H = U + pV = \frac{5}{2} N k_B T. \tag{21.34}$$

Before proceeding to the entropy, it is going to be necessary to worry about what the constants are in eqn 21.30. Returning to eqn 21.29, we write

$$\begin{aligned}\ln Z_N &= N \ln V - 3N \ln \lambda_{\text{th}} - N \ln N + N \\ &= N \ln\left(\frac{Ve}{N\lambda_{\text{th}}^3}\right),\end{aligned} \tag{21.35}$$

where we have used Stirling's approximation, $\ln N! \approx N \ln N - N$ (see eqn 1.17). Hence we can obtain the following expression for the Helmholtz function F:

$$\begin{aligned}F &= -N k_B T \ln\left(\frac{Ve}{N\lambda_{\text{th}}^3}\right) \\ &= N k_B T [\ln(n\lambda_{\text{th}}^3) - 1].\end{aligned} \tag{21.36}$$

This allows us to derive the entropy S:

$$\begin{aligned}S &= \frac{U - F}{T} = \frac{3}{2} N k_B + N k_B \ln\left(\frac{Ve}{N\lambda_{\text{th}}^3}\right) \\ &= N k_B \ln\left(\frac{Ve^{5/2}}{N\lambda_{\text{th}}^3}\right) \\ &= N k_B \left[\frac{5}{2} - \ln(n\lambda_{\text{th}}^3)\right],\end{aligned} \tag{21.37}$$

and hence the entropy is expressed in terms of the thermal wavelength of the molecules. We can also derive the Gibbs function G

$$\begin{aligned}G &= H - TS = \frac{5}{2} N k_B T - N k_B T \ln\left(\frac{Ve^{5/2}}{N\lambda_{\text{th}}^3}\right) \\ &= N k_B T \ln(n\lambda_{\text{th}}^3).\end{aligned} \tag{21.38}$$

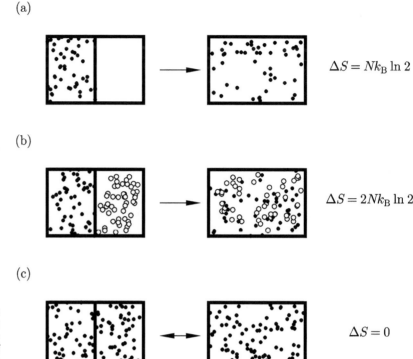

Fig. 21.4 (a) Joule expansion of an ideal gas (an irreversible process). (b) Mixing of two different gases, equivalent to the Joule expansion of each of the gases (an irreversible process). (c) Mixing of two identical gases, which is clearly a reversible process – how can you tell if they have been mixed?

21.5 Gibbs paradox

The expression for the entropy in eqn 21.37 is called the **Sackur–Tetrode equation** and can be used to demonstrate the **Gibbs paradox**. Consider the process shown in Fig. 21.4(a), namely the Joule expansion of N molecules of an ideal gas. This is an irreversible process which halves the number density n so that the increase in entropy is given by

$$\begin{aligned} \Delta S &= S_{\text{final}} - S_{\text{initial}} \\ &= Nk_{\text{B}}\left[\frac{5}{2} - \ln(\frac{n}{2}\lambda_{\text{th}}^3)\right] - Nk_{\text{B}}\left[\frac{5}{2} - \ln(n\lambda_{\text{th}}^3)\right] \\ &= Nk_{\text{B}}\ln 2, \end{aligned} \qquad (21.39)$$

in agreement with eqn 14.29. This reflects the fact that, following the Joule expansion, we have an uncertainty about each molecule as to whether it is on the left- or right-hand side of the chamber, whereas beforehand there was no uncertainty (all molecules were on the left-hand side). Hence the uncertainty is one bit per molecule, and hence $\Delta S/k_{\text{B}} = N\ln 2$.

Now consider the situation depicted in Fig. 21.4(b) in which two *different* gases are allowed to mix following the removal of a partition which separated them. This is clearly an irreversible process and is equivalent to the Joule expansion of each gas. Thus the entropy increase is

$$\Delta S = 2Nk_B \ln 2. \qquad (21.40)$$

An apparently similar case is shown in Fig. 21.4(c), but this time the two gases on either side of the partition are *indistinguishable*. Removing the partition is now an eminently reversible operation so $\Delta S = 0$. Yet, it might be argued, is it not the case that the removal of the partition simply allows the gases which were initially on either side of the partition to each undergo a Joule expansion? Surely, the change of entropy would then be $\Delta S = 2Nk_B \ln 2$. This apparent paradox is resolved by understanding that indistinguishable really means indistinguishable! In other words, the case shown in Fig. 21.4(c) is *fundamentally different* from that shown in Fig. 21.4(b). Removing the partition in the case of Fig. 21.4(c) is a reversible operation since we have no way of losing information about which side of the partition certain bits of gas are; this is because all molecules of this gas look the same to us and we never had such information in the first place. Hence $\Delta S = 0$.

Gibbs resolved this paradox himself by realizing that indistinguishability was fundamental and that all states of the system that differ only by a permutation of identical molecules should be considered as the same state. Failure to do this results in an expression for the entropy that is not extensive (see Exercise 21.2, which was the original manifestation of the Gibbs paradox).

21.6 Heat capacity of a diatomic gas

The energy of a diatomic molecule in a gas can be written using eqn 19.19 as the sum of three translational, two rotational, and two vibrational terms, giving seven modes in total. The equipartition theorem shows that the mean energy per molecule at high temperature is therefore $\frac{7}{2}k_B T$ (see eqn 19.20). Because the modes are independent, the partition function of a diatomic molecule, Z, can be written as the product of partition functions for the translational, rotational and vibrational modes as

$$Z = Z_{\text{trans}} Z_{\text{vib}} Z_{\text{rot}}, \qquad (21.41)$$

where $Z_{\text{trans}} = V/\lambda_{\text{th}}^3$ from eqn 21.19, $Z_{\text{vib}} = e^{-\frac{1}{2}\beta\hbar\omega}/(1 - e^{-\beta\hbar\omega})$, from eqn 20.3, and Z_{rot} is the rotational partition function

$$Z_{\text{rot}} = \sum_\alpha e^{-\beta E_\alpha} = \sum_{J=0}^{\infty} (2J+1)e^{-\beta\hbar^2 J(J+1)/2I}, \qquad (21.42)$$

from eqn 20.6. Thus the mean energy U of such a diatomic molecule is given by $U = -d \ln Z/d\beta$ and is the *sum* of the energies of the individual modes. Similarly, the heat capacity C_V is the *sum* of the heat capacities of the individual modes. This gives rise to the behaviour shown in Fig. 21.5 in which the heat capacity goes through a series of plateaus:

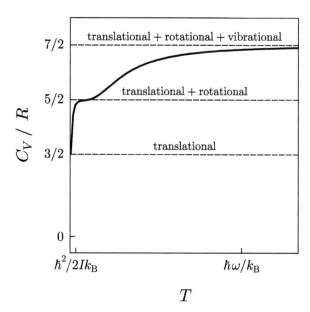

Fig. 21.5 The molar heat capacity at constant volume of a diatomic gas as a function of temperature.

at any non-zero temperature, all the translational modes are excited (a failure of the ideal gas model, because C_V should go to zero as $T \to 0$, see Chapter 18) and $C_V = \frac{3}{2}R$ (for one mole of gas); above $T \approx \hbar^2/2Ik_B$ the rotational modes are also excited and C_V rises to $\frac{5}{2}R$; above $T \approx \hbar\omega/k_B$, the vibrational modes are excited and hence C_V rises to $\frac{7}{2}R$.

Chapter summary

- For an ideal gas, the partition function can be written
$$Z = V/\lambda_{\text{th}}^3,$$
where $\lambda_{\text{th}} = h/\sqrt{2\pi m k_B T}$ is the thermal wavelength.
- The quantum concentration $n_Q = 1/\lambda_{\text{th}}^3$.
- The N-particle partition function is given by
$$Z_N = \frac{(Z_1)^N}{N!}$$
for indistinguishable particles in the low-density case when $n/n_Q \ll 1$ so that $n\lambda_{\text{th}}^3 \ll 1$.

Exercises

(21.1) Show that the single-partition function Z_1 of a two-dimensional gas confined in an area A is given by
$$Z_1 = \frac{A}{\lambda_{\text{th}}^2}, \qquad (21.43)$$
where $\lambda_{\text{th}} = h/\sqrt{2\pi m k_B T}$.

(21.2) Show that S as given by eqn 21.37 (the Sackur–Tetrode equation) is an extensive quantity, but that the entropy of a gas of distinguishable particles is given by
$$S = Nk_B \left[\frac{3}{2} - \ln(\lambda_{\text{th}}^3/V) \right], \qquad (21.44)$$
and show that this quantity is not extensive. This non-extensive entropy provided the original version of the Gibbs paradox.

(21.3) Show that the number of states in a gas with energies below $E_{\max}$ is
$$\int_0^{\sqrt{2mE_{\max}/\hbar^2}} g(k)\,dk = \frac{V}{6\pi^2}\left(\frac{2mE_{\max}}{\hbar^2}\right)^{3/2}. \qquad (21.45)$$
Putting $E_{\max} = \frac{3}{2}k_B T$, show that the number of states is $\Xi V n_Q$ where Ξ is a numerical constant of order unity.

(21.4) An atom in a solid has two energy levels: a ground state of degeneracy g_1 and an excited state of degeneracy g_2 at an energy Δ above the ground state. Show that the partition function Z_{atom} is
$$Z_{\text{atom}} = g_1 + g_2 e^{-\beta\Delta}. \qquad (21.46)$$
Show that the heat capacity of the atom is given by
$$C = \frac{g_1 g_2 \Delta^2 e^{-\beta\Delta}}{k_B T^2 (g_1 + g_2 e^{-\beta\Delta})^2}. \qquad (21.47)$$
A monatomic gas of such atoms has a partition function given by
$$Z_N = Z_1^N/N!, \qquad (21.48)$$
where $Z_1 = Z_{\text{atoms}} Z_{\text{translation}}$, and $Z_{\text{translation}} = V/\lambda_{\text{th}}^3$ is the partition function due to the translational motion of the gas atoms. Show that the heat capacity of such as gas is
$$C = N\left[\frac{3}{2}k_B + \frac{g_1 g_2 \Delta^2 e^{-\beta\Delta}}{k_B T^2 (g_1 + g_2 e^{-\beta\Delta})^2}\right]. \qquad (21.49)$$

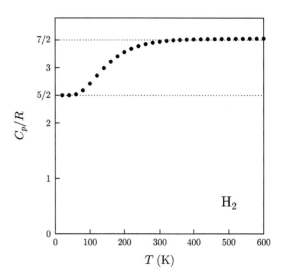

Fig. 21.6 The heat capacity of hydrogen gas as a function of temperature.

(21.5) Explain the behaviour of the experimental heat capacity (measured at constant pressure) of hydrogen (H_2) gas shown in Fig. 21.6.

(21.6) Show that the single–particle partition function Z_1 of a gas of hydrogen atoms is given approximately by
$$Z_1 = \frac{Ve^{\beta R}}{\lambda_{\text{th}}^3}, \qquad (21.50)$$
where $R = 13.6\,\text{eV}$ and the contribution due to excited states has been neglected.

22 The chemical potential

22.1 A definition of the chemical potential	244
22.2 The meaning of the chemical potential	245
22.3 Grand partition function	247
22.4 Grand potential	248
22.5 Chemical potential as Gibbs function per particle	250
22.6 Many types of particle	250
22.7 Particle number conservation laws	251
22.8 Chemical potential and chemical reactions	252
22.9 Osmosis	257
Chapter summary	261
Further reading	261
Exercises	262

We now want to consider systems that can exchange particles with their surroundings and we will show in this chapter that this feature leads to a new concept, known as the *chemical potential*. Differences in the chemical potential drive the flow of particles from one place to another in much the same way as differences in temperature drive the flow of heat. The chemical potential turns up in chemical reactions (hence the name) because in a reaction such as

$$2\text{H}_2 + \text{O}_2 \to 2\text{H}_2\text{O}, \tag{22.1}$$

you are changing the number of particles in your system (three molecules on the left, two on the right). However, as we shall see, the chemical potential applies to more than just chemical systems. It is connected with conservation laws, so that particles such as electrons (which are conserved) and photons (which are not) have different chemical potentials and this has consequences for their behaviour.

22.1 A definition of the chemical potential

If you add a particle to a system without changing the system's volume or entropy,[1] then the internal energy will change by an amount which we call the **chemical potential** μ. Thus the first and second laws of thermodynamics expressed in eqn 14.18 must, in the case of changing numbers of particles, be modified to contain an extra term, so that

$$dU = TdS - pdV + \mu dN, \tag{22.2}$$

where N is the number of particles in the system.[2] This means that we can write an expression for μ as a partial differential of U as follows:

$$\mu = \left(\frac{\partial U}{\partial N}\right)_{S,V}. \tag{22.3}$$

However, keeping S and V constant is a difficult constraint to apply, so it is convenient to consider other thermodynamic potentials. Equation 22.2, together with the definitions $F = U - TS$ and $G = U + pV - TS$, implies that

$$dF = -pdV - SdT + \mu dN, \tag{22.4}$$
$$dG = Vdp - SdT + \mu dN, \tag{22.5}$$

[1] Volume and entropy are the natural variables for U.

[2] If we are dealing with discrete particles, then N is an integer and can only change by integer amounts; hence using calculus expressions like dN is a bit sloppy, but this is an indiscretion for which we may be excused if N is large. However, there exist systems such as **quantum dots**, which are semiconductor nanocrystals whose size is a few nanometres. Quantum dots are so small that μ jumps discontinuously when you add one electron to the quantum dot.

and hence we can make the more useful definitions:

$$\mu = \left(\frac{\partial F}{\partial N}\right)_{V,T} \quad \text{or} \quad (22.6)$$

$$\mu = \left(\frac{\partial G}{\partial N}\right)_{p,T}. \quad (22.7)$$

The constraints of constant p and T are experimentally convenient for chemical systems and so eqn 22.7 will be particularly useful.

22.2 The meaning of the chemical potential

What drives a system to form a particular equilibrium state? As we have seen in Chapter 14, it is the second law of thermodynamics which states that entropy always increases. The entropy of a system can be considered to be a function of U, V, and N, so that $S = S(U, V, N)$. Therefore, we can immediately write down

$$dS = \left(\frac{\partial S}{\partial U}\right)_{N,V} dU + \left(\frac{\partial S}{\partial V}\right)_{N,U} dV + \left(\frac{\partial S}{\partial N}\right)_{U,V} dN. \quad (22.8)$$

Equation 22.2 implies that

$$dS = \frac{dU}{T} + \frac{pdV}{T} - \frac{\mu dN}{T}. \quad (22.9)$$

Comparison of eqn 22.8 and 22.9 implies that we can therefore make the following identifications:

$$\left(\frac{\partial S}{\partial U}\right)_{N,V} = \frac{1}{T}, \quad \left(\frac{\partial S}{\partial V}\right)_{N,U} = \frac{p}{T}, \quad \left(\frac{\partial S}{\partial N}\right)_{U,V} = -\frac{\mu}{T}. \quad (22.10)$$

Now consider two systems which are able to exchange heat or particles between them. If we write down an expression for dS, then we can use the second law of thermodynamics in the form $dS \geq 0$ to determine the equilibrium state. We repeat this analysis for two cases as follows:

- **The case of heat flow**
  Consider two systems which are able to exchange heat with each other while remaining thermally isolated from their surroundings (see Fig. 22.1). If system 1 loses internal energy dU, system 2 must gain internal energy dU. Thus the change of entropy is

$$\begin{aligned} dS &= \left(\frac{\partial S_1}{\partial U_1}\right)_{N,V} dU_1 + \left(\frac{\partial S_2}{\partial U_2}\right)_{N,V} dU_2 \\ &= \left(\frac{\partial S_1}{\partial U_1}\right)_{N,V} (-dU) + \left(\frac{\partial S_2}{\partial U_2}\right)_{N,V} (dU) \\ &= \left(-\frac{1}{T_1} + \frac{1}{T_2}\right) dU \geq 0. \quad (22.11) \end{aligned}$$

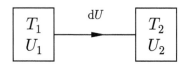

Fig. 22.1 Two systems which are able to exchange heat with each other.

So $dU > 0$, i.e., energy flows from 1 to 2, when $T_1 > T_2$. As expected, equilibrium is found when $T_1 = T_2$, i.e., when the temperatures of the two systems are equal.

- **The case of particle exchange**
  Now consider two systems which are able to exchange particles with each other, but remain isolated from their surroundings (see Fig. 22.2). If system 1 loses dN particles, system 2 must gain dN particles. Thus the change of entropy is

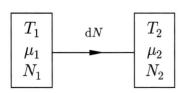

Fig. 22.2 Two systems which are able to exchange particles with each other.

$$\begin{aligned} dS &= \left(\frac{\partial S_1}{\partial N_1}\right)_{U,V} dN_1 + \left(\frac{\partial S_2}{\partial N_2}\right)_{U,V} dN_2 \\ &= \left(\frac{\partial S_1}{\partial N_1}\right)_{U,V} (-dN) + \left(\frac{\partial S_2}{\partial N_2}\right)_{U,V} (dN) \\ &= \left(\frac{\mu_1}{T_1} - \frac{\mu_2}{T_2}\right) dN \geq 0 \end{aligned} \qquad (22.12)$$

Assuming that $T_1 = T_2$, we find that $dN > 0$ (so that particles flow from 1 to 2) when $\mu_1 > \mu_2$. Similarly, if $\mu_1 < \mu_2$, then $dN < 0$. Hence equilibrium is found when $\mu_1 = \mu_2$, i.e., when the chemical potentials are the same for each system. This demonstrates that chemical potential plays a similar rôle in particle exchange as $1/\text{temperature}$ does in heat exchange.

Example 22.1

Find the chemical potential for an ideal gas.
Solution:
We use eqn 22.6 ($\mu = (\partial F/\partial N)_{V,T}$), which relates μ to F, together with eqn 21.36, which gives an expression for F, namely

$$F = Nk_BT[\ln(n\lambda_{th}^3) - 1]. \qquad (22.13)$$

Recalling also that $n = N/V$, we find that

$$\mu = k_BT[\ln(n\lambda_{th}^3) - 1] + Nk_BT\left(\frac{1}{N}\right), \qquad (22.14)$$

and hence

$$\mu = k_BT \ln(n\lambda_{th}^3). \qquad (22.15)$$

In this case, comparison with eqn 21.38 shows that $\mu = G/N$. We will see in Section 22.5 that this property has more general applicability than just this specific case.

22.3 Grand partition function

In this section we will introduce a version of the partition function we met in Chapter 20 but now generalized to include the effect of variable numbers of particle. To do this, we have to generalize the canonical ensemble we met in Chapter 4 to the case of both energy and particle exchange.

Let us write the entropy S as a function of internal energy U and particle number N. Consider a small system with fixed volume V and with energy ϵ and containing N particles, connected to a reservoir with energy $U - \epsilon$ and $\mathcal{N} - N$ particles (see Fig. 22.3). We assume that $U \gg \epsilon$ and $\mathcal{N} \gg N$. Using a Taylor expansion, we can write the entropy of the reservoir as

$$S(U - \epsilon, \mathcal{N} - N) = S(U, \mathcal{N}) - \epsilon \left(\frac{\mathrm{d}S}{\mathrm{d}U}\right)_{\mathcal{N},V} - N \left(\frac{\mathrm{d}S}{\mathrm{d}\mathcal{N}}\right)_{U,V}, \quad (22.16)$$

and using the differentials defined in eqn 22.10, we have that

$$S(U - \epsilon, \mathcal{N} - N) = S(U, \mathcal{N}) - \frac{1}{T}(\epsilon - \mu N). \quad (22.17)$$

The probability $P(\epsilon, N)$ that the system chooses a particular macrostate is proportional to the number Ω of microstates corresponding to that macrostate, and using $S = k_\mathrm{B} \ln \Omega$ we have that

$$P(\epsilon, N) \propto \mathrm{e}^{S(U-\epsilon, \mathcal{N}-N)/k_\mathrm{B}} \propto \mathrm{e}^{\beta(\mu N - \epsilon)}. \quad (22.18)$$

This is known as the **Gibbs distribution** and the situation is known as the **grand canonical ensemble**. In the case in which $\mu = 0$, this reverts to the Boltzmann distribution (the canonical ensemble). Normalizing this distribution, we have that the probability of a state of the system with energy E_i and with N_i particles is given by

$$\boxed{P_i = \frac{\mathrm{e}^{\beta(\mu N_i - E_i)}}{\mathcal{Z}},} \quad (22.19)$$

where $\mathcal{Z}$ is a normalization constant. The normalization constant is known as the **grand partition function** $\mathcal{Z}$, which we write as follows:

$$\boxed{\mathcal{Z} = \sum_i \mathrm{e}^{\beta(\mu N_i - E_i)},} \quad (22.20)$$

which is a sum over all states of the system. The grand partition function $\mathcal{Z}$ can be used to derive many thermodynamic quantities, and we write down the most useful equations here without detailed proof.[3]

$$N = \sum_i N_i P_i = k_\mathrm{B} T \left(\frac{\partial \ln \mathcal{Z}}{\partial \mu}\right)_\beta, \quad (22.21)$$

$$U = \sum_i E_i P_i = -\left(\frac{\partial \ln \mathcal{Z}}{\partial \beta}\right)_\mu + \mu N, \quad (22.22)$$

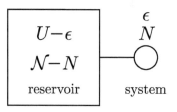

Fig. 22.3 A small system with energy ϵ and containing N particles, connected to a reservoir with energy $U - \epsilon$ and $\mathcal{N} - N$ particles.

[3]See Exercise 22.4.

and
$$S = -k_{\rm B} \sum_i P_i \ln P_i = \frac{U - \mu N + k_{\rm B} T \ln \mathcal{Z}}{T}. \tag{22.23}$$

For convenience, let us summarize the various ensembles considered in statistical mechanics.

(1) The **microcanonical ensemble**: an ensemble of systems, all of which have the same fixed energy. The entropy S is related to the number of microstates by $S = k_{\rm B} \ln \Omega$, and hence by
$$\Omega = e^{\beta TS}. \tag{22.24}$$

(2) The **canonical ensemble**: an ensemble of systems, each of which can exchange its energy with a large reservoir of heat. As we shall see, this fixes (and defines) the temperature of the system. Since $F = -k_{\rm B} T \ln Z$, the partition function is given by
$$Z = e^{-\beta F}, \tag{22.25}$$
where F is the Helmholtz function.

(3) The **grand canonical ensemble**: an ensemble of systems, each of which can exchange both energy and particles with a large reservoir. This fixes the system's temperature and chemical potential. By analogy with the canonical ensemble, we write the grand partition function as
$$\mathcal{Z} = e^{-\beta \Phi_{\rm G}}, \tag{22.26}$$
where $\Phi_{\rm G}$ is the **grand potential**, which we discuss in the next section.

22.4 Grand potential

Using eqn 22.26, we have defined a new state function, the grand potential $\Phi_{\rm G}$, by
$$\boxed{\Phi_{\rm G} = -k_{\rm B} T \ln \mathcal{Z}.} \tag{22.27}$$
Rearranging eqn 22.23, we have that
$$-k_{\rm B} T \ln \mathcal{Z} = U - TS - \mu N, \tag{22.28}$$
and hence
$$\Phi_{\rm G} = U - TS - \mu N = F - \mu N. \tag{22.29}$$
The grand potential has differential $d\Phi_{\rm G}$ given by
$$d\Phi_{\rm G} = dF - \mu \, dN - N \, d\mu, \tag{22.30}$$
and, substituting in eqn 22.4, we therefore have
$$d\Phi_{\rm G} = -S \, dT - p \, dV - N \, d\mu, \tag{22.31}$$

and this leads to the following equations for S, p and N:

$$S = -\left(\frac{\partial \Phi_G}{\partial T}\right)_{V,\mu}, \quad (22.32)$$

$$p = -\left(\frac{\partial \Phi_G}{\partial V}\right)_{T,\mu}, \quad (22.33)$$

$$N = -\left(\frac{\partial \Phi_G}{\partial \mu}\right)_{T,V}. \quad (22.34)$$

Example 22.2

Find the grand potential for an ideal gas, and show that eqns 22.33 and 22.34 lead to the correct expressions for p and N.

Solution:

Using eqns 21.36 and 22.15 we have that

$$\begin{aligned}\Phi_G &= Nk_BT[\ln(n\lambda_{th}^3) - 1] - Nk_BT\ln(n\lambda_{th}^3) \\ &= -Nk_BT, \quad (22.35)\end{aligned}$$

and using the ideal gas equation ($pV = Nk_BT$) this becomes

$$\Phi_G = -pV. \quad (22.36)$$

We can check that eqn 22.34 leads to the correct value of N by evaluating

$$\left(\frac{\partial \Phi_G}{\partial \mu}\right)_{T,V} = \left(\frac{\partial \Phi_G}{\partial N}\right)_{T,V}\left(\frac{\partial N}{\partial \mu}\right)_{T,V}, \quad (22.37)$$

and since $\left(\frac{\partial \Phi_G}{\partial N}\right)_{T,V} = -k_BT$ (from eqn 22.35) and $\left(\frac{\partial \mu}{\partial N}\right)_{T,V} = k_BT/N$ we have that

$$\left(\frac{\partial \Phi_G}{\partial \mu}\right)_{T,V} = -k_BT \times \frac{N}{k_BT} = -N, \quad (22.38)$$

justifying eqn 22.34. Similarly,[4]

$$\left(\frac{\partial \Phi_G}{\partial V}\right)_{T,\mu} = -\left(\frac{\partial \Phi_G}{\partial \mu}\right)_{T,V}\left(\frac{\partial \mu}{\partial V}\right)_{T,\Phi_G} = N\left(\frac{\partial \mu}{\partial V}\right)_{T,\Phi_G} \quad (22.39)$$

[4]Using the reciprocity theorem, with T held constant for all terms.

and since the constraint of constant T and constant $\Phi_G = -Nk_BT$ means constant T and N, and using $N = nV$; we can use eqn 22.15 to obtain

$$\left(\frac{\partial \mu}{\partial V}\right)_{T,N} = k_BT\left(\frac{\partial \ln(N\lambda_{th}^3/V)}{\partial V}\right)_{T,N} = -\frac{k_BT}{V}, \quad (22.40)$$

and eqn 22.39 becomes

$$\left(\frac{\partial \Phi_G}{\partial V}\right)_{T,\mu} = -\frac{Nk_BT}{V} = -p, \quad (22.41)$$

thus justifying eqn 22.33.

22.5 Chemical potential as Gibbs function per particle

If we scale a system by a factor λ, then we expect all the extensive[5] variables will scale with λ, thus

[5] The distinction between intensive and extensive variables is discussed in Section 11.1.2.

$$U \to \lambda U, \qquad S \to \lambda S, \qquad V \to \lambda V, \qquad N \to \lambda N, \qquad (22.42)$$

and writing the entropy S as a function of U, V, and N, we have

$$\lambda S(U, V, N) = S(\lambda U, \lambda V, \lambda N), \qquad (22.43)$$

so that differentiating with respect to λ we have

$$S = \frac{\partial S}{\partial(\lambda U)} \frac{\partial(\lambda U)}{\partial \lambda} + \frac{\partial S}{\partial(\lambda V)} \frac{\partial(\lambda V)}{\partial \lambda} + \frac{\partial S}{\partial(\lambda N)} \frac{\partial(\lambda N)}{\partial \lambda}, \qquad (22.44)$$

so that setting $\lambda = 1$ and using eqn 22.10, we have that

$$S = \frac{U}{T} + \frac{pV}{T} - \frac{\mu N}{T}, \qquad (22.45)$$

and hence

$$U - TS + pV = \mu N. \qquad (22.46)$$

We recognize the left-hand side of this equation as the Gibbs function, and so we have

$$G = \mu N. \qquad (22.47)$$

This gives a new interpretation for the chemical potential: by rearranging the above equation, one has that

$$\boxed{\mu = \frac{G}{N},} \qquad (22.48)$$

so that the chemical potential μ can be thought of as the Gibbs function per particle.

This analysis also implies that the grand potential $\Phi_G = F - \mu N$ can be rewritten (using eqn 22.46 and $F = U - TS$) as

$$\boxed{\Phi_G = -pV.} \qquad (22.49)$$

This equation has been demonstrated to be correct for the specific example of the ideal gas (see eqn 22.36), but we have now shown that it is always correct if entropy is an extensive property.

22.6 Many types of particle

If there is more than one type of particle, then one can generalize the treatment in Section 22.5, and write

$$dU = T dS - p dV + \sum_i \mu_i dN_i, \qquad (22.50)$$

where N_i is the number of particles of species i and μ_i is the chemical potential of species i. Correspondingly, we have the equations

$$dF = -pdV - SdT + \sum_i \mu_i dN_i, \quad (22.51)$$

$$dG = Vdp - SdT + \sum_i \mu_i dN_i, \quad (22.52)$$

and in particular, when the pressure and temperature are held constant we have that

$$\boxed{dG = \sum_i \mu_i dN_i.} \quad (22.53)$$

This generalization will be useful in our treatment of chemical reactions in Section 22.8. In the following section, we make the connection between μ and the conservation of particle number.

22.7 Particle number conservation laws

Imagine that one has a set of particles in a box in which particle number is not conserved. This means that we are free to create or destroy particles at will. There might be an energy cost associated with doing this, but provided we have energy to "pay" for the particles, no conservation laws would be broken. In this case, the system will try to minimize its availability (see Section 16.5) and if the constraints are that the box has fixed volume and fixed temperature, then the appropriate availability is the Helmholtz function[6] F. The system will therefore choose a number of particles N by minimizing F with respect to N, i.e.,

$$\left(\frac{\partial F}{\partial N}\right)_{V,T} = 0. \quad (22.54)$$

This means that, from eqn 22.6,

$$\mu = 0. \quad (22.55)$$

We arrive at the important result that, for a set of particles with no conservation law concerning particle number, the chemical potential μ is zero. One example of such a particle is the photon.[7]

To understand this further, let us consider a set of particles for which particle number is a conserved quantity. Consider a gas of electrons. Electrons do have a conservation law: electron number has to be conserved, so the only way of annihilating an electron is by reacting it with a positron[8] via the reaction

$$e^- + e^+ \rightleftharpoons \gamma + \gamma, \quad (22.56)$$

where γ denotes a photon. Thus imagine that our box contains N_- electrons and N_+ positrons. We are constrained by our conservation law to fix the number $N = N_+ - N_-$, which also serves to ensure that

[6] If the constraints were constant pressure and temperature, we would be dealing with G not F; see Section 16.5.

[7] Strictly this is only for photons in a vacuum, which the following example will assume. Photons can have a non-zero chemical potential under some circumstances. For example, if electrons and holes combine in a light-emitting diode, it may be that the chemical potential of the electrons μ_e, from the conduction band, is not balanced by the chemical potential of the holes μ_h, from the valence band, and this leads to light with a non-zero chemical potential $\mu_\gamma = \mu_e + \mu_h$.

[8] A positron e^+ is an antielectron.

charge is conserved. The system is at fixed T and V, and hence we should minimize F with respect to any variable, so let us choose N_- as a variable to vary. Thus

$$\left(\frac{\partial F}{\partial N_-}\right)_{V,T,N} = 0. \tag{22.57}$$

In this case, F is the sum of a term due to the Helmholtz function for the electrons and one for the positrons. Thus

$$\left(\frac{\partial F}{\partial N_-}\right)_{V,T,N_+} + \left(\frac{\partial F}{\partial N_+}\right)_{V,T,N_-} \frac{dN_+}{dN_-} = 0. \tag{22.58}$$

Now we have that

$$\left(\frac{\partial F}{\partial N_-}\right)_{V,T,N_+} = \mu_-, \tag{22.59}$$

the chemical potential of the electrons, while

$$\left(\frac{\partial F}{\partial N_+}\right)_{V,T,N_-} = \mu_+, \tag{22.60}$$

the chemical potential of the positrons. Moreover, since

$$\frac{dN_-}{dN_+} = 1, \tag{22.61}$$

we have that

$$\mu_+ + \mu_- = 0. \tag{22.62}$$

We are ignoring the chemical potential of the photons, since this is zero because photons do not have a conservation law.[9]

[9] Again, this is true for most circumstances, the photons from a light-emitting diode being a rather notable counterexample.

22.8 Chemical potential and chemical reactions

We next want to consider how the chemical potential can be used to determine the equilibrium position of a chemical reaction. Before proceeding, we will prove an important result concerning the way the chemical potential of an ideal gas depends on pressure.

Example 22.3

Derive an expression for the dependence of the chemical potential of an ideal gas on pressure at fixed temperature.
Solution:
Equation 22.15 and the ideal gas equation ($p = nk_BT$) imply that

$$\mu = k_B T \ln\left(\frac{\lambda_{th}^3}{k_B T}\right) + k_B T \ln p. \tag{22.63}$$

It is useful to compare the chemical potential at standard temperature (273 K) and pressure ($p^{\ominus} = 1\,\text{bar} = 10^5\,\text{Pa}$), which we denote by $\mu^{\ominus}$, with the chemical potential measured at some other pressure p. Here the symbol $\ominus$ denotes the value of a function measured at standard temperature and pressure. The chemical potential $\mu(p)$ at pressure p is then given by

$$\mu(p) = \mu^{\ominus} + k_B T \ln \frac{p}{p^{\ominus}}. \tag{22.64}$$

Chemists often define their chemical potentials as the Gibbs function per mole, rather than per particle. In those units, one would have

$$\mu(p) = \mu^{\ominus} + RT \ln \frac{p}{p^{\ominus}}. \tag{22.65}$$

Another way of solving this is to use the equation for the change in Gibbs function $dG = V\,dp - S\,dT$, which when the temperature is constant is $dG = V\,dp$. This can be integrated to give

$$G(p) = G^{\ominus} + \int_{p^{\ominus}}^{p} V\,dp$$

and hence

$$G(p) = G^{\ominus} + n_m RT \ln \frac{p}{p^{\ominus}}$$

for n_m moles of gas. Equation 22.65 then follows.

We are now ready to think about a simple chemical reaction. Consider the chemical reaction

$$A \rightleftharpoons B. \tag{22.66}$$

The symbol $\rightleftharpoons$ indicates that in this reaction it is possible to have both the forwards reaction A→B and the backwards reaction B→A. If we have a container filled with a mixture of A and B, and we leave it to react for a while, then depending on whether A→B is more or less important than B→A, we can determine the equilibrium concentrations of A and B. For gaseous reactions, the concentration of A (or B) is related to that species' partial pressure[10] p_A (or p_B). We define the **equilibrium constant** K as the ratio of these two partial pressures at equilibrium, i.e.,

$$K = \frac{p_B}{p_A}. \tag{22.67}$$

When $K \ll 1$, the backwards reaction dominates and our container will be mainly filled with A. When $K \gg 1$, the forwards reaction dominates and our container will be mainly filled with B.

The change in Gibbs function as this reaction proceeds is

$$dG = \mu_A\,dN_A + \mu_B\,dN_B. \tag{22.68}$$

However, since an increase in B is always accompanied by a corresponding decrease in A, we have that

$$dN_B = -dN_A, \tag{22.69}$$

and hence

$$dG = (\mu_B - \mu_A)\,dN_B. \tag{22.70}$$

Let us now denote the total molar Gibbs function[11] difference in a reaction by the symbol $\Delta_r G$. For a gaseous reaction, eqn 22.65 implies that

$$\Delta_r G = \Delta_r G^{\ominus} + RT \ln \frac{p_B}{p_A}, \tag{22.71}$$

[10] The partial pressure of a gas in a mixture is what the pressure of that gas would be if all other components suddenly vanished. Dalton's law states that the total pressure of a mixture of gases is equal to the sum of the individual partial pressures of the gases in the mixture (see Section 6.3).

[11] Thus $\Delta_r G = N_B(\mu_B - \mu_A)$. In many chemistry books it is conventional to define chemical potential as Gibbs function per mole, rather than Gibbs function per particle. Under this definition, one would have $\Delta_r G = \mu_B - \mu_A$.

where $\Delta_r G^\ominus$ is the difference between the molar chemical potentials of the two species. When $\Delta_r G < 0$, the forwards reaction $A \to B$ occurs spontanously. When $\Delta_r G > 0$, the backwards reaction $B \to A$ occurs spontanously. Equilibrium occurs when $\Delta_r G = 0$, and substituting this into eqn 22.71 and using eqn 22.67 shows that

$$\ln K = -\frac{\Delta_r G^\ominus}{RT}. \qquad (22.72)$$

Hence there is a direct relationship between the equilibrium constant of a reaction and the difference in chemical potentials (measured under standard conditions) of the product and reactant.[12]

[12] The reactant is defined to be the chemical on the left-hand side of the reaction; the product is defined to be the chemical on the right-hand side of the reaction.

It is useful to generalize these ideas to the case in which the chemical reaction is a bit more complicated than $A \rightleftharpoons B$. A general chemical reaction, with p reactants and q products, can be written in the form

$$\sum_{j=1}^{p}(-\nu_j)A_j \to \sum_{j=p+1}^{p+q}(+\nu_j)A_j, \qquad (22.73)$$

where the ν_j coefficients are here defined to be negative for the reactants and where A_j represents the jth substance. This can be rearranged to give

$$0 \to \sum_{j=1}^{p+q} \nu_j A_j. \qquad (22.74)$$

Example 22.4

Equation 22.53 can be applied to chemical reactions, such as

$$N_2 + 3H_2 \to 2NH_3. \qquad (22.75)$$

This can be cast into the general form of eqn 22.74 by writing

$$\nu_1 = -1, \qquad \nu_2 = -3, \qquad \nu_3 = 2. \qquad (22.76)$$

In a chemical system in equilibrium at constant temperature and pressure we have that the Gibbs function is minimized and so eqn 22.53 gives

$$\sum_{j=1}^{p+q} \mu_j dN_j = 0, \qquad (22.77)$$

where N_j is the number of molecules of type A_j. To keep the reaction balanced, the dN_j must be proportional to ν_j and hence

$$\sum_{j=1}^{p+q} \nu_j \mu_j = 0. \qquad (22.78)$$

This equation is very general.

Example 22.5
For the chemical reaction
$$N_2 + 3H_2 \rightarrow 2NH_3,$$
eqn 22.78 implies that
$$-\mu_{N_2} - 3\mu_{H_2} + 2\mu_{NH_3} = 0. \tag{22.79}$$

One can generalize the previous definition of the equilibrium constant for a gaseous reaction in eqn 22.67 (for a simple $A \rightleftharpoons B$ reaction) to the following expression (for our general reaction in eqn 22.74):

$$K = \prod_{j=1}^{p+q} \left(\frac{p_j}{p^\ominus}\right)^{\nu_j}. \tag{22.80}$$

Example 22.6
For the chemical reaction
$$N_2 + 3H_2 \rightarrow 2NH_3,$$
the equilibrium constant is
$$K = \frac{(p_{NH_3}/p^\ominus)^2}{(p_{N_2}/p^\ominus)(p_{H_2}/p^\ominus)^3} = \frac{p_{NH_3}^2 p^{\ominus 2}}{p_{N_2} p_{H_2}^3}. \tag{22.81}$$

Equilibrium, given by eqn 22.78, implies that
$$\sum_{j=1}^{p+q} \nu_j \left(\mu_j^\ominus + RT \ln \frac{p_j}{p^\ominus}\right) = 0 \tag{22.82}$$

and writing
$$\Delta_r G^\ominus = \sum_{j=1}^{p+q} \nu_j \mu_j^\ominus, \tag{22.83}$$

we have that
$$\Delta_r G^\ominus + RT \sum_{j=1}^{p+q} \nu_j \ln \frac{p_j}{p^\ominus} = 0 \tag{22.84}$$

and hence
$$\Delta_r G^\ominus + RT \ln K = 0, \tag{22.85}$$

or equivalently

$$\ln K = -\frac{\Delta_r G^\ominus}{RT}, \tag{22.86}$$

in agreement with eqn 22.72 (which was proved only for the simple reaction $A \rightleftharpoons B$).

Since $\ln K = -\Delta_r G^\ominus / RT$, we have that

$$\frac{d \ln K}{dT} = -\frac{1}{R} \frac{d(\Delta_r G^\ominus / T)}{dT}, \tag{22.87}$$

and using the Gibbs–Helmholtz relation (eqn 16.26) this becomes

$$\frac{d \ln K}{dT} = \frac{\Delta_r H^\ominus}{RT^2}. \tag{22.88}$$

Note that if the reaction is exothermic under standard conditions, then $\Delta_r H^\ominus < 0$ and hence K decreases as temperature increases. Equilibrium therefore shifts away from the products of the reaction.

If on the other hand the reaction is endothermic under standard conditions, then $\Delta_r H^\ominus > 0$ and hence K increases as temperature increases. Equilibrium therefore shifts towards the products of the reaction.

This observation agrees with **Le Chatelier's principle**, which states that "a system at equilibrium, when subjected to a disturbance, responds in such a way as to minimize that disturbance". In this case an exothermic reaction produces heat and this can raise the temperature, which then slows the forwards reaction towards the products. In the case of an endothermic reaction, heat is absorbed by the reactants and this can lower the temperature which would speed up the forwards reaction towards the products.

Equation 22.88 can be written in the following form:

$$\frac{d \ln K}{d(1/T)} = -\frac{\Delta_r H^\ominus}{R}, \tag{22.89}$$

which is known as the **van 't Hoff equation**.[13] This implies that a graph of $\ln K$ against $1/T$ should yield a straight line whose gradient is $-\Delta_r H^\ominus / R$. This fact is used in the following example.

[13] Jacobus Henricus van 't Hoff (1852–1911).

Example 22.7

Consider the dissociation reaction of molecular hydrogen into atomic hydrogen, i.e., the reaction

$$H_2 \rightarrow H\cdot + H\cdot \tag{22.90}$$

The equilibrium constant for this reaction is plotted in Fig. 22.4. The plot of K against T emphasizes that the "equilibrium for this reaction is well and truly on the left", meaning that the main constituent is H_2; molecular hydrogen is only very slightly dissociated even at 2000 K. Plotting the same data as $\ln K$ against $1/T$ yields a straight-line graph whose gradient yields $-\Delta H^\ominus/R$ for this reaction. For these data we find that $\Delta H^\ominus$ is about $440\,\text{kJ}\,\text{mol}^{-1}$. This is positive and hence the reaction is endothermic, which makes sense because you need to heat H_2 to break the molecular bond. This corresponds to a bond enthalpy per hydrogen molecule of $(440\,\text{kJ}\,\text{mol}^{-1}/N_A e) \approx 4.5\,\text{eV}$.

22.9 Osmosis

We have seen in Section 22.2 that differences in chemical potential can drive a flow of particles from one reservoir to another. This is driven by entropy, as the joint system finds its most likely macrostate (maximizing the entropy). This flow of particles can give rise to a force (sometimes described as an **entropic force** because it is determined by entropy, rather than energy, considerations). A good example of this is the phenomenon of **osmosis**.

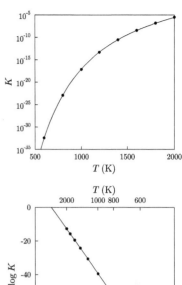

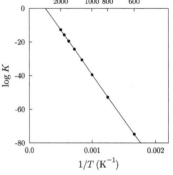

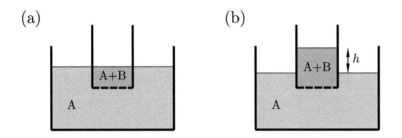

Fig. 22.4 The equilibrium constant for the reaction $H_2 \to H\cdot + H\cdot$, as a function of temperature. The same data are plotted in two different ways.

Fig. 22.5 (a) A bath of solvent A surrounds a solution of B dissolved in A. A semipermeable membrane allows solvent to flow through it but not the solute B. (b) Osmotic flow of solvent into the solution and in equilibrium the level of the solution is higher than that of the pure solvent.

Consider the situation shown in Fig. 22.5(a), in which a solution of some solute is contained inside a **semipermeable membrane** which allows the smaller solvent molecules to flow through it, but not the larger solute molecules. This is placed in a large bath of pure solvent. For example, the solvent could be water and the solute could be sugar. What is observed is that pure water is drawn through the semipermeable membrane and into the sugar solution (this is called **osmotic flow**), causing the level to rise up until equilibrium is obtained as shown in Fig. 22.5(b). The height h reached by the solution is proportional to the **osmotic pressure**, $\Pi = \rho_\text{solution} g h$, the additional pressure needed to

The liquid forming the main component is known as the **solvent**, while the material dissolved in it is known as the **solute**.

stop the osmotic flow and bring about equilibrium, and this pressure is provided by the column of solution (density ρ_{solution}).

It is osmosis that is responsible for giving cells their internal pressure and that provides the structural stability of many plants; water is absorbed into a plant's cells and provides **turgor pressure**. Water in the ground seems to be "sucked up" from the roots up to the top of the plant, but no suction is involved; the water flows upwards because it is driven by osmotic pressure. The water draws small nutrients from the soil with it as it flows up into the plant, and some water evaporates from the leaves (**transpiration**, a process that occurs through tiny pores in the leaves, called stomata), and provides the plant with cooling. Failure to water a plant results in wilting because the turgor pressure falls; wilting can also be produced by watering with salty water, which would cause osmotic flow of water out of the plant.

Freshwater and saltwater fish are adapted for their respective environments and can be harmed by placing them in water with the wrong salinity, precisely because this upsets the osmotic balance of their cells. Amoebae swimming in fresh water have to contend with fluid continually passing through their cell's wall, and need a pump (the **contractile vacuole**) to discharge water periodically and prevent bursting.

Even our own blood cells are adapted for a particular osmotic pressure. Therefore, in blood tranfusions and intravenous feeding it is important that the liquid injected is **isotonic** with blood; if instead it is too concentrated, and therefore has a higher osmotic pressure than blood, it is said to be **hypertonic** and will draw water out of the blood cells, causing them to shrivel; if it is too diluted, and therefore has a lower osmotic pressure than blood, it is said to be **hypotonic** and water will flow into the blood cells, causing them to burst.

What causes the osmotic pressure? The answer is that it is nothing more than the tendency to maximize entropy, the drive of a system to approach equilibrium because it is the most likely state of the system. Temperature gradients are equalized by the flow of heat and concentration gradients are equalized by the flow of particles. The difference in concentration between the solvent on one side of the semipermeable membrane and the other therefore drives an osmotic flow (see Fig. 22.6(a)), and its effects can only be countered by providing a pressure-driven flow in the opposite direction (and this is what happens in equilibrium, see Fig. 22.6(b)). In fact, if you apply a greater pressure to the solution than the osmotic pressure you can cause the pressure-driven flow to be larger than the osmotic flow, leading to the phenomenon of **reverse osmosis** (see Fig. 22.6(c)). This is used in some water filtration and desalination processes, whereby for example pure water can be caused to flow out of a quantity of sea water contained inside a suitable membrane by applying mechanical pressure.

The osmotic pressure is an example of an entropic force, but there are many others. In Section 17.1 we found that if you hang a weight by a piece of rubber and heat the rubber, it will contract, thus raising the weight. This is another entropic force because the contraction of the

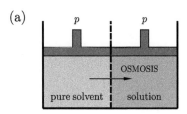
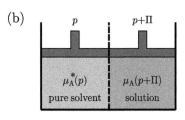
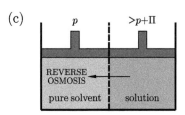

Fig. 22.6 (a) Osmotic flow into a solution from pure solvent, through a semipermeable membrane. Both sides are subjected to an identical pressure p, shown schematically by the pistons. (b) In equilibrium, the chemical potential of the solvent and solution are equal. This occurs when the solution is subjected to an additional pressure equal to Π, the osmotic pressure, over and above the pressure p. (c) If a pressure difference is provided that is larger than this, flow of solvent molecules from the solution to the solvent occurs (reverse osmosis).

rubber that raises the weight is due to maximizing entropy (there are many more disordered configurations of the rubber molecules in which their mean random-walk length is shorter than there are in which the rubber molecules are fully extended in a long line, see Fig. 17.2).

Example 22.8

Find the chemical potential of a solvent A with a solute B dissolved in it. The mole fraction of the solvent is x_A.

Solution:
Recall from eqn 22.65 that the chemical potential[14] of a gas (let us call it a gas of molecules of A) with pressure p_A^* is given by

[14] We are here again using the chemistry definition of chemical potential as Gibbs function *per mole*.

$$\mu_A^{(g)*} = \mu_A^\ominus + RT \ln \frac{p_A^*}{p^\ominus}, \tag{22.91}$$

where the superscript (g) indicates gas and the superscript * indicates that we are dealing with a pure substance. If this is in equilibrium with the liquid form of A, then we also have

$$\mu_A^{(\ell)*} = \mu_A^\ominus + RT \ln \frac{p_A^*}{p^\ominus}, \tag{22.92}$$

where the superscript (ℓ) indicates liquid. Now imagine that we mix some B molecules into the liquid. The mole fraction of A, x_A, is now less than one. The chemical potential of A in the liquid is now still equal to the chemical potential of A in the gas, but the gas has a different vapour pressure p_A (no asterisk because we are no longer dealing with pure substances). Thus

$$\mu_A^{(\ell)} = \mu_A^{(g)} = \mu_A^\ominus + RT \ln \frac{p_A}{p^\ominus}. \tag{22.93}$$

Equations 22.92 and 22.93 give that

$$\mu_A^{(\ell)} = \mu_A^{(\ell)*} + RT \ln \frac{p_A}{p_A^*}. \tag{22.94}$$

The vapour pressure of A in the mixed system can be estimated using **Raoult's law**,[15] which states that $p_A = x_A p_A^*$ (i.e., that the vapour pressure of A is proportional to its mole fraction). Hence eqn 22.94 becomes

[15] François-Marie Raoult (1830–1901).

$$\mu_A^{(g)} = \mu_A^{(\ell)} = \mu_A^{(\ell)*} + RT \ln x_A. \tag{22.95}$$

Since $x_A < 1$, we find that $\mu_A^{(\ell)} < \mu_A^{(\ell)*}$ and so the chemical potential of A is depressed compared to the pure case.[16]

[16] This result will be used to explain colligative properties in Section 28.6.

We can use the result in the previous example to derive an equation which is useful in describing osmosis for dilute solutions. Let us consider

the equilibrium between (i) the pure solvent, A, held at pressure p (provided by the atmosphere) and (ii) the solution, which contains a small number of B molecules dissolved in solvent A, held at pressure $p + \Pi$, where Π is the osmotic pressure (see Fig. 22.6(b)). As before, (i) and (ii) are separated by a semipermeable membrane which only allows passage of the solvent A molecules. Equilibrium between the A molecules implies

$$\mu_A^*(p) = \mu_A(p + \Pi), \tag{22.96}$$

or equivalently

$$\mu_A^*(p) = \mu_A^*(p + \Pi) + RT \ln x_A, \tag{22.97}$$

where the second term in this equation uses the result from eqn 22.95. The pressure dependence of μ_A can be accounted for by remembering that $(\partial G/\partial p)_T = V$ and hence $\mu_A^*(p+\Pi) = \mu_A^*(p) + \int_p^{p+\Pi} V_A \, dp$ where V_A is the partial molar volume of solvent, which we can assume is constant over this pressure range. Hence

$$\mu_A^*(p) = \mu_A^*(p) + \Pi V_A + RT \ln x_A, \tag{22.98}$$

and therefore

$$\Pi V_A = -RT \ln x_A. \tag{22.99}$$

Remembering that $x_A + x_B = 1$ and $x_B \ll 1$, we can write $-\ln x_A \approx x_B$ and hence

$$\Pi V_A = RT x_B. \tag{22.100}$$

Further writing $x_B = n_B/(n_A + n_B)$ and the total volume $V \approx n_A V_A$ and $n_B \ll n_A$, we have

$$\Pi V = n_B RT. \tag{22.101}$$

This equation applies only to dilute, ideal solutions.[17] The ratio n_B/V is the concentration of the solution, expressed as the number of moles of solute per unit volume.[18]

We close by trying to examine what causes the osmotic flow at a microscopic level. As shown in Fig. 22.7, although the small solvent molecules can pass through the membrane, the large solute molecules cannot. The number density of solvent molecules on the right-hand side is lower than on the left-hand side, and this concentration gradient has an effect close to the pores in the membrane (since solute molecules are excluded from this region, their centres never being able to cross to the left of the vertical dashed line). All the molecules bounce off each other and therefore momentum is transferred between molecules of all types. However, close to the membrane, the solute molecules receive a kick from the membrane only in the rightwards direction. This momentum is eventually distributed throughout all the molecules on the right-hand side of the diagram, since the small solvent molecules colliding with these solute molecules receive a net transfer of rightwards momentum. This drags solvent through the pores and causes the osmotic flow. Equilibrium is only obtained when a pressure equal to the osmotic pressure is provided on the right-hand side and causes an equal and opposite flow of solvent molecules in a leftwards direction.

---

[17] In more advanced treatments, osmotic pressure can be modelled using a virial-type expression

$$\Pi V = n_B RT(1 + \alpha(n_B/V) + \cdots),$$

where α is a constant. Measuring Π as a function of the mass concentration of the solute in the solution (proportional to n_B/V) can allow this constant to be determined, as well as the molar mass of the solvent and this is useful in the mass determination of macromolecules (the technique is called **osmometry**).

[18] Chemists write the concentration n_B/V using the symbol [B]. Using this notation, eqn 22.101 becomes

$$\Pi = [B]RT.$$

It is therefore no surprise that we find an ideal-gas type expression for the osmotic pressure in eqn 22.101 since this would be the pressure exerted by the solute molecules on the semipermeable membrane, treating those molecules as an ideal gas; by Newton's third law, that pressure is also exerted by the semipermeable membrane on the solute molecules and through collisions with the solvent leads to the rightwards osmotic flow. Only by applying an equal and opposite pressure on the right-hand side (the osmotic pressure) can equilibrium be obtained.

Chapter summary

- An extra term is appropriately introduced into the combined first and second laws to give $dU = TdS - pdV + \mu dN$, and this allows for cases in which the number of particles can vary.
- μ is the chemical potential, which can be expressed as $\mu = \left(\frac{\partial G}{\partial N}\right)_{p,T}$. It is also the Gibbs function per particle.
- For a system that can exchange particles with its surroundings, the chemical potential plays a similar rôle in particle exchange as temperature does in heat exchange.
- The grand partition function $\mathcal{Z}$ is given by $\mathcal{Z} = \sum_i e^{\beta(\mu N_i - E_i)}$.
- The grand potential is $\Phi_G = -k_B T \ln \mathcal{Z} = U - TS - \mu N = -pV$.
- $\mu = 0$ for particles with no conservation law.
- For a chemical reaction $dG = \sum \mu_j dN_j = 0$ and hence $\sum \nu_j \mu_j = 0$.
- The equilibrium constant K can be written as $\ln K = -\Delta_r G^\ominus / RT$.
- The temperature dependence of K follows $d \ln K / dT = \Delta_r H^\ominus / RT^2$.
- Osmosis is an example of an entropic force. An osmotic flow occurs from a solvent into a solution through a semipermeable membrane. The osmotic pressure Π obeys $\Pi V = n_B RT$ where n_B/V is the concentration of the solute.

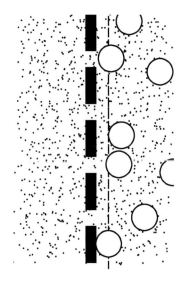

Fig. 22.7 A microscopic view of osmosis. The large solute molecules are kept on one side of the semipermeable membrane and their centres are confined in the region right of the vertical dashed line (drawn a radius away from the right-hand side of the semipermeable membrane).

Further reading

- Baierlein (2001) and Cook and Dickerson (1995) are both excellent articles concerning the nature of the chemical potential.
- Atkins and de Paulo (2006) contains a treatment of the chemical potential from the perspective of chemistry.

Exercises

(22.1) Maximize the entropy $S = -k_B \sum_i P_i \ln P_i$, where P_i is the probability of the ith level being occupied, subject to the constraints that $\sum P_i = 1$, $\sum P_i E_i = U$, and $\sum P_i N_i = N$ to rederive the grand canonical ensemble.

(22.2) The **fugacity** z is defined as $z = e^{\beta\mu}$. Using eqn 22.15, show that

$$z = n\lambda_{\text{th}}^3 \quad (22.102)$$

for an ideal gas, and comment on the limits $z \ll 1$ and $z \gg 1$.

(22.3) Estimate the bond enthalpy of Br_2 using the data plotted in Fig. 22.8.

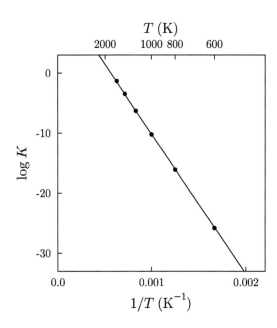

Fig. 22.8 The equilibrium constant for the reaction $Br_2 \to Br\cdot + Br\cdot$, as a function of temperature.

(22.4) Derive eqns 22.21, 22.22, and 22.23.

(22.5) If the partition function Z_N of a gas of N indistinguishable particles is given by $Z_N = Z_1^N/N!$, where Z_1 is the single–particle partition function, show that the chemical potential is given by

$$\mu = -k_B T \ln \frac{Z_1}{N}. \quad (22.103)$$

(22.6) (a) Consider the ionization of atomic hydrogen, governed by the equation

$$H \rightleftharpoons p^+ + e^-, \quad (22.104)$$

where p^+ is a proton (equivalently a positively ionized hydrogen) and e^- is an electron. Explain why

$$\mu_H = \mu_p + \mu_e. \quad (22.105)$$

Using the partition function for hydrogen atoms from eqn 21.50, and using eqn 22.103, show that

$$-k_B T \ln \frac{Z_1^p}{N_p} - k_B T \ln \frac{Z_1^e}{N_e} = -k_B T \ln \frac{Z_1^H}{N_H} e^{\beta R}, \quad (22.106)$$

where Z_1^x and N_x are the single–particle partition function and number of particles for species x respectively, and where $R = 13.6\,\text{eV}$. Hence show that

$$\frac{n_e n_p}{n_H} = \frac{(2\pi m_e k_B T)^{3/2}}{h^3} e^{-\beta R}, \quad (22.107)$$

where $n_x = N_x/V$ is the number density of species x, stating any approximations you make. Equation 22.107 is known as the **Saha equation**.

(b) Explain why charge neutrality implies that $n_e = n_p$ and conservation of nucleons implies $n_H + n_p = n$, where n is the total number density of hydrogen (neutral and ionized). Writing $y = n_p/n$ as the degree of ionization, show that

$$\frac{y^2}{1-y} = \frac{e^{-\beta R}}{n\lambda_{\text{th}}^3}, \quad (22.108)$$

where λ_{th} is the thermal wavelength for the electrons. Find the degree of ionization of a cloud of atomic hydrogen at 1000 K and density $10^{20}\,\text{m}^{-3}$.

(c) Equation 22.108 shows that the degree of ionization goes up when the density n goes down. Why is that?

(22.7) A solution of NaCl dissolved in water, which is 0.9% by weight NaCl, is isotonic with blood. Show that the osmotic pressure of blood is nearly eight times atmospheric pressure.

Photons

23

In this chapter, we will consider the thermodynamics of electromagnetic radiation. It was Maxwell who realized that light was an electromagnetic wave and that the speed of light, c, could be expressed in terms of fundamental constants taken from the theories of electricity and magnetism. In modern notation, this relation is

$$c = 1/\sqrt{\epsilon_0 \mu_0}, \qquad (23.1)$$

where ϵ_0 and μ_0 are the permittivity and permeability of free space respectively. Later, Planck realized that light behaved not only like a *wave* but also like a *particle*. In the language of quantum mechanics, electromagnetic waves can be *quantized* as a set of particles, which are known as **photons**. Each photon has an energy $\hbar\omega$ where $\omega = 2\pi\nu$ is the angular frequency.[1] Each photon has a momentum $\hbar k$ where k is the wave vector.[2] The ratio of the energy to the momentum of a photon is

$$\frac{\omega}{k} = 2\pi\nu \times \frac{\lambda}{2\pi} = \nu\lambda = c. \qquad (23.2)$$

Electromagnetic radiation is emitted from any substance at non-zero temperature. This is known as **thermal radiation**. For objects at room temperature, you may not have noticed this effect because the frequency of the electromagnetic radiation is low and most of the emission is in the infrared region of the electromagnetic spectrum. Our eyes are only sensitive to electromagnetic radiation in the visible region. However, you may have noticed that a piece of metal in a furnace glows "red hot" so that, for such objects at higher temperature, your eyes are able to pick up some of the thermal radiation.[3]

This chapter is all about the properties of this thermal radiation. We will begin in Sections 23.1–23.4 by restricting ourselves to simple thermodynamics arguments to derive as much as we can about thermal radiation without going into the gory details, in much the same way as was originally done in the nineteenth century. This approach doesn't get us the whole way, but provides a lot of insight. Then in Sections 23.5 and 23.6, we will use the more advanced statistical mechanical techniques introduced in the previous chapters to do the job properly. The final sections concern the thermal radiation that exists in the Universe as a remnant of the hot big bang and the effect of thermal radiation on the behaviour of atoms and hence the operation of the laser.

23.1	The classical thermodynamics of electromagnetic radiation 264
23.2	Spectral energy density 265
23.3	Kirchhoff's law 266
23.4	Radiation pressure 268
23.5	The statistical mechanics of the photon gas 269
23.6	Black-body distribution 270
23.7	Cosmic microwave background radiation 273
23.8	The Einstein A and B coefficients 274
	Chapter summary 277
	Further reading 277
	Exercises 278

[1] ν is the frequency. The energy can also be expressed as $h\nu$. Recall also that $\hbar = h/(2\pi)$.

[2] The wave vector $k = 2\pi/\lambda$ where λ is the wavelength.

[3] Your eyes can pick up a lot of the thermal radiation if they are assisted by infrared goggles.

23.1 The classical thermodynamics of electromagnetic radiation

In this section, we will consider the thermodynamics of electromagnetic radiation from a classical standpoint, although we will allow ourselves the post-nineteenth century luxury of considering the electromagnetic radiation to consist of a gas of photons. First we will consider the effect of a collection of photons on the surroundings that contain it. Let us consider the surroundings to be a container of volume V, which in this subject is termed a "cavity", which is held at temperature T. The photons inside the cavity are in thermal equilibrium with the cavity walls, and form electromagnetic standing waves. The walls of the cavity, shown in Fig. 23.1, are made of diathermal material (i.e., they transmit heat between the gas of photons inside the cavity and the surroundings). If n photons per unit volume comprise the gas of photons in the cavity then the energy density u of the gas may be written as:

$$u = \frac{U}{V} = n\hbar\omega, \tag{23.3}$$

where $\hbar\omega$ is the mean energy of a photon. From kinetic theory (eqn 6.15), the pressure p of a gas of particles is $\frac{1}{3}nm\langle v^2\rangle$. For photons, we replace $\langle v^2\rangle$ in this formula by c^2, the square of the speed of light. Interpreting mc^2 as the energy of a photon, we then have that p is one-third of the energy density. Thus

$$p = \frac{u}{3}, \tag{23.4}$$

which is different from the expression in eqn 6.25 ($p = 2u/3$) from the kinetic theory of gases, a point which we will return to in Section 25.2 (see eqn 25.21).[4] Equation 23.4 gives an expression for the radiation pressure due to the electromagnetic radiation. Also from kinetic theory (eqn 7.6), the flux Φ of photons on the walls of their container, that is to say the number of photons striking unit area of their container per second, is given by

$$\Phi = \frac{1}{4}nc, \tag{23.5}$$

where c is the speed of light. From this, and eqn 23.3, we can write the power incident per unit area of cavity wall, due to the photons, as

$$F = \hbar\omega\Phi = \frac{1}{4}uc. \tag{23.6}$$

This relation will be important as we now derive the Stefan–Boltzmann law, which relates the temperature of a body to the energy flux radiating from it in the form of electromagnetic radiation. We can derive this using the first law of thermodynamics in the form $dU = TdS - pdV$ to give

$$\begin{aligned}\left(\frac{\partial U}{\partial V}\right)_T &= T\left(\frac{\partial S}{\partial V}\right)_T - p \\ &= T\left(\frac{\partial p}{\partial T}\right)_V - p, \end{aligned} \tag{23.7}$$

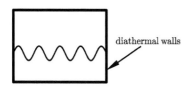

Fig. 23.1 A cavity of photons whose walls are diathermal, meaning they are in thermal contact with their surroundings, so that the temperature within may be controlled.

[4] The factor of two difference arises from writing the kinetic energy as mc^2 and not as $\frac{1}{2}m\langle v^2\rangle$, and thus reflects the difference in form between the equation for the relativistic energy of a photon and that for the kinetic energy of a non-relativistic particle.

where the last equality follows from using a Maxwell relation. The left-hand side of eqn 23.7 is simply[5] the energy density u. Hence, using eqn 23.7, together with eqn 23.4, we obtain

$$u = \frac{1}{3}T\left(\frac{\partial u}{\partial T}\right)_V - \frac{u}{3}. \tag{23.8}$$

Rearranging gives:

$$4u = T\left(\frac{\partial u}{\partial T}\right)_V, \tag{23.9}$$

from which follows

$$4\frac{dT}{T} = \frac{du}{u}. \tag{23.10}$$

Equation 23.10 may be integrated to give:

$$u = AT^4, \tag{23.11}$$

where A is a constant of the integration with units $\mathrm{J\,K^{-4}\,m^{-3}}$. We can now use eqn 23.6 to give us the power incident[6] per unit area.[7]

$$F = \frac{1}{4}uc = \left(\frac{1}{4}Ac\right)T^4 = \sigma T^4, \tag{23.12}$$

where the term in brackets, $\sigma = \frac{1}{4}Ac$, is the **Stefan–Boltzmann constant**. Equation 23.12 is known as the **Stefan–Boltzmann law** or sometimes as **Stefan's law**. For the moment, we have no idea what value the constant σ takes and this is something that was originally determined from experiment. In Section 23.5, using the techniques of statistical mechanics, we will derive an expression for this constant.

[5]This should be obvious since it is the definition of energy density. However, if you want to convince yourself, notice that differentiating $U = uV$ with respect to V yields

$$\left(\frac{\partial U}{\partial V}\right)_T = u + V\left(\frac{\partial u}{\partial V}\right)_T = u,$$

because $\left(\frac{\partial u}{\partial V}\right)_T = 0$ since u, an energy density, is independent of volume.

[6]Note that when the cavity is in equilibrium with the radiation inside it, the power incident is equal to the power emitted; hence the expression for F expresses the power emitted by the surface and the power incident on the surface.

[7]The power per unit area is equal to an energy flux, sometimes called a radiative flux or irradiance, see Section 37.3.

23.2 Spectral energy density

The energy density u of electromagnetic radiation is a quantity that tells you how many Joules are stored in a cubic metre of cavity. What we want to do now is to specify in which frequency ranges that energy is stored. All of this will fall out of the statistical mechanical treatment in Section 23.5, but we want to continue to apply a classical treatment to see how far we can get. To do this, consider two containers, each in contact with thermal reservoirs at temperature T and joined to one another by a tube, as illustrated schematically in Fig. 23.2. The system is allowed to come to equilibrium.

The thermal reservoirs are at the same temperature T and so we know from the second law of thermodynamics that there can be no net heat flow from either one of the bodies to the other. Therefore there can be no net energy flux along the tube, so that the energy flux from the soot-lined cavity along the tube from left to right must be balanced by the energy flux from the mirror-lined cavity along the tube from right to left. Equation 23.11 thus tells us that each cavity must have the same energy density u. This argument can be repeated for cavities of different

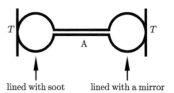

Fig. 23.2 Two cavities at temperature T: one is lined with soot and the other with a mirror coating.

shape and size as well as different coatings. Hence we conclude that u is independent of shape, size, or material of the cavity. But maybe one cavity might have more energy density than the other at certain wavelengths, even if it has to have the same energy density overall? This is not the case, as we shall now prove. First, we make a definition.

- The **spectral energy density** u_λ is defined as follows: $u_\lambda \mathrm{d}\lambda$ is the energy density due to those photons which have wavelengths between λ and $\lambda + \mathrm{d}\lambda$. The total energy density is then

$$u = \int u_\lambda \, \mathrm{d}\lambda. \tag{23.13}$$

u_λ has units $\mathrm{J\,m^{-3}\,m^{-1}}$.

We can also define a spectral density in terms of frequency ν, so that $u_\nu \mathrm{d}\nu$ is the energy density due to those photons which have frequencies between ν and $\nu + \mathrm{d}\nu$.

Now imagine that a filter, which only allows a narrow band of radiation at wavelength λ to pass, is inserted at point A in Fig. 23.2 and the system is left to come to equilibrium. The same arguments listed above apply in this case: there is no net energy flux from one cavity to the other and hence the specific internal energy within a narrow wavelength range is the same for each case:

$$u_\lambda^{\mathrm{soot}}(T) = u_\lambda^{\mathrm{mirror}}(T). \tag{23.14}$$

This demonstrates that the spectral internal energy has no dependence on the material, shape, size, or nature of a cavity. The spectral energy density is thus a universal function of λ and T only.

23.3 Kirchhoff's law

We now wish to discuss how well particular surfaces of a cavity will absorb or emit electromagnetic radiation of a particular frequency or wavelength. We therefore make the following additional definitions:

- The **spectral absorptivity** α_λ is the fraction of the incident radiation that is absorbed at wavelength λ.
- The **spectral emissive power** e_λ of a surface is a function such that $e_\lambda \mathrm{d}\lambda$ is the power emitted per unit area by the electromagnetic radiation having wavelengths between λ and $\lambda + \mathrm{d}\lambda$.

α_λ is dimensionless.

e_λ has units $\mathrm{W\,m^{-2}\,m^{-1}}$.

Using these definitions, we may now write down the form for the power per unit area absorbed by a surface, if the incident spectral energy density is $u_\lambda \mathrm{d}\lambda$, as follows:

$$\left(\frac{1}{4} u_\lambda \mathrm{d}\lambda \, c\right) \alpha_\lambda. \tag{23.15}$$

The power per unit area emitted by a surface is given by

$$e_\lambda \mathrm{d}\lambda. \tag{23.16}$$

In equilibrium, the expressions in eqns 23.15 and 23.16 must be equal, and hence

$$\frac{e_\lambda}{\alpha_\lambda} = \frac{c}{4} u_\lambda. \tag{23.17}$$

Equation 23.17 expresses **Kirchhoff's law**, which states that the ratio e_λ/α_λ is a universal function of λ and T. Therefore, if you fix λ and T, the ratio e_λ/α_λ is fixed and hence $e_\lambda \propto \alpha_\lambda$. In other words "good absorbers are good emitters" and "bad absorbers are bad emitters".

Example 23.1

Dark-coloured objects which absorb most of the light that falls on them will be good at emitting thermal radiation. One has to be a bit careful here because you have to be sure about which wavelength you are talking about. A better statement of Kirchhoff's laws would be "good absorbers at one wavelength are good emitters at the same wavelength".

For example, a white coffee mug absorbs poorly in visible wavelengths so looks white. A black, but otherwise identical, coffee mug absorbs well in visible wavelengths so looks black. Which one is best at keeping your coffee warm? You might conclude that it is the white mug because "poor absorbers are poor emitters" and that the mug will lose less heat by thermal radiation. However, a hot mug emits radiation mainly in the infrared region of the electromagnetic spectrum,[8] and so the mug being white in the visible is immaterial; what you need to know is what "colour" each mug is in the infrared, i.e., measuring their absorption spectra at infrared wavelengths will tell you about their emission properties there.

[8]See Appendix D.

A perfect **black body** is an object that is defined to have $\alpha_\lambda = 1$ for all λ. Kirchhoff's law expressed in eqn 23.17 tells us that for this maximum value of α, a black body is the best possible emitter. It is often useful to think of a **black-body cavity**, which is an enclosure whose walls have $\alpha_\lambda = 1$ for all λ and which contains a gas of photons at the same temperature as the walls, due to emission and absorption of photons by the atoms in the walls. The gas of photons contained in the black-body cavity is known as **black-body radiation**.

Example 23.2

The temperature of the Earth's surface is maintained by radiation from the Sun. By making the approximation that the Sun and the Earth behave as black bodies, show that the ratio of the Earth's temperature to that of the Sun is given by

$$\frac{T_{\text{Earth}}}{T_{\text{Sun}}} = \sqrt{\frac{R_{\text{Sun}}}{2D}}, \qquad (23.18)$$

where R_{Sun} is the radius of the Sun and the Earth–Sun separation is D.

Solution:
The Sun emits a power equal to its surface area $4\pi R_{\text{Sun}}^2$ multiplied by σT_{Sun}^4. This power is known as its **luminosity** L (measured in watts):

$$L = 4\pi R_{\text{Sun}}^2 \sigma T_{\text{Sun}}^4. \tag{23.19}$$

At a distance D from the Sun, this power is uniformly distributed over a sphere with surface area $4\pi D^2$, and the Earth is only able to "catch" this power over its projected area πR_{Earth}^2. Thus the power incident on the Earth is

$$\text{power incident} = L\left(\frac{\pi R_{\text{Earth}}^2}{4\pi D^2}\right). \tag{23.20}$$

The power emitted by the Earth, assuming that it has a uniform temperature T_{Earth} and behaves as a black body, is simply $\sigma T_{\text{Earth}}^4$ multiplied by the Earth's surface area $4\pi R_{\text{Earth}}^2$, so that

$$\text{power emitted} = 4\pi R_{\text{Earth}}^2 \sigma T_{\text{Earth}}^4. \tag{23.21}$$

Equating eqn 23.20 and eqn 23.21 yields the desired result.

Putting in the numbers $R_{\text{Sun}} = 7 \times 10^8$ m, $D = 1.5 \times 10^{11}$ m, and $T_{\text{Sun}} = 5800$ K yields $T_{\text{Earth}} = 280$ K, which is not bad given the crudeness of the assumptions.

23.4 Radiation pressure

To summarize the results of the earlier sections in this chapter, for black-body radiation we have:

$$\text{power radiated per unit area} \quad F = \frac{1}{4}uc = \sigma T^4, \tag{23.22}$$

$$\text{energy density in radiation} \quad u = \left(\frac{4\sigma}{c}\right)T^4, \tag{23.23}$$

$$\text{pressure on cavity walls} \quad p = \frac{u}{3} = \frac{4\sigma T^4}{3c}. \tag{23.24}$$

If, however, one is dealing with a beam of light, in which all the photons are going in the same direction (rather than in each and every direction as we have in a gas of photons) then these results need to be modified. The pressure exerted by a collimated beam of light can be calculated as follows: a cubic metre of this beam has momentum $n\hbar k = n\hbar\omega/c$, and this momentum is absorbed by a unit area of surface, normal to the beam, in a time $1/c$. Thus the pressure is $p = [n\hbar\omega/c]/[1/c] = n\hbar\omega = u$. A cubic metre of the beam has energy $n\hbar\omega$, so the power F incident on unit area of surface is $F = n\hbar\omega/(1/c) = uc$. Hence, we have[9]

$$\text{power radiated per unit area} \quad F = uc = \sigma T^4, \tag{23.25}$$

$$\text{energy density in radiation} \quad u = \left(\frac{\sigma}{c}\right)T^4, \tag{23.26}$$

$$\text{pressure on cavity walls} \quad p = u = \frac{\sigma T^4}{c}. \tag{23.27}$$

[9] Equations 23.25–23.27 are appropriate for a situation in which some black-body radiation, emitted by a cavity at temperature T, has been collimated into a beam.

It is worth emphasizing that electromagnetic radiation exerts a real pressure on a surface and this can be calculated using eqn 23.24 or eqn 23.27 as appropriate. An example of a calculation of radiation pressure is given below.

Example 23.3

Sunlight falls on the surface of the Earth with a power per unit area equal to $F = 1370\,\text{W}\,\text{m}^{-2}$. Calculate the radiation pressure and compare it with atmospheric pressure.

Solution:

Sunlight on the Earth's surface consists of photons all going in the same direction,[10] and hence we can use

$$p = \frac{F}{c} = 4.6\,\mu\text{Pa}, \qquad (23.28)$$

which is more than ten orders of magnitude lower than atmospheric pressure (which is $\sim 10^5\,\text{Pa}$).

[10] We make this approximation because the Sun is sufficiently far from the Earth, so that all the rays of light arriving on Earth are parallel.

23.5 The statistical mechanics of the photon gas

Our argument so far has only used classical thermodynamics. We have been able to predict that the energy density u of a photon gas behaves as AT^4 but we have been able to say nothing about the constant A. It was only through the development of quantum theory that it was possible to derive what A is, and we will present this in what follows. The crucial insight is that electromagnetic waves in a cavity can be described by simple harmonic oscillators. The angular frequency ω of each mode of oscillation is related to the wave vector k by

$$\omega = ck \qquad (23.29)$$

(see Fig. 23.3) and hence the density of states[11] of electromagnetic waves as a function of wave vector k is given by

$$g(k)\,\text{d}k = \frac{4\pi k^2\,\text{d}k}{(2\pi/L)^3} \times 2, \qquad (23.30)$$

where the cavity is assumed to be a cube of volume $V = L^3$ and the factor two corresponds to the two possible polarizations of the electromagnetic waves. Thus

$$g(k)\,\text{d}k = \frac{Vk^2\,\text{d}k}{\pi^2}, \qquad (23.31)$$

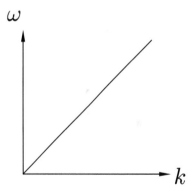

Fig. 23.3 The relation between ω and k, for example that in eqn 23.29, is known as a **dispersion relation**. For light (plotted here) this relation is very simple and is called non-dispersive because both the phase velocity (ω/k) and the group velocity ($\text{d}\omega/\text{d}k$) are equal.

[11] This treatment is similar to the analysis in Section 21.1 for the ideal gas.

and hence the density of states $g(\omega)$, now written as a function of frequency using eqn 23.29, is

$$g(\omega) = g(k)\frac{\mathrm{d}k}{\mathrm{d}\omega} = \frac{g(k)}{c}, \qquad (23.32)$$

and hence

$$g(\omega)\,\mathrm{d}\omega = \frac{V\omega^2\,\mathrm{d}\omega}{\pi^2 c^3}. \qquad (23.33)$$

We can derive U for the photon gas by using the expression for U for a single simple harmonic oscillator in eqn 20.29 to give

$$U = \int_0^\infty g(\omega)\,\mathrm{d}\omega\,\hbar\omega\left(\frac{1}{2} + \frac{1}{e^{\beta\hbar\omega}-1}\right). \qquad (23.34)$$

This presents us with a problem since the first part of this expression, due to the sum of all the zero-point energies, diverges:

$$\int_0^\infty g(\omega)\,\mathrm{d}\omega\,\frac{1}{2}\hbar\omega \to \infty. \qquad (23.35)$$

This must correspond to the energy of the vacuum, so after swallowing hard we redefine our zero of energy so that this infinite contribution is swept conveniently under the carpet. We are therefore left with

$$U = \int_0^\infty g(\omega)\,\mathrm{d}\omega\,\frac{\hbar\omega}{e^{\beta\hbar\omega}-1} = \frac{V\hbar}{\pi^2 c^3}\int_0^\infty \frac{\omega^3\,\mathrm{d}\omega}{e^{\beta\hbar\omega}-1}. \qquad (23.36)$$

If we make the substitution $x = \hbar\beta\omega$, we can rewrite this as

$$U = \frac{V\hbar}{\pi^2 c^3}\left(\frac{1}{\hbar\beta}\right)^4 \int_0^\infty \frac{x^3\,\mathrm{d}x}{e^x - 1} = \left(\frac{V\pi^2 k_B^4}{15 c^3 \hbar^3}\right) T^4, \qquad (23.37)$$

and hence $u = U/V = AT^4$. Here, use has been made of the integral

$$\int_0^\infty \frac{x^3\,\mathrm{d}x}{e^x - 1} = \zeta(4)\Gamma(4) = \frac{\pi^4}{15}, \qquad (23.38)$$

which is proved in Appendix C.4 (see eqn C.25). This therefore establishes that the constant $A = 4\sigma/c$ is given by

$$A = \frac{\pi^2 k_B^4}{15 c^3 \hbar^3}, \qquad (23.39)$$

and hence the Stefan–Boltzmann constant[12] σ is

$$\sigma = \frac{\pi^2 k_B^4}{60 c^2 \hbar^3} = 5.67 \times 10^{-8}\,\mathrm{W\,m^{-2}\,K^{-4}}. \qquad (23.40)$$

23.6 Black-body distribution

The expression in eqn 23.36 can be rewritten as

$$u = \frac{U}{V} = \int u_\omega\,\mathrm{d}\omega, \qquad (23.41)$$

---

[12] If you prefer to use h, rather than $\hbar$, the **Stefan–Boltzmann constant** is written as

$$\sigma = \frac{2\pi^5 k_B^4}{15 c^2 h^3}.$$

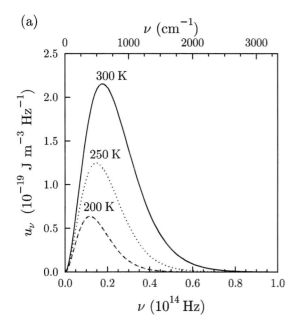
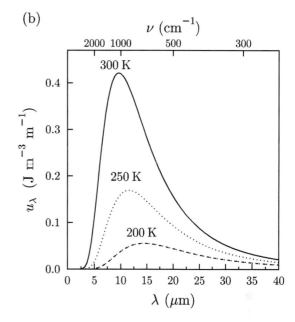

Fig. 23.4 The black-body distribution of spectral energy density, plotted for 200 K, 250 K, and 300 K as a function of (a) frequency and (b) wavelength. The upper scale shows the frequency in inverse centimetres, a unit beloved of spectroscopists.

where u_ω is a different form of the spectral energy density (written this time as a function of angular frequency $\omega = 2\pi\nu$). It thus takes the form

$$u_\omega = \frac{\hbar}{\pi^2 c^3} \frac{\omega^3}{e^{\beta\hbar\omega} - 1}. \quad (23.42)$$

This spectral energy density function is known as a **black-body distribution**. We can also express this in terms of frequency ν by writing $u_\omega \, d\omega = u_\nu \, d\nu$, and using $\omega = 2\pi\nu$ and hence $d\omega/d\nu = 2\pi$. This yields

$$u_\nu = \frac{8\pi h}{c^3} \frac{\nu^3}{e^{\beta h\nu} - 1}. \quad (23.43)$$

This function is plotted in Fig. 23.4(a). Similarly, we can transform this into wavelength, by writing[13] $u_\nu \, |d\nu| = u_\lambda \, |d\lambda|$, and using $\nu = c/\lambda$ and hence $d\nu/d\lambda = -c/\lambda^2$. This yields an expression for u_λ as follows:

$$u_\lambda = \frac{8\pi hc}{\lambda^5} \frac{1}{e^{\beta hc/\lambda} - 1}. \quad (23.44)$$

This is shown in Fig. 23.4(b).

We note several features of this black-body distribution.

- At low frequency (i.e., long wavelength), when $h\nu/k_\mathrm{B} T \ll 1$, the exponential term can be written as

$$e^{\beta h\nu} \approx 1 + \frac{h\nu}{k_\mathrm{B} T}, \quad (23.45)$$

[13] There are modulus signs in the expression since we are counting the energy density within either a frequency or a wavelength interval. Because $\nu = c/\lambda$, a positive $d\nu$ corresponds to a negative $d\lambda$, but we do not care in which direction the interval is defined.

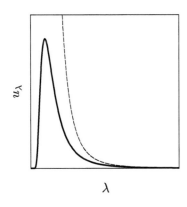

Fig. 23.5 The black-body energy density u_λ (thick solid line), together with the Rayleigh–Jeans expression ,eqn 23.47 (dashed line), which is the long-wavelength limit of the black-body distribution.

and hence
$$u_\nu \to \frac{8\pi k_B T \nu^2}{c^3}, \qquad (23.46)$$
and equivalently
$$u_\lambda \to \frac{8\pi k_B T}{\lambda^4}. \qquad (23.47)$$

These two expressions are different forms of the **Rayleigh–Jeans law**, and were derived without using quantum mechanics. As that might imply, Planck's constant h does not appear in them. These expressions are the correct limit of the black-body distribution, as shown in Fig. 23.5. They created problems before quantum mechanics was fully appreciated, because if you take the Rayleigh–Jeans form of u_λ and assume that it is true for all wavelengths, and then try and integrate it to get the total internal energy density u, you find that

$$u = \int_0^\infty u_\lambda \, d\lambda = \int_0^\infty \frac{8\pi k_B T \, d\lambda}{\lambda^4} \to \infty. \qquad (23.48)$$

This apparent divergence in u was called the **ultraviolet catastrophe**, because integrating down to small wavelengths (towards the ultraviolet) produced a divergence. In fact, such high–energy electromagnetic waves are not excited because light is quantized and it costs too much energy to produce an ultraviolet photon when the temperature is too low. Of course, using the correct black-body u_λ from eqn 23.44, the correct form

$$u = \int_0^\infty u_\lambda \, d\lambda = \frac{4\sigma}{c} T^4 \qquad (23.49)$$

is obtained.

- One can also define the **radiance** (or **surface brightness**) B_ν as the flux of radiation per steradian (the unit of solid angle, abbreviated to sr) in a unit frequency interval. This function gives the power *through* an element of unit area, per unit frequency, *from* an element of solid angle. The units of radiance are $\mathrm{W\,m^{-2}\,Hz^{-1}\,sr^{-1}}$. Because there are a total of 4π steradians, we have that[14]

$$B_\nu(T) = \frac{c}{4\pi} u_\nu(T) = \frac{2h}{c^2} \frac{\nu^3}{e^{\beta h\nu} - 1}. \qquad (23.50)$$

By analogy, B_λ, with units $\mathrm{W\,m^{-2}\,m^{-1}\,sr^{-1}}$, is defined by

$$B_\lambda(T) = \frac{c}{4\pi} u_\lambda(T) = \frac{2hc^2}{\lambda^5} \frac{1}{e^{\beta hc/\lambda} - 1}. \qquad (23.51)$$

- Wilhelm Wien[15] found experimentally in 1896, before the advent of quantum mechanics, that the product of the temperature and the wavelength at which the maximum of the black-body distribution u_λ is found is a constant. This is a statement of what is known as **Wien's law**. The constant can be given as follows:

$$\lambda_{\max} T = \text{a constant}. \qquad (23.52)$$

[14] Note that if we divide the energy density by the time taken for unit volume of photons to pass through unit area of surface, namely $1/c$, we have the energy flux.

[15] His full name was Wilhelm Carl Werner Otto Fritz Franz Wien (1864–1928).

Wien's law follows from the fact that $\lambda_{\max}$ can be determined by the condition $du_\lambda/d\lambda = 0$, and applying this to eqn 23.44 leads to $\beta hc/\lambda_{\max} =$ a constant. Hence $\lambda_{\max} T$ is a constant, which is Wien's law. The law tells us that at room temperature, objects that are approximately black bodies will radiate the most at wavelength $\lambda_{\max} \approx 10\,\mu\mathrm{m}$, which is in the infrared region of the electromagnetic spectrum, as demonstrated in Fig. 23.4(b). One can easily show[16] that the maximum in u_ν occurs at a frequency given by

[16] See Exercise 23.2.

$$\frac{h\nu}{k_\mathrm{B} T} = 2.82144 \qquad (23.53)$$

and the maximum in u_λ occurs at a wavelength given by

$$\frac{hc}{\lambda k_\mathrm{B} T} = 4.96511. \qquad (23.54)$$

This can be used to show that the product λT is given by

$$\lambda T = \begin{cases} 5.1\,\mathrm{mm\,K} & \text{at the maximum of } u_\nu(T), \\ 2.9\,\mathrm{mm\,K} & \text{at the maximum of } u_\lambda(T). \end{cases} \qquad (23.55)$$

These maxima do not occur at the same place for each distribution because one is measured per unit frequency interval and the other per unit wavelength interval, and these are different.[17]

Figure 23.6(a) shows how the shape of the distribution changes with temperature for u_ν and Fig. 23.6(b) for u_λ on log–log scales. These diagrams show how the peak of the black-body distribution lies in the optical region of the spectrum for temperatures of several thousand kelvin, but in the microwave region for a temperature of a few kelvin. This fact is very relevant for the black-body radiation in the Universe, which we describe in the following section.

[17] The difference between $d\nu$ and $d\lambda$ is derived as follows. The relationship between frequency and wavelength is given by
$$c = \nu\lambda,$$
and hence
$$\nu = c/\lambda,$$
so that
$$d\nu = -\frac{c}{\lambda^2} d\lambda.$$

23.7 Cosmic microwave background radiation

In 1978, Penzias and Wilson of Bell Labs, New Jersey, USA won the Nobel Prize for their serendipitous discovery (in 1963–1965) of seemingly uniform microwave emission coming from all directions in the sky, which has come to be known as the **cosmic microwave background** (CMB). Remarkably, the spectral shape of this emission exhibits, to high precision, the distribution for black-body radiation of temperature 2.7 K (see Fig. 23.7) with a peak in the emission spectrum at a wavelength of about 1 mm. It is startling that the radiation is uniform, or isotropic, to better than 1 part in 10^5 (meaning that its spectrum and intensity are almost the same if you measure in different directions in the sky). This is one of the key pieces of evidence in favour of the hot big bang model for the origin of the Universe. It implies that there was a time when all of the Universe we see now was in thermal equilibrium.[18]

[18] Note that different black-body distributions, that is multiple curves corresponding to regions at a variety of different temperatures, do not superpose to form a single black-body distribution.

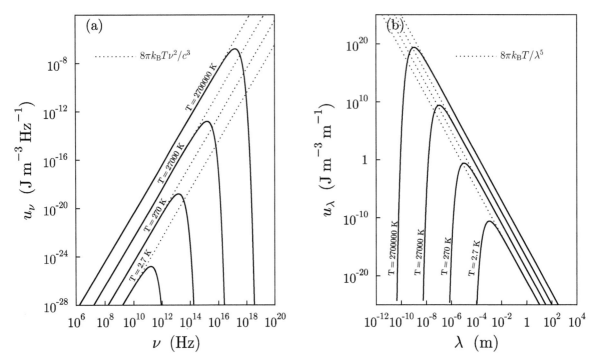

Fig. 23.6 The black-body distribution of spectral energy density, plotted on a logarithmic scale for four different temperatures as a function of (a) frequency and (b) wavelength. The dotted lines show the Rayleigh–Jeans law (eqns 23.46 and 23.47) which are valid in the low frequency (long wavelength) limit.

We can make various inferences about the origin of the Universe from observations of the cosmic microwave background. It can be shown that the energy density of radiation in the expanding Universe falls off as one over the fourth power of the scale factor (which you can think of as the linear magnification factor describing the separation of a pair of marker galaxies in the Universe, a quantity that increases with cosmic time). From the Stefan–Boltzmann law, the energy density of radiation is proportional to T^4, so temperature and scale factor are inversely proportional to one another, so the Universe cools as it expands. Conversely, when the Universe was much younger, it was much smaller and much hotter. Extrapolating back in time, one finds that temperatures were such that physical conditions were very different. For example, it was too hot for matter to exist as atoms, and everything was ionized. Further back in cosmic time still, even quarks and hadrons, the sub-structure of protons and neutrons were thought to be dissociated.

23.8 The Einstein A and B coefficients

If a gas of atoms is subjected to thermal radiation, the atoms can respond by making transitions between different energy levels. We can

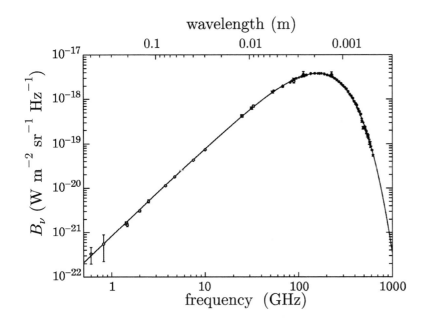

Fig. 23.7 The experimentally determined spectrum of the cosmic microwave background (data courtesy of NASA).

think about this effect in terms of absorption and emission of photons by the atom. The atoms are sitting in a bath of photons which we call the **radiation field** and it has an energy density u_ω given by eqn 23.42. In this section, we will consider the effect of this radiation field on the transitions between atomic energy levels by modelling the atom as a simple two-level system. Consider the two-level system shown in Fig. 23.8, which comprises two energy levels, a lower level 1 and an upper level 2, separated by an energy $\hbar\omega$. In the absence of the radiation field, atoms in the upper level can decay to the lower level by the process of **spontaneous emission** of a photon [Fig. 23.8(a)]. The number of atoms in the upper level, N_2, is given by solving a simple differential equation

$$\frac{dN_2}{dt} = -A_{21} N_2, \qquad (23.56)$$

where A_{21} is a constant. This expresses simply that the decay rate depends on the number of atoms in the upper level. The solution of this equation is

$$N_2(t) = N_2(0) e^{-t/\tau}, \qquad (23.57)$$

where $\tau \equiv 1/A_{21}$ is the **natural radiative lifetime** of the upper level.

In the presence of a radiation field of energy density u_ω, two further processes are possible:

- An atom in level 1 can absorb a photon of energy $\hbar\omega$ and will end up in level 2 [Fig. 23.8(b)]. This process is called **absorption**, and will occur at a rate that is proportional both to u_ω and to the number of atoms in level 1. Thus the rate can be written as $N_1 B_{12} u_\omega$, where B_{12} is a constant.

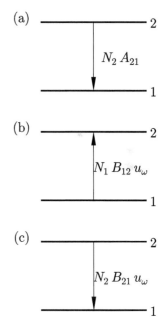

Fig. 23.8 Transitions for a two-level system: (a) spontaneous emission of a photon; (b) absorption of a photon; (c) stimulated emission of a photon.

- Quantum mechanics allows the reverse process to occur. Thus an atom in level 2 can emit a photon of energy $\hbar\omega$ as a direct result of the radiation field, and the atom will end up in level 1 [Fig. 23.8(c)]. In terms of individual photons, this process involves two photons: the presence of a first photon in the radiation field (which is absorbed and then re-emitted) stimulates the emission by the atom of an additional photon. This process is called **stimulated emission**, and will occur at a rate which is proportional both to u_ω and to the number of atoms in level 2. Thus the rate can be written as $N_2 B_{21} u_\omega$ where B_{21} is a constant.

The constants A_{21}, B_{12} and B_{21} are called the **Einstein A and B coefficients**. To summarize, our three processes are:

(1) spontaneous emission (one photon emitted);

(2) absorption (one photon absorbed);

(3) stimulated emission (one photon absorbed, two photons emitted).

In the steady state, with all three processes occurring simultaneously, we must have

$$N_2 B_{21} u_\omega + N_2 A_{21} = N_1 B_{12} u_\omega. \tag{23.58}$$

This can be rearranged to give

$$u_\omega = \frac{A_{21}/B_{21}}{(N_1 B_{12}/N_2 B_{21}) - 1}. \tag{23.59}$$

If the system is in thermal equilibrium, then the relative populations of the two levels must be given by a Boltzmann factor, i.e.,

$$\frac{N_2}{N_1} = \frac{g_2}{g_1} e^{-\beta\hbar\omega}, \tag{23.60}$$

where g_1 and g_2 are the degeneracies of levels 1 and 2 respectively. Substitution of eqn 23.60 into eqn 23.59 yields

$$u_\omega = \frac{A_{21}/B_{21}}{(g_1 B_{12}/g_2 B_{21}) e^{\beta\hbar\omega} - 1}, \tag{23.61}$$

and comparison with eqn 23.42 yields the following relations between the Einstein A and B coefficients:

$$\frac{B_{21}}{B_{12}} = \frac{g_1}{g_2} \quad \text{and} \quad A_{21} = \frac{\hbar\omega^3}{\pi^2 c^3} B_{21}. \tag{23.62}$$

Example 23.4

When will a system of atoms in a radiation field exhibit **gain**, i.e., produce more photons than they absorb?
Solution: The atoms will produce more photons than they absorb if the rate of stimulated emission is greater than the absorption rate, and this will occur if

$$N_2 B_{21} u_\omega > N_1 B_{12} u_\omega, \tag{23.63}$$

which implies that
$$\frac{N_2}{g_2} > \frac{N_1}{g_1}. \tag{23.64}$$

This means that we need to have a **population inversion**, so that the number of atoms ("the population") in the upper state (per degenerate level) exceeds that in the lower state. This is the principle behind the operation of the **laser** (a word that stands for light amplification by stimulated emission of radiation). However, in our two-level system such a population inversion is not possible in thermal equilibrium. For laser operation, it is necessary to have further energy levels to provide additional transitions: these can provide a mechanism to ensure that level 2 is *pumped* (fed by transitions from another level, keeping its population high) and that level 1 can *drain away* (into another lower level, so that level 1 has a low population).

Chapter summary

- The power emitted per unit area of a black-body surface at temperature T is given by σT^4, where
$$\sigma = \frac{\pi^2 k_B^4}{60 c^2 \hbar^3} = 5.67 \times 10^{-8} \, \text{W m}^{-2} \, \text{K}^{-4}.$$

- Radiation pressure p due to black-body photons is equal to $u/3$ where u is the energy density. Radiation pressure due to a collimated beam of light is equal to u.

- The spectral energy density u_ω takes the form of a black-body distribution. This form fits well to the experimentally measured form of the cosmic microwave background. It is also important in the theory of lasers.

Further reading

- A discussion of lasers may be found in Foot (2004), Chapters 1 and 7.
- More information concerning the cosmic microwave background is in Liddle (2003) Chapter 10 and Carroll and Ostlie (1996) Chapter 27.

Exercises

(23.1) The temperature of the Earth's surface is maintained by radiation from the Sun. By making the approximation that the Sun is a black body, but now assuming that the Earth is a grey body with albedo A (this means that it reflects a fraction A of the incident energy), show that the ratio of the Earth's temperature to that of the Sun is given by

$$T_{\text{Earth}} = T_{\text{Sun}}(1-A)^{1/4}\sqrt{\frac{R_{\text{Sun}}}{2D}}, \qquad (23.65)$$

where R_{Sun} is the radius of the Sun and the Earth–Sun separation is D.

(23.2) Show that the maxima in the functions u_ν and u_λ can be computed by maximizing the function $x^\alpha/(e^x-1)$ for $\alpha = 3$ and $\alpha = 5$ respectively. Show that this implies that

$$x = \alpha(1-e^{-x}). \qquad (23.66)$$

This equation can be solved by iterating

$$x_n = \alpha(1-e^{-x_{n-1}}); \qquad (23.67)$$

now show that (using an initial guess of $x_1 = 1$) this leads to the values given in eqns 23.53 and 23.54.

(23.3) The cosmic microwave background (CMB) radiation has a temperature of 2.73 K.
(a) What is the photon energy density in the Universe?
(b) Estimate the number of CMB photons that fall on the outstretched palm of your hand every second.
(c) What is the average energy due to CMB radiation that lands on your outstretched palm every second?
(d) What radiation pressure do you feel from CMB radiation?

(23.4) What is the ratio of the number of photons from the Sun to the number of CMB photons that irradiate your outstretched hand every second (during the daytime!)?

(23.5) Thermal radiation can be treated thermodynamically as a gas of photons with internal energy $U = u(T)V$ and pressure $p = u(T)/3$, where $u(T)$ is the energy density. Show that:
(a) the entropy density s is given by $s = 4p/T$;
(b) the Gibbs function $G = 0$;
(c) the heat capacity at constant volume $C_V = 3s$ per unit volume;
(d) the heat capacity at constant pressure, C_p, is infinite. (What on earth does that mean?)

(23.6) Ignoring the zero-point energy, show that the partition function Z for a gas of photons in volume V is given by

$$\ln Z = -\frac{V}{\pi^2 c^3}\int_0^\infty \omega^2 \ln(1-e^{-\hbar\omega\beta})\,d\omega, \qquad (23.68)$$

and hence, by integrating by parts, that

$$\ln Z = \frac{V\pi^2(k_\text{B}T)^3}{45\hbar^3 c^3}. \qquad (23.69)$$

Hence show that

$$F = -\frac{4\sigma V T^4}{3c}, \qquad (23.70)$$

$$S = \frac{16\sigma V T^3}{3c}, \qquad (23.71)$$

$$U = \frac{4\sigma V T^4}{c}, \qquad (23.72)$$

$$p = \frac{4\sigma T^4}{3c}, \qquad (23.73)$$

and hence that $U = -3F$, $pV = U/3$, and $S = 4U/3T$.

(23.7) Show that the total number N of photons in blackbody radiation contained in a volume V is

$$N = \int_0^\infty \frac{g(\omega)\,d\omega}{e^{\hbar\omega/k_\text{B}T}-1} = \frac{2\zeta(3)}{\pi^2}\left(\frac{k_\text{B}T}{\hbar c}\right)^3 V, \qquad (23.74)$$

where $\zeta(3) = 1.20206$ is a Riemann zeta function (see Appendix C.4). Hence show that the average energy per photon is

$$\frac{U}{N} = \frac{\pi^4}{30\zeta(3)}k_\text{B}T = 2.701k_\text{B}T, \qquad (23.75)$$

and that the average entropy per photon is

$$\frac{S}{N} = \frac{2\pi^4}{45\zeta(3)}k_\text{B} = 3.602k_\text{B}. \qquad (23.76)$$

The result for the internal energy of a photon gas is therefore $U = 2.701Nk_\text{B}T$, whereas for a classical ideal gas one obtains $U = \frac{3}{2}Nk_\text{B}T$. Why should the two results be different? Compare the expression for the entropy of a photon gas with that for an ideal gas (the Sackur–Tetrode equation); what is the *physical* reason for the difference?

Phonons

24

In a solid, energy can be stored in vibrations of the atoms which are arranged in a **lattice**.[1] In the same way that photons are quantized electromagnetic waves that describe the elementary excitations of the electromagnetic field, **phonons** are the quantized lattice waves that describe the elementary excitations of vibrations of the lattice. Rather than treating the vibration of each individual atom, our focus is on the normal modes of the system, which oscillate independently of each other. Each normal mode can be treated as a simple harmonic oscillator, and thus can contain an integer number of energy quanta. These energy quanta can be considered discrete "particles", known as phonons. The thermodynamic properties of a solid can therefore be calculated in much the same way as was done for photons in the previous chapter – by evaluating the statistical mechanics of a set of simple harmonic oscillators. The problem here is more complex because of the dispersive nature of lattice waves, but two models (the Einstein model and the Debye model) are commonly used to describe solids and we evaluate each in turn in the following two sections.

24.1 The Einstein model	279
24.2 The Debye model	281
24.3 Phonon dispersion	284
Chapter summary	287
Further reading	287
Exercises	287

[1] We assume a crystalline solid, though analogous results can be derived for non-crystalline solids. A lattice is a three-dimensional array of regularly spaced points, each point coinciding with the mean position of the atoms in the crystal.

24.1 The Einstein model

The **Einstein model** treats the problem by making the assumption that all vibrational modes of the solid have the same frequency ω_E. There are $3N$ such modes[2] (each atom of the solid has three vibrational degrees of freedom). We will assume that these normal modes are independent and do not interact with each other. In this case, the partition function Z can be written as the product

$$Z = \prod_{k=1}^{3N} Z_k, \quad (24.1)$$

where Z_k is the partition function of a single mode. Hence, the logarithm of the partition function is a simple sum over all the modes of the system:

$$\ln Z = \sum_{k=1}^{3N} \ln Z_k. \quad (24.2)$$

Each mode can be modelled as a simple harmonic oscillator, so we can use the expression in eqn 20.3 to write down the partition function of a

[2] Strictly speaking, a solid has $3N - 6$ vibrational modes, since although each atom can move in one of three directions (hence $3N$ degrees of freedom) one has to subtract six modes, which correspond to translation and rotation of the solid as a whole. When N is large, as it will be for any macroscopic sample, a correction of six modes is irrelevant.

single mode as

$$Z_k = \sum_{n=0}^{\infty} e^{-(n+\frac{1}{2})\hbar\omega_E \beta} = \frac{e^{-\frac{1}{2}\hbar\omega_E \beta}}{1 - e^{-\hbar\omega_E \beta}}. \quad (24.3)$$

This expression is independent of k because all the modes are identical, and so the partition function is $Z = (Z_k)^{3N}$ and hence

$$\ln Z = 3N\left[-\frac{1}{2}\hbar\omega_E \beta - \ln(1 - e^{-\hbar\omega_E \beta})\right], \quad (24.4)$$

and the internal energy U is

$$U = -\left(\frac{\partial \ln Z}{\partial \beta}\right) = \frac{3N}{2}\hbar\omega_E + \frac{3N}{1 - e^{-\hbar\omega_E \beta}}\hbar\omega_E e^{-\hbar\omega_E \beta}$$

$$= \frac{3N}{2}\hbar\omega_E + \frac{3N\hbar\omega_E}{e^{\hbar\omega_E \beta} - 1}. \quad (24.5)$$

In fact, we could have got immediately to eqn 24.5 simply by multiplying $3N$ by the expression in eqn 20.29, but we have taken a longer route to reiterate the basic principles. Writing $\hbar\omega_E = k_B \Theta_E$ defines a temperature Θ_E which scales with the vibrational frequency in the Einstein model. This allows us to rewrite eqn 24.5 as

Recall that $N_A k_B = R$.

$$U = 3R\Theta_E \left[\frac{1}{2} + \frac{1}{e^{\Theta_E/T} - 1}\right], \quad (24.6)$$

where U is now per mole of solid. In the high–temperature limit, $U \to 3RT$ because

$$\frac{1}{e^{\Theta_E/T} - 1} \to \frac{T}{\Theta_E} \quad \text{as } T \to \infty. \quad (24.7)$$

Example 24.1

Derive the molar heat capacity of an Einstein solid as a function of temperature, and show how it behaves in the low- and high-temperature limits.

Solution:

Using the expression for the molar internal energy in eqn 24.6, one can use $C = \left(\frac{\partial U}{\partial T}\right)$ to show that[3]

[3] For a solid, $C_V \approx C_p$, and so the subscript will be omitted.

$$C = 3R\Theta_E \frac{-1}{(e^{\Theta_E/T} - 1)^2} e^{\Theta_E/T} \left[-\frac{\Theta_E}{T^2}\right],$$

$$= 3R\frac{x^2 e^x}{(e^x - 1)^2}, \quad (24.8)$$

where $x = \Theta_E/T$.

- As $T \to 0$, $x \to \infty$ and $C \to 3Rx^2 e^{-x}$.
- As $T \to \infty$, $x \to 0$ and $C \to 3R$.

The high–temperature result is known as the **Dulong–Petit rule**.[4]

In summary, the molar heat capacity of an Einstein solid falls off very fast at low temperature (because it will be dominated by the $e^{-\Theta_E/T}$ term), but saturates to a value of $3R$ at high temperature.

[4] The Dulong–Petit rule is named after P. L. Dulong and A. T. Petit who measured it in 1819. It agrees with our expectations based on the equipartition theorem, see eqn 19.25.

24.2 The Debye model

The Einstein model makes a rather gross assumption that the normal modes of a solid all have the same frequency. It is clearly better to assume a *distribution* of frequencies. Hence, we would like to choose a function $g(\omega)$, which is the density of vibrational states. The number of vibrational states with frequencies between ω and $\omega + d\omega$ should be given by $g(\omega)\,d\omega$ and we require that the total number of normal modes be given by

$$\int g(\omega)\,d\omega = 3N. \quad (24.9)$$

The Einstein model took the density of states to be simply a delta function, i.e.,

$$g_{\text{Einstein}}(\omega) = 3N\delta(\omega - \omega_E), \quad (24.10)$$

as shown in Fig. 24.1, but we would now like to do better.

The next simplest approximation is to assume that lattice vibrations correspond to waves, all with the same speed v_s, which is the speed of sound in the solid. Thus we assume that

$$\omega = v_s q, \quad (24.11)$$

where q is the wave vector of the lattice vibration.[5] The density of states of lattice vibrations in three dimensions as a function of q is given by

$$g(q)\,dq = \frac{4\pi q^2\,dq}{(2\pi/L)^3} \times 3, \quad (24.12)$$

where the solid is assumed to be a cube of volume $V = L^3$ and the factor three corresponds to the three possible "polarizations" of the lattice vibration (one longitudinal and two transverse polarizations are possible for each value of q). Thus

$$g(q)\,dq = \frac{3Vq^2\,dq}{2\pi^2}, \quad (24.13)$$

and hence

$$g(\omega)\,d\omega = \frac{3V\omega^2\,d\omega}{2\pi^2 v_s^3}. \quad (24.14)$$

Because there is a limit ($3N$) on the total number of modes, we will now assume that lattice vibrations are possible up to a maximum frequency ω_D known as the **Debye frequency**. This is defined by

$$\int_0^{\omega_D} g(\omega)\,d\omega = 3N, \quad (24.15)$$

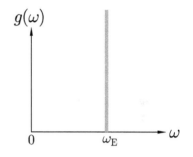

Fig. 24.1 The density of states for the Einstein model, using eqn 24.10.

[5] P. Debye (1884–1966) introduced this model in 1912, but assumed that a solid was a continuous elastic medium with a linear dispersion relation. We will improve on this dispersion relation in Section 24.3.

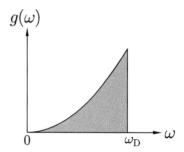

Fig. 24.2 The density of states for the Debye model, using eqn 24.14.

which, using eqn 24.14, implies that

$$\omega_D = \left(\frac{6N\pi^2 v_s^3}{V}\right)^{1/3}. \tag{24.16}$$

This allows us to rewrite eqn 24.14 as

$$g(\omega)\,d\omega = \frac{9N\omega^2\,d\omega}{\omega_D^3}. \tag{24.17}$$

The density of states for the Debye model is shown in Fig. 24.2. We also define the **Debye temperature**[6] Θ_D by

$$\Theta_D = \frac{\hbar\omega_D}{k_B}, \tag{24.18}$$

which gives the temperature scale corresponding to the Debye frequency. We are now ready to roll up our sleeves and tackle the statistical mechanics of this model.

---

[6]Some example Debye temperatures are shown in the following table:

Material	Θ_D (K)
Ne	63
Na	150
NaCl	321
Al	394
Si	625
C (diamond)	1860

The Debye temperature is higher for harder materials, since the bonds are stiffer and the phonon frequencies correspondingly higher.

Example 24.2

Derive the molar heat capacity of a Debye solid as a function of temperature.

Solution:
To obtain $C = (\partial U/\partial T)$, we first need to obtain U, which we can do by one of two methods.

Method 1 (Starting from the partition function)
We begin by writing down the logarithm of the partition function as follows:

$$\ln Z = \int_0^{\omega_D} d\omega\, g(\omega) \ln\left[\frac{e^{-\frac{1}{2}\hbar\omega\beta}}{1 - e^{-\hbar\omega\beta}}\right]. \tag{24.19}$$

This integral looks a bit daunting, but we can do it by first expanding the logarithm term (using $\ln(a/b) = \ln a - \ln b$):

$$\ln Z = -\int_0^{\omega_D} \frac{1}{2}\hbar\omega\beta g(\omega)\,d\omega - \int_0^{\omega_D} g(\omega)\ln(1 - e^{-\hbar\omega\beta})\,d\omega. \tag{24.20}$$

The first term of eqn 24.20 is easily evaluated to be $-\frac{9}{8}N\hbar\omega_D\beta$ while the second term we will leave unevaluated for the moment. Thus we have

$$\ln Z = -\frac{9}{8}N\hbar\omega_D\beta - \frac{9N}{\omega_D^3}\int_0^{\omega_D} \omega^2 \ln(1 - e^{-\hbar\omega\beta})\,d\omega. \tag{24.21}$$

Now we can use $U = -\partial \ln Z/\partial \beta$, and hence we find that

$$U = \frac{9}{8}N\hbar\omega_D + \frac{9N\hbar}{\omega_D^3}\int_0^{\omega_D} \frac{\omega^3\,d\omega}{e^{\hbar\omega\beta} - 1}. \tag{24.22}$$

24.2 The Debye model

Method 2 (Using the expression for U of a simple harmonic oscillator)
We can derive the internal energy U by using the expression for U for a single simple harmonic oscillator in eqn 20.29 to give

$$U = \int_0^{\omega_D} g(\omega) \, d\omega \, \hbar\omega \left(\frac{1}{2} + \frac{1}{e^{\beta\hbar\omega} - 1} \right), \quad (24.23)$$

which results in eqn 24.22 after substituting in eqn 24.17 and integrating.

Obtaining C

The heat capacity can be derived from $C = \left(\frac{\partial U}{\partial T}\right)$ and hence, using eqn 24.22, we have that

$$C = \frac{9N\hbar}{\omega_D^3} \int_0^{\omega_D} \frac{-\omega^3 \, d\omega}{(e^{\hbar\omega\beta} - 1)^2} e^{\hbar\omega\beta} \left(-\frac{\hbar\omega}{k_B T^2} \right). \quad (24.24)$$

Making the substitution $x = \hbar\beta\omega$, and hence $x_D = \hbar\beta\omega_D$, eqn 24.24 can be rewritten as

$$C = \frac{9R}{x_D^3} \int_0^{x_D} \frac{x^4 e^x \, dx}{(e^x - 1)^2}. \quad (24.25)$$

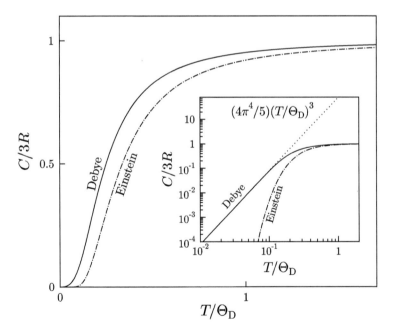

Fig. 24.3 The molar specific heat capacity for the Einstein solid and the Debye solid, according to eqn 24.8 and eqn 24.25 respectively. The inset shows the same information on a log–log scale, illustrating the difference between the low-temperature-specific heat capacities of the two models. The Debye model predicts a cubic temperature dependence at low temperature according to eqn 24.28, as shown by the dotted line. The figure is drawn with $\Theta_E = \Theta_D$.

The expression in eqn 24.25 is quite complicated and it is not obvious, just by looking at the equation, what the temperature dependence of the heat capacity will be. This is because $x_D = \hbar\omega_D \beta$ is temperature dependent and hence both the prefactor $9/x_D^3$ and the integral are temperature

Example 24.3

Show how the molar heat capacity of a Debye solid, derived in eqn 24.25, behaves in the low- and high-temperature limits.
Solution:

- At high temperature, $x \to 0$ and hence $e^x - 1 \to x$. Hence, the heat capacity C behaves as

$$C \to \frac{9R}{x_D^3} \int_0^{x_D} \frac{x^4}{x^2}\, dx = 3R, \quad (24.26)$$

which is the equipartition result (Dulong–Petit rule, eqn 19.25) again.

- At low temperature, x becomes very large and $e^x \gg 1$. The heat capacity is given by

$$C \to \frac{9R}{x_D^3} \int_0^\infty \frac{x^4 e^x\, dx}{(e^x - 1)^2} = \frac{12R\pi^4}{5 x_D^3}. \quad (24.27)$$

Here we can use the integral

$$\int_0^\infty \frac{x^4 e^x\, dx}{(e^x - 1)^2} = \frac{4\pi^4}{15}$$

which is derived in the Appendix B, see eqn C.31.

Thus an expression for the low temperature heat capacity of a solid is

$$C = 3R \times \frac{4\pi^4}{5} \left(\frac{T}{\Theta_D}\right)^3. \quad (24.28)$$

This demonstrates that the molar heat capacity of a Debye solid saturates to a value of $3R$ at high temperature and is proportional to T^3 at low temperature.

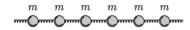

Fig. 24.4 A monatomic linear chain.

24.3 Phonon dispersion

We have so far assumed that the phonon dispersion relation is given by eqn 24.11. In this section, we will improve on this substantially. Let us first consider the vibrations on a monatomic linear chain of atoms, each atom with mass m connected to its nearest neighbour by a spring with force constant K (see Fig. 24.4). The displacement from the equilibrium position of the nth mass is given the symbol u_n. Hence, the equation of motion of the nth mass is given by

$$m\ddot{u}_n = K(u_{n+1} - u_n) - K(u_n - u_{n-1}) = K(u_{n+1} - 2u_n + u_{n-1}). \quad (24.29)$$

To solve this equation, we must attempt to look for a wave-like solution. A trial normal mode solution $u_n = \exp[i(qna - \omega t)]$ yields

$$-m\omega^2 = K(e^{iqa} - 2 + e^{-iqa}), \quad (24.30)$$

and hence
$$mw^2 = 2K(1 - \cos qa), \quad (24.31)$$
which simplifies to
$$\omega^2 = \frac{4K}{m} \sin^2(qa/2), \quad (24.32)$$
and hence
$$\omega = \left(\frac{4K}{m}\right)^{1/2} |\sin(qa/2)|. \quad (24.33)$$

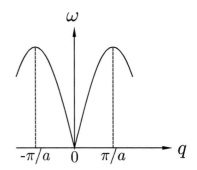

Fig. 24.5 The dispersion relation for a monatomic linear chain, given in eqn 24.33.

This result is plotted in Fig. 24.5. In the long-wavelength limit when $qa \to 0$, we have that $\omega \to v_s q$, where
$$v_s = a\left(\frac{K}{m}\right)^{1/2}, \quad (24.34)$$
and hence eqn 24.11 is obtained in this limit.

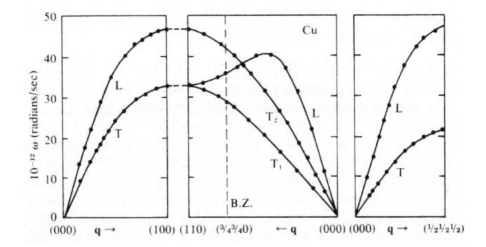

Fig. 24.6 The phonon dispersion in copper (Cu). Because Cu is a three-dimensional metal, the phonon dispersion has to be evaluated in three dimensions, and it is here shown as a function of wave vector in different directions. Along the (100) direction, bands can be seen, which look somewhat like the simple monatomic chain. Both longitudinal (L) and transverse (T) modes are present. The wave vector q is plotted in units of π/a where a is the lattice spacing. The data shown are from Svensson et al., *Phys. Rev.* **155**, 619 (1967), and are obtained using inelastic neutron scattering. In this technique, a beam of slow neutrons is scattered from the sample and the changes in both the energy and momentum of the neutrons are measured. This can be used to infer the energy $\hbar\omega$ and momentum $\hbar q$ of the phonons. Copyright (1967) by the American Physical Society.

The measured phonon dispersion for copper (Cu), which is a monatomic metal with a face-centred cubic structure, is shown in Fig. 24.6 and demonstrates that for small wave vectors (long wavelengths) the angular frequency ω is indeed proportional to the wave vector q and hence $\omega = v_s q$ is a good approximation in this limit. However, there are both longitudinal and transverse modes present and so in the Debye model one would need to use a suitably modified sound speed.[7] Where the bands flatten over, peaks can be seen in the phonon density of states because states are uniformly distributed in wave vector and so will be

[7] Usually what is used in the expression for the Debye frequency is
$$\frac{3}{v_s^3} = \frac{2}{v_{s,T}^3} + \frac{1}{v_{s,L}^3}$$
where $v_{s,T}$ and $v_{s,L}$ are the transverse and longitudinal sound speeds respectively. The weighting of 2:1 is because there are two orthogonal transverse modes and only one longitudinal mode.

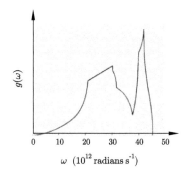

Fig. 24.7 The density of states $g(\omega)$ for the phonons in copper. The curve is obtained by numerical analysis of the measured phonon dispersion relation. Data from Svensson et al., *Phys. Rev.* **155**, 619 (1967). Copyright (1967) by the American Physical Society.

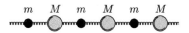

Fig. 24.8 A diatomic linear chain.

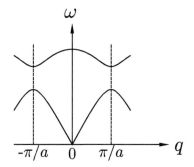

Fig. 24.9 The dispersion relation for a diatomic linear chain. The lower curve is the acoustic branch, the upper curve is the optic branch.

Fig. 24.10 The measured phonon dispersion relation in germanium. B. N. Brockhouse, *Rev. Mod. Phys.* **67**, 735 (1995). Copyright (1995) by the American Physical Society.

concentrated at energies corresponding to regions in the dispersion relation which are horizontal. This is illustrated in Fig. 24.7, and you can compare each peak in this graph with a flattening of the band in some part of the dispersion relation shown in Fig. 24.6. The phonon density of states clearly follows a quadratic dependence at low frequency, corresponding to the non-dispersive parts of the dispersion relation.

If the solid contains more than one crystallographically independent atom per unit cell, the situation is a little more complicated. To gain insight into this problem, one can solve the diatomic linear chain problem (see Fig. 24.8) which is composed of an alternating series of two different atoms. The dispersion relation for this is plotted in Fig. 24.9, and shows two branches. The **acoustic branch** is very similar to the monatomic linear chain dispersion and near $q = 0$ corresponds to neighbouring atoms vibrating almost in phase (and the group velocity near $q = 0$ is the speed of sound in the material, hence the adjective "acoustic"). The modes of vibration in the acoustic branch are called **acoustic modes**. The **optic branch** has non-zero ω at $q = 0$ and near $q = 0$ corresponds to vibrations in which neighbouring atoms vibrate almost out of phase. The modes of vibration in the optic branch are called **optic modes**. It is called an optic branch because, if the chain contains ions of different charges, an oscillation with small q causes an oscillating electric dipole moment, which can couple with electromagnetic radiation.

An example of such a phonon dispersion in which optic modes are present is provided by germanium (Ge), shown in Fig. 24.10. Although all the atoms in Ge are identical, there are two crystallographically distinct atomic sites and hence an optic branch is observed in the phonon dispersion. These data have been measured by inelastic neutron scattering.

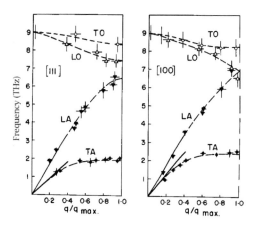

Though the phonon dispersion relations of real solids are more complicated than the linear relation, $\omega = v_s q$, assumed by the Debye model, they are linear at low frequency. With this relationship we have that the phonon density of states is approximately quadratic at low frequency.

At low temperature (where only low-energy, i.e., low-frequency, phonons can be excited), the heat capacity of most solids therefore shows the Debye T^3 behaviour. In practice, the acoustic modes of a solid can be well described by the Debye model, while the optic modes (whose frequencies do not vary much with wave vector) are quite well described by the Einstein model.

Chapter summary

- A phonon is a quantized lattice vibration.
- The Einstein model of a solid assumes that all phonons have the same frequency.
- The Debye model allows a range of phonon frequencies up to a maximum frequency called the Debye frequency. The density of states is quadratic in frequency, and this assumes that $\omega = v_s q$.
- The dispersion relation of a real solid is more complicated and may contain acoustic and optic branches. It can be experimentally determined using inelastic neutron scattering.
- The heat capacity of a three-dimensional solid is proportional to T^3 at low temperature and saturates to a value of $3R$ at high temperature.

Further reading

A wonderful introduction to waves in periodic structures may be found in Brillouin (1953). Useful information about phonons may be found in Ashcroft and Mermin (1976) Chapters 22–24, Dove (2003) Chapters 8 and 9 and Singleton (2001) Appendix D.

Exercises

(24.1) A primitive cubic crystal has lattice parameter 0.3 nm and Debye temperature 100 K. Estimate the maximum phonon frequency in Hz and the speed of sound in m s^{-1}.

(24.2) Show that eqn 24.22 can be rewritten as

$$U = \frac{9}{8}N_A \hbar \omega_D + \frac{9RT}{x_D^3} \int_0^{x_D} \frac{x^3 \, dx}{e^x - 1}, \quad (24.35)$$

by making the substitution $x = \hbar \beta \omega$, and hence $x_D = \hbar \beta \omega_D$.

(24.3) Show that the Debye model of a d-dimensional crystal predicts that the low temperature heat capacity is proportional to T^d.

(24.4) Show that the density of states of lattice vibrations on a monatomic linear chain (see Section 24.3) is given by $g(\omega) = (2L/\pi a)[4K/m - \omega^2]^{-1/2}$.

Sketch $g(\omega)$ and comment on the singularity at $\omega = \sqrt{4K/m}$.

(24.5) Generalize the treatment of a monatomic linear chain to the transverse vibrations of atoms on a (two-dimensional) square lattice of atoms and show that

$$\omega = \left(\frac{2K}{m}\right)^{1/2} [2 - \cos q_x a - \cos q_y a]^{1/2}, \quad (24.36)$$

and derive an expression for the speed of sound.

(24.6) Show that the dispersion relation for the diatomic chain shown in Fig. 24.8 is

$$\frac{\omega^2}{K} = \left(\frac{1}{M} + \frac{1}{m}\right) \pm \left[\left(\frac{1}{M} + \frac{1}{m}\right)^2 - \frac{4}{Mm}\sin^2\frac{qa}{2}\right]^{1/2}. \quad (24.37)$$

(24.7) The treatment of the monatomic linear chain in Section 24.3 included only nearest-neighbour interactions. Show that if a force constant K_j links an atom with one j atoms away, then the dispersion relation becomes

$$\omega^2 = \frac{4}{m}\sum_j K_j \sin^2 \frac{jqa}{2}. \quad (24.38)$$

A measurement is made of $\omega(q)$. Show that the force constants can be obtained from $\omega(q)$ using

$$K_j = -\frac{ma}{2\pi}\int_{-\frac{\pi}{a}}^{\frac{\pi}{a}} dq\, \omega^2(q) \cos jqa. \quad (24.39)$$

Part VIII

Beyond the ideal gas

In this part we introduce various extensions to the ideal gas model which allow us to take account of various complications that make the subject of thermal physics more rich and interesting, but of course also slightly more complicated! This part is structured as follows:

- In Chapter 25, we study the consequences of allowing the *dispersion relation*, the equation that connects energy and momentum, to be relativistic. We examine the differences between the relativistic and non-relativistic cases.

- In Chapter 26, we introduce several equations of state, which take into account the *interactions* between molecules in a gas. These include the *van der Waals* model, the *Dieterici model*, and the *virial expansion*. We discuss the *law of corresponding states*.

- In Chapter 27 we discuss how to cool real gases using the *Joule–Kelvin expansion* and the operation of a *liquefier*.

- In Chapter 28 we discuss *phase transitions*, discussing *latent heat* and deriving the *Clausius–Clapeyron equation*. We discuss the criteria for *stability* and *metastability* and derive the *Gibbs phase rule*. We introduce *colligative properties* and classify the different types of phase transition.

- In Chapter 29 we examine the effect that exchange symmetry has on the quantum wave functions of collections of identical particles. This allows us to introduce *bosons* and *fermions*, which can be used to describe the *Bose–Einstein distribution* and *Fermi–Dirac distribution* respectively.

- In Chapter 30, we show how the results of the previous chapter can be applied to *quantum gases*, and we consider non-interacting fermion and boson gases and discuss *Bose–Einstein condensation*.

25 Relativistic gases

25.1 Relativistic dispersion relation for massive particles 290
25.2 The ultrarelativistic gas 290
25.3 Adiabatic expansion of an ultrarelativistic gas 293
Chapter summary 295
Exercises 295

In this chapter we will repeat our derivation of the partition function for a gas, and hence of the other thermodynamic properties that can be obtained from it, but this time include relativistic effects. We will see that this leads to some subtle changes in these properties which have profound consequences. First we will review the full relativistic dispersion relation for particles with non-zero mass and then derive the partition function for ultrarelativistic particles.

25.1 Relativistic dispersion relation for massive particles

In deriving the partition function for a gas, we assumed that the kinetic energy E of a molecule of mass m was equal to $p^2/2m$, where p is the momentum (and using $p = \hbar k$, we wrote down $E(k) = \hbar^2 k^2/2m$; see eqn 21.15). This is a classical approximation valid only when $p/m \ll c$ (where c is the speed of light), and in general we should use the relativistic formula

$$E^2 = p^2 c^2 + m^2 c^4, \tag{25.1}$$

where m is now taken to be the **rest mass**, i.e., the mass of the molecule in its rest frame. This is plotted in Fig. 25.1. When $p \ll mc$ (the **non-relativistic limit**) this reduces to

$$E = \frac{p^2}{2m} + mc^2, \tag{25.2}$$

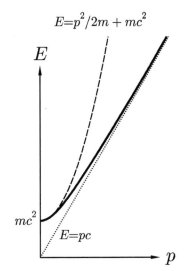

Fig. 25.1 The dispersion relation of a particle with mass m (thick solid line) according to eqn 25.1. The dashed line is the non-relativistic limit ($p \ll mc$). The dotted line is the ultrarelativistic limit ($p \gg mc$).

[1] The relation between E and p is known as a **dispersion relation**. By scaling $E = \hbar \omega$ and $p = \hbar k$ by a factor $\hbar$ we have a relation between ω and k, which is perhaps more familiar as a dispersion relation from wave physics.

which is identical to our classical approximation $E = p^2/2m$ apart from the extra constant mc^2 (the rest mass energy), which just defines a new "zero" for the energy (see Fig. 25.1). In the case $p \gg mc$ (the **ultrarelativistic limit**), eqn 25.1 reduces to

$$E = pc, \tag{25.3}$$

which is the appropriate relation[1] for photons (this is the straight line in Fig. 25.1).

25.2 The ultrarelativistic gas

Let us now consider a gas of particles with non-zero mass in the ultrarelativistic limit, which means that $E = pc$. Such a linear dispersion

25.2 The ultrarelativistic gas

relation means that some of the algebra in this chapter is actually much simpler than we had to deal with for the partition function in the non-relativistic case where the dispersion relation is quadratic. Using the ultrarelativistic limit means that all the particles (or at the very least, the vast majority of them), will be moving so quickly that their kinetic energy is much greater than their rest mass energy.[2] Using the ultrarelativistic limit $E = pc = \hbar kc$, we can write down the single-particle partition function

$$Z_1 = \int_0^\infty e^{-\beta \hbar k c} g(k) \, dk, \qquad (25.4)$$

[2] Note however that we are ignoring any quantum effects that may come into play; these will be considered in Chapter 30.

where we recall that (eqn 21.6)

$$g(k) \, dk = \frac{V k^2 \, dk}{2\pi^2}, \qquad (25.5)$$

and so, using the substitution $x = \beta \hbar k c$, we have

$$Z_1 = \frac{V}{2\pi^2} \left(\frac{1}{\beta \hbar c} \right)^3 \int_0^\infty e^{-x} x^2 \, dx, \qquad (25.6)$$

and recognizing that the integral is 2!, we have finally that

$$Z_1 = \frac{V}{\pi^2} \left(\frac{k_B T}{\hbar c} \right)^3. \qquad (25.7)$$

Notice immediately that we find that $Z_1 \propto VT^3$, whereas in the non-relativistic case we had that $Z_1 \propto VT^{3/2}$. We can also write eqn 25.7 in a familiar form

$$Z_1 = \frac{V}{\Lambda^3}, \qquad (25.8)$$

where Λ is not the same as the expression for the thermal wavelength in eqn 21.18, but is given by

$$\Lambda = \frac{\hbar c \pi^{2/3}}{k_B T}, \qquad (25.9)$$

Equivalently, one can write

$$\Lambda = \frac{hc}{2\pi^{1/3} k_B T},$$

It now becomes a simple exercise to determine all the properties of the ultrarelativistic gas using our practised methods of partition functions.

Example 25.1

Find U, C_V, F, p, S, H, and G for an ultrarelativistic gas of indistinguishable particles.

Solution:
The N-particle partition function Z_N is given by[3]

$$Z_N = \frac{Z_1^N}{N!}, \qquad (25.10)$$

[3] This is assuming the density is not so high that this approximation breaks down.

and hence

$$\ln Z_N = N \ln V + 3N \ln T + \text{constants}. \qquad (25.11)$$

The internal energy U is given by

$$U = -\frac{d \ln Z_N}{d\beta} = 3Nk_BT, \quad (25.12)$$

which is different from the non-relativistic case (which gave $U = \frac{3}{2}Nk_BT$). The heat capacity C_V is

$$C_V = \left(\frac{\partial U}{\partial T}\right)_V, \quad (25.13)$$

and hence is given by[4] $C_V = 3Nk_B$. The Helmholtz function is

$$F = -k_BT \ln Z_N = -k_BTN \ln V - 3Nk_BT \ln T - k_BT \times \text{constants}, \quad (25.14)$$

so that

$$p = -\left(\frac{\partial F}{\partial V}\right)_T = \frac{Nk_BT}{V} = nk_BT, \quad (25.15)$$

which is the ideal gas equation,[5] as for the non-relativistic case. This also gives the enthalpy H via

$$H = U + pV = 4Nk_BT. \quad (25.16)$$

As we found for the non-relativistic case, getting the entropy involves bothering with what the constants are in eqn 25.11. Hence, let us write this equation as

$$\begin{aligned}\ln Z_N &= N \ln V - 3N \ln \Lambda - N \ln N + N \\ &= N \ln\left(\frac{1}{n\Lambda^3}\right) + N, \end{aligned} \quad (25.17)$$

where $n = N/V$, so we immediately have (using the usual statistical mechanics manipulations listed in Table 20.1):

$$\begin{aligned} F &= -k_BT \ln Z_N \\ &= Nk_BT[\ln(n\Lambda^3) - 1], \quad (25.18)\\ S &= \frac{U - F}{T} \\ &= Nk_B[4 - \ln(n\Lambda^3)], \quad (25.19)\\ G &= H - TS = 4Nk_BT - Nk_BT[4 - \ln(n\Lambda^3)]\\ &= Nk_BT \ln(n\Lambda^3). \quad (25.20)\end{aligned}$$

The results from this problem are summarized in Table 25.1.

---

[4] Notice that this does not agree with the equipartition theorem, which would predict $C_V = \frac{3}{2}Nk_B$, half of the value that we have found. Why does the equipartition theorem fail? Because the dispersion relation is not a quadratic one (i.e., $E \propto p^2$), as is needed for the equipartition theorem to hold, but instead is a linear one ($E \propto p$).

[5] Note that we have $p = nk_BT$ for both the non-relativistic and ultrarelativistic cases. This is because $Z_1 \propto V$ in both cases; hence $Z_N \propto V^N$ and $F = -k_BTN \ln V +$(other terms not involving V), so that $p = -(\partial F/\partial V)_T = nk_BT$.

One consequence of these results is that the pressure p is related to the energy density $u = U/V$ using

$$p = \frac{u}{3}, \quad (25.21)$$

which is very different from the non-relativistic case $p = 2u/3$ (see eqn 6.25). This has some rather dramatic consequences for the structure of stars (see Section 35.1.3).

Property	Non-relativistic	Ultrarelativistic
Z_1	$\dfrac{V}{\lambda_{\text{th}}^3}$ $\lambda_{\text{th}} = \dfrac{h}{\sqrt{2\pi m k_B T}}$	$\dfrac{V}{\Lambda^3}$ $\Lambda = \dfrac{\hbar c \pi^{2/3}}{k_B T}$
U	$\tfrac{3}{2} N k_B T$	$3 N k_B T$
H	$\tfrac{5}{2} N k_B T$	$4 N k_B T$
p	$\dfrac{N k_B T}{V} = \dfrac{2u}{3}$	$\dfrac{N k_B T}{V} = \dfrac{u}{3}$
F	$N k_B T [\ln(n\lambda_{\text{th}}^3) - 1]$	$N k_B T [\ln(n\Lambda^3) - 1]$
S	$N k_B [\tfrac{5}{2} - \ln(n\lambda_{\text{th}}^3)]$	$N k_B [4 - \ln(n\Lambda^3)]$
G	$N k_B T \ln(n\lambda_{\text{th}}^3)$	$N k_B T \ln(n\Lambda^3)$
Adiabatic expansion	$VT^{3/2} = \text{constant}$ $pV^{5/3} = \text{constant}$	$VT^3 = \text{constant}$ $pV^{4/3} = \text{constant}$

Table 25.1 The properties of non-relativistic and ultrarelativistic monatomic gases of indistinguishable particles of mass m.

25.3 Adiabatic expansion of an ultrarelativistic gas

We will now consider the adiabatic expansion of an ultrarelativistic monatomic gas. This means that we will keep the gas thermally isolated from its surroundings and no heat will enter or leave. The entropy stays constant in such a process, and hence (from Table 25.1) so does $n\Lambda^3$, which implies that

$$VT^3 = \text{constant}, \qquad (25.22)$$

or equivalently (using $pV \propto T$)

$$pV^{4/3} = \text{constant}. \qquad (25.23)$$

This implies that the adiabatic index $\gamma = 4/3$. This contrasts with the non-relativistic cases (for which $VT^{3/2}$ and $pV^{5/3}$ are constants, and $\gamma = 5/3$).

Example 25.2

An example of the adiabatic expansion of an ultrarelativistic gas relates to the expansion of the Universe. If the Universe expands adiabatically (how can heat enter or leave it when it presumably doesn't have any "surroundings" by definition?) then we expect that an ultrarelativistic gas inside the Universe, such as the cosmic microwave background photons,[6] behaves according to

$$VT^3 = \text{constant}, \quad (25.24)$$

where T is the temperature of the Universe and V is its volume. Hence

$$T \propto V^{-1/3} \propto a^{-1}, \quad (25.25)$$

where a is the scale factor[7] of the Universe ($V \propto a^3$). Thus the temperature of the cosmic microwave background is inversely proportional to the scale factor of the Universe.

A non-relativistic gas in the Universe would behave according to

$$VT^{3/2} = \text{constant}, \quad (25.26)$$

in which case

$$T \propto V^{-2/3} \propto a^{-2}, \quad (25.27)$$

so the non-relativistic gas would cool faster than the cosmic microwave background as the Universe expands.

We can also work out the density ρ of both types of gas as a function of the scale factor a. For the adiabatic expansion of a gas of non-relativistic particles, the density $\rho \propto V^{-1}$ (because the mass stays constant) and hence

$$\rho \propto a^{-3}. \quad (25.28)$$

For relativistic particles,

$$\rho = \frac{u}{c^2}, \quad (25.29)$$

where $u = U/V$ is the energy density. Now $u = 3p$ (by eqn 25.21) and since $p \propto V^{-4/3}$ for relativistic particles, we have that

$$\rho \propto a^{-4}. \quad (25.30)$$

Thus the density drops off faster for a gas of relativistic particles than it does for non-relativistic particles, as the Universe expands.[8]

The Universe contains both matter (mostly non-relativistic) and photons (clearly ultrarelativistic). This simple analysis shows that as the Universe expanded, the matter cooled faster than the photons, but the density of the matter decreases less quickly than that of the photons. The density of the early Universe is said to be **radiation dominated** but as time has passed the Universe has become **matter dominated** as far as its density (and hence expansion dynamics) is concerned.

---

[6] See Section 23.7.

[7] See Section 23.7.

[8] This is because, for both cases, you have the effect of volume dilution, due to the Universe expanding, which goes as a^3; but only for the relativistic case do you have an energy loss (and hence a density loss) due to the Universe expanding, giving an extra factor of a.

Chapter summary

- Using the ultrarelativistic dispersion relation $E = pc$, rather than the non-relativistic dispersion relation $E = p^2/2m$, leads to changes in various thermodynamic functions (Table 25.1).

Exercises

(25.1) Find the phase velocity and the group velocity for a relativistic particle whose energy E is $E^2 = p^2c^2 + m_0^2c^4$ and examine the limit $p \ll mc$ and $p \gg mc$.

(25.2) In D dimensions, show that the density of states of particles with spin degeneracy g in a volume V is
$$g(k)\,dk = \frac{gVD\pi^{D/2}k^{D-1}\,dk}{\Gamma(\frac{D}{2}+1)(2\pi)^D}. \quad (25.31)$$
You may need to use the fact that the volume of a sphere of radius r in D dimensions is (see Appendix C.8)
$$\frac{\pi^{D/2}r^D}{\Gamma(\frac{D}{2}+1)}. \quad (25.32)$$

(25.3) Consider a general dispersion relation of the form
$$E = \alpha p^s, \quad (25.33)$$
where p is the momentum and α and s are constants. Using the result of the previous question, show that the density of states as a function of energy is
$$g(E)\,dE = \frac{gVD\pi^{D/2}}{h^D \alpha^{D/s} s \Gamma(\frac{D}{2}+1)} E^{\frac{D}{s}-1}\,dE. \quad (25.34)$$
Hence show that the single-particle partition function takes the form
$$Z_1 = \frac{V}{\lambda^D}, \quad (25.35)$$
where λ is given by
$$\lambda = \frac{h}{\pi^{1/2}}\left(\frac{\alpha}{k_B T}\right)^{1/s}\left[\frac{\Gamma(\frac{D}{2}+1)}{\Gamma(\frac{D}{s}+1)}\right]^{1/D}. \quad (25.36)$$
Show that this result for three dimensions ($D=3$) agrees with (i) the non-relativistic case when $s=2$ and (ii) the ultrarelativistic case when $s=1$.

26 Real gases

26.1 The van der Waals gas 296
26.2 The Dieterici equation 304
26.3 Virial expansion 306
26.4 The law of corresponding states 310
Chapter summary 311
Exercises 312

In this book we have spent a lot of time considering the so-called ideal (sometimes called "perfect") gas, which has an equation of state given by

$$pV = n_{\text{moles}} RT, \qquad (26.1)$$

where n_{moles} is the number of moles, or equivalently by

$$pV_{\text{m}} = RT, \qquad (26.2)$$

where $V_{\text{m}} = V/n_{\text{moles}}$ is the molar volume (i.e., the volume occupied by 1 mole). This equation of state leads to isotherms as plotted in Fig. 26.1. However, real gases don't behave quite like this, particularly when the pressure is high and the volume is small. For a start, if you get a real gas cold enough it will liquefy, and this is something that the ideal gas equation does not predict or describe. In a liquid, the intermolecular attractions, which we have so far preferred to ignore, are really significant. In fact, even before the gas liquefies, there are departures from ideal-gas behaviour. This chapter deals with how this additional element of real behaviour can be modelled, by introducing various extensions to the ideal gas model, including those introduced by *van der Waals* (Section 26.1) and *Dieterici* (Section 26.2). An alternative series expansion approach is the so-called *virial expansion* in Section 26.3. Many similar systems behave in analogous ways once the differences in the magnitude of the intermolecular interactions have been factored out by some appropriate scaling. This forms the basis of the *law of corresponding states* in Section 26.4.

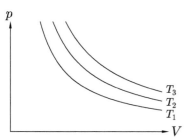

Fig. 26.1 Isotherms of the ideal gas for three different temperatures $T_3 > T_2 > T_1$.

26.1 The van der Waals gas

The most commonly used model of real gas behaviour is the **van der Waals gas**. This is the simplest real gas model which includes the two crucial ingredients we need: (i) intermolecular interactions (gas molecules actually weakly attract one another) and (ii) the non-zero size of molecules (gas molecules don't have freedom to move around in *all* the volume of the container, because some of the volume is occupied by the other gas molecules!). Like the ideal gas, the van der Waals gas is only a *model* of real behaviour, but by being a slightly more complicated description than the ideal gas (more complicated in the right way!) it is able to describe more of the physical properties exhibited by real gases.

26.1 The van der Waals gas

The equation of state for a van der Waals gas is

$$\left(p + \frac{a}{V_m^2}\right)(V_m - b) = RT. \qquad (26.3)$$

In this equation, the constant a parameterizes the strength of the intermolecular interactions, while the constant b accounts for the volume excluded owing to the finite size of molecules. If a and b are both set to zero, we recover the equation of state for an ideal gas, $pV_m = RT$. Moreover, in the low-density limit (when $V_m \gg b$ and $V_m \gg (a/p)^{1/2}$) we also recover the ideal gas behaviour. However, when the density is high, and we try to make V_m approach b, the pressure p shoots up.[1] The motivation for the a/V_m^2 term in the van der Waals model is outlined in the box on page 298.

[1] To understand this, consider eqn 26.3 at some fixed T. When $V_m \to b$, the term $(V_m - b)$ is very small, and hence $(p + a/V_m^2) \approx (p + a/b^2)$ is very big and hence p increases.

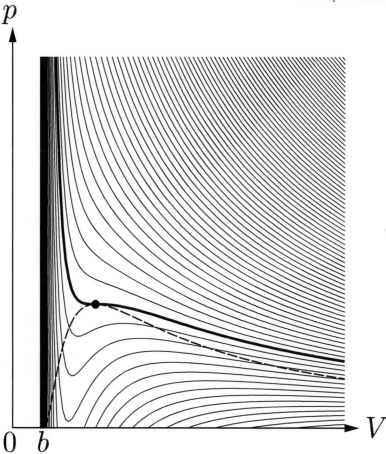

Fig. 26.2 Isotherms of the van der Waals gas. Isotherms towards to the top right of the graph correspond to higher temperatures. The dashed line shows the region in which liquid and vapour are in equilibrium (see the end of Section 26.1). The thick line is the critical isotherm and the dot marks the critical point.

Origin of the a/V_m^2 term

Assume n_moles moles of gas in volume V. The number of nearest neighbours is proportional to n_moles/V, and so attractive intermolecular interactions lower the total potential energy by an amount proportional to the number of atoms multiplied by the number of nearest neighbours, i.e., we can write the energy change as

$$-\frac{an_\mathrm{moles}^2}{V}, \tag{26.4}$$

where a is a constant. Hence, if you change V, the energy changes by an amount

$$\frac{an_\mathrm{moles}^2 \mathrm{d}V}{V^2}, \tag{26.5}$$

but this energy change can be thought of as being due to an effective pressure p_eff, so that the energy change would be $-p_\mathrm{eff}\,\mathrm{d}V$. Hence

$$p_\mathrm{eff} = -a\frac{n_\mathrm{moles}^2}{V^2} = -\frac{a}{V_\mathrm{m}^2}. \tag{26.6}$$

The pressure p that you measure is the sum of the pressure p_ideal neglecting intermolecular interactions and p_eff. Therefore

$$p_\mathrm{ideal} = p - p_\mathrm{eff} = p + \frac{a}{V_\mathrm{m}^2} \tag{26.7}$$

is the pressure that you have to enter into the formula for the ideal gas,

$$p_\mathrm{ideal}V_\mathrm{m} = RT, \tag{26.8}$$

making the correction $V_\mathrm{m} \to V_\mathrm{m} - b$ to take account of the excluded volume. This yields

$$\left(p + \frac{a}{V_\mathrm{m}^2}\right)(V_\mathrm{m} - b) = RT, \tag{26.9}$$

in agreement with eqn 26.3. This equation of state can also be justified from statistical mechanics as follows: taking the expression for the partition function of N molecules in a gas, $Z_N = (1/N!)(V/\lambda_\mathrm{th}^3)^N$, we replace the volume V by $V - n_\mathrm{moles}b$, the volume actually available for molecules to move around in; we also include a Boltzmann factor $\mathrm{e}^{-\beta(-an_\mathrm{moles}^2/V)}$ to give

$$Z_N = \frac{1}{N!}\left(\frac{V - n_\mathrm{moles}b}{\lambda_\mathrm{th}^3}\right)^N \mathrm{e}^{\beta an_\mathrm{moles}^2/V}, \tag{26.10}$$

which after using $F = -k_\mathrm{B}T \ln Z_N$ and $p = -(\partial F/\partial V)_T$ yields the van der Waals equation of state.

Multiplying eqn 26.3 by V^2 for one mole of van der Waals gas (where $V_m = V$), we have

$$pV^3 - (pb + RT)V^2 + aV - ab = 0, \qquad (26.11)$$

which is a cubic equation in V. The equation of state of the van der Waals gas is plotted in Fig. 26.2 for various isotherms. As the temperature is lowered, the isotherms change from being somewhat ideal-gas like, at the top right of the figure, to exhibiting an S-shape with a minimum and a maximum (as expected for a general cubic equation) in the lower left of the figure. This provides us with a complication: the isothermal compressibility (eqn 16.68) is $\kappa_T = -\frac{1}{V}(\partial V/\partial p)_T$, and for the ideal gas this is always positive (and equal to the inverse pressure of the gas). However, for the van der Waals gas, when the isotherms become S-shaped, there is a region when the gradient $(\partial V/\partial p)_T$ is positive and hence the compressibility κ_T will be negative. This is not a stable situation: a negative compressibility means that when you try to compress the gas it gets bigger! If a pressure fluctuation momentarily increases the pressure, the volume increases (rather than decreases) and negative work is done on the gas, providing energy to amplify the pressure fluctuation; thus a negative compressibility means that the system is unstable with respect to fluctuations. The problem starts when the isotherms become S-shaped, and this happens when the temperature is lower than a certain critical temperature. This temperature is that of the **critical isotherm**, which is indicated by the thick solid line in Fig. 26.2. This does not have a maximum or minimum but shows a point of inflection, known as the **critical point**, which is marked by the dot on Fig. 26.2.

Example 26.1

Find the temperature T_c, pressure p_c, and volume V_c at the critical point of a van der Waals gas, and calculate the ratio $p_c V_c / RT_c$.
Solution:
The equation of state for one mole of van der Waals gas can be rewritten with p as the subject as follows:

$$p = \frac{RT}{V-b} - \frac{a}{V^2}. \qquad (26.12)$$

The point of inflection can be found by using

$$\left(\frac{\partial p}{\partial V}\right)_T = -\frac{RT}{(V-b)^2} + \frac{2a}{V^3} = 0 \qquad (26.13)$$

and

$$\left(\frac{\partial^2 p}{\partial V^2}\right)_T = \frac{2RT}{(V-b)^3} - \frac{6a}{V^4} = 0. \qquad (26.14)$$

Equation 26.13 implies that
$$RT = \frac{2a(V-b)^2}{V^3}, \tag{26.15}$$

while eqn 26.14 implies that
$$RT = \frac{3a(V-b)^3}{V^4}, \tag{26.16}$$

and equating these last two equations gives
$$\frac{3(V-b)}{V} = 2, \tag{26.17}$$

which implies that $V = V_c$, where V_c is the **critical volume** given by
$$V_c = 3b. \tag{26.18}$$

Substituting this back into eqn 26.14 yields $RT = 8a/27b$ and hence $T = T_c$ where T_c is the **critical temperature** given by
$$T_c = \frac{8a}{27Rb}. \tag{26.19}$$

Substituting our expressions for V_c and T_c back into the equation of state for a van der Waals gas gives the **critical pressure** p_c as
$$p_c = \frac{a}{27b^2}. \tag{26.20}$$

We then have that
$$\frac{p_c V_c}{RT_c} = \frac{3}{8} = 0.375, \tag{26.21}$$

independent of both a and b. At the critical point,
$$\left(\frac{\partial p}{\partial V}\right)_{T_c} = 0, \tag{26.22}$$

and hence the isothermal compressibility diverges since
$$\kappa_T = -\frac{1}{V}(\partial V/\partial p)_T \to \infty. \tag{26.23}$$

We have found that the compressibility κ_T is negative when $T < T_c$ and so the system is then unstable. Let us now examine the isotherms below the critical temperature. Since the constraints in an experiment are often those of constant pressure and temperature, it is instructive to examine the Gibbs function for the van der Waals gas, which we can obtain as follows. The Helmholtz function F is related to p by $p = -(\partial F/\partial V)_T$ and so (for 1 mole)
$$F = f(T) - RT\ln(V-b) - \frac{a}{V}, \tag{26.24}$$

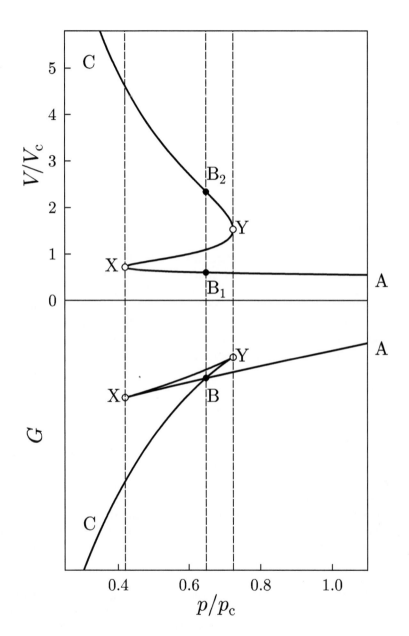

Fig. 26.3 The behaviour of the volume V and Gibbs function G of a van der Waals gas as a function of pressure at $T = 0.9T_c$.

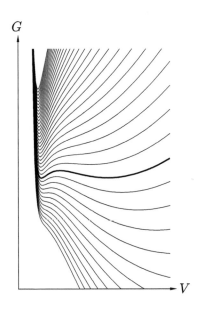

Fig. 26.4 The Gibbs function for different pressures for the van der Waals gas with $T/T_c = 0.9$. The line corresponding to highest (lowest) pressure is at the top (bottom) of the figure. The thick solid line corresponds to the critical pressure p_c.

where $f(T)$ is a function of temperature. Hence the Gibbs function is

$$G = F + pV = f(T) - RT\ln(V-b) - \frac{a}{V} + pV, \quad (26.25)$$

and this is plotted as a function of pressure p in the lower half of Fig. 26.3 for a temperature $T = 0.9T_c$, i.e., below the critical temperature. What is found is that the Gibbs function becomes multiply valued for certain values of pressure. Since a system held at constant temperature and pressure will minimize its Gibbs function, the system will normally ignore the upper loop of the Gibbs function, i.e., the path BXYB in Fig. 26.3, and proceed from A to B to C as the pressure is reduced. The upper part of Fig. 26.3 also shows the corresponding behaviour of the volume as a function of pressure for this same temperature. We see here that the two points B_1 and B_2 on the curve representing the volume correspond to the single point B on the curve representing the Gibbs function. Since the Gibbs function is the same for these two points, phases corresponding to these two points can be in equilibrium with each other. The point B is thus a two-phase region, in which gas and liquid coexist together. Thus liquid (with a much smaller compressibility) is stable in the region A→B and gas (with a much larger compressibility) is stable in the region B→C.

The line BX represents a **metastable** state, in this case superheated liquid. The line BY represents another metastable state, supercooled gas. These metastable states are not the phase corresponding to the lowest Gibbs function of the system for the given conditions of temperature and pressure. They can, however, exist for limited periods of time. The dependence of the Gibbs function on volume for various pressures, expressed in eqn 26.25 and plotted Fig. 26.4, helps us to understand why. At high pressure, there is a single minimum in the Gibbs function corresponding to a low-volume state (the liquid). At low pressure, there is a single minimum in the Gibbs function corresponding to a high-volume state (the gas). At the critical pressure (the thick solid line in Fig. 26.4) there are two minima, corresponding to the coexisting liquid and gas states. If you take gas initially at low pressure and raise the pressure, then when you reach the critical pressure, the system will still be in the right-hand minimum of the Gibbs function. Raising the pressure above p_c would make the left-hand minimum (liquid state) the more stable state, but the system might be stuck in the right-hand minimum (gaseous state) because there is a small energy barrier to surmount to achieve the true stable state. The system is thus, at least temporarily, stuck in a metastable state.

Of course, the triangle BXY in Fig. 26.3 vanishes for temperatures above the critical temperature and then there is simply a crossover from a system with low compressibility to one with progressively higher compressibility as the pressure is reduced. When $T > T_c$, the sharp distinction between liquid and gas is lost and you cannot really tell precisely where the system stops being liquid and starts being a gas. This is a point we will return to in Section 28.7.

We have noted that at points B_1 and B_2 in Fig. 26.3 we have phase

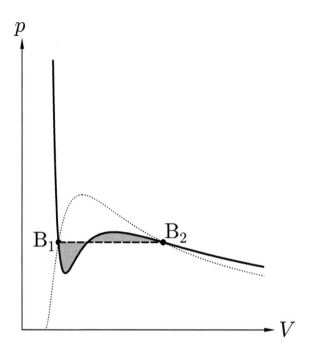

Fig. 26.5 The Maxwell construction for the van der Waals gas. Phase coexistence occurs between points B$_1$ and B$_2$ when the shaded areas are equal. The dotted line shows the locus of such points for different temperatures (and is identical to the dashed line in Fig. 26.2).

coexistence because the Gibbs function is equal at these points. In general, we can always write that the Gibbs function at some pressure p_1 is related to the Gibbs function at some pressure p_0 by

$$G(p_1, T) = G(p_0, T) + \int_{p_0}^{p_1} \left(\frac{\partial G}{\partial p}\right)_T dp, \qquad (26.26)$$

and since

$$\left(\frac{\partial G}{\partial p}\right)_T = V, \qquad (26.27)$$

we have

$$G(p_1, T) = G(p_0, T) + \int_{p_0}^{p_1} V \, dp. \qquad (26.28)$$

Applying this equation between the points B$_1$ and B$_2$ we have that

$$G(p_{B_2}, T) = G(p_{B_1}, T) + \int_{B_1}^{B_2} V \, dp \qquad (26.29)$$

and since $G(p_{B_1}, T) = G(p_{B_2}, T)$, we have that

$$\int_{B_1}^{B_2} V \, dp = 0. \qquad (26.30)$$

This result gives us a useful way of identifying the points B$_1$ and B$_2$, as illustrated in Fig. 26.5. These two points show phase coexistence when the two shaded areas are equal, and this follows directly from eqn 26.30.

304 Real gases

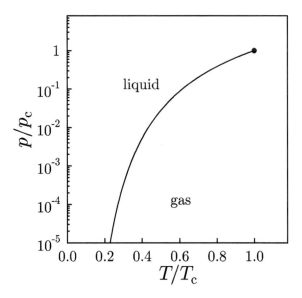

Fig. 26.6 The p–T phase diagram for a van der Waals gas.

The horizontal dashed line separating the two equal shaded areas in Fig. 26.5 is known as the **Maxwell construction**.

The dotted line in Fig. 26.5 shows the locus of such points of coexistence for different temperatures (and is identical to the dashed line in Fig. 26.2). This allows us to plot the phase diagram shown in Fig. 26.6, which shows p against T. The line of phase coexistence is shown, ending in the critical point at $T = T_c$ and $p = p_c$. At fixed pressure, the stable low-temperature state is the liquid, while the stable high-temperature state is the gas. Note that when $p > p_c$ and $T > T_c$, there is no sharp phase boundary separating gas from liquid. Thus it is possible to "avoid" a sharp phase transition between liquid and gas by, for example, starting with a gas, heating it at low pressure to above T_c, isothermally pressurizing above p_c, and then isobarically cooling to below T_c and obtaining a liquid. We will consider these transitions between different phases in more detail in Chapter 28.

26.2 The Dieterici equation

The van der Waals equation of state can be written in the form

$$p = p_{\text{repulsive}} + p_{\text{attractive}}, \qquad (26.31)$$

where the first term is a repulsive hard-sphere interaction

$$p_{\text{repulsive}} = \frac{RT}{V-b}, \qquad (26.32)$$

which is an ideal-gas-like term but with the denominator being the volume available to gas molecules, namely that of the container V minus that of the molecules, b. The second term is the attractive interaction

$$p_{\text{attractive}} = -\frac{a}{V^2}. \tag{26.33}$$

There have been other attempts to model non-ideal gases. In the **Berthelot equation**, the attractive force is made temperature dependent by writing[2]

$$p_{\text{attractive}} = -\frac{a}{TV^2}. \tag{26.34}$$

Another approach is given by Dieterici,[3] who in 1899 proposed an alternative equation of state,[4] in which he wrote that

$$p = p_{\text{repulsive}} \exp\left(-\frac{a}{RTV}\right), \tag{26.35}$$

and using eqn 26.32 this leads to

$$\boxed{p(V_{\text{m}} - b) = RT \exp\left(-\frac{a}{RTV_{\text{m}}}\right),} \tag{26.36}$$

which is the **Dieterici equation**, here written in terms of the molar volume. The constant a is, again, a parameter that controls the strength of attractive interactions. Isotherms of the Dieterici equation of state are shown in Fig. 26.7; they are similar to those for the van der Waals gas (Fig. 26.2), showing a very sudden increase in pressure as V approaches b.

The critical point can be identified for this model by evaluating

$$\left(\frac{\partial^2 p}{\partial V^2}\right)_T = \left(\frac{\partial p}{\partial V}\right)_T = 0, \tag{26.37}$$

and this yields (after a little algebra)

$$T_{\text{c}} = \frac{a}{4Rb}, \quad p_{\text{c}} = \frac{a}{4e^2 b^2}, \quad V_{\text{c}} = 2b, \tag{26.38}$$

for the critical temperature, pressure, and volume, and hence

$$\frac{p_{\text{c}} V_{\text{c}}}{RT_{\text{c}}} = \frac{2}{e^2} = 0.271. \tag{26.39}$$

This value agrees well with those listed in Table 26.1 (and is better than the van der Waals result, which is 0.375, as shown in eqn 26.21).

[2] Here a is a different constant, with different units, from that used in the van der Waals equation.

[3] Conrad Dieterici (1858–1929)

[4] The constant a is different again. The motivation for the Dieterici equation is an assumption that the pressure should be corrected by a Boltzmann factor, reflecting the fact that wall-striking molecules have reduced kinetic energy and number density compared with the bulk of the gas. (Pressure, kinetic energy, and number density are all related to each other.) The Boltzmann factor should take the form $e^{-\Delta E/k_{\text{B}} T}$ where the energy reduction ΔE is due to the attractive interaction, which is proportional to $1/V$. This motivates eqn 26.35, though it should be stressed that these equations of state are all phenomenological.

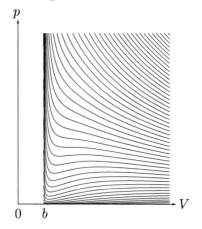

Fig. 26.7 Isotherms for the Dieterici equation of state.

	Ne	Ar	Kr	Xe
$p_{\text{c}} V_{\text{c}} / RT_{\text{c}}$	0.287	0.292	0.291	0.290

Table 26.1 The values of $p_{\text{c}} V_{\text{c}} / RT_{\text{c}}$ for various noble gases.

26.3 Virial expansion

Another method of modelling real gases is to take the ideal gas equation and modify it using a power series in $1/V_m$ (where V_m is the molar volume). This leads to the following **virial expansion**:

$$\frac{pV_m}{RT} = 1 + \frac{B}{V_m} + \frac{C}{V_m^2} + \cdots \qquad (26.40)$$

In this equation, the parameters B, C, etc., are called **virial coefficients** and can be made to be temperature dependent (so that we will denote them by $B(T)$ and $C(T)$). The temperature at which the virial coefficient $B(T)$ goes to zero is called the **Boyle temperature** T_B since it is the temperature at which Boyle's law is approximately obeyed (neglecting the higher-order virial coefficients), as shown in Fig. 26.8.

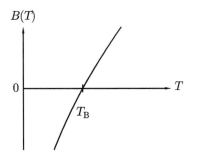

Fig. 26.8 The temperature dependence of the virial coefficient B.

Example 26.2

Express the van der Waals equation of state in terms of a virial expansion and hence find the Boyle temperature in terms of the critical temperature.

Solution:
The van der Waals equation of state can be rewritten as

$$p = \frac{RT}{V-b} - \frac{a}{V^2} = \frac{RT}{V}\left(1 - \frac{b}{V}\right)^{-1} - \frac{a}{V^2}, \qquad (26.41)$$

and using the binomial expansion, the term in brackets can be expanded into a series, resulting in

$$\frac{pV}{RT} = 1 + \frac{1}{V}\left(b - \frac{a}{RT}\right) + \left(\frac{b}{V}\right)^2 + \left(\frac{b}{V}\right)^3 + \cdots, \qquad (26.42)$$

which is in the same form as the virial expansion in eqn 26.40 with

$$B(T) = b - \frac{a}{RT}. \qquad (26.43)$$

The Boyle temperature T_B is defined by $B(T_B) = 0$ and hence

$$T_B = \frac{a}{bR}, \qquad (26.44)$$

and hence using eqn 26.19 we have that

$$T_B = \frac{27 T_c}{8}. \qquad (26.45)$$

The additional terms in the virial expansion give information about the nature of the intermolecular interactions. We can show this using the following argument, which shows how to model intermolecular interactions in the dilute gas limit using a statistical mechanical argument.[5] The total internal energy U of the molecules (each with mass m) in a gas can be written as

$$U = U_{\text{K.E.}} + U_{\text{P.E.}}, \tag{26.46}$$

where the kinetic energy $U_{\text{K.E.}}$ is given by a sum over all N molecules

$$U_{\text{K.E.}} = \sum_{i=1}^{N} \frac{p_i^2}{2m}, \tag{26.47}$$

where p_i is the momentum of the ith molecule, and the potential energy is given by

$$U_{\text{P.E.}} = \sum_{i \neq j} \frac{1}{2} \mathcal{V}(|\mathbf{r}_i - \mathbf{r}_j|), \tag{26.48}$$

where $\mathcal{V}(|\mathbf{r}_i - \mathbf{r}_j|)$ is the potential energy between the ith and jth molecules and the factor $\frac{1}{2}$ is to avoid double counting of pairs of molecules in the sum. The partition function Z is then given by

$$\begin{aligned} Z &= \int \cdots \int d^3 r_1 \cdots d^3 r_N \, d^3 p_1 \cdots d^3 p_N \, e^{-\beta [U_{\text{K.E.}}(\{p_i\}) + U_{\text{P.E.}}(\{r_i\})]} \\ &= Z_{\text{K.E.}} Z_{\text{P.E.}}, \end{aligned} \tag{26.49}$$

where the last equality follows because the integrals in momentum and in position variables are separable. Now $Z_{\text{K.E.}}$ is the partition function for the ideal gas which we have already derived (Chapter 21) and which leads to the ideal gas equation $pV = RT$, so we focus entirely on $Z_{\text{P.E.}}$ which is given by

$$Z_{\text{P.E.}} = \frac{1}{V^N} \int \cdots \int d^3 r_1 \cdots d^3 r_N \, e^{-\beta U_{\text{P.E.}}}, \tag{26.50}$$

where we have included the factor $1/V^N$ so that when $U_{\text{P.E.}} = 0$ then $Z_{\text{P.E.}} = 1$. Hence

$$Z_{\text{P.E.}} = \frac{1}{V^N} \int \cdots \int d^3 r_1 \cdots d^3 r_N \, e^{-\frac{\beta}{2} \sum_{i \neq j} \mathcal{V}(|\mathbf{r}_i - \mathbf{r}_j|)}, \tag{26.51}$$

and adding one and subtracting one from this equation,[6] we have

$$Z_{\text{P.E.}} = 1 + \frac{1}{V^N} \int \cdots \int d^3 r_1 \cdots d^3 r_N \left[e^{-\frac{\beta}{2} \sum_{i \neq j} \mathcal{V}(|\mathbf{r}_i - \mathbf{r}_j|)} - 1 \right]. \tag{26.52}$$

We presume that the intermolecular interactions are only significant for molecules that are virtually touching, so the integrand is appreciably different from zero only when two molecules are very close together. If the gas is dilute this condition of two molecules being close will only happen relatively rarely, and so we will assume that this condition occurs only for one pair of molecules at any one time. There are N ways of

[5] This argument is a little more technical than the material in the rest of this chapter and can be skipped at first reading.

[6] This trick is done because we know that later we will want to play with the log of the partition function, and $\ln(1 + x) \approx x$ for $x \ll 1$, so having Z in the form one plus something small is convenient.

picking the first molecule for a collision, and $N-1$ ways of picking the second molecule for a collision, and since we don't care which is the "first" molecule and which is the "second", the number of ways to select a pair of molecules from N molecules is

$$\frac{N(N-1)}{2}, \tag{26.53}$$

which is approximately $N^2/2$ when N is large. Writing r for the coordinate separating these two molecules, we then have

$$Z_{\text{P.E.}} \approx 1 + \frac{N^2}{2V^N} \int \cdots \int d^3r_1 \cdots d^3r_N \left[e^{-\beta \mathcal{V}(r)} - 1\right]. \tag{26.54}$$

Since the integral depends only on the separation r of these two molecules, we can integrate out the other $N-1$ volume coordinates (resulting in integrals equal to unity multiplied by the volume V) and obtain

$$Z_{\text{P.E.}} \approx 1 + \frac{N^2}{2V} \int d^3r \left[e^{-\beta \mathcal{V}(r)} - 1\right], \tag{26.55}$$

and writing $B(T)$ (the virial coefficient) as

$$B(T) = \frac{N}{2} \int d^3r \left[1 - e^{-\beta \mathcal{V}(r)}\right], \tag{26.56}$$

we have that

$$Z_{\text{P.E.}} \approx 1 - \frac{N B(T)}{V}, \tag{26.57}$$

and hence

$$\begin{aligned} F &= -k_{\text{B}} T \ln Z \\ &= -k_{\text{B}} T \ln(Z_{\text{K.E.}} Z_{\text{P.E.}}) \\ &= F_0 + \frac{N k_{\text{B}} T B(T)}{V}, \end{aligned} \tag{26.58}$$

where F_0 is the Helmholtz function of the ideal gas and the last equality is accomplished using $\ln(1+x) \approx x$ for $x \ll 1$. Hence, we can evaluate the pressure p as follows:

$$p = -\left(\frac{\partial F}{\partial V}\right)_T = \frac{N k_{\text{B}} T}{V} + \frac{N k_{\text{B}} T B(T)}{V^2}. \tag{26.59}$$

Rearranging, we have that for one mole of gas

$$\frac{pV}{RT} = 1 + \frac{B(T)}{V}, \tag{26.60}$$

which is of the form of the virial expansion in eqn 26.40 but with only a single non-ideal term.

The temperature dependence of the virial coefficient $B(T)$ for argon is shown in Fig. 26.9. It is large and negative at low temperatures but changes sign (at the Boyle temperature) and then becomes small

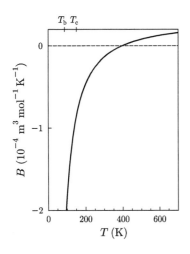

Fig. 26.9 The temperature dependence of the virial coefficient $B(T)$ in argon. Argon has a boiling point at atmospheric pressure of $T_{\text{b}} = 87\,\text{K}$ and the critical point is at $T_{\text{c}} = 151\,\text{K}$ and $p_{\text{c}} = 4.86\,\text{MPa}$.

and positive at higher temperatures. We can understand this from the expression for $B(T)$ given in eqn 26.56. This is an integral of the function $1 - e^{-\beta \mathcal{V}(r)}$, which is shown in Fig. 26.10(b). At low temperatures, the integral of this function is dominated by the negative peak which is centred around $r_{\min}$, the minimum in the potential well (corresponding to the particles spending more time with this intermolecular spacing). Hence $B(T)$ is negative and large at low temperatures. As temperature increases, the peak in this function broadens out as molecules spend more time a long way from each other, resulting in a weakened average potential energy. Here, the effect of the positive plateau below $r_{\min}$ begins to dominate the integral and B changes sign.

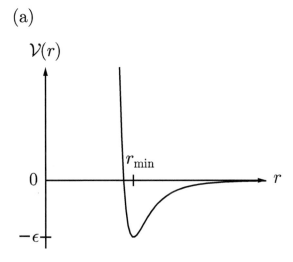

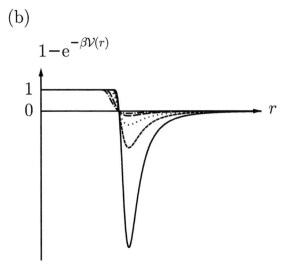

Fig. 26.10 The upper graph, (a), shows the intermolecular potential energy $\mathcal{V}(r)$. The integrand in eqn 26.56 is $1 - e^{-\beta \mathcal{V}(r)}$ and this is plotted in the lower graph, (b), for different values of β. The solid curve shows a value with large β (low temperature) and the other curves show the effect of reducing β (raising the temperature), where in order of decreasing β the lines are dashed, dotted, long-dashed, and dot-dashed.

26.4 The law of corresponding states

For different substances, the size of the molecules (which controls b in the van der Waals model) and the strength of the intermolecular interactions (which controls a in the van der Waals model) will vary, and hence their phase diagrams will be different. For example, the critical temperatures and pressures for different gases are different. However, the phase diagram of substances should be the same when plotted in **reduced coordinates**, which can be obtained by dividing a quantity by its value at the critical point. Hence, if we replace the quantities p, V, T by their reduced coordinates $\tilde{p}, \tilde{V}, \tilde{T}$ defined by

$$\tilde{p} = \frac{p}{p_c}, \qquad \tilde{V} = \frac{V}{V_c}, \qquad \tilde{T} = \frac{T}{T_c}, \qquad (26.61)$$

then phase diagrams of materials that are not wholly different from one another should lie on top of each other. This is called the **law of corresponding states**.

Example 26.3

Express the equation of state of the van der Waals gas in reduced coordinates.

Solution:
Substituting eqns 26.61 into eqn 26.3 we find that

$$p_c \tilde{p} = \frac{RT_c \tilde{T}}{V_c \tilde{V} - b} - \frac{a}{V_c^2 \tilde{V}^2}, \qquad (26.62)$$

and this can be rearranged to give

$$\left(\tilde{p} + \frac{3}{\tilde{V}^2}\right) = \frac{8\tilde{T}}{3\tilde{V} - 1}. \qquad (26.63)$$

The law of corresponding states works well in practice for real experimental data, since the intermolecular potential energies are usually of a similar form in different substances, as shown in Fig. 26.10(a). There is a repulsive region at small distances, a stable minimum at a separation $r_{\min}$ corresponding to a potential well depth of $-\epsilon$, and then a long-range attractive region at larger distances. For different molecules, the length scale $r_{\min}$ and the energy scale ϵ may be different, but these two parameters together are sufficient to give a reasonable description of the intermolecular potential energy. The parameter $r_{\min}$ sets the scale of the molecular size and the parameter ϵ sets the scale of the intermolecular interactions. Dividing p, V and T by their values at the critical point removes these scales and allows the different phase diagrams to be superimposed.

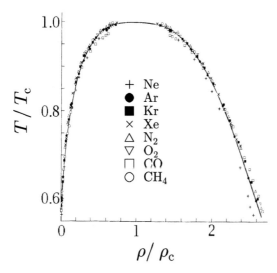

Fig. 26.11 The liquid–gas coexistence for a number of different substances can be superimposed once they are plotted in reduced coordinates. The solid line is a scaling relation. This plot is adapted from E. A. Guggenheim. *Thermodynamics*, (8th edition), North-Holland, Amsterdam (1945).

An example of this for real data is shown in Fig. 26.11. The form of the liquid–gas coexistence is different in detail from that predicted by the van der Waals equation, but shows that the underlying behaviour in different real systems is similar and shows "universal" features.

Chapter summary

- Attractive intermolecular interactions and the non-zero size of molecules lead to departures from ideal gas behaviour.
- The van der Waals equation of state is

$$\left(p + \frac{a}{V_m^2}\right)(V_m - b) = RT.$$

- The Dieterici equation of state is

$$p(V_m - b) = RT e^{-a/RTV_m}.$$

- The virial expansion of a gas can be written as

$$\frac{pV_m}{RT} = 1 + \frac{B}{V_m} + \frac{C}{V_m^2} + \cdots$$

- The law of corresponding states implies that if the variables p, V, and T are scaled by their values at the critical point, the behaviour of different gases in these scaled variables is often very similar to that of other gases scaled in the same way.

Exercises

(26.1) Show that the isothermal compressibility κ_T of a van der Waals gas at $V = V_c$ can be written as
$$\kappa_T = \frac{4b}{3R}(T - T_c)^{-1}. \qquad (26.64)$$
Sketch the temperature dependence of κ_T and explain what happens to the properties of the gas when the temperature is lowered through the critical temperature.

(26.2) The equation of state of a certain gas is $p(V - b) = RT$, where b is a constant. What order of magnitude do you expect b to be? Show that the internal energy of this gas is a function of temperature only.

(26.3) Show that the Dieterici equation of state,
$$p(V - b) = RTe^{-a/RTV},$$
can be written in reduced units as
$$\tilde{p}(2\tilde{V} - 1) = \tilde{T}\exp\left[2\left(1 - \frac{1}{\tilde{T}\tilde{V}}\right)\right],$$
where $\tilde{p} = p/p_c$, $\tilde{T} = T/T_c$, $\tilde{V} = V/V_c$, and (p_c, T_c, V_c) is the critical point.

(26.4) Show that the isobaric expansivity β_p of the van der Waals gas is given by
$$\beta_p = \frac{1}{T}\left(1 + \frac{b}{V - b} - \frac{2a}{pV^2 + a}\right)^{-1}. \qquad (26.65)$$
What happens to this quantity close to the critical point?

(26.5) Show that eqn 26.10 leads to
$$U = \frac{3}{2}RT - \frac{a}{V}, \qquad (26.66)$$
for one mole of gas.

(26.6) The total energy of one mole of a van der Waals gas can be written as
$$U = \frac{f}{2}RT - \frac{a}{V}, \qquad (26.67)$$
where f is the number of degrees of freedom (see eqn 19.22). Show that
$$C_V = \frac{f}{2}R \qquad (26.68)$$
and
$$C_p - C_V \approx R + \frac{2a}{TV}. \qquad (26.69)$$

Cooling real gases

27

In Chapter 26, we considered how to model the properties of real gases using various corrections to the ideal gas model. In this chapter, we will use these results to explore some of the deviations from ideal gas behaviour that can be observed in practice, in particular with changes in the behaviour of a Joule expansion. Then we will introduce the Joule–Kelvin throttling process (which has no effect on an ideal gas, but which can lead to cooling of a real gas) and discuss how real gases can be liquefied.

27.1 The Joule expansion	313
27.2 Isothermal expansion	315
27.3 Joule–Kelvin expansion	316
27.4 Liquefaction of gases	318
Chapter summary	319
Exercises	320

27.1 The Joule expansion

We have discussed the properties of non-ideal gases in some detail. In this section, we will see how the intermolecular interactions in such gases lead to departures from ideal-gas behaviour for the Joule expansion. Recall from Section 14.4 that a Joule expansion is an irreversible expansion of a gas into a vacuum which can be accomplished by opening a tap connecting the vessel containing gas and an evacuated vessel (see Fig. 27.1). The entire system is isolated from its surroundings and so no heat enters or leaves. No work is done, so the internal energy U is unchanged. We are interested in finding out whether the gas warms, cools, or remains at constant temperature in this expansion.

To answer this, we define the **Joule coefficient** μ_J using

$$\mu_J = \left(\frac{\partial T}{\partial V}\right)_U, \tag{27.1}$$

where the constraint of constant U is relevant for the Joule expansion. This partial differential can be transformed using eqn 16.64 and the definition of C_V to give

$$\mu_J = -\left(\frac{\partial T}{\partial U}\right)_V \left(\frac{\partial U}{\partial V}\right)_T = -\frac{1}{C_V}\left(\frac{\partial U}{\partial V}\right)_T. \tag{27.2}$$

Now the first law, $dU = TdS - pdV$, implies that

$$\left(\frac{\partial U}{\partial V}\right)_T = T\left(\frac{\partial S}{\partial V}\right)_T - p, \tag{27.3}$$

and using a Maxwell relation (eqn 16.50) this becomes

$$\left(\frac{\partial U}{\partial V}\right)_T = T\left(\frac{\partial p}{\partial T}\right)_V - p, \tag{27.4}$$

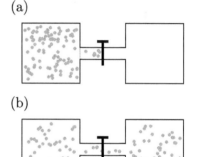

Fig. 27.1 The Joule expansion: (a) before opening the tap and (b) after opening the tap.

and hence
$$\mu_J = -\frac{1}{C_V}\left[T\left(\frac{\partial p}{\partial T}\right)_V - p\right]. \tag{27.5}$$

For an ideal gas, $p = RT/V$, $(\partial p/\partial T)_V = R/V$, and hence $\mu_J = 0$. Hence, as we found in Section 14.4, the temperature of an ideal gas is unchanged in a Joule expansion. For real gases, you always get cooling because of the attractive effect of interactions. This is because $C_V > 0$ and $(\partial U/\partial V)_T > 0$ and so $\mu_J = -\frac{1}{C_V}(\partial U/\partial V)_T < 0$. This can be understood physically as follows: when a gas freely expands into a vacuum, the time-averaged distance between neighbouring molecules increases and the magnitude of the potential energy resulting from the attractive intermolecular interactions is reduced. However, this potential energy is a negative quantity (because the interactions are attractive) and so the potential energy is actually *increased* (because it is made less negative).[1] Since U must be unchanged in a Joule expansion (no heat enters or leaves and no work is done), the kinetic energy must be reduced (by the same amount by which the potential energy rises) and hence the temperature falls.

[1] Of course, at very high densities, the intermolecular interactions become repulsive rather than attractive, but at such a density one is probably dealing with a solid rather than a gas.

Example 27.1

Evaluate the Joule coefficient for a van der Waals gas.
Solution: The equation of state is $p = RT/(V-b) - a/V^2$ and so
$$\left(\frac{\partial p}{\partial T}\right)_V = \frac{R}{V-b}, \tag{27.6}$$
and hence
$$\mu_J = -\frac{1}{C_V}\left[\frac{RT}{V-b} - \frac{RT}{V-b} + \frac{a}{V^2}\right] = -\frac{a}{C_V V^2}. \tag{27.7}$$

The temperature change in a Joule expansion from V_1 to V_2 can be evaluated simply by integrating the Joule coefficient as follows:
$$\Delta T = \int_{V_1}^{V_2} \mu_J \, dV = -\int_{V_1}^{V_2} \frac{1}{C_V}\left[T\left(\frac{\partial p}{\partial T}\right)_V - p\right] dV. \tag{27.8}$$

Example 27.2

Evaluate the change in temperature for a van der Waals gas which undergoes a Joule expansion from volume V_1 to volume V_2.
Solution: Using eqn 27.8, we have that
$$\Delta T = -\frac{a}{C_V}\int_{V_1}^{V_2} \frac{dV}{V^2} = -\frac{a}{C_V}\left(\frac{1}{V_1} - \frac{1}{V_2}\right) < 0 \tag{27.9}$$
since $V_2 > V_1$ in an expansion.

27.2 Isothermal expansion

Consider the isothermal expansion of a non-ideal gas. Equation 27.4 states that

$$\left(\frac{\partial U}{\partial V}\right)_T = T\left(\frac{\partial p}{\partial T}\right)_V - p, \qquad (27.10)$$

so that the change of U in an isothermal expansion is

$$\Delta U = \int_{V_1}^{V_2}\left[T\left(\frac{\partial p}{\partial T}\right)_V - p\right]\mathrm{d}V. \qquad (27.11)$$

- For an ideal gas,[2] $\Delta U = 0$.
- For a van der Waals gas, $\Delta U = \int_{V_1}^{V_2}\frac{a}{V^2}\,\mathrm{d}V = a(1/V_1 - 1/V_2)$.

[2] For an ideal gas, $p = RT/V$, $(\partial p/\partial T)_V = R/V$, and hence $T\left(\frac{\partial p}{\partial T}\right)_V - p = 0$.

Note that U depends on a, not b (it is influenced by the intermolecular interactions but does not "care" that they have non-zero size). Note also that for large volumes, U becomes independent of V and one recovers the ideal gas limit.

Example 27.3

Calculate the entropy of a van der Waals gas.

Solution:

The entropy S can be written as a function of T and V so that $S = S(T, V)$. Hence

$$\begin{aligned}\mathrm{d}S &= \left(\frac{\partial S}{\partial T}\right)_V \mathrm{d}T + \left(\frac{\partial S}{\partial V}\right)_T \mathrm{d}V \\ &= \frac{C_V}{T}\mathrm{d}T + \left(\frac{\partial p}{\partial T}\right)_V \mathrm{d}V, \end{aligned} \qquad (27.12)$$

where eqns 16.65 and 16.50 have been used to obtain the second line. For the van der Waals gas, we can write $(\partial p/\partial T)_V = R/(V-b)$, and hence we have that

$$S = C_V \ln T + R\ln(V - b) + \text{constant}. \qquad (27.13)$$

Note that the entropy depends on the constant b, but not a. Entropy "cares" about the volume occupied by the molecules in the gas (because this determines how much available space there is for the molecules to move around in, and this in turn determines the number of possible microstates of the system) but not about the intermolecular interactions.

27.3 Joule–Kelvin expansion

The Joule expansion is a useful conceptual process, but it is not much practical use for cooling gases. Gas slightly cools when it is expanded into a second evacuated vessel, but what do you do with it then? What is wanted is some kind of flow process where warm gas can be fed into some kind of a "cooling machine" and cold gas (or better still, cold liquid) emerges from the other end. Such a process was discovered by James Joule and William Thomson (later Lord Kelvin) and is known as a **Joule–Thomson expansion** or a **Joule–Kelvin expansion**.

Consider a steady flow process in which gas at high pressure p_1 is forced through a throttle valve or a porous plug to a lower pressure p_2. This is illustrated in Fig. 27.2. Consider a volume V_1 of gas on the high-pressure side. Its internal energy is U_1. To push the gas through the constriction, the high pressure gas behind it has to do work on it equal to $p_1 V_1$ (since the pressure p_1 is maintained on the high-pressure side of the constriction). The gas expands as it passes through to the low-pressure region and now occupies volume V_2 which is larger than V_1. It has to do work on the low-pressure gas in front of it which is at pressure p_2 and hence this work is $p_2 V_2$. The gas may change its temperature in the process and hence its new internal energy is U_2. The change in internal energy $(U_2 - U_1)$ must be equal to the work done on the gas $(p_1 V_1)$ minus the work done by the gas $(p_2 V_2)$. Thus

$$U_1 + p_1 V_1 = U_2 + p_2 V_2 \tag{27.14}$$

or equivalently

$$H_1 = H_2, \tag{27.15}$$

so that it is *enthalpy* that is conserved in this flow process.

Since we are now interested in how much the gas changes temperature when we reduce its pressure at constant enthalpy, we define the Joule–Kelvin coefficient by

$$\mu_{\text{JK}} = \left(\frac{\partial T}{\partial p}\right)_H. \tag{27.16}$$

This can be transformed using the reciprocity theorem (eqn C.42) and the definition of C_p to give

$$\mu_{\text{JK}} = -\left(\frac{\partial T}{\partial H}\right)_p \left(\frac{\partial H}{\partial p}\right)_T = -\frac{1}{C_p}\left(\frac{\partial H}{\partial p}\right)_T. \tag{27.17}$$

Now the relation $dH = T dS + V dp$ implies that

$$\left(\frac{\partial H}{\partial p}\right)_T = T\left(\frac{\partial S}{\partial p}\right)_T + V, \tag{27.18}$$

and using a Maxwell relation (eqn 16.51) this becomes

$$\left(\frac{\partial H}{\partial p}\right)_T = -T\left(\frac{\partial V}{\partial T}\right)_p + V, \tag{27.19}$$

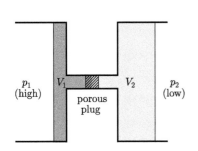

Fig. 27.2 A throttling process.

and hence
$$\mu_{JK} = \frac{1}{C_p}\left[T\left(\frac{\partial V}{\partial T}\right)_p - V\right]. \qquad (27.20)$$

The change in temperature for a gas following a Joule–Kelvin expansion from pressure p_1 to pressure p_2 is given by
$$\Delta T = \int_{p_1}^{p_2} \frac{1}{C_p}\left[T\left(\frac{\partial V}{\partial T}\right)_p - V\right] dp. \qquad (27.21)$$

Since $dH = TdS + Vdp = 0$, the entropy change is
$$\Delta S = -\int_{p_1}^{p_2} \frac{V}{T} dp, \qquad (27.22)$$

and for an ideal gas this is $R\ln(p_1/p_2) > 0$. Thus this is an irreversible process.

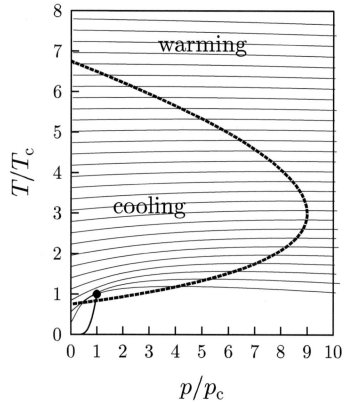

Fig. 27.3 The inversion curve of the van der Waals gas is shown as the heavy dashed line. The isenthalps (lines of constant enthalpy) are shown as thin solid lines. When the gradients of the isenthalps on this diagram are positive, then cooling can be obtained when pressure is reduced at constant enthalpy (i.e., in a Joule–Kelvin expansion). Also shown (as a solid line near the bottom left-hand corner of the graph which terminates at the dot) is the line of coexistence of liquid and gas (from Fig. 26.6) ending in the critical point ($p = p_c$, $T = T_c$, shown by the dot).

Whether the Joule–Kelvin expansion results in heating or cooling is more subtle and in fact μ_{JK} can take either sign. It is convenient to consider when μ_{JK} changes sign, and this will occur when $\mu_{JK} = 0$, i.e., when $T(\partial V/\partial T)_p - V = 0$, or equivalently
$$\left(\frac{\partial V}{\partial T}\right)_p = \frac{V}{T}. \qquad (27.23)$$

^{4}He	H$_2$	N$_2$	Ar	CO$_2$
43	204	607	794	1275

Table 27.1 The maximum inversion temperature in kelvin for several gases.

This equation defines the so-called **inversion curve** in the T–p plane. This is plotted for the van der Waals gas in Fig. 27.3 as a heavy dashed line. The lines of constant enthalpy are also shown and their gradients change sign when they cross the inversion curve. When the gradient of the isenthalps on this diagram are positive, then cooling can be obtained when pressure is reduced at constant enthalpy (i.e., in a Joule–Kelvin expansion).

A crucial parameter is the maximum inversion temperature, below which the Joule–Kelvin expansion can result in cooling. These are listed for several real gases in Table 27.1. In the case of helium, this temperature is 43 K, so helium gas must be cooled to below this temperature by some other means before it can be liquefied using the Joule–Kelvin process.

27.4 Liquefaction of gases

For achieving the liquefaction of gases, the Joule–Kelvin process is extremely useful, though it must be carried out below the maximum inversion temperature of the particular gas in question. A schematic diagram of a liquefier is shown in Fig. 27.4. High–pressure gas is forced through a throttle valve, resulting in cooling by the Joule–Kelvin process. Low–pressure gas plus liquid results, and the process is made more efficient by use of a counter-current heat exchanger by which the outgoing cold low–pressure gas is used to precool the incoming warm high pressure gas, helping to ensure that by the time it reaches the throttle valve the incoming high–pressure gas is already as cool as possible and at least at a temperature such that the Joule–Kelvin effect will result in cooling.

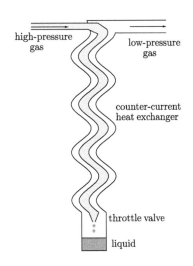

Fig. 27.4 A schematic diagram of a gas liquefier.

We can consider the liquefier as a "black box" into which you put 1 kg of warm gas and get out y kg of liquid, as well as $(1 - y)$ kg of exhaust gas (see Fig. 27.5). The variable y is the efficiency y of a liquefier, i.e., the mass fraction of incoming gas which is liquefied. Since enthalpy is conserved in a Joule–Kelvin process, we have that

$$h_i = y h_L + (1 - y) h_f, \tag{27.24}$$

where h_i is the specific enthalpy of the incoming gas, h_L is the specific enthalpy of the liquid, and h_f is the specific enthalpy of the outgoing gas. Hence the efficiency y is given by

$$y = \frac{h_f - h_i}{h_f - h_L}. \tag{27.25}$$

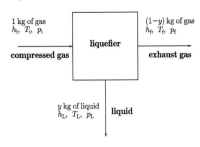

Fig. 27.5 A block diagram showing the liquefaction process.

For an efficient heat exchanger, the temperature of the compressed gas T_i and the exhaust gas T_f will be the same. We also have that $p_f = 1\,\text{atm}$, and T_L is fixed (because the liquid will be in equilibrium with its vapour). Therefore h_f and h_L are fixed. The only parameter to vary is then h_i and to maximize y we must minimize h_i, i.e.,

$$\left(\frac{\partial h_i}{\partial p_i}\right)_{T_i} = 0 \tag{27.26}$$

and since $(\partial h/\partial p)_T = -C_p \mu_{\text{JK}}$, we therefore require that

$$\mu_{\text{JK}} = 0. \tag{27.27}$$

This means that it is best to work the liquefier right on the inversion curve ($\mu_{\text{JK}} = 0$) for maximum efficiency.

Most gases could be liquefied by the end of the nineteenth century, but the modern form of gas liquefier dates back to the work of the German chemist Karl von Linde (1842–1934) who commercialized liquid-air production in 1895 using the Joule–Kelvin effect with a counter-current heat exchanger (as shown in Fig. 27.4; this is known as the **Linde process**) and discovered various uses of liquid nitrogen. The British scientist James Dewar (1842–1923) was the first to liquefy hydrogen using the Linde process in 1898, and in 1899 succeeded in solidifying it. Dewar was also the first, in 1891, to study the magnetic properties of liquid oxygen. The Dutch physicist Heike Kamerlingh Onnes (1853–1926) was the first to produce liquid helium in 1908 by a similar process, precooling the helium gas using liquid hydrogen. Using liquid helium, he then discovered superconductivity in 1911, and was awarded the Nobel Prize in 1913 for 'his investigations on the properties of matter at low temperatures which led, inter alia, to the production of liquid helium."

Chapter summary

- The Joule expansion results in cooling for non-ideal gases because of the attractive interactions between molecules.
- The entropy of a gas depends on the non-zero size of molecules.
- The Joule–Kelvin expansion is a steady flow process in which enthalpy is conserved. It can result in either warming or cooling of a gas. It forms the basis of many gas liquefaction techniques.

Exercises

(27.1) (a) Derive the following general relations

(i) $\left(\dfrac{\partial T}{\partial V}\right)_U = -\dfrac{1}{C_V}\left[T\left(\dfrac{\partial p}{\partial T}\right)_V - p\right],$

(ii) $\left(\dfrac{\partial T}{\partial V}\right)_S = -\dfrac{1}{C_V}T\left(\dfrac{\partial p}{\partial T}\right)_V,$

(iii) $\left(\dfrac{\partial T}{\partial p}\right)_H = \dfrac{1}{C_p}\left[T\left(\dfrac{\partial V}{\partial T}\right)_p - V\right].$

In each case the quantity on the left-hand side is the appropriate thing to consider for a particular type of expansion. State what type of expansion each refers to.

(b) Using these relations, verify that for an ideal gas $(\partial T/\partial V)_U = 0$ and $(\partial T/\partial p)_H = 0$, and that $(\partial T/\partial V)_S$ leads to the familiar relation $pV^\gamma =$ constant along an isentrope.

(27.2) In a Joule–Kelvin liquefier, gas is cooled by expansion through an insulated throttle – a simple but inefficient process with no moving parts at low temperature. Explain why enthalpy is conserved in this process. Deduce that

$$\left(\dfrac{\partial T}{\partial P}\right)_H = \dfrac{1}{C_P}\left[T\left(\dfrac{\partial V}{\partial T}\right)_P - V\right].$$

Estimate the highest starting temperature at which the process will work for helium at low densities, on the following assumptions:

(i) The pressure is given at low densities by a virial expansion of the form

$$\dfrac{PV}{RT} = 1 + \left(b - \dfrac{a}{RT}\right)\left(\dfrac{1}{V}\right) + \cdots,$$

and

(ii) The Boyle temperature a/bR (the temperature at which the second virial coefficient vanishes) is known from experiment to be 19 K for helium.

[Hint: One method of solving this problem is to remember that p is easily made the subject of the equation of state and one can then use $(\partial V/\partial T)_p = -(\partial p/\partial T)_V/(\partial p/\partial V)_T$.]

(27.3) For a gas obeying Dieterici's equation of state

$$p(V - b) = RTe^{-a/RTV},$$

for 1 mole, prove that the equation of the inversion curve is

$$p = \left(\dfrac{2a}{b^2} - \dfrac{RT}{b}\right)\exp\left(\dfrac{1}{2} - \dfrac{a}{RTb}\right),$$

and hence find the maximum inversion temperature $T_{\max}$.

(27.4) Show that the equation for the inversion curve of the Dieterici gas in reduced units is

$$\tilde{P} = (8 - \tilde{T})\exp\left[\dfrac{5}{2} - \dfrac{4}{\tilde{T}}\right],$$

and sketch it in the $\tilde{T}$–$\tilde{P}$ plane.

(27.5) Why is enthalpy conserved in steady flow processes? A helium liquefier in its final stage of liquefaction takes in compressed helium gas at 14 K, liquefies a fraction y, and rejects the rest at 14 K and atmospheric pressure. Use the values of specific enthalpy h of helium gas at 14 K as a function of pressure p in the table below to determine the input pressure which allows y to take its maximum value, and determine what this value is.

p (atm)	0	10	20	30	40
h (kJ kg^{-1})	87.4	78.5	73.1	71.8	72.6

[Enthalpy of liquid helium at atmospheric pressure $= 10.1$ kJ kg^{-1}].

Phase transitions

28

In this chapter we will consider **phase transitions**, in which one thermodynamic phase changes into another. An example would be the transition from liquid water to gaseous steam, which occurs when you boil a kettle of water. If you start with cold water, and warm it in the kettle, all that happens initially is that the water gets progressively hotter. However, when the temperature of the water reaches $100\,°$C interesting things begin to happen. Bubbles of gas form with different sizes, making the kettle considerably noisier, and water molecules begin to leave the liquid surface in large quantities, and steam is emitted. The transition between different phases is very sudden. It is only when the boiling point is reached that liquid water becomes thermodynamically unstable and gaseous water, steam, becomes thermodynamically stable. In this chapter, we will look in detail at the thermodynamics of this and other phase transitions.

28.1	Latent heat	321
28.2	Chemical potential and phase changes	324
28.3	The Clausius–Clapeyron equation	324
28.4	Stability and metastability	329
28.5	The Gibbs phase rule	332
28.6	Colligative properties	334
28.7	Classification of phase transitions	335
28.8	The Ising model	338
Chapter summary		343
Further reading		343
Exercises		343

28.1 Latent heat

To increase the temperature of a substance, one needs to apply heat, and how much heat is needed can be calculated from the heat capacity because adding heat to the substance increases its entropy. The gradient of entropy with temperature is related to the heat capacity via

$$C_x = T \left(\frac{\partial S}{\partial T}\right)_x, \quad (28.1)$$

where x is the appropriate constraint (e.g., p, V, B, etc). Now consider two phases, which are in thermodynamic equilibrium at a critical temperature T_c. Very often, it is found that to change from phase 1 to phase 2 at a constant temperature T_c, you need to supply some extra heat, known as the **latent heat** L, which is given by

$$L = \Delta Q_{\text{rev}} = T_c(S_2 - S_1), \quad (28.2)$$

where S_1 is the entropy of phase 1 and S_2 is the entropy of phase 2. This, together with eqn 28.1, implies that there will be a spike in the heat capacity C_x as a function of temperature.

An example of a phase transition involving a latent heat is the liquid–gas transition. The entropy as a function of temperature for H_2O is shown in Fig. 28.1. The entropy is shown to change discontinuously at the phase transition. The heat capacity[1] C_p of the liquid phase,

[1] We use C_p because the constraint usually applied in the laboratory is that of constant pressure.

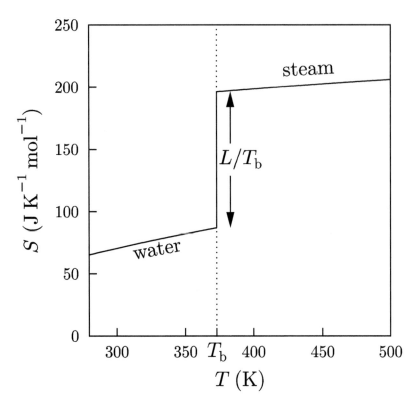

Fig. 28.1 The entropy of H$_2$O as a function of temperature. The boiling point is $T_b = 373$ K.

water, is about $75\,\text{J}\,\text{K}^{-1}\,\text{mol}^{-1}$ (equivalent to about $4.2\,\text{kJ}\,\text{kg}^{-1}\,\text{K}^{-1}$) at temperatures below the boiling point T_b, and this is responsible for the gradient of $S(T)$ below the transition (because $\Delta S = \int C_p\, dT/T$), while the heat capacity of the gaseous phase, steam, is about $34\,\text{J}\,\text{K}^{-1}\,\text{mol}^{-1}$, and this is responsible for the gradient in $S(T)$ above the transition. The sudden, discontinuous change in S which occurs at T_b is a jump of magnitude L/T_b, where L is the latent heat, equal to $40.7\,\text{kJ}\,\text{mol}^{-1}$ (or equivalently $2.26\,\text{MJ}\,\text{kg}^{-1}$).

Example 28.1

If it takes 3 minutes to boil a kettle of water, which was initially at $20\,°\text{C}$, how much longer will it take to boil the kettle dry?
Solution:
Using the data above, the energy required to raise water from $20\,°\text{C}$ to $100\,°\text{C}$ is $80 \times 4.2 = 336\,\text{kJ}\,\text{kg}^{-1}$. The energy required to turn it into steam at $100\,°\text{C}$ is $2.26\,\text{MJ}\,\text{kg}^{-1}$, which is 6.7 times as much. Therefore it would take $6.7 \times 3 \approx 20$ minutes to boil the kettle dry (though, of course, having an automatic switch-off mechanism saves this from happening!).

Let us now perform a rough estimate for the entropy discontinuity at a vapour[2]–liquid transition. The number of microstates Ω available to a single gas molecule is proportional to its volume,[3] and hence the ratio of Ω for one mole of vapour and one mole of liquid is

$$\frac{\Omega_{\text{vapour}}}{\Omega_{\text{liquid}}} = \left(\frac{V_{\text{vapour}}}{V_{\text{liquid}}}\right)^{N_A}. \tag{28.3}$$

Hence

$$\frac{\Omega_{\text{vapour}}}{\Omega_{\text{liquid}}} = \left(\frac{\rho_{\text{liquid}}}{\rho_{\text{vapour}}}\right)^{N_A} \sim (10^3)^{N_A}, \tag{28.4}$$

since the density of the vapour is roughly 10^3 times smaller than the density of the liquid. Hence, using $S = k_B \ln \Omega$, we have that the entropy discontinuity is approximately

$$\Delta S = \Delta(k_B \ln \Omega) = k_B \ln(10^3)^{N_A} = R \ln 10^3 \approx 7R, \tag{28.5}$$

so that

$$L \approx 7RT_b. \tag{28.6}$$

This relationship is known as **Trouton's rule**, and is an empirical relationship, which has been noticed for many systems, although it is usually stated with a slightly different prefactor:

$$L \approx 10RT_b. \tag{28.7}$$

The fact that the latent heat is slightly larger than expected from our simple argument stems from the fact that the latent heat also involves a contribution from the attractive intermolecular potential. However, the law of corresponding states[4] implies that if substances have similarly shaped intermolecular potentials then certain properties should scale in the same way, so we do expect L/RT_b to be a constant.

[2] The word *vapour* is a synonym for gas, but is often used when conditions are such that the substance in the gas can also exist as a liquid or solid; if $T < T_c$, the vapour can be condensed into a liquid or solid with the application of pressure.

[3] Recall from Section 21.1 that one microstate occupies a volume in k-space equal to $(2\pi/L)^3 \propto V^{-1}$, and hence the density of states is proportional to the system volume V.

Remember that $R = N_A k_B$.

[4] See Section 26.4.

	Ne	Ar	Kr	Xe	He	H$_2$O	CH$_4$	C$_6$H$_6$
T_b (K)	27.1	87.3	119.8	165.0	4.22	373.15	111.7	353.9
L (kJ mol^{-1})	1.77	6.52	9.03	12.64	0.084	40.7	8.18	30.7
L/RT_b	7.85	8.98	9.06	9.21	2.39	13.1	8.80	10.5

Table 28.1 The values of T_b, L and L/RT_b for several common substances.

This can be tested for various real substances (see Table 28.1) and indeed it is found that for many substances the ratio L/RT_b is in the range 8–10, confirming Trouton's empirical rule. Notable outliers include helium (He) for which quantum effects are very important (see Chapter 30) and water[5] (H$_2$O), which is a polar liquid (because the water molecule has a dipole moment) and which therefore possesses a rather different intermolecular potential.

[5] Water, being a special case, has a lot of consequences; see, for example, Section 37.4.

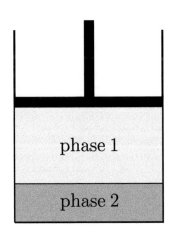

Fig. 28.2 Two phases in equilibrium at constant pressure (the constraint of constant pressure is maintained by the piston).

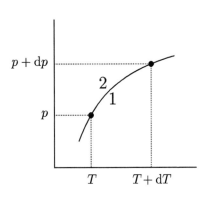

Fig. 28.3 Two phases in the p–T plane coexist at the phase boundary, shown by the solid line.

28.2 Chemical potential and phase changes

We have seen in Section 16.5 that the Gibbs function is the quantity that must be minimized when systems are held at constant pressure and temperature. In Section 22.5 we found that the chemical potential is the Gibbs function per particle. We were also able to write in eqn 22.52 that

$$dG = V\,dp - S\,dT + \sum_i \mu_i\,dN_i. \tag{28.8}$$

Now consider the situation in Fig. 28.2, in which N_1 particles of phase 1 are in equilibrium with N_2 particles of phase 2. Then the total Gibbs free energy is

$$G_{\text{tot}} = N_1\mu_1 + N_2\mu_2, \tag{28.9}$$

and since we are in equilibrium we must have

$$dG_{\text{tot}} = 0, \tag{28.10}$$

and hence

$$dG_{\text{tot}} = dN_1\mu_1 + dN_2\mu_2 = 0. \tag{28.11}$$

But if we increase the number of particles in phase 1, the number of particles in phase 2 must decrease by the same amount, so that $dN_1 = -dN_2$. Hence we have that

$$\mu_1 = \mu_2. \tag{28.12}$$

Thus in phase equilibrium, each coexisting phase has the same chemical potential. The lowest μ phase is the stable phase. Along a line of coexistence, $\mu_1 = \mu_2$.

28.3 The Clausius–Clapeyron equation

We now want to find the equation that describes the phase boundary in the p–T plane (see Fig. 28.3). This line of coexistence of the two phases is determined by the equation

$$\mu_1(p, T) = \mu_2(p, T). \tag{28.13}$$

If we move along this phase boundary, we must also have

$$\mu_1(p + dp, T + dT) = \mu_2(p + dp, T + dT), \tag{28.14}$$

so that when we change p to $p + dp$ and T to $T + dT$ we must have

$$d\mu_1 = d\mu_2. \tag{28.15}$$

This implies that (using eqns 16.22 and 22.48)

$$-s_1 dT + v_1 dp = -s_2 dT + v_2 dp, \tag{28.16}$$

where s_1 and s_2 are the entropy per particle in phases 1 and 2, and v_1 and v_2 are the volume per particle in phases 1 and 2. Rearranging this equation therefore gives that

$$\frac{dp}{dT} = \frac{s_2 - s_1}{v_2 - v_1}. \qquad (28.17)$$

If we define the latent heat per particle as $l = T\Delta s$, we then have that

$$\frac{dp}{dT} = \frac{l}{T(v_2 - v_1)}, \qquad (28.18)$$

or equivalently

$$\boxed{\frac{dp}{dT} = \frac{L}{T(V_2 - V_1)},} \qquad (28.19)$$

which is known as the **Clausius–Clapeyron equation**. This shows that the gradient of the phase boundary of the p–T plane is purely determined by the latent heat, the temperature at the phase boundary, and the difference in volume between the two phases.[6]

[6] This can be obtained from the difference in *densities*.

Example 28.2

Derive an equation for the phase boundary of the liquid and gas phases under the assumptions that the latent heat L is temperature independent, that the vapour can be treated as an ideal gas, and that $V_{\text{vapour}} = V \gg V_{\text{liquid}}$.

Solution:

Assuming that $V_{\text{vapour}} = V \gg V_{\text{liquid}}$ and that $pV = RT$ for one mole, the Clausius–Clapeyron equation becomes

$$\frac{dp}{dT} = \frac{Lp}{RT^2}. \qquad (28.20)$$

This can be rearranged to give

$$\frac{dp}{p} = \frac{L\,dT}{RT^2}, \qquad (28.21)$$

and hence integrating we obtain

$$\ln p = -\frac{L}{RT} + \text{constant}. \qquad (28.22)$$

Hence the equation of the phase boundary is

$$p(T) = p_0 \exp\left(-\frac{L}{RT}\right), \qquad (28.23)$$

where the exponential looks like a Boltzmann factor $e^{-\beta l}$ with $l = L/N_A$, the latent heat per particle.

In this case, the constant p_0 is given by $p_0 = p(\infty)$, the pressure at infinite temperature.

Remember again that $R = N_A k_B$.

The temperature dependence of the latent heat of many substances cannot be neglected. As an example, the temperature dependence of the latent heat of water is shown in Fig. 28.4 and this shows a weak temperature dependence. A method of treating this is outlined in the following example.

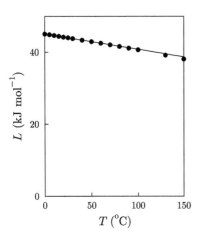

Fig. 28.4 The temperature dependence of the latent heat of water. The solid line is according to eqn 28.30.

Example 28.3

Evaluate the temperature dependence of the latent heat along the phase boundary in a liquid–gas transition and hence deduce the equation of the phase boundary including this temperature dependence.

Solution:

Along the phase boundary, we can write that the gradient in the temperature is (see Fig. 28.5)

$$\frac{\mathrm{d}}{\mathrm{d}T} = \left(\frac{\partial}{\partial T}\right)_p + \frac{\mathrm{d}p}{\mathrm{d}T}\left(\frac{\partial}{\partial p}\right)_T. \tag{28.24}$$

Hence, applying this to the quantity $\Delta S = S_v - S_L = L/T$ where the subscripts v and L refer to vapour and liquid respectively, we have that

$$\frac{\mathrm{d}}{\mathrm{d}T}\left(\frac{L}{T}\right) = \left(\frac{\partial(\Delta S)}{\partial T}\right)_p + \frac{\mathrm{d}p}{\mathrm{d}T}\left(\frac{\partial(\Delta S)}{\partial p}\right)_T$$

$$= \frac{C_{pv} - C_{pL}}{T} + \left[\left(\frac{\partial S_v}{\partial p}\right)_T - \left(\frac{\partial S_L}{\partial p}\right)_T\right]\frac{\mathrm{d}p}{\mathrm{d}T}, \tag{28.25}$$

so that

$$\frac{\mathrm{d}}{\mathrm{d}T}\left(\frac{L}{T}\right) = \frac{C_{pv} - C_{pL}}{T} - \left[\frac{\partial}{\partial T}(V_v - V_L)\right]\frac{\mathrm{d}p}{\mathrm{d}T}. \tag{28.26}$$

Using $V_v \gg V_L$ and $pV_v = RT$, we have that

$$\frac{\mathrm{d}}{\mathrm{d}T}\left(\frac{L}{T}\right) = \frac{C_{pv} - C_{pL}}{T} - \frac{R}{p} \times \frac{Lp}{RT^2}, \tag{28.27}$$

and expanding

$$\frac{\mathrm{d}}{\mathrm{d}T}\left(\frac{L}{T}\right) = \frac{1}{T}\frac{\mathrm{d}L}{\mathrm{d}T} - \frac{L}{T^2}, \tag{28.28}$$

yields

$$\mathrm{d}L = (C_{pv} - C_{pL})\,\mathrm{d}T, \tag{28.29}$$

so that

$$L = L_0 + (C_{pv} - C_{pL})T. \tag{28.30}$$

Thus the latent heat contains a linear temperature dependence and this is shown by the solid line in Fig. 28.4. The slope is negative because $C_{pL} > C_{pv}$. Substituting this value of L into eqn 28.20 yields the equation of the phase boundary:

$$p = p_0 \exp\left(-\frac{L_0}{RT} + \frac{(C_{pv} - C_{pL})\ln T}{R}\right). \tag{28.31}$$

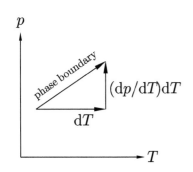

Fig. 28.5 The phase boundary.

We can also use the Clausius–Clapeyron equation to derive the phase boundary of the liquid–solid coexistence line, as shown in the following example.

Example 28.4

Find the equation in the p–T plane for the phase boundary between the liquid and solid phases of a substance.
Solution:
The Clausius–Clapeyron equation (eqn 28.19) can be rearranged to give

$$\mathrm{d}p = \frac{L\,\mathrm{d}T}{T\Delta V}, \tag{28.32}$$

and neglecting the temperature dependence of L and ΔV, we find that this integrates to

$$p = p_0 + \frac{L}{\Delta V}\ln\left(\frac{T}{T_0}\right), \tag{28.33}$$

where T_0 and p_0 are constants such that $(T,p) = (T_0, p_0)$ is a point on the phase boundary. The volume change ΔV on melting is relatively small, so that the gradient of the phase boundary in the p–T plane is very steep.

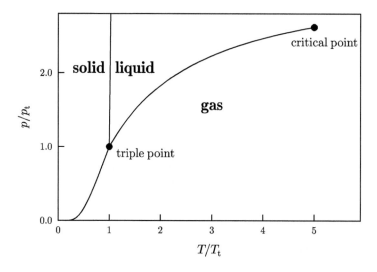

Fig. 28.6 A schematic phase diagram of a (hypothetical) pure substance.

A phase diagram of a hypothetical pure substance is shown in Fig. 28.6 and shows the solid, liquid, and gaseous phases coexisting with the phase boundaries calculated from the Clausius–Clapeyron equation. The three phases coexist at the **triple point**. The solid–liquid phase boundary is very steep, reflecting the large change in entropy in going from liquid to solid and the very small change in volume. This phase boundary does not terminate, but continues indefinitely. By way of contrast, the phase boundary between liquid and gas terminates at the **critical point**, as we have seen in Section 26.1. (We will have more to say about this observation in Section 28.7.) Note also that, at temperatures close to the triple point, the latent heat of **sublimation** (changing from solid to gas) is equal to the sum of the latent heat of melting (solid→liquid)[7] and the latent heat of vaporization (liquid→gas).[8]

[7] The latent heat of melting is sometimes known as the latent heat of fusion.

[8] This fact will be used in Exercise 28.5.

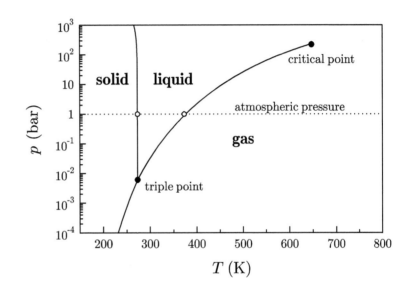

Fig. 28.7 The phase diagram of H_2O showing the solid (ice), liquid (water), and gaseous (steam) phases. The horizontal dashed line corresponds to atmospheric pressure, and the normally experienced freezing and boiling points of water are indicated by the open circles.

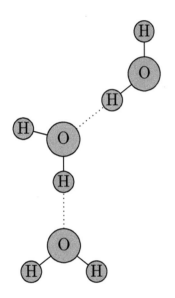

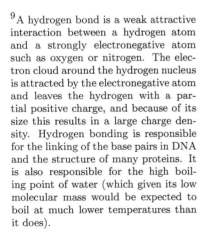

Fig. 28.8 Schematic diagram of hydrogen bonding in water.

[9] A hydrogen bond is a weak attractive interaction between a hydrogen atom and a strongly electronegative atom such as oxygen or nitrogen. The electron cloud around the hydrogen nucleus is attracted by the electronegative atom and leaves the hydrogen with a partial positive charge, and because of its size this results in a large charge density. Hydrogen bonding is responsible for the linking of the base pairs in DNA and the structure of many proteins. It is also responsible for the high boiling point of water (which given its low molecular mass would be expected to boil at much lower temperatures than it does).

The gradient of the liquid–solid coexistence line is normally positive because most substances expand when they melt. A notable counterexample is water, which slightly shrinks when it melts. Hence, the gradient of the ice–water coexistence line is negative (see Fig. 28.7; because the solid–liquid line is so steep, it is not easy to see that the slope is negative). This effect occurs because of the **hydrogen bonding**[9] in water (see Fig. 28.8), which results in a rather open structure of the ice crystal lattice. This collapses on melting, resulting in a slightly denser liquid. This result has many consequences: for example, icebergs float on the ocean and ice cubes float in your gin and tonic. Lakes and oceans only freeze from the top, preserving life living in the cool liquid beneath. Since water close to 4 °C has the largest density, this sinks in the oceans, fixing the temperature of the deep oceans, which facilitates the survival of life in the oceans. The pressure dependence of the coexistence line means that pressing ice can cause it to melt, an effect that is responsible for the movement of glaciers, which can press against rock, melt near the region of contact with the rock, and slowly creep downhill.

28.4 Stability and metastability

We have seen in Section 28.2 that the phase with the lowest chemical potential μ is the most stable. Let us see how the phase transition varies as a function of pressure. Since μ is the Gibbs function per particle, eqn 16.24 implies that

$$\left(\frac{\partial \mu}{\partial p}\right)_T = v, \qquad (28.34)$$

where v is the volume per particle. Since $v > 0$, the gradient of the chemical potential with pressure must always be positive. The behaviour of μ as a function of pressure as one crosses the phase transition between the liquid and gas phases is shown in Fig. 28.9. This figure shows that the phase that is stable at the highest pressure must therefore have the smallest volume. This of course makes sense since, when you apply large pressure, you expect the smallest space-occupying phase to be the most stable.

We can also think about μ as a function of temperature. Equation 16.23 implies that

$$\left(\frac{\partial \mu}{\partial T}\right)_p = -s, \qquad (28.35)$$

where s is the entropy per particle. Since $s > 0$, the gradient of μ as a function of temperature must always be negative. The behaviour of μ as a function of temperature as you cross the phase transition between the liquid and gas phases is shown in Fig. 28.10. This figure shows that the phase that is stable at the highest temperature must therefore have the highest entropy. This makes sense because $G = H - TS$ and so at higher temperature, you minimize G by maximizing S.

This also shows that as you warm a substance through its boiling point, it is possible to continue momentarily on the curve corresponding to $\mu_{\rm liq}$ and to form **superheated liquid**, which is a metastable state. Although for temperatures above the boiling point it is the gaseous state which is thermodynamically stable (i.e., has the lowest Gibbs function), there may be reasons why this state cannot be formed immediately and the liquid state persists. Similarly, if you cool a gas below the boiling point, it is possible to continue momentarily on the curve corresponding to $\mu_{\rm gas}$ and to form **supercooled vapour**, which is a metastable state. Again, this is not the thermodynamically stable state of the system but there may be reasons why the liquid state cannot nucleate immediately and the gaseous state persists.

Let us now try and fathom the reason why the thermodynamically most stable state sometimes doesn't form. Consider a liquid with pressure $p_{\rm liq}$ in equilibrium with a vapour at pressure p. The chemical potentials of the liquid and vapour must be equal. Now imagine that the

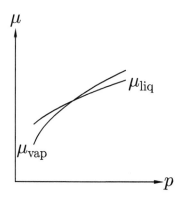

Fig. 28.9 The chemical potential as a function of pressure.

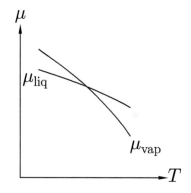

Fig. 28.10 The chemical potential as a function of temperature.

liquid pressure increases slightly to $p_{\text{liq}} + dp_{\text{liq}}$. If the vapour is still in equilibrium with the liquid, then its pressure must increase to $p + dp$ and we must have that

$$\left(\frac{\partial \mu_{\text{liq}}}{\partial p_{\text{liq}}}\right)_T dp_{\text{liq}} = \left(\frac{\partial \mu_{\text{vap}}}{\partial p}\right)_T dp, \qquad (28.36)$$

so that the chemical potentials of the liquid and vapour are still equal. Using eqn 28.34, this implies that

$$v_{\text{liq}} dp_{\text{liq}} = v_{\text{vap}} dp, \qquad (28.37)$$

where v_{liq} is the volume per particle occupied by the liquid and v_{vap} is the volume per particle occupied by the vapour. Hence multiplying this by N_A and using $pV = RT$ for one mole of the gas, we find that

$$V_{\text{liq}} dp_{\text{liq}} = \frac{RT dp}{p}, \qquad (28.38)$$

where V_{liq} is the molar volume of the liquid. We can use this to find the dependence of the vapour pressure[10] on the pressure in the liquid at constant temperature. Integrating eqn 28.38 leads to

$$p = p_0 \exp\left(\frac{V_{\text{liq}} \Delta p_{\text{liq}}}{RT}\right), \qquad (28.39)$$

where Δp_{liq} is the extra pressure applied to the liquid, p_0 is the vapour pressure of the gas with no excess pressure applied to the liquid and p is the vapour pressure of the gas with excess pressure Δp_{liq} in the liquid.

This result can be used to derive the vapour pressure of a droplet of liquid. Recall from eqn 17.18 that the excess pressure in a droplet of liquid of radius r can be obtained as

$$\Delta p_{\text{liq}} = \frac{2\gamma}{r}, \qquad (28.40)$$

where γ is the surface tension. Hence we find that

$$\boxed{p = p_0 \exp\left(\frac{2\gamma V_{\text{liq}}}{rRT}\right),} \qquad (28.41)$$

which is known as **Kelvin's formula**. This formula shows that small droplets have a very high vapour pressure, and this gives some understanding about why the vapour sometimes doesn't condense when you cool it through the boiling temperature. Small droplets initially begin to nucleate, but have a very high vapour pressure and therefore instead of growing can evaporate. This stabilizes the vapour, even though it is the thermodynamically unstable phase. The thermodynamic driving force to condense is overcome by the tendency to evaporate. This effect occurs very often in the atmosphere, which contains water vapour that has risen to an altitude where it is sufficiently cold to condense into water droplets, but the droplets cannot form owing to this tendency to evaporate. Clouds do form through the nucleation of droplets on minute dust

[10] The **vapour pressure** of a liquid (or a solid) is the pressure of vapour in equilibrium with the liquid (or the solid).

particles, which have sufficient surface area for the liquid to condense and then grow above the critical size.

A similar effect occurs for superheated liquids. The pressure of liquid near a vapour-filled cavity of radius r is less than that in the bulk liquid according to

$$\Delta p_{\text{liq}} = -\frac{2\gamma}{r}, \qquad (28.42)$$

and hence the vapour pressure inside the cavity follows:

$$p = p_0 \exp\left(-\frac{2\gamma V_{\text{liq}}}{rRT}\right). \qquad (28.43)$$

Thus the vapour pressure inside a cavity is lower than one might expect. As you boil a liquid, any bubble of vapour which does form tends to collapse. This means the liquid can become superheated and kinetically stable above its boiling point, even though the vapour is the true thermodynamic ground state. The only bubbles which then do survive are very large ones, and this causes the violent bumping that can be observed in boiling liquids. This can be avoided by boiling liquids with small pieces of glass or ceramic, so that there are plenty of nucleation centres for small bubbles to form.

Example 28.5

A **bubble chamber** is used in particle physics to detect electrically charged subatomic particles. It consists of a container filled with a superheated transparent liquid such as liquid hydrogen, at a temperature just above its boiling point. The motion of the charged particle is sufficient to nucleate a string of bubbles of vapour, which display the track of the particle. A magnetic field can be applied to the chamber so that the shape of the curved tracks of the particle can be used to infer its charge to mass ratio. Its invention in 1952 earned Donald Glaser (1926–) the 1960 Nobel Prize for Physics.

Example 28.6

Calculate the Gibbs function for a droplet of liquid of radius r (and hence surface area $A = 4\pi r^2$) in equilibrium with vapour. Assume that the temperature is such that the liquid is the thermodynamically stable phase.

Solution:

Writing the number of molecules (of mass m) in the liquid and vapour as N_{liq} and N_{vap} respectively, the change in Gibbs function is

$$dG = \mu_{\text{liq}}\, dN_{\text{liq}} + \mu_{\text{vap}}\, dN_{\text{vap}} + \gamma\, dA, \qquad (28.44)$$

where γ is the surface tension. Since molecules must be conserved, $dN_{vap} = -dN_{liq}$. Differentiating $A = 4\pi r^2$ yields $dA = 8\pi r\, dr$, and writing $\Delta\mu = \mu_{vap} - \mu_{liq}$ (which will be positive since the liquid is the thermodynamically stable phase) we have

$$dG = \left(8\pi\gamma r - \frac{4\pi r^2 \Delta\mu \rho_{liq}}{m}\right) dr, \qquad (28.45)$$

where ρ_{liq} is the density of the liquid. This can be integrated to yield

$$G(r) = G(0) + 4\pi\gamma r^2 - \frac{4\pi\Delta\mu\rho_{liq}}{3m} r^3, \qquad (28.46)$$

and hence equilibrium is established when $dG/dr = 0$, and this occurs at the critical radius r^*, given by

$$r^* = \frac{2\gamma m}{\rho_{liq}\Delta\mu}. \qquad (28.47)$$

This function is sketched in Fig. 28.11 and shows that r^* is indeed a stationary point, but is a maximum in G, not a minimum! Thus $r = r^*$ is a point of unstable equilibrium. If $r < r^*$, the system can minimize G by shrinking r to zero, i.e., the droplet evaporates. If $r > r^*$, the system can minimize G by the droplet growing to infinite size.

This effect occurs as water condenses in a cloud. The large droplets keep the partial pressure of the water vapour low. The smaller droplets therefore evaporate and the water can transfer from the smaller to the larger droplets.

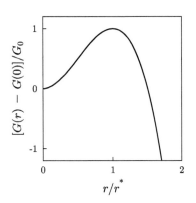

Fig. 28.11 The Gibbs function of the droplet as a function of radius, plotted in units of G_0, where

$$G_0 = \frac{16\pi\gamma^3 m^2}{3(\rho_{liq}\Delta\mu)^2}.$$

28.5 The Gibbs phase rule

In this section, we want to find out how much freedom a system has to change its internal parameters while keeping the different substances in various combinations of phases in equilibrium with each other. We want to include the possibility of having mixtures of different substances, and we will call the different substances *components*. A component is a chemically independent constituent of the system. To keep track of the number of molecules in these different components, we introduce the **mole fraction** x_i, which is defined to be the ratio of the number of moles, n_i, of the ith substance, divided by the total number of moles n, so that

$$x_i = \frac{n_i}{n}. \qquad (28.48)$$

By definition, we have that

$$\sum x_i = 1. \qquad (28.49)$$

Each of the components can be in different *phases* (where here we mean phases such as "solid", "liquid" and "gas", but we might also wish to include other possibilities, such as "ferromagnetic" and "paramagnetic" phases, or "superconducting" and "non-superconducting" phases). We denote by the symbol F the number of degrees of freedom the system has while keeping the different phases in equilibrium, and it is this quantity we now want to calculate, following a method introduced by Gibbs.

Consider a multicomponent system, containing C components. Each component can be in any one of P different phases. The system is characterized by the intensive variables, the pressure p, the temperature T and the mole fractions of $C-1$ of the components (we don't need all C of them, since $\sum_{i=1}^{C} x_i = 1$) for each of the P phases, so that is

$$2 + P(C-1) \tag{28.50}$$

variables. If the phases of each component are in equilibrium with one another, then we must have, as i runs from 1 to C,

$$\mu_i(\text{phase 1}) = \mu_i(\text{phase 2}) = \cdots = \mu_i(\text{phase } P), \tag{28.51}$$

which gives us $P-1$ equations to solve for each component, and hence $C(P-1)$ equations to solve for each of the C components.

The number of degrees of freedom F the system has is given by the difference between the number of variables and the number of constraining equations to solve. Hence $F = [P(C-1) + 2] - C(P-1)$, and thus

$$\boxed{F = C - P + 2} \tag{28.52}$$

which is known as the **Gibbs phase rule**.[11]

[11] The Gibbs phase rule is of great use in interpreting complex phase diagrams of mixtures of substances.

Example 28.7

For a single-component system, $C = 1$ and hence $F = 3 - P$. Thus:

- If there is one phase, $F = 2$, and the whole p–T plane is accessible.
- If there are two phases, $F = 1$, and these two phases can only coexist at a common *line* of coexistence in the p–T plane.
- If there are three phases, $F = 0$, and these three phases can only coexist at a common *point* of coexistence in the p–T plane (the triple point).

For a two-component system, $C = 2$ and hence $F = 4 - P$. If we fix the pressure, then the number of remaining degrees of freedom $F' = F - 1 = 3 - P$. Having fixed the pressure, we have two variables which are temperature T and the mole fraction x_1 of the first component (the mole fraction of the second component being given by $1 - x_1$). Thus:

- If there is one phase, $F = 2$, and the whole x_1–T plane is accessible.
- If there are two phases, $F = 1$, and these two phases can only coexist at a common *line* of coexistence in the x_1–T plane.[12]
- If there are three phases, $F = 0$, and these three phases can only coexist at a common *point* of coexistence in the x_1–T plane.

[12] An example would be a mixture of two liquids in equilibrium with their vapour (thus two components and two phases). Note that this analysis can be complicated by the presence of a **solubility gap**: there may be concentrations x_1 which are unobtainable since, for example, the system "phase separates" into two distinct phases, a phase rich in one component and a phase rich in the other component.

28.6 Colligative properties

When a liquid of a particular material (we will call it A) has another species, B, dissolved in it, the chemical potential of A is decreased. The result of this is that the boiling point of the liquid A is elevated and the freezing point of the liquid is depressed compared with the pure liquid. These effects are known as **colligative properties**.[13] The magnitude of the effect can be worked out from the reduction of the chemical potential.

[13] The word colligative means a collection of things fastened together.

We can now derive formulae to describe the colligative properties. When the solution A is in equilibrium with its vapour $\mu_A^{(g)} = \mu_A^{(\ell)}$ and equation 22.95 states that $\mu_A^{(\ell)} = \mu_A^{(\ell)*} + RT \ln x_A$ and hence

$$\ln x_A = \frac{\Delta G_{\text{vap}}}{RT}, \tag{28.53}$$

Recall that the liquid which is the main component is known as the **solvent**, while the material which is dissolved in it is known as the **solute**. See Section 22.9.

where $\Delta G_{\text{vap}} = \mu_A^{(g)} - \mu_A^{(\ell)*}$. When $x_A = 1$, then equilibrium between vapour and liquid occurs at a temperature T^* given by (using eqn 28.53 with $x_A = 1$)

$$\frac{\Delta G_{\text{vap}}(T^*)}{RT^*} = 0, \tag{28.54}$$

which implies that (recall that $G = H - TS$)

$$\Delta H_{\text{vap}}(T^*) - T^* \Delta S_{\text{vap}}(T^*) = 0. \tag{28.55}$$

When $x_B = 1 - x_A$ is very small, then we have that

$$\ln x_A = \ln(1 - x_B) \approx -x_B, \tag{28.56}$$

and hence eqn 28.53 implies that

$$-x_B = \frac{\Delta G_{\text{vap}}}{RT} = \frac{1}{RT}\left[\Delta H_{\text{vap}}(T) - T\Delta S_{\text{vap}}(T)\right]. \tag{28.57}$$

Assuming that ΔH_{vap} and ΔS_{vap} are only weakly temperature-dependent, eqn 28.55 implies $\Delta H_{\text{vap}} - T\Delta S_{\text{vap}} \approx \Delta H_{\text{vap}}(1 - T/T^*)$ and this yields

$$-x_B = \frac{\Delta H_{\text{vap}}}{R}\left(\frac{1}{T} - \frac{1}{T^*}\right) \approx \frac{\Delta H_{\text{vap}}}{RT^{*2}}(T^* - T). \tag{28.58}$$

Hence $T - T^*$, the elevation in boiling point, is given approximately by

$$T - T^* \approx \frac{RT^{*2} x_B}{\Delta H_{\text{vap}}}. \tag{28.59}$$

It is often written $T - T^* = K_b x_B$, where $K_b \approx RT^{*2}/\Delta H_{\text{vap}}$ is known as the **ebullioscopic constant**. For water, $K_b = 0.51$ K mol^{-1} kg^{-1}. There is a similar effect on the depression of the freezing point. One can show similarly that the freezing point is depressed by $T^* - T = K_f x_B$, where $K_f \approx RT^{*2}/\Delta H_{\text{fus}}$ is the **cryoscopic constant** and ΔH_{fus} is the latent heat of fusion. The salt water in the oceans freezes at a lower temperature than fresh water. The effect is also relevant for salt being put on pavements (sidewalks) in winter to stop them becoming icy.

Adding a small quantity of solute to a solvent increases the entropy of the solvent because the solute atoms are randomly located in the solvent. This means that there is a weaker tendency to form a gas (which would increase the solvent's entropy) because the entropy of the solvent has been increased anyway. This results in an elevation of the boiling point. Similarly, this additional entropy opposes the tendency to freeze and the freezing point is depressed.

28.7 Classification of phase transitions

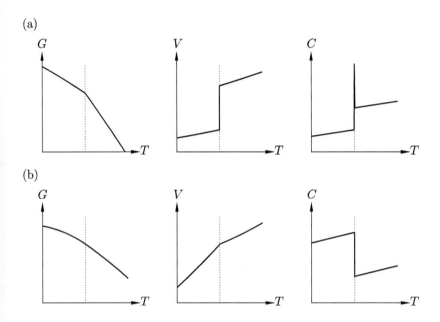

Fig. 28.12 Ehrenfest's classification of phase transitions. (a) First-order phase transition. (b) Second-order phase transition. The critical temperature T_c is marked by a vertical dotted line in each case.

Paul Ehrenfest (1880–1933) proposed a classification of phase transitions which goes as follows: the **order** of a phase transition is the order of the lowest differential of G (or μ) that shows a discontinuity at T_c. Thus **first-order phase transitions** involve a latent heat because the entropy (a first differential of G) shows a discontinuity. The volume is also a first differential of G and this also shows a discontinuous jump. The heat capacity is a second differential of G and thus it shows a sharp spike, as does the compressibility. This is illustrated in Fig. 28.12(a). Examples of first-order phase transitions include the solid–liquid transition, the solid–vapour transition, and the liquid–vapour transition.

By Ehrenfest's classification, a **second-order phase transition** has no latent heat because the entropy does not show a discontinuity (and neither does the volume – both are first differentials of G), but quantities like the heat capacity and compressibility (second differentials of G) do. This is illustrated in Fig. 28.12(b). Examples of second-order phase transitions include the superconducting transition, or the order–disorder transition in β-brass.

See Section 16.4 for the differentials of the Gibbs function G.

However, a big problem with the approach we have been using so far in studying phase transitions is that one key approximation made in thermodynamics, namely that the number of particles is so large that average properties such as pressure and density are well defined, breaks down at a phase transition. Fluctuations build up near a phase transition and so the behaviour of the system does not follow the expectations of our analysis very close to the phase transition temperature. This critical region is characterized by fluctuations, at all length scales. For example, when a saucepan of water is heated, the water warms quite quietly and unobtrusively until near the boiling point, when it makes a great deal of noise and bubbles violently.[14] We have already analysed the behaviour of the formation of bubbles in Section 28.4. Therefore, it has been found that Ehrenfest's approach is rather too simple. We will have more to say concerning fluctuations in Chapters 33 and 34.

A more modern approach to classifying phase transitions simply distinguishes between those that show a latent heat, for which Ehrenfest's term "first-order phase transition" is retained, and those that do not, which are called a **continuous phase transition** (and include Ehrenfest's phase transitions of second order, third order, fourth order, etc., all lumped together).

[14] A visual demonstration of this is found in the phenomenon known as **critical opalescence** which is the blurring and clouding of images seen through a volume of gas near its critical point. This occurs because density fluctuations are strong near the critical point and give rise to large variations in refractive index.

Example 28.8

- The liquid–gas phase transition is a first-order transition, except at the critical point where the phase transition involves no latent heat and is a continuous phase transition.
- A ferromagnet[15] such as iron loses its ferromagnetism when heated to the Curie temperature, T_C (a particular example of a critical temperature). This phase transition is a continuous phase transition, since there is no latent heat. The magnetization is a first differential of the Gibbs function and does not change discontinuously at T_C. The specific heat C_B, at constant magnetic field B, has a finite peak at T_C.

[15] A ferromagnet is a material containing magnetic moments that are all aligned in parallel below a transition temperature called the Curie temperature. Above this temperature, the magnetic moments become randomly aligned. This state is known as the paramagnetic state.

A further classification of phase transitions involves the notion of **symmetry breaking**. Figure 28.13 shows atoms in a liquid and in a solid. As a liquid cools there is a very slight contraction of the system but it retains a very high degree of symmetry. However, below the melting temperature, the liquid becomes a solid and that symmetry is broken. This may at first sight seem surprising because the picture of the solid "looks" more symmetrical than that of the liquid. The atoms in the solid are all symmetrically lined up while in the liquid they are all over the place. The crucial observation is that any point in a liquid is, on average, exactly the same as any other. If you average the system over time, each position is visited by atoms as often as any other. There are no unique directions or axes along which atoms line up. In short, the system possesses complete translational and rotational symmetry. In the solid, however, this high degree of symmetry is nearly all lost. The solid drawn in Fig. 28.13 still possesses some residual symmetry: rather than being invariant under arbitrary rotations, it is invariant under four-fold rotations ($\pi/2$, π, $3\pi/2$, 2π); rather than being invariant under arbitrary translations, it is now invariant under a translation of an integer combination of lattice basis vectors. Therefore not all symmetry has been lost but the high symmetry of the liquid state has been, to use the technical term, "broken". It is impossible to change symmetry *gradually*. Either a particular symmetry is present or it is not. Hence, phase transitions are sharp and there is a clear delineation between the ordered and disordered states.

Not all phase transitions involve a change of symmetry. Consider the liquid–gas coexistence line again (see Fig. 28.7). The boundary line between the liquid and gas regions is terminated by a critical point. Hence it is possible to "cheat" the sharp phase transition by taking a path through the phase diagram that avoids a discontinuous change. For temperatures above the critical temperature (647 K for water) the gaseous and liquid states are distinguished only by their density. The transition between a gas and a liquid involves no change of symmetry and therefore it is possible to avoid it by working round the critical end point. In contrast, the solid–liquid transition involves a change of symmetry and consequently there is no critical point for the melting curve.

Symmetry-breaking phase transitions include those between the ferromagnetic and paramagnetic states (in which the low-temperature state does not possess the rotational symmetries of the high-temperature state) and those between the superconducting and normal metal states of certain materials (in which the low-temperature state does not possess the same symmetry in the phase of the wavefunction as the high-temperature state).

The concept of broken symmetry is very wide ranging and is used to explain how the electromagnetic and weak forces originated. In the early Universe, when the temperature was very high, it is believed that the electromagnetic and weak forces were part of the same, unified, electroweak force. When the temperature cooled[16] to below about 10^{11} eV a symmetry was broken and a phase transition occured, via what is known as the **Higgs mechanism**, and the W and Z bosons (mediating the weak force) acquired mass while the photon (mediating the electromagnetic force) remained massless. It is suggested that, at even earlier times, when the temperature of the Universe was around 10^{21} eV, the electroweak and strong forces were unified, and as the Universe expanded and its temperature lowered, another symmetry-breaking transition caused them to appear as different forces.

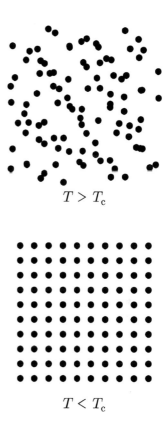

Fig. 28.13 The liquid–solid phase transition. *Top*: the high temperature state (statistically averaged) has complete translational and rotational symmetry. *Bottom*: these symmetries are broken as the system becomes a solid below the critical temperature T_c.

[16] In other words when $k_B T$ was lower than this energy, corresponding to a temperature $T \sim 10^{15}$ K.

28.8 The Ising model

To appreciate some of the intriguing properties of phase transitions, it is helpful to consider a very simple model that can exhibit a phase transition. Imagine a set of atoms arranged on a regular lattice (it could be in one, two, or three dimensions) and imagine further that each atom can be in one of only two states, which we will denote by the values $+1$ and -1. This could be achieved by allowing the atom to have two possible spin states, up and down. Let us also suppose that the interaction between these atoms is such that it costs less energy for a site to be in the same state as its nearest neighbours (there is an energetic "peer pressure" for neighbouring sites to conform). The lowest energy configuration of such a system is for all atoms to be in state $+1$ (shown in Fig. 28.14(a)) or for all to be in state -1 (shown in Fig. 28.14(b)). In this way, all the nearest-neighbour interactions will save energy and so these states "win" over configurations in which there is a completely random distribution of atoms being in a $+1$ or -1 state (as shown in Fig. 28.14(c)). However, the ordered configuration isn't necessarily adopted by the system in equilibrium, even though it minimizes energy. This is because there are only two ways for the atoms to be arranged with identical states (all $+1$ or all -1), whereas there are lots of ways of making disordered configurations with half the sites $+1$ and half the sites -1. Equilibrium is determined[17] by minimizing the Helmholtz function F. Therefore, while the ordered configuration will be favoured at low temperature (when $F = U - TS$ is dominated by U and minimizing U is therefore what counts), the disordered configuration will be favoured at high temperature (when F is dominated by $-TS$ and maximizing S is what counts).

What we have been discussing is the **Ising model**,[18] which can be expressed by the following Hamiltonian

$$\hat{H} = -J \sum_{\langle i,j \rangle} S_i S_j, \tag{28.60}$$

where $S_i = \pm 1$ is the state of the atom at site i, the symbol $\sum_{\langle i,j \rangle}$ denotes a sum over nearest-neighbour sites i and j, and J is a constant.[19] The notation is chosen to emphasize the connection with magnetism. In magnetism, J is interpreted as the exchange constant, S_i is the z-component of the spin on the ith site, and $J > 0$ corresponds to **ferromagnetic** interactions (all the spins line up), while $J < 0$ corresponds to **antiferromagnetic** interactions (where neighbouring spins are oppositely aligned). However, the Ising model can be relevant in many other situations. For example, the Ising model has been used to describe the behaviour of β-brass, an alloy of copper (Cu) and zinc (Zn) with formula CuZn; at low temperature this alloy has a body-centred cubic crystal structure, with Cu atoms sitting at the corners of the cube and Zn atoms sitting at the centres. In this case, it is energetically favourable for Zn atoms to be surrounded by Cu atoms (and vice versa); however, above a critical temperature the structure changes so that Cu atoms and Zn

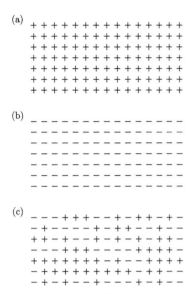

Fig. 28.14 A two-dimensional array of Ising "spins", atoms which can exist in either a $+$ or $-$ state. (a) An ordered configuration with all atoms in the $+$ state. (b) An ordered configuration with all atoms in the $-$ state. (c) A disordered configuration.

[17] We assume that lattice is rigid, so changes in external pressure make negligible difference. In this case, we minimize F and not the G.

[18] Ernst Ising (1900–1998), a doctoral student of Wilhelm Lenz, formulated the problem in his doctoral work which was published in 1925.

[19] The constant J is assumed positive for nearest-neighbour interactions in order to cause the state on neighbouring sites to be the same.

atoms occupy each site with equal probability. This is a phase transition and can be described by the Ising model.

Example 28.9

Consider four spins, located on the corners of a square, and interacting according to the Ising model with a ferromagnetic interaction. There are $2^4 = 16$ different configurations of these atoms and the energy of each configuration is shown in Fig. 28.15. The ground state has energy $E = -4J$ (all four nearest-neighbour interactions saving energy J) and is two-fold degenerate, the two configurations minimizing the energy being the case with all four spins $+$ and the case with all four spins $-$. The magnetic moment m (in units of the magnetic moment due to a single spin) is the number of spins that are in the $+$ state minus the number of spins that are in the $-$ state, and so the ground state corresponds to $m = \pm 4$. The other fourteen states correspond to $m = 0, \pm 2$ and are arranged as shown in Fig. 28.15; twelve have energy $E = 0$ and two have energy $E = +4J$. The partition function Z is therefore given by

$$Z = 2e^{-\beta(-4J)} + 12e^{-\beta \times 0} + 2e^{-\beta(4J)} = 12 + 4\cosh 4\beta J, \quad (28.61)$$

and the mean energy $\langle E \rangle$ is given by $-d\ln Z/d\beta$ and hence

$$\langle E \rangle = -\frac{4J \sinh 4\beta J}{3 + \cosh 4\beta J}, \quad (28.62)$$

so that at high temperature $\langle E \rangle \to 0$ (all states are occupied with equal probability) and at low temperature $\langle E \rangle \to -4J$ (only the ground state is occupied). Equation 28.62 is plotted in Fig. 28.16. There are a couple of important lessons to learn from this exercise.

(1) There is a crossover from a low temperature "ordered" state to a high-temperature disordered state. However, the crossover is very broad and not a true phase transition. If the number of atoms in the model increases, the sharpness of the crossover would increase and it would start to look like a transition. Phase transitions "emerge" as the number of atoms gets large.

(2) The ground state is degenerate, containing the configurations for $m = +4$ and $m = -4$. This feature would remain as the number of atoms became very large, because it reflects the symmetry of the problem. The ground state will always be degenerate since you can either have all the Ising spins in the $+1$ state or all of them in the -1 state. However, a real system can get "stuck" in one or other of those ground states, simply because it is unlikely at low temperature to be able to simultaneously reverse all the spins as a result of a fluctuation of the system's properties.[20] Thus working out "thermal averages", using the conventional machinery of statistical mechanics, might not be the best way to understand the real behaviour of a complex system.[21]

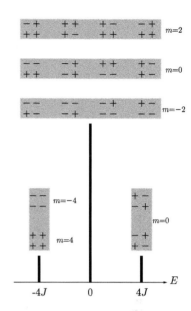

Fig. 28.15 The energy distribution of the $2^4 = 16$ configurations of the system of four Ising spins arranged on a square. Two states have energy $-4J$, twelve have energy 0 and two have energy $4J$. The configurations are shown in the shaded regions and labelled according to their net magnetic moments.

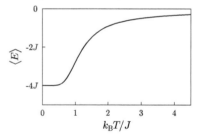

Fig. 28.16 The mean energy of the system of four Ising spins, according to eqn 28.62.

[20] This is symmetry breaking again, see Section 28.7.

[21] This is where Monte-Carlo simulations come in, as shown next.

The simplest version of the Ising model that we can consider is when the spins are arranged in a one-dimensional chain. This is called the one-dimensional Ising model. The ground state occurs when all the sites align, i.e., we have the state

$$\cdots{+}\cdots$$

We can ask what happens if we insert a break into the chain, so that we have the state

$$\cdots{+}{+}{+}{+}{+}{+}{+}{+}{+}{+}{+}{+}{+}{+}{+}{+}{-}{-}{-}{-}{-}{-}{-}{-}{-}{-}{-}{-}{-}{-}{-}{-}{-}\cdots$$

This costs an extra amount of energy $2J$ because we have made one interaction unfavourable (the one which occurs across the boundary, which has changed from energy $-J$ to $+J$). However, there is an entropy gain equal to $S = k_B \ln N$ because we can put this break in any one of N places. The Helmholtz-free-energy cost of the break is therefore $\Delta F = 2J - k_B T \ln N$. As we let the system get very large ($N \to \infty$) the energy cost of a break in the chain remains the same ($2J$) but the entropy gain becomes infinite. The free-energy cost ΔF of a break in the chain is always negative as $N \to \infty$ as long as $T > 0$. This means that chain breaks can spontaneously form and so no long-range order occurs for $T > 0$. Another way of saying this is that the critical temperature is zero. The system stays disordered at all temperatures.[22] This consideration is valid for all models on one-dimensional lattices (because entropy always wins in one dimension) and we conclude that long-range order is not possible in one-dimension.

In two dimensions, you cannot play the same trick. You need more than a single defect in order to switch a region of spins a different way from that of surrounding spins.[23] It turns out that long-range order is stable in two dimensions for the Ising model. However, the exact solution is rather complicated and it took almost twenty years for anyone to work it out.[24] It has been shown that the two-dimensional Ising model does exhibit a phase transition.

Since an exact calculation is complicated, why not use a computer to work it out directly? It turns out that there is a practical problem. The number of distinct configurations for the Ising model on a lattice of $N \times N$ sites is 2^{N^2} and this grows very rapidly[25] with N. For large N, direct calculation of the partition function is clearly going to be a hopeless task. One way to attack this problem is on the basis of what is known as a **Monte-Carlo method**. The name of this model evokes the gambling casinos of Monte Carlo, where people roll dice and try their luck.[26] But there is a method behind the madness of this particular type of gambling. A real system adopts a particular configuration and, as it fluctuates randomly, explores the configuration space defined by the 2^{N^2} states. It periodically gets "stuck" in particular configurations or evolves between certain arrangements, and of course its speed of fluctuation depends on the temperature. Rather than computing all the configurations directly, why not let the computer simply take a similar path through the configuration space?

---

[22] Spins do start to correlate with their neighbours, and it is possible to show that the **correlation length**, the length scale over which fluctuations are correlated, increases as the system is cooled. It only becomes infinite (perfect order) at absolute zero.

[23] For one dimension, a break in the chain stops the message "line up this way" getting through. In two or higher dimensions, a region of flipped spins does not do this since the message can pass around it.

[24] The feat was accomplished in 1944 by Lars Onsager (1903–1976).

[25] The factor 2^{N^2} increases very fast with N, as shown in the table below:

N	N^2	2^{N^2}
2	4	16
3	9	512
4	16	65 536
5	25	33 554 432
6	36	68 719 476 736
⋮	⋮	⋮

[26] The name originates from Los Alamos physicists working on nuclear weapons in the 1940s. Legend has it that one of them, Stanislaw Ulam, had an uncle who was a regular gambler at the casinos in Monte Carlo, Monaco.

One promising way to do this for the two-dimensional Ising model is the **Metropolis algorithm**,[27] which can be very simply implemented on a computer. In this algorithm, a spin is chosen at random and the state is flipped. If this lowers the energy of the system as a whole, you always leave it flipped. If instead it raises the energy (let us say by an amount ΔE), you return it to its original unflipped state with probability $e^{-\beta \Delta E}$ or leave it flipped with probability $1 - e^{-\beta \Delta E}$. You then choose another spin at random and repeat the process. The process is repeated until equilibrium is reached for that temperature.

To understand how the Metropolis algorithm works, notice that (i) at $T = 0$, $e^{-\beta \Delta E} = 1$ and so you only flip a spin if it lowers the system's energy; (ii) at high temperature, $e^{-\beta \Delta E} \ll 1$, and therefore you always end up flipping the spin. Thus the algorithm at low temperature will always drive the system towards its lowest energy state; at high temperature, the algorithm tends to randomize all the states, which flip repeatedly and randomly. At a particular β one needs to iterate the algorithm repeatedly until the system parameters settle down. Once this occurs, you can then take averages of system properties.

It is easy to work out the energy of a particular configuration using the Ising Hamiltonian (eqn 28.60) and to average it over time. One can then obtain $\langle E \rangle$ and then $\langle E^2 \rangle$. Now we have that

$$\langle E \rangle = \frac{1}{Z} \sum_i E_i e^{-\beta E_i} = -\frac{1}{Z} \frac{dZ}{d\beta} = -\frac{d \ln Z}{d\beta}, \qquad (28.63)$$

since $Z = \sum_i e^{-\beta E_i}$. Similarly

$$\langle E^2 \rangle = \frac{1}{Z} \sum_i E_i^2 e^{-\beta E_i} = \frac{1}{Z} \frac{d^2 Z}{d\beta^2}. \qquad (28.64)$$

The heat capacity $C = d\langle E \rangle/dT$ is then, after a couple of lines of algebra, given by

$$C = k_B \beta^2 (\langle E^2 \rangle - \langle E \rangle^2), \qquad (28.65)$$

and this provides a useful way to obtain the heat capacity from the measured $\langle E \rangle$ and $\langle E^2 \rangle$.

If the energy of the system depended linearly on some quantity X, such that $\Delta E = -BX$, where B is a constant, then

$$\langle X \rangle = \frac{1}{Z} \sum_i X_i e^{-\beta(E_i - BX)} = \frac{1}{\beta} \left(\frac{\partial \ln Z}{\partial B} \right)_T = -\left(\frac{\partial F}{\partial B} \right)_T, \qquad (28.66)$$

where $F = -k_B T \ln Z$ is the Helmholtz function. A further differentiation gives

$$\langle X^2 \rangle - \langle X \rangle^2 = k_B T \chi, \qquad (28.67)$$

where $\chi = \partial \langle X \rangle / \partial B$. For the magnetic case, we can interpret X as the magnetic moment, B as a magnetic field, and χ as a magnetic susceptibility. This shows that fluctuations in the magnetic moment are related to the system's susceptibility. It also provides a method for extracting susceptibility from Monte-Carlo simulations.

[27] Named after Nicholas Metropolis, who invented it in 1953. N. Metropolis, A.W. Rosenbluth, M.N. Rosenbluth, A.H. Teller, and E. Teller, J. Chem. Phys. **21**, 1087 (1953).

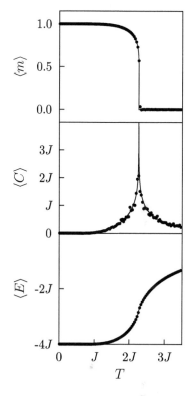

Fig. 28.17 The average magnetic moment, heat capacity, and energy per site for the two-dimensional Ising model. The solid lines are the exact solution while the points are obtained from a Monte-Carlo simulation using the Metropolis algorithm. The phase transition T_c occurs at a temperature given by $k_B T_c / J = 2/\ln(1 + \sqrt{2}) = 2.269185\ldots$, predicted by Onsager.

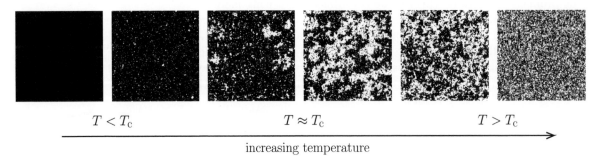

Fig. 28.18 A Monte-Carlo simulation of spins in the two-dimensional Ising model at various temperatures. Spins are coloured black or white according to whether they are in the +1 or −1 state respectively.

Results from Monte-Carlo simulations of the two-dimensional Ising model are shown in Fig. 28.17. They follow the behaviour of the exact solution very well. The disappearance of the average magnetic moment per site is apparent at the critical temperature T_c, as is a sharp peak in the heat capacity. The average energy climbs up from $-4J$ at $T = 0$ in a rather leisurely way with temperature, in a manner similar to that shown in Fig. 28.16, though the marked change in gradient seen in Fig. 28.17 signifies a phase transition in the present case. It is noticeable though that the region in which the agreement of the simulated heat capacity with the exact solution is poorest is close to the phase transition. To understand this, examine snapshots of configurations of the system which are taken after the system has had time to settle down. These are shown in Fig. 28.18. At low temperature all the spins are aligned and the region is uniformly coloured black, representing identical spin states pointing up. As the temperature increases, a few white blobs begin to appear as small fluctuations start to occur more readily. Close to the transition, the fluctuating regions are very large and the system starts to lose its net magnetic moment. Above the transition, fluctuations occur more readily, but the size of correlated regions of spins begins to decrease and, at the highest temperature shown, the state of each individual spin is only weakly correlated with its neighbour.

Close to T_c, the fluctuations become very slow owing to the correlation length of fluctuating spins becoming very large. This is because to cause a fluctuation it is necessary to flip a large, correlated block of spins, which is a slow process. Both the correlation length and the fluctuation time diverge at T_c (the latter phenomenon is known as **critical slowing down**). This slowing down affects the simulations as much as in the real system, increasing the convergence time of calculations[28] near T_c.

In this section, we have demonstrated Monte-Carlo methods for a problem that actually has an exact solution. The real benefit of this approach is in studying statistical systems for which the solution is not known and, in this respect, Monte-Carlo techniques form a mainstay of many calculations in statistical physics.

[28] One way around this is to use the Wolff algorithm, which involves building up clusters of spins which are then flipped together and this procedure proves more effective than the Metropolis algorithm. This is described in U. Wolff, *Phys. Rev. Lett.* **62**, 361 (1989).

Chapter summary

- The latent heat is related to the change in entropy at a first-order phase transition.
- The Clausius–Clapeyron equation states that

$$\frac{\mathrm{d}p}{\mathrm{d}T} = \frac{L}{T(V_2 - V_1)},$$

and this can be used to determine the shape of the phase boundary.
- The Kelvin formula states that the pressure in a droplet is given by

$$p = p_0 \exp\left(\frac{2\gamma V_{\text{liq}}}{rRT}\right).$$

- The Gibbs phase rule states that $F = C - P + 2$.
- Dissolving a solute in a solvent results in the elevation of the solvent's boiling point and a depression of its freezing point.
- A first-order phase transition involves a latent heat while a continuous phase transition does not.
- Certain phase transitions involve the breaking of symmetry.
- Statistical systems can be simulated using Monte-Carlo techniques.

Further reading

More information on phase transitions may be found in Binney *et al.* (1992), Yeomans (1992), Le Bellac et al. (2004), Blundell (2001) and Anderson (1984). A good discussion of Monte-Carlo methods may be found in Binney et al. (1992) and Krauth (2006). Onsager's solution of the two-dimensional Ising model is described in Plischke and Bergersen (1989).

Exercises

(28.1) When lead is melted at atmospheric pressure, the melting point is 327.0 °C, the density decreases from 1.101×10^4 to 1.065×10^4 kg m^{-3} and the latent heat is 24.5 kJ kg^{-1}. Estimate the melting point of lead at a pressure of 100 atm.

(28.2) Some tea connoisseurs claim that a good cup of tea cannot be brewed with water at a temperature less than 97 °C. Assuming this to be the case, is it possible for an astronomer, working on the summit of Mauna Kea in Hawaii (elevation 4207 m, though you don't need to know this to solve the problem) where the air pressure is 615 mbar, to make a good cup of tea without the aid of a pressure vessel?

(28.3) The gradient of the melting line of water on a

p–T diagram close to $0\,°C$ is $-1.4 \times 10^7\,\mathrm{Pa\,K^{-1}}$. At $0\,°C$, the specific volume of water is $1.00 \times 10^{-3}\,\mathrm{m^3\,kg^{-1}}$ and of ice is $1.09 \times 10^{-3}\,\mathrm{m^3\,kg^{-1}}$. Using this information, deduce the latent heat of fusion of ice.

In winter, a lake of water is covered initially by a uniform layer of ice of thickness $1\,\mathrm{cm}$. The air temperature at the surface of the ice is $-0.5\,°C$. Estimate the rate at which the layer of ice begins to thicken, assuming that the temperature of the water just below the ice is $0\,°C$. You can also assume steady-state conditions and ignore convection.

The temperature of the water at the bottom of the lake, depth $1\,\mathrm{m}$, is maintained at $2\,°C$. Find the thickness of ice that will eventually be formed. [The thermal conductivity of ice is $2.3\,\mathrm{W\,m^{-1}\,K^{-1}}$ and of water is $0.56\,\mathrm{W\,m^{-1}\,K^{-1}}$.]

(28.4) (a) Show that the temperature dependence of the latent heat of vaporization L is given by the following expression:

$$\frac{\mathrm{d}}{\mathrm{d}T}\left(\frac{L}{T}\right) = \frac{C_{pv} - C_{pL}}{T} \quad (28.68)$$
$$+ \left[\left(\frac{\partial S_v}{\partial p}\right)_T - \left(\frac{\partial S_L}{\partial p}\right)_T\right]\frac{\mathrm{d}p}{\mathrm{d}T}.$$

In this equation, S_v and S_L are the entropies of the vapour and liquid and C_{pv} and C_{pL} are the heat capacities of the vapour and liquid. Hence show that $L = L_0 + L_1 T$, where L_0 and L_1 are constants.

(b) Show further that when the saturated vapour of an incompressible liquid is expanded adiabatically, some liquid condenses out if

$$C_{pL} + T\frac{\mathrm{d}}{\mathrm{d}T}\left(\frac{L}{T}\right) < 0, \quad (28.69)$$

where C_{pL} is the heat capacity of the liquid (which is assumed constant) and L is the latent heat of vaporization.

(Hint: consider the gradient of the phase boundary in the p–T plane and the corresponding curve for adiabatic expansion.)

(28.5) The equilibrium vapour pressure p of water as a function of temperature is given in the following table:

T (°C)	p (Pa)
0	611
10	1228
20	2339
30	4246
40	7384
50	12349

Deduce a value for the latent heat of evaporation L_v of water. State clearly any simplifying assumptions that you make.

Estimate the pressure at which ice and water are in equilibrium at $-2\,°C$, given that ice cubes float with $4/5$ of their volume submerged in water at the triple point $(0.01\,°C, 612\,\mathrm{Pa})$.
[Latent heat of sublimation of ice at the triple point, $L_s = 2776 \times 10^3\,\mathrm{J\,kg^{-1}}$.]

(28.6) It is sometimes stated that the weight of a skater pressing down on thin skates is enough to melt ice, so that the skater can glide around on a thin film of liquid water. Assuming an ice rink at $-5\,°C$, make some estimates and show that this mechanism won't work. [In fact, frictional heating of ice is much more important, see S. C. Colbeck, *Am. J. Phys.* **63**, 888 (1995) and S. C. Colbeck, L. Najarian, and H. B. Smith *Am. J. Phys.* **65**, 488 (1997).]

(28.7) Write a computer program that will implement the Metropolis algorithm for the two-dimensional Ising model.

Bose–Einstein and Fermi–Dirac distributions

29

In this chapter, we are going to consider the way in which *quantum mechanics* changes the statistical properties of gases. The crucial ingredient is the concept of *identical particles*. The results of quantum mechanics show that there are two types of identical particle: **bosons** and **fermions**. Bosons can share quantum states, while fermions cannot share quantum states. Another way of stating this is to say that bosons are not subject to the **Pauli exclusion principle**, while fermions are. This difference in ability to share quantum states (arising from what we shall call *exchange symmetry*) has a profound effect on the statistical distribution of these particles over the energy states of the system. This distribution over energy states is called the **statistics** of these particles, and we will demonstrate the effect of exchange symmetry on statistics. However, it can also be shown that another difference between bosons and fermions is the type of spin angular momentum that they may possess. This is enshrined in the **spin-statistics theorem**, which we will not prove but which states that bosons have integer spin while fermions have half-integer spin.

29.1 Exchange and symmetry	345
29.2 Wave functions of identical particles	346
29.3 The statistics of identical particles	349
Chapter summary	353
Further reading	353
Exercises	354

Example 29.1

- Examples of bosons include: photons (spin 1), ^{4}He atoms (spin 0).
- Examples of fermions include: electrons (spin $\frac{1}{2}$), neutrons (spin $\frac{1}{2}$), protons (spin $\frac{1}{2}$), ^{3}He atoms (spin $\frac{1}{2}$), ^{7}Li nuclei (spin $\frac{3}{2}$).

29.1 Exchange and symmetry

In this section, we will argue why a two-particle wave function can be either symmetric or antisymmetric under exchange of particles. Consider two identical particles, one at position $\boldsymbol{r}_1$ and the other at position $\boldsymbol{r}_2$. The wave function describing this is $\psi(\boldsymbol{r}_1, \boldsymbol{r}_2)$. We now define an exchange operator $\hat{P}_{12}$, which exchanges particles 1 and 2. Thus

$$\hat{P}_{12}\psi(\boldsymbol{r}_1, \boldsymbol{r}_2) = \psi(\boldsymbol{r}_2, \boldsymbol{r}_1). \qquad (29.1)$$

Since the particles are identical, we also expect that the Hamiltonian $\hat{\mathcal{H}}$, which describes this two-particle system must commute with $\hat{P}_{12}$, i.e.,

$$[\hat{\mathcal{H}}, \hat{P}_{12}] = 0, \tag{29.2}$$

so that the energy eigenfunctions must be simultaneously eigenfunctions of the exchange operator. However, because the particles are identical, swapping them over must have no effect on the probability density. Thus

$$|\psi(\mathbf{r}_1, \mathbf{r}_2)|^2 = |\psi(\mathbf{r}_2, \mathbf{r}_1)|^2. \tag{29.3}$$

[1] A Hermitian operator has real eigenvalues, so is useful for representing real physical quantities in quantum mechanics.

If $\hat{P}_{12}$ is a Hermitian[1] operator, it must have real eigenvalues, so we expect that $\hat{P}_{12}\psi = \lambda\psi$, where λ is a real eigenvalue. Equation 29.3 shows that the only solution to this is $\lambda = \pm 1$, i.e.,

$$\hat{P}_{12}\psi(\mathbf{r}_1, \mathbf{r}_2) = \psi(\mathbf{r}_2, \mathbf{r}_1) = \pm\psi(\mathbf{r}_1, \mathbf{r}_2). \tag{29.4}$$

The wave function must therefore have one of two types of **exchange symmetry**, as follows:

- The wave function is **symmetric** under exchange of particles:

$$\psi(\mathbf{r}_2, \mathbf{r}_1) = \psi(\mathbf{r}_1, \mathbf{r}_2), \tag{29.5}$$

and the particles are called bosons.

- The wave function is **antisymmetric** under exchange of particles:

$$\psi(\mathbf{r}_2, \mathbf{r}_1) = -\psi(\mathbf{r}_1, \mathbf{r}_2), \tag{29.6}$$

and the particles are called fermions.

This argument is valid for particles in three dimensions, the situation we usually encounter in our three-dimensional world, but fails in two dimensions. This occurs because you have to be a bit more careful than we've been here about how you exchange two particles. This point is rather an esoteric one, but the interested reader can follow up this point in the box on page 347 and in the further reading.

29.2 Wave functions of identical particles

In the previous section, we wrote down a two-particle wave function $\psi(\mathbf{r}_2, \mathbf{r}_1)$, which labelled the particles according to their position. However, there are lots more ways in which one could label a particle, such as which orbital state it is in, or what its momentum is. To keep things completely general, we will label the particles according to their state in a more abstract way. The effect on the statistics will then be more transparent, and is demonstrated by Example 29.2 (on page 348).

Anyons

The argument that we have used to describe exchange symmetry is, in fact, only strictly valid in three dimensions. In two dimensions, there are further possibilities other than fermions and bosons. For the interested reader, we give a more detailed description in this box.

We begin by noticing that eqn 29.3 allows the solution $\psi(r_2, r_1) = e^{i\theta}\psi(r_1, r_2)$, where θ is a phase factor. Thus exchanging identical particles means that the wave function acquires a phase θ. Defining $r = r_2 - r_1$, the action of exchanging the position coordinates of two particles involves letting this vector execute some path from r to $-r$, but avoiding the origin so that the two particles do not ever occupy the same position.

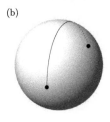

Fig. 29.1 Paths in r-space, for the three-dimensional case, corresponding to (a) no exchange of particles and (b) exchange of particles.

We can therefore imagine the exchange of particles as a path in r-space. Without loss of generality, we can keep $|r|$ fixed, so that in the process of exchanging the two particles, they move relative to each other at a fixed separation. Thus, for the case of three dimensions, the path is on the surface of a sphere in r-space. Since the two particles are identical, opposite points on the surface of the sphere are equivalent and must be identified (giving r-space the topology of what is known as real, two-dimensional projective space). It turns out that all paths on this surface fall into two classes: those which are contractible to a point [and thus correspond to no exchange of particles, yielding $\theta = 0$ to ensure the wave function is single-valued; see Fig. 29.1(a)] and those which are not [and thus correspond to exchange of particles; see Fig. 29.1(b)]. For this latter case we have to assign $\theta = \pi$, so that two exchanges correspond to no exchange, i.e.,

$e^{i\theta}e^{i\theta} = 1$, so that $\theta = \pi$. This argument thus justifies that the phase factor $e^{i\theta} = \pm 1$, giving rise to bosons ($e^{i\theta} = +1$) and fermions ($e^{i\theta} = -1$).

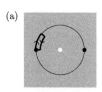

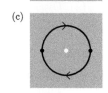

Fig. 29.2 Paths in r-space, for the two-dimensional case, corresponding to (a) no exchange, (b) a single exchange and (c) two exchanges of particles.

However, the argument fails in two dimensions. In the two-dimensional case, the path is on a circle in r-space in which opposite points on the circle are equivalent and are identified. In this case, the paths in r-space can wind round the origin an integer number of times. This means that two successive exchanges of the particles [as shown in Fig. 29.2(c)] are not topologically equivalent to zero exchanges [if performed by winding round the origin in the same direction, as shown in Fig. 29.2(c)] and thus the phase θ can take any value. (In this case, r-space has the topology of real one-dimensional projective space, which is the same as that of a circle.) The resulting particles have more complicated statistical properties than either bosons or fermions and are called **anyons** (because θ can take "any" value). Since θ/π is no longer forced to be ± 1, and can take any *fractional* value in between, anyons can have **fractional statistics**. The crucial distinction between r-space in two and three dimensions is that the removal of the origin in two-dimensional space makes the space multiply connected (allowing paths that wind around the origin), whereas three-dimensional space remains singly connected (and a path that tries to wind round the origin can be deformed into one which does not).

We live in a three-dimensional world, so is any of this relevant? In fact, anyons turn out to be important in the **fractional quantum Hall effect**, which occurs in certain two-dimensional electron systems under high magnetic field. For more details concerning anyons, see the further reading.

Example 29.2

Imagine that a particle can exist in one of two states, which we will label $|0\rangle$ and $|1\rangle$. We now consider two such particles, and describe their joint behaviour by a product state. Thus

$$|0\rangle|1\rangle \tag{29.7}$$

describes the state in which the first particle is in state 0 and the second particle is in state 1. What are the possible states for this system if the particles are (a) distinguishable, (b) indistinguishable, but classical, (c) indistinguishable bosons, and (d) indistinguishable fermions?
Solution:

(a) *Distinguishable particles*
There are four possible states, which are

$$|0\rangle|0\rangle, \quad |1\rangle|0\rangle, \quad |0\rangle|1\rangle, \quad |1\rangle|1\rangle. \tag{29.8}$$

(b) *Indistinguishable, but classical, particles*
There are now only three possible states, which are

$$|0\rangle|0\rangle, \quad |1\rangle|0\rangle, \quad |1\rangle|1\rangle. \tag{29.9}$$

Since the particles are indistinguishable, there is no way of distinguishing the state $|1\rangle|0\rangle$ and $|0\rangle|1\rangle$.

(c) *Indistinguishable bosons*
There are also only three possible states. Clearly both $|0\rangle|0\rangle$ and $|1\rangle|1\rangle$ are eigenstates of the exchange operator, but $|1\rangle|0\rangle$ and $|0\rangle|1\rangle$ are not. However, if we make a linear combination[2]

$$\frac{1}{\sqrt{2}}\left(|1\rangle|0\rangle + |0\rangle|1\rangle\right), \tag{29.10}$$

this will be an eigenstate of the exchange operator with eigenvalue 1. Thus the three possible states are (all symmetric under exchange):

$$|0\rangle|0\rangle, \quad |1\rangle|1\rangle, \quad \frac{1}{\sqrt{2}}\left(|1\rangle|0\rangle + |0\rangle|1\rangle\right). \tag{29.11}$$

(d) *Indistinguishable fermions*
No two fermions can be in the same quantum state (by the Pauli exclusion principle), so $|0\rangle|0\rangle$ and $|1\rangle|1\rangle$ are not allowed. Thus only one state is possible, which is antisymmetric under exchange:

$$\frac{1}{\sqrt{2}}\left(|1\rangle|0\rangle - |0\rangle|1\rangle\right). \tag{29.12}$$

This wave function is an eigenstate of the exchange operator with eigenvalue -1.

---

[2]This example demonstrates quantum-mechanical **entanglement**. The states of the two particles are entangled because they cannot be separated into a product of two single-particle states. In this case, if one particle is in the $|0\rangle$ state, the other particle has to be in the $|1\rangle$ state, and vice versa. Thus the behaviour of the two particles are correlated. An example of a non-entangled state is $\frac{1}{\sqrt{2}}(|0\rangle|0\rangle+|0\rangle|1\rangle)$, which can be separated into $\frac{1}{\sqrt{2}}|0\rangle(|0\rangle+|1\rangle)$, a product state.

In general, for fermions, the requirement that $\hat{P}_{12}|\psi\rangle = -|\psi\rangle$ means that if $|\psi\rangle$ is a two-particle state consisting of two particles in the *same* quantum state, i.e., if $\psi = |\varphi\rangle|\varphi\rangle$, then

$$\hat{P}_{12}|\varphi\rangle|\varphi\rangle = |\varphi\rangle|\varphi\rangle = -|\varphi\rangle|\varphi\rangle, \quad (29.13)$$

so that

$$|\varphi\rangle|\varphi\rangle = 0, \quad (29.14)$$

i.e., the doubly-occupied state cannot exist. This, again, illustrates the Pauli exclusion principle, namely that two identical fermions cannot coexist in the same quantum state.

29.3 The statistics of identical particles

In the last section, we have demonstrated that exchange symmetry has an important effect on the statistics of two identical particles. Now we want to do the same for cases in which we have many more than two identical particles. Our derivation will be easiest if we do this by finding the grand partition function $\mathcal{Z}$ (see Section 22.3) for a system comprised either of fermions or bosons. In this approach, the total number of particles is not fixed, and this is an easy constraint to apply, as we shall see. If one is treating a system in which the number of particles *is* fixed, we can always fix it at the end of our calculation. Our method will be to use the expression $\mathcal{Z} = \sum_\alpha e^{\beta(\mu N_\alpha - E_\alpha)}$ (from eqn 22.20). Here, α denotes a particular state of the system.

Let us begin to think about this problem by first studying a very simple case. Assume that there is only one state in which we can put particles and that the energy cost of each particle is E. In this case, the grand partition function $\mathcal{Z}$ is simply a sum over configurations, each with a different number n of particles in that state, so that

$$\mathcal{Z} = \sum_n e^{n\beta(\mu - E)}. \quad (29.15)$$

Note that one can extract the mean number of particles, $\langle n \rangle$, in this state using

$$\langle n \rangle = \frac{\sum_n n e^{n\beta(\mu - E)}}{\sum_n e^{n\beta(\mu - E)}} = -\frac{1}{\beta \mathcal{Z}} \frac{\partial \mathcal{Z}}{\partial E} = -\frac{1}{\beta} \frac{\partial \ln \mathcal{Z}}{\partial E}. \quad (29.16)$$

For fermions, the sum over n in this expression only includes $n = 0$ and $n = 1$ because of the Pauli exclusion principle. Hence

$$\mathcal{Z} = \sum_{n=0}^{1} e^{n\beta(\mu - E)} = 1 + e^{\beta(\mu - E)}, \quad (29.17)$$

and hence

$$\ln \mathcal{Z} = \ln(1 + e^{\beta(\mu - E)}). \quad (29.18)$$

For bosons, the Pauli exclusion principle does not apply and hence n runs from 0 to ∞, and so $\mathcal{Z}$ is

$$\mathcal{Z} = \sum_{n=0}^{\infty} e^{n\beta(\mu-E)} = \frac{1}{1 - e^{\beta(\mu-E)}}, \tag{29.19}$$

and therefore

$$\ln \mathcal{Z} = -\ln(1 - e^{\beta(\mu-E)}). \tag{29.20}$$

In general then, instead of eqns 29.18 and 29.20, one can write

$$\boxed{\ln \mathcal{Z} = \pm \ln(1 \pm e^{\beta(\mu-E)}),} \tag{29.21}$$

where the $\pm$ sign means $+$ for fermions and $-$ for bosons. The mean number of particles in the state is then, using eqn 29.16, given by

$$\langle n \rangle = -\frac{1}{\beta} \frac{\partial \ln \mathcal{Z}}{\partial E} = \frac{e^{\beta(\mu-E)}}{1 \pm e^{\beta(\mu-E)}}, \tag{29.22}$$

and hence dividing top and bottom by $e^{\beta(\mu-E)}$ gives

$$\boxed{\langle n \rangle = \frac{1}{e^{\beta(E-\mu)} \pm 1},} \tag{29.23}$$

where, again, the $\pm$ sign means $+$ for fermions and $-$ for bosons. In the following example,[3] we will derive the general case in which there are many more quantum states accessible to our particles.

[3] This example may be skipped on first reading. It turns out that the expression we have derived in eqn 29.23 holds in the general case.

Example 29.3

Let us assume that the energy cost of putting a particle into the ith single-particle state of a system is given by E_i. We will put n_i particles into the ith state; here n_i is called the **occupation number** of the ith state. A particular configuration of the system is then described by the product

$$\left[e^{\beta(\mu-E_1)}\right]^{n_1} \times \left[e^{\beta(\mu-E_2)}\right]^{n_2} \times \cdots = \prod_i e^{n_i \beta(\mu-E_i)}. \tag{29.24}$$

The grand partition function is the sum of such products for all sets of occupation numbers that are allowed by the symmetry of the particles. Hence

$$\mathcal{Z} = \sum_{\{n_i\}} \prod_i e^{n_i \beta(\mu-E_i)}, \tag{29.25}$$

where the symbol $\{n_i\}$ denotes a set of occupation numbers allowed by the symmetry of the particles.

Fortunately, the total number of particles $\sum_i n_i$ does not have to be fixed,[4] because that would have been a fiddly constraint to apply to this expression. In fact, we will only be considering two cases: fermions, for which $\{n_i\} = \{0,1\}$ (independent of i), and bosons, for which $\{n_i\} = \{0,1,2,3,\ldots\}$ (independent of i). This allows us to factor out the terms in the product for each state i and hence write

[4]The total number of particles is not fixed in the grand canonical ensemble, which is the one we are using here.

$$\mathcal{Z} = \prod_i \sum_{\{n_i\}} e^{n_i \beta (\mu - E_i)}. \qquad (29.26)$$

We now proceed to evaluate $\ln \mathcal{Z}$ for a gas of (a) fermions and (b) bosons, and the derivation mirrors what we did for a single quantum state.

(a) For fermions, each state can either be empty or singly occupied, so that $\{n_i\} = \{0,1\}$, and hence eqn 29.26 becomes

$$\mathcal{Z} = \prod_i 1 + e^{\beta(\mu - E_i)}. \qquad (29.27)$$

Hence

$$\ln \mathcal{Z} = \sum_i \ln(1 + e^{\beta(\mu - E_i)}). \qquad (29.28)$$

(b) For bosons, each state can contain any integer number of particles, so that $\{n_i\} = \{0,1,2,3,\ldots\}$, and hence eqn 29.26 becomes

$$\mathcal{Z} = \prod_i 1 + e^{\beta(\mu - E_i)} + e^{2\beta(\mu - E_i)} + \cdots \qquad (29.29)$$

and therefore, by summing this geometric series, we have that

$$\mathcal{Z} = \prod_i \frac{1}{1 - e^{\beta(\mu - E_i)}}, \qquad (29.30)$$

and hence

$$\ln \mathcal{Z} = -\sum_i \ln(1 - e^{\beta(\mu - E_i)}). \qquad (29.31)$$

Summarizing these, we have

$$\ln \mathcal{Z} = \pm \sum_i \ln(1 \pm e^{\beta(\mu - E_i)}), \qquad (29.32)$$

where again the $\pm$ sign means $+$ for fermions and $-$ for bosons. The number of particles in each energy level is given by

$$\langle n_i \rangle = -\frac{1}{\beta}\left(\frac{\partial \ln \mathcal{Z}}{\partial E_i}\right) = \frac{e^{\beta(\mu - E_i)}}{1 \pm e^{\beta(\mu - E_i)}}, \qquad (29.33)$$

and hence dividing top and bottom by $e^{\beta(\mu - E_i)}$ gives

$$\langle n_i \rangle = \frac{1}{e^{\beta(E_i - \mu)} \pm 1}, \qquad (29.34)$$

where, again, the $\pm$ sign means $+$ for fermions and $-$ for bosons. Thus the results that we have obtained in the general case follow precisely what we found for a single quantum state in eqn 29.23. If μ and T are fixed for a particular system, eqn 29.34 shows that the mean occupation of the ith state, $\langle n_i \rangle$, is a function only of the energy E_i.

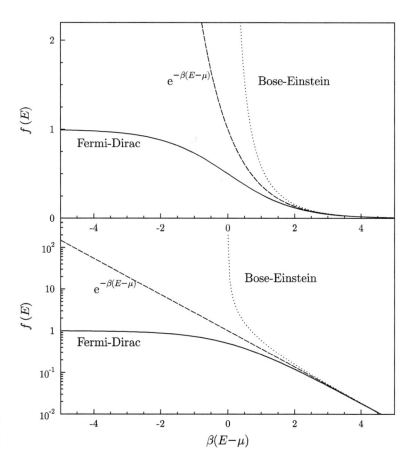

Fig. 29.3 The Fermi–Dirac and Bose–Einstein distribution functions, together with the Boltzmann factor $e^{-\beta(E-\mu)}$. The upper panel shows the distributions with a linear vertical axis; in the lower panel, the vertical axis has a logarithmic scale.

It is convenient to consider the **distribution function** $f(E)$ for fermions and bosons, which is defined to be the mean occupation of a single-particle state with energy E (following eqn 29.23, so that $f(E) = \langle n \rangle$). We can therefore immediately write down the distribution function $f(E)$ for fermions as

$$f(E) = \frac{1}{e^{\beta(E-\mu)} + 1}, \tag{29.35}$$

which is known as the **Fermi–Dirac distribution function**, and for bosons as

$$f(E) = \frac{1}{e^{\beta(E-\mu)} - 1}, \tag{29.36}$$

which is known as the **Bose–Einstein distribution function**. Sometimes the term on the right-hand side of eqn 29.35 is referred to as the **Fermi factor** and the term on the right-hand side of eqn 29.36 is referred to as the **Bose factor**. These are sketched in Fig. 29.3. Note that in the limit $\beta(E - \mu) \gg 1$, both functions tend to the Boltzmann distribution $e^{-\beta(E-\mu)}$. This is because this limit corresponds to low density (μ small) and here there are many more states thermally accessible to the particles than there are particles; thus double occupancy never occurs and the requirements of exchange symmetry become irrelevant and both fermions and bosons behave like classical particles. The differences, however, are particularly felt at high density. In particular, note that the distribution function for bosons diverges when $\mu = E$. Thus for bosons, the chemical potential must always be below, even if only slightly, the lowest-energy state. If it is not, then the lowest-energy state would become occupied with an infinite number of particles, which is unphysical. The implications for the properties of quantum gases will be considered in the next chapter.

Chapter summary

- The wave function of a pair of bosons is symmetric under exchange of particles, while the wave function of a pair of fermions is antisymmetric under exchange of particles.
- Bosons can share quantum states, while fermions cannot share quantum states.
- Bosons obey Bose–Einstein statistics, given by

$$f(E) = \frac{1}{e^{\beta(E-\mu)} - 1},$$

while fermions obey Fermi–Dirac statistics, given by

$$f(E) = \frac{1}{e^{\beta(E-\mu)} + 1}.$$

Further reading

More information about anyons may be found in Canright and Girvin (1990), Rao (1992), and in the collection of articles in Shapere and Wilczek (1989).

Exercises

(29.1) Differentiate between particles that obey Bose–Einstein and Fermi–Dirac statistics, giving two examples of each.

(29.2) For the particles considered in Example 29.2, what is the probability that both particles are in the $|0\rangle$ state when (a) distinguishable, (b) indistinguishable, but classical, (c) indistinguishable bosons, and (d) indistinguishable fermions?

(29.3) By rewriting the Fermi–Dirac function

$$f(E) = \frac{1}{e^{\beta(E-\mu)} + 1} \quad (29.37)$$

as

$$f(E) = \frac{1}{2}\left(1 - \tanh\frac{1}{2}\beta(E-\mu)\right), \quad (29.38)$$

show that $f(E)$ is symmetric about $E = \mu$ and sketch it. Find simplified expressions for $f(E)$ when (i) $E \ll \mu$, (ii) $E \gg \mu$, and (iii) E is very close to μ.

(29.4) Are identical particles always indistinguishable?

(29.5) Hydrogen (H$_2$) gas can exist in two forms. If the proton spins are in an exchange symmetric triplet ($S = 1$) state, it is known as **ortho-hydrogen**. If the proton spins are in an exchange antisymmetric singlet ($S = 0$) state, it is known as **para-hydrogen**. The symmetry of the total wave function must be antisymmetric overall, so that the rotational part of the wave function must be antisymmetric for ortho-hydrogen (so that the angular momentum quantum number J is 1, 3, 5, ...) or symmetric for para-hydrogen (so that $J = 0, 2, 4, ...$) The proton separation in H$_2$ is 7.4×10^{-11} m. Estimate the spacing in kelvin between the ground state and first excited state in para-hydrogen.
Show that the ratio f of ortho-hydrogen to para-hydrogen is given by

$$f = 3\frac{\sum_{J=1,3,5,...}(2J+1)e^{-J(J+1)\hbar^2/2Ik_BT}}{\sum_{J=0,2,4,...}(2J+1)e^{-J(J+1)\hbar^2/2Ik_BT}}, \quad (29.39)$$

and find f at 50 K.

(29.6) In this exercise, we derive Fermi–Dirac and Bose–Einstein statistics using the microcanonical ensemble.

(a) Show that the number of ways of distributing n_j fermions among g_j states with not more than one particle in each is

$$\Omega_j = \frac{g_j!}{n_j!(g_j - n_j)!}. \quad (29.40)$$

Here, j labels a particular group of states. Hence the entropy S is given by

$$S = k_B \ln\left[\prod_j \frac{g_j!}{n_j!(g_j - n_j)!}\right]. \quad (29.41)$$

Hence show (using Stirling's approximation) that

$$S = -k_B \sum_j g_j[\bar{n}_j \ln \bar{n}_j + (1 - \bar{n}_j)\ln(1 - \bar{n}_j)], \quad (29.42)$$

where $\bar{n}_j = n_j/g_j$ are the mean occupation numbers of the quantum states. Maximize this expression subject to the constraint that the total energy E and number of particles N are constant, and hence show that

$$\bar{n}_j = \frac{1}{e^{\alpha + \beta E_j} + 1}. \quad (29.43)$$

(b) Show that the number of ways of distributing n_j bosons among g_j states with any number of particles in each is

$$\Omega_j = \frac{(g_j + n_j - 1)!}{n_j!(g_j - 1)!}. \quad (29.44)$$

Hence show that

$$S = k_B \sum_j g_j[(1 + \bar{n}_j)\ln(1 + \bar{n}_j) - \bar{n}_j \ln \bar{n}_j].$$

$$(29.45)$$

Maximize this expression subject to the constraint that the total energy E and number of particles N are constant, and hence show that

$$\bar{n}_j = \frac{1}{e^{\alpha + \beta E_j} - 1}. \quad (29.46)$$

Albert Einstein (1879–1955)

Fig. 29.4 Albert Einstein

Albert Einstein's academic career began badly. In 1895, he failed to get into the prestigious Eidgenössische Technische Hochschule (ETH) in Zürich, and was sent to nearby Aarau to finish secondary school. He enrolled at ETH the following year, but failed to get a teaching assistant job there after his degree. After teaching maths at technical schools in Winterthur and Schaffhausen, Einstein finally landed a job at a patent office in Bern in 1902 and was to stay there for seven years. Though Einstein was present in the office, his mind was elsewhere and he combined the day job with doctoral studies at the University of Zürich.

In 1905 this unknown patent clerk submitted his doctoral thesis (which derived a relationship between diffusion and frictional forces, and which contained a new method to determine molecular radii) and also published four revolutionary papers in the journal *Annalen der Physik*. The first paper proposed that Planck's energy quanta were real entities and would show up in the photoelectric effect, work for which he was awarded the 1921 Nobel Prize. The citation stated that the prize was "for his services to Theoretical Physics, and especially for his discovery of the law of the photoelectric effect". The second paper explained Brownian motion on the basis of statistical mechanical fluctuations of atoms. The third and fourth papers introduced his special theory of relativity and his famous equation $E = mc^2$. Any one of these developments alone was sufficient to earn him a major place in the history of physics; the combined achievement led to more modest immediate rewards: the following year, Einstein was promoted by the patent office to "technical examiner second class". Einstein only became a professor (at Zürich) in 1909, moving to Prague in 1911, ETH in 1912, and Berlin in 1914.

In 1915, Einstein presented his general theory of relativity, which included gravity. These ideas led to the prediction of gravitational lensing and gravitational waves, and the general theory of relativity is of fundamental importance in modern astrophysics. In the 1920's, Einstein battled with Bohr on the interpretation of quantum theory, a subject which he had helped found through his work on the photoelectric effect. Einstein did not believe quantum theory to be complete, while completeness was the central thesis of Bohr's Copenhagen interpretation. Einstein seemed to lose the battles, but his criticisms illuminated the understanding of quantum mechanics, particularly concerning the nature of quantum entanglement. Einstein also contributed to quantum statistical mechanics through his work on Bose–Einstein statistics (see the biography of Bose).

The rise of Nazi Germany led to Einstein's departure in 1933 from the country of his birth, and after receiving offers from Jerusalem, Leiden, Oxford, Madrid, and Paris, he settled on Princeton where he remained for the rest of his life. When he arrived there in 1935, and was asked what he would require for his study, he is reported to have replied "A desk, some pads and a pencil, and a large wastebasket to hold all of my mistakes."

In 1939, following persuasion from Szilárd, he played a crucial rôle in alerting President Roosevelt to the theoretical possibility of nuclear weapons being developed based on the discovery of nuclear fission and the need for the Allies to have this before the Nazis; this eventually led to the Manhattan project and the development of the atomic bomb. Einstein's final years were spent in an unsuccessful search for a grand unified theory, which would combine the fundamental forces into a single theory.

Interstingly, Einstein said that his search for the principle of relativity had been motivated by his yearning for a grand universal principle, which was on the same level of the second law of thermodynamics. He saw many theories of physics as constructive, such as the kinetic theory of gases, which build up a description of complex behaviour from a simple scheme of mechanical and diffusional processes. Instead, he was after something much grander, in which many subtle consequences followed from a single universal principle. His model was thermodynamics, in which everything flowed from a fundamental principle about increase of entropy. Thus in some sense, thermodynamics was the template for relativity.

Satyendra Nath Bose (1894–1974)

Satyendra Nath Bose was born in Calcutta and graduated from the Presidency College there in 1915. He was appointed to Calcutta's new research institute, University College, in 1917 along with M. N. Saha and, the following year, C. V. Raman. All three were to make pioneering contributions to physics. Four years later, Bose moved to the University of Dhaka as Reader of Physics (though he returned to Calcutta in 1945). Bose had a prodigious memory and was legendary for giving highly polished lectures without consulting any notes.

Fig. 29.5 S. N. Bose

In 1924, Bose sent a paper to Einstein in Berlin, together with a handwritten covering letter:

> Respected Sir: I have ventured to send you the accompanying article for your perusal and opinion. I am anxious to know what you think of it. You will see that I have tried to deduce the coefficient of $8\pi\nu^2/c^3$ in Planck's Law independent of the classical electrodynamics, only assuming that the ultimate elementary regions in the phase space has the content h^3.

Bose had treated black-body radiation as a photon gas, using phase-space arguments; Planck's distribution came simply from maximizing the entropy. Einstein was impressed and translated Bose's paper into German and submitted it to *Zeitschrift für Physik* on Bose's behalf. Einstein followed up Bose's work in 1924 by generalizing it to non-relativistic particles with non-zero mass and in 1925 he deduced the phenomenon now known as Bose–Einstein condensation. This purely theoretical proposal was a full thirteen years before Fritz London proposed interpreting the superfluid transition in ^{4}He as just such a Bose–Einstein condensation.

Enrico Fermi (1901–1954)

Enrico Fermi was born in Rome and gained a degree at the University of Pisa in 1922. He spent a brief period working with Born and then returned to Italy, first as a lecturer in Florence (where he worked out "Fermi statistics", the statistical mechanics of particles subject to the Pauli exclusion principle) and then as a professor of physics at Rome in 1927. In Rome, Fermi made important contributions, including the theory of beta decay, the demonstration of nuclear transformation in elements subjected to neutron bombardment, and the discovery of slow neutrons. These results demonstrate Fermi's extraordinary ability to excel in both theory and experiment. Though extremely adept at detailed mathematical analysis, Fermi disliked complicated theories and had an aptitude for getting the right answer simply and quickly using the most efficient method possible.

Fig. 29.6 E. Fermi

Fermi was awarded the Nobel Prize in 1938 for his "demonstrations of the existence of new radioactive elements produced by neutron irradiation, and for his related discovery of nuclear reactions brought about by slow neutrons". After picking up his prize in Stockholm, he emigrated to the United States. He was one of the first to realize the possibility of a chain reaction in uranium, demonstrating the first self-sustaining nuclear reaction in a squash court near the University of Chicago in December 1942. Following this event, a coded phone call was sent to the leaders of the Manhattan project, with the message: "The Italian navigator has landed in the new world... The natives were very friendly".

Fermi became a major player in the Manhattan project, and following the end of World War II he remained in Chicago, working in high energy physics and cosmic rays until his untimely death due to stomach cancer.

Paul Dirac (1902–1984)

Paul Adrien Maurice Dirac was brought up in Bristol by his English mother and Swiss father. His father insisted that only French was spoken at the dinner table, a stipulation that left Dirac with something of a distaste for speaking at all.

Fig. 29.7 P. A. M. Dirac

He read engineering at Bristol University, graduating in 1921, and then took another degree in maths and got a first in 1923. This led him to doctoral research in Cambridge under the supervision (if one can use such a word of what was rather a tenuous relationship) of Fowler. During this period, Dirac's brother committed suicide and Dirac broke off contact with his father; this all contributed to making Dirac even more socially withdrawn. In 1925, he read Heisenberg's paper on commutators and realized the connection with Poisson brackets from classical mechanics. His Ph.D. thesis, submitted the following year, was entitled simply *Quantum Mechanics*. In 1926, Dirac showed how the antisymmetry of the wave function under particle exchange led to statistics that were identical to those derived by Fermi. Particles obeying such Fermi–Dirac statistics Dirac called (generously) "fermions", while those obeying Bose–Einstein statistics were "bosons".

After having spent time with Bohr in Copenhagen, Born in Göttingen, and Ehrenfest in Leiden, Dirac returned to Cambridge in 1927 to take up a fellowship at St John's College. His famous Dirac equation (which predicted the existence of the positron) appeared in 1928 and his book, *The Principles of Quantum Mechanics* (still highly readable, and in print), in 1930. In 1932 he was appointed to the Lucasian chair (held before by Newton, Airy, Babbage, Stokes, and Larmor, and later by Hawking) and the following year he shared the Nobel Prize with Schrödinger "for the discovery of new productive forms of atomic theory". Following a sabbatical visit to work with Eugene Wigner at Princeton, Dirac married Wigner's sister Margrit in 1937. In 1969, Dirac retired from Cambridge and moved to Tallahassee, Florida, where he became a professor at FSU.

Dirac had a very high view of mathematics, stating in the preface to his 1930 book that it was "the tool specially suited for dealing with abstract concepts of any kind and there is no limit to its power in this field." Later he remarked that in science "one tries to tell people, in such a way as to be understood by everyone, something that no one ever knew before. But in poetry, it's the exact opposite." Clarity for Dirac was fundamental, as was beauty, as it was "more important to have beauty in one's equations than to have them fit experiment." Failure to match the results of experimental data can be rectified by further experiment, or by the sorting out of some minor feature not taken into account that subsequent theoretical development will resolve; but for Dirac, an ugly theory could never be right.

Dirac said "I was taught at school never to start a sentence without knowing the end of it." This explains a lot. Dirac's famously taciturn and precise nature spawned many "Dirac stories". Dirac once fell asleep during someone else's lecture, but woke during a moment when the speaker was getting stuck in a mathematical derivation, muttering: "Here is a minus sign where there should be a plus. I seem to have dropped a minus sign somewhere." Dirac opened one eye and interjected: "Or an odd number of them." One further example concerns a conference lecture he himself gave, following which a questioner indicated that he had not followed a particular part of Dirac's argument. A long silence ensued, broken finally by the chairman asking if Professor Dirac would deal with the question. Dirac responded, "It was a statement, not a question."

30 Quantum gases and condensates

- 30.1 The non-interacting quantum fluid — 358
- 30.2 The Fermi gas — 361
- 30.3 The Bose gas — 366
- 30.4 Bose–Einstein condensation (BEC) — 367
- Chapter summary — 372
- Further reading — 373
- Exercises — 373

Exchange symmetry affects the occupation of allowed states in quantum gases. If the density of the gas is very low, such that $n\lambda_{\rm th}^3 \ll 1$, we can ignore this and forget about exchange symmetry; this is what we do for gases at room temperature. But if the density is high, the effects of exchange symmetry become very important and it really starts to matter whether the particles you are considering are fermions or bosons. In this chapter, we consider quantum gases in detail and explore the possible effects that one can observe.

30.1 The non-interacting quantum fluid

We first consider a fluid composed of non-interacting particles. To keep things completely general for the moment, we will consider particles with spin S. This means that each allowed momentum state is associated with $2S+1$ possible spin states.[1] If we can ignore interactions between particles, the grand partition function $\mathcal{Z}$ is simply the product of partition functions for each state, so that

$$\mathcal{Z} = \prod_{\bm{k}} \mathcal{Z}_{\bm{k}}^{2S+1}, \tag{30.1}$$

where

$$\mathcal{Z}_{\bm{k}} = (1 \pm e^{-\beta(E_{\bm{k}}-\mu)})^{\pm 1} \tag{30.2}$$

is a partition function for state $\bm{k}$ and where the $\pm$ sign is $+$ for fermions and $-$ for bosons.[2]

[1] If the spin is S, there are $2S+1$ possible states corresponding to the z-component of angular momentum being $-S, -S+1, \ldots S$.

[2] These results follow directly from eqns 29.27 and 29.30.

Example 30.1

Find the grand potential for a three-dimensional gas of non-interacting bosons and fermions with spin S.
Solution:
The grand potential $\Phi_{\rm G}$ is obtained from eqn 30.1 as follows:

$$\begin{aligned}
\Phi_{\rm G} &= -k_{\rm B} T \ln \mathcal{Z} \\
&= \mp k_{\rm B} T (2S+1) \sum_{\bm{k}} \ln(1 \pm e^{-\beta(E_{\bm{k}}-\mu)}) \\
&= \mp k_{\rm B} T \int_0^\infty \ln(1 \pm e^{-\beta(E-\mu)}) \, g(E) \, {\rm d}E,
\end{aligned} \tag{30.3}$$

where $g(E)$ is the density of states (containing the spin degeneracy factor $2S+1$), which can be derived as follows. States in k-space are uniformly distributed, and so

$$g(k)\,\mathrm{d}k = \frac{4\pi k^2\,\mathrm{d}k}{(2\pi/L)^3} \times (2S+1) = \frac{(2S+1)Vk^2\,\mathrm{d}k}{2\pi^2}, \qquad (30.4)$$

where $V = L^3$ is the volume. Using $E = \hbar^2 k^2/2m$ we can transform this into

$$g(E)\,\mathrm{d}E = \frac{(2S+1)VE^{1/2}\,\mathrm{d}E}{(2\pi)^2}\left(\frac{2m}{\hbar^2}\right)^{3/2}, \qquad (30.5)$$

and hence

$$\Phi_\mathrm{G} = \mp k_\mathrm{B}T\frac{(2S+1)V}{(2\pi)^2}\left(\frac{2m}{\hbar^2}\right)^{3/2}\int_0^\infty \ln(1\pm e^{-\beta(E-\mu)})\,E^{1/2}\,\mathrm{d}E, \qquad (30.6)$$

which after integrating by parts yields

$$\Phi_\mathrm{G} = -\frac{2}{3}\frac{(2S+1)V}{(2\pi)^2}\left(\frac{2m}{\hbar^2}\right)^{3/2}\int_0^\infty \frac{E^{3/2}\,\mathrm{d}E}{e^{\beta(E-\mu)}\pm 1}. \qquad (30.7)$$

The grand potential evaluated in the previous example can be used to derive various thermodynamic functions for fermions and bosons.[3] Another way to get to the same result is to evaluate the mean occupation $n_{\boldsymbol{k}}$ of a state with wave vector $\boldsymbol{k}$, which is given by

[3] Note that in the derived expressions, the $\pm$ sign means $+$ for fermions and $-$ for bosons.

$$n_{\boldsymbol{k}} = k_\mathrm{B}T\frac{\partial}{\partial \mu}\ln \mathcal{Z}_{\boldsymbol{k}} = \frac{1}{e^{\beta(E_{\boldsymbol{k}}-\mu)}\pm 1}, \qquad (30.8)$$

and then use this expression to derive directly quantities such as

$$N = \sum_{\boldsymbol{k}} n_{\boldsymbol{k}} = \int_0^\infty \frac{g(E)\,\mathrm{d}E}{e^{\beta(E-\mu)}\pm 1}, \qquad (30.9)$$

and

$$U = \sum_{\boldsymbol{k}} n_{\boldsymbol{k}} E_{\boldsymbol{k}} = \int_0^\infty \frac{E\,g(E)\,\mathrm{d}E}{e^{\beta(E-\mu)}\pm 1}. \qquad (30.10)$$

For reasons which will become more clear below, we will write $e^{\beta\mu}$ as the **fugacity** z, i.e.,

$$\boxed{z = e^{\beta\mu}.} \qquad (30.11)$$

These give expressions for N and U as follows:

$$N = \left[\frac{(2S+1)V}{(2\pi)^2}\left(\frac{2m}{\hbar^2}\right)^{3/2}\right]\int_0^\infty \frac{E^{1/2}\,\mathrm{d}E}{z^{-1}e^{\beta E}\pm 1} \qquad (30.12)$$

and

$$U = \left[\frac{(2S+1)V}{(2\pi)^2}\left(\frac{2m}{\hbar^2}\right)^{3/2}\right]\int_0^\infty \frac{E^{3/2}\,\mathrm{d}E}{z^{-1}e^{\beta E}\pm 1}. \qquad (30.13)$$

One problem with all these types of formula, such as eqns 30.7, 30.12 and 30.13, is that to simplify them any further, you have to evaluate a difficult integral. Fortunately, we can show that these integrals are related to the **polylogarithm** function $\text{Li}_n(x)$ (see Appendix C.5), so that

$$\int_0^\infty \frac{E^{n-1}\,\mathrm{d}E}{z^{-1}\mathrm{e}^{\beta E} \pm 1} = (k_\mathrm{B}T)^n \Gamma(n)[\mp \text{Li}_n(\mp z)], \qquad (30.14)$$

where $\Gamma(n)$ is a gamma function. This result is proved in the appendix (eqn C.36). The crucial thing to realize is that $\text{Li}_n(z)$ is just a numerical function of z, i.e., of the temperature and the chemical potential. This integral then allows us to establish, after a small amount of algebra, that the number N of particles is given by

$$N = \frac{(2S+1)V}{\lambda_\mathrm{th}^3}[\mp \text{Li}_{3/2}(\mp z)], \qquad (30.15)$$

and the internal energy U is given by

$$\begin{aligned} U &= \frac{3}{2}k_\mathrm{B}T \frac{(2S+1)V}{\lambda_\mathrm{th}^3}[\mp \text{Li}_{5/2}(\mp z)] \\ &= \frac{3}{2}N k_\mathrm{B}T \frac{\text{Li}_{5/2}(\mp z)}{\text{Li}_{3/2}(\mp z)}. \end{aligned} \qquad (30.16)$$

We will use these equations in subsequent sections. Note also that we have from eqns 30.7 and 30.13 that

$$\Phi_\mathrm{G} = -\frac{2}{3}U. \qquad (30.17)$$

Example 30.2

Evaluate N, U and Φ_G (from eqns 30.15, 30.16 and 30.17) in the ideal-gas limit.

Solution:
We choose $z = \mathrm{e}^{\beta\mu} \ll 1$ (corresponding to $(N/V)\lambda_\mathrm{th}^3 \ll 1$) and use the fact that $\text{Li}_n(z) \approx z$ when $z \ll 1$. Hence elementary substitution yields

$$N \approx \frac{(2S+1)Vz}{\lambda_\mathrm{th}^3}, \qquad (30.18)$$

$$U \approx \frac{3}{2}Nk_\mathrm{B}T, \qquad (30.19)$$

$$\Phi_\mathrm{G} \approx -Nk_\mathrm{B}T. \qquad (30.20)$$

The equation for N shows that the number density of particles N/V is such that, on average, $2S+1$ particles (one for each spin state) occupy a volume λ_th^3/z. Since $z \ll 1$, this means that particle wave functions do not overlap. The other two equations are reassuringly familiar. The equation for U asserts that the energy per particle is the well-known equipartition result $\frac{3}{2}k_\mathrm{B}T$. The equation for Φ_G, together with $\Phi_\mathrm{G} = -pV$ (from eqn 22.49) yields the ideal gas law $pV = Nk_\mathrm{B}T$.

30.2 The Fermi gas

What we have done so far is to consider bosons and fermions on an equal footing. Let us now restrict our attention to a gas of fermions (known as a **Fermi gas**) and to get a feel for what is going on, let us also consider $T = 0$. Fermions will occupy the lowest-energy states, but we can only put one fermion in each state, and thus only $2S+1$ in each energy level. The fermions will fill up the energy levels until they get to an energy E_F, known as the **Fermi energy**, which is the energy of the highest occupied state at a temperature of absolute zero.[4] Thus we define

$$E_F = \mu(T=0). \tag{30.21}$$

This makes sense because $\mu(T=0) = \partial E/\partial N$, which gives $\mu(T=0) = E(N) - E(N-1) = E_F$. At absolute zero, we have that $\beta \to \infty$, and hence the occupation n_k is given by

$$n_k = \frac{1}{e^{\beta(E_k - \mu)} + 1} = \theta(\mu - E_k) = \theta(E_F - E_k), \tag{30.22}$$

where $\theta(x)$ is a Heaviside step function.[5] At absolute zero, therefore, the number of states is given by

$$N = \int_0^{k_F} g(\mathbf{k}) \, \mathrm{d}^3 k, \tag{30.23}$$

where k_F is the **Fermi wave vector**, defined by

$$E_F = \frac{\hbar^2 k_F^2}{2m}. \tag{30.24}$$

Hence the number of fermions N is given by

$$N = \frac{(2S+1)V}{2\pi^2} \frac{k_F^3}{3}, \tag{30.25}$$

so that writing $n = N/V$, we have

$$k_F = \left[\frac{6\pi^2 n}{2S+1} \right]^{1/3}, \tag{30.26}$$

and hence

$$E_F = \frac{\hbar^2}{2m} \left[\frac{6\pi^2 n}{2S+1} \right]^{2/3}. \tag{30.27}$$

[4]The chemical potential is sometimes known as the **Fermi level**, though this can be a misleading term as, for example in semiconductors, there may not be any states at the chemical potential (which lies somewhere in the energy gap) so in this case there would be "no actual occupied levels at the Fermi level."

[5]The Heaviside step function $\theta(x)$ is defined by

$$\theta(x) = \begin{cases} 0 & x < 0 \\ 1 & x > 0 \end{cases}$$

It is plotted in Fig. 30.1.

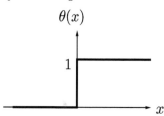

Fig. 30.1 The Heaviside step function.

Example 30.3

Evaluate k_F and E_F for spin-$\frac{1}{2}$ particles.
Solution:
When $S = \frac{1}{2}$, $2S+1 = 2$ and hence eqns 30.26 and 30.27 become

$$k_F = \left[3\pi^2 n \right]^{1/3}, \tag{30.28}$$

and

$$E_F = \frac{\hbar^2}{2m} \left[3\pi^2 n \right]^{2/3}. \tag{30.29}$$

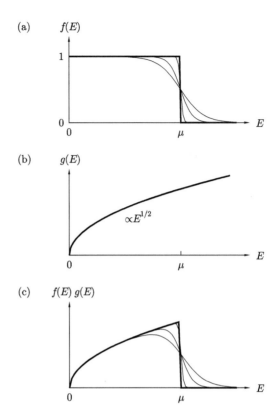

Fig. 30.2 (a) The Fermi function $f(E)$ defined by eqn 29.35. The thick line is for $T = 0$. The step function is smoothed out as the temperature is increased (shown as thinner lines). The temperatures shown are $T = 0$, $T = 0.01\mu/k_\text{B}$, $T = 0.05\mu/k_\text{B}$, and $T = 0.1\mu/k_\text{B}$. (b) The density of states $g(E)$ for a non-interacting fermion gas in three dimensions is proportional to $E^{1/2}$. (c) $f(E)g(E)$ for the same temperatures as in (a).

At $T = 0$, the distribution function $f(E)$ is a Heaviside step function, taking the value 1 for $E < \mu$ and 0 for $E > \mu$. This step is smoothed out as the temperature T increases, as shown in Fig. 30.2(a). The density of states $g(E)$ for a non-interacting fermion gas in three dimensions is proportional to $E^{1/2}$ (as shown in eqn 30.4) and this is plotted in Fig. 30.2(b). The product of $f(E)g(E)$ gives the actual number distribution of fermions, and this is shown in Fig. 30.2(c). The sharp cut-off you would expect at $T = 0$ is smoothed over an energy scale $k_\text{B}T$ around the chemical potential μ.

The electrons in a metal can be treated as a non-interacting gas of fermions. Using the number density n of electrons in a metal, one can calculate the Fermi energy using eqn 30.29, and some example results are shown in Table 30.1. The Fermi energies are all several eV; converting each number into a temperature, the so-called **Fermi temperature** $T_\text{F} = E_\text{F}/k_\text{B}$, yields values of several tens of thousands of kelvin. Thus the Fermi energy is a large energy scale, and hence for most metals the Fermi function is close to a step function, at pretty much all temperatures below their melting temperature. In this case, the electrons in a metal are said to be in the **degenerate limit**.

The pressure of these electrons is given (by using eqns 22.49 and 30.17) as

$$p = \frac{2U}{3V}, \tag{30.30}$$

Table 30.1 Properties of selected metals

	n (10^{28} m^{-3})	E_F (eV)	$\tfrac{2}{3}nE_F$ (10^9 N m^{-2})	B (10^9 N m^{-2})
Li	4.70	4.74	23.8	11.1
Na	2.65	3.24	9.2	6.3
K	1.40	2.12	3.2	3.1
Cu	8.47	7.00	63.3	137.8
Ag	5.86	5.49	34.3	103.6

as is appropriate for non-relativistic electrons (see Table 25.1). The mean energy of the electrons at $T=0$ is given by

$$\langle E \rangle = \frac{\int_0^{E_F} E g(E)\,dE}{\int_0^{E_F} g(E)\,dE}, \tag{30.31}$$

which with $g(E) \propto E^{1/2}$ gives $\langle E \rangle = \tfrac{3}{5} E_F$. Writing $U = N\langle E \rangle$, we have that the bulk modulus B is

$$B = -V \frac{\partial p}{\partial V} = \frac{10U}{9V} = \frac{2}{3} n E_F. \tag{30.32}$$

This expression is evaluated in Table 30.1 and gives results that are of the same order of magnitude as experimental values.

The next example computes an integral that is useful for considering analytically the effect of finite temperature.

Example 30.4

Evaluate the integral $I = \int_0^\infty \phi(E) f(E)\,dE$ as a power series in temperature.

Solution:
Consider the function $\psi(E) = \int_0^E \phi(E')\,dE'$, which is defined so that $\phi(E) = d\psi/dE$ and therefore

$$I = \int_0^\infty \frac{d\psi}{dE} f(E)\,dE = [\psi(E) f(E)]_0^\infty - \int_0^\infty \psi(E) \frac{df}{dE}\,dE$$

$$= -\int_0^\infty \psi(E) \frac{df}{dE}\,dE. \tag{30.33}$$

Now put $x = (E-\mu)/k_B T$ and hence

$$\frac{df}{dE} = -\frac{1}{k_B T} \frac{e^x}{(e^x+1)^2}. \tag{30.34}$$

Writing $\psi(E)$ as a power series in x as

$$\psi(E) = \sum_{s=0}^{\infty} \frac{x^s}{s!} \left(\frac{\mathrm{d}^s\psi}{\mathrm{d}x^s}\right)_{x=0}, \qquad (30.35)$$

we can express I as a power series of integrals as follows:

$$I = \sum_{s=0}^{\infty} \frac{1}{s!}\left(\frac{\mathrm{d}^s\psi}{\mathrm{d}x^s}\right)_{x=0} \int_{-\mu/k_\mathrm{B}T}^{\infty} \frac{x^s e^x\,\mathrm{d}x}{(e^x+1)^2}. \qquad (30.36)$$

[6] This approximation is valid when $k_\mathrm{B}T \ll E_\mathrm{F}$.

The integral is rewritten using
$$\frac{1}{1+z^2} = 1 - 2z + 3z^2 - 4z^3 + \cdots$$
$$= \sum_{n=0}^{\infty} (-1)^n (n+1) z^n$$

The sum is reordered using $n = m + 1$.

The integral part of this can be simplified by replacing[6] the lower limit with $-\infty$. It vanishes for odd s, but for even s

$$\int_{-\infty}^{\infty} \frac{x^s e^x\,\mathrm{d}x}{(e^x+1)^2} = 2\int_0^{\infty} \frac{x^s e^x\,\mathrm{d}x}{(e^x+1)^2} = 2\int_0^{\infty} \frac{x^s e^{-x}\,\mathrm{d}x}{(e^{-x}+1)^2}$$

$$= 2\int_0^{\infty} \mathrm{d}x\, x^s e^{-x} \sum_{m=0}^{\infty} (-1)^m (m+1) e^{-mx}$$

$$= 2\sum_{n=1}^{\infty} (-1)^{n+1} n \int_0^{\infty} x^s e^{-nx}\,\mathrm{d}x$$

$$= 2(s!)\sum_{n=1}^{\infty} \frac{(-1)^{n+1}}{n^s}$$

$$= 2(s!)(1 - 2^{1-s})\zeta(s), \qquad (30.37)$$

where $\zeta(s)$ is the Riemann zeta function.

Thus the integral is

$$I = \sum_{s=0,\,s\,\mathrm{even}}^{\infty} 2\left(\frac{\mathrm{d}^s\psi}{\mathrm{d}x^s}\right)_{x=0} (1-2^{1-s})\zeta(s)$$

$$= \psi(x=0) + \frac{\pi^2}{6}\left(\frac{\mathrm{d}^2\psi}{\mathrm{d}x^2}\right)_{x=0} + \frac{7\pi^4}{360}\left(\frac{\mathrm{d}^4\psi}{\mathrm{d}x^4}\right)_{x=0} + \cdots$$

$$= \int_0^{\mu} \phi(E)\,\mathrm{d}E + \frac{\pi^2}{6}(k_\mathrm{B}T)^2 \left(\frac{\mathrm{d}\phi}{\mathrm{d}E}\right)_{E=\mu} \qquad (30.38)$$

$$+ \frac{7\pi^4}{360}(k_\mathrm{B}T)^4 \left(\frac{\mathrm{d}^3\phi}{\mathrm{d}E^3}\right)_{E=\mu} + \cdots$$

This expression is known as the **Sommerfeld formula**.

Having derived the Sommerfeld formula, we can now evaluate N and U quite easily. Let us choose $S = \frac{1}{2}$, just to make the equations a little less cumbersome. Then

$$N = \frac{V}{2\pi^2}\left(\frac{2m}{\hbar^2}\right)^{3/2} \int_0^{\infty} E^{1/2} f(E)\,\mathrm{d}E$$

$$= \frac{V}{3\pi^2}\left(\frac{2m}{\hbar^2}\right)^{3/2} \mu^{3/2} \left[1 + \frac{\pi^2}{8}\left(\frac{k_\mathrm{B}T}{\mu}\right)^2 + \cdots\right], \qquad (30.39)$$

which implies that

$$\mu(T) = \mu(0)\left[1 - \frac{\pi^2}{12}\left(\frac{k_B T}{\mu(0)}\right)^2 + \cdots\right]. \tag{30.40}$$

In fact, equating E_F and μ is good to 0.01% for typical metals even at room temperature, although it is worthwhile keeping in the back of one's mind that the two quantities are not the same.

We can also compute the heat capacity of electrons in a metal by a similar technique, as shown in the following example.

Example 30.5

Compute the heat capacity of non-interacting free electrons in a three-dimensional metal.
Solution:

$$\begin{aligned}
U &= \frac{V}{2\pi^2}\left(\frac{2m}{\hbar^2}\right)^{3/2}\int_0^\infty E^{3/2} f(E)\,dE \\
&= \frac{V}{5\pi^2}\left(\frac{2m}{\hbar^2}\right)^{3/2}\mu(T)^{5/2}\left[1 + \frac{5\pi^2}{8}\left(\frac{k_B T}{\mu(0)}\right)^2 + \cdots\right] \\
&= \frac{3}{5}N\mu(T)\left[1 + \frac{\pi^2}{2}\left(\frac{k_B T}{\mu(0)}\right)^2 + \cdots\right] \\
&= \frac{3}{5}N\mu(0)\left[1 + \frac{5\pi^2}{12}\left(\frac{k_B T}{\mu(0)}\right)^2 + \cdots\right] \tag{30.41}
\end{aligned}$$

and hence

$$C_V = \frac{3}{2}Nk_B\left(\frac{\pi^2}{3}\frac{k_B T}{\mu(0)}\right) + O(T^3). \tag{30.42}$$

Thus the contribution to the heat capacity from electrons is linear in temperature (recall from Chapter 24 that the heat capacity from lattice vibrations (phonons) is proportional to T^3 at low temperature) and will therefore dominate the heat capacity of a metal at very low temperatures.

The **Fermi surface** is the set of points in k-space whose energy is equal to the chemical potential. If the chemical potential lies in a gap[7] between energy bands, then the material is a semiconductor or an insulator and there will be no Fermi surface. Thus a metal is a material with a Fermi surface.

[7]The periodic potential that exists in crystalline metals can lead to the formation of energy gaps, i.e., intervals in energy in which there are no allowed states.

30.3 The Bose gas

For the **Bose gas** (a gas composed of bosons), we can use our expressions for N and U in eqns 30.15 and 30.16 to give

$$N = \frac{(2S+1)V}{\lambda_{\text{th}}^3} \text{Li}_{3/2}(z) \tag{30.43}$$

and

$$U = \frac{3}{2} N k_B T \frac{\text{Li}_{5/2}(z)}{\text{Li}_{3/2}(z)}. \tag{30.44}$$

Example 30.6

Evaluate eqns 30.43 and 30.44 for the case $\mu = 0$.
Solution:
If $\mu = 0$ then $z = 1$. Now $\text{Li}_n(1) = \zeta(n)$ where $\zeta(n)$ is the Riemann zeta function. Therefore

$$N = \frac{(2S+1)V}{\lambda_{\text{th}}^3} \zeta\left(\frac{3}{2}\right) \tag{30.45}$$

and

$$U = \frac{3}{2} N k_B T \frac{\zeta(\frac{5}{2})}{\zeta(\frac{3}{2})}. \tag{30.46}$$

The numerical values are $\zeta(\frac{3}{2}) = 2.612$, $\zeta(\frac{5}{2}) = 1.341$, and hence we have that $\zeta(\frac{5}{2})/\zeta(\frac{3}{2}) = 0.513$.

Note that these results will not apply to photons because we have assumed at the beginning that $E = \hbar^2 k^2/2m$, whereas for a photon $E = \hbar k c$. This is worked through in the following example.

Example 30.7

Rederive the equation for U for a gas of photons using the formalism of this chapter.
Solution:
The density of states is $g(k)\,dk = (2S+1)Vk^2\,dk/(2\pi^2)$. A photon has a spin of 1, but the 0 state is not allowed, so the spin-degeneracy factor $(2S+1)$ is in this case only 2. Using $E = \hbar k c$ we arrive at

$$g(E)\,dE = \frac{V}{\pi^2 \hbar^3 c^3} E^2\,dE, \tag{30.47}$$

and hence

$$U = \int_0^\infty \frac{E\,g(E)\,dE}{z^{-1}e^{\beta E} - 1} = \frac{V}{\pi^2 \hbar^3 c^3} \int_0^\infty \frac{E^3\,dE}{z^{-1}e^{\beta E} - 1}, \tag{30.48}$$

and using

$$\int_0^\infty \frac{E^3 \, dE}{z^{-1}e^{\beta E} - 1} = (k_B T)^4 \Gamma(4) \text{Li}_4(z), \quad (30.49)$$

and recognizing that $z = 1$ because $\mu = 0$ and hence $\text{Li}_4(z) = \zeta(4) = \pi^4/90$, and using $\Gamma(4) = 3! = 6$, we have that

$$U = \frac{V\pi^2}{15\hbar^3 c^3}(k_B T)^4, \quad (30.50)$$

which agrees with eqn 23.37.

For Bose systems with a dispersion relation like $E = \hbar^2 k^2/2m$ (i.e., for a gapless dispersion, where the lowest-energy level, corresponding to $k = 0$ or infinite wavelength, is at zero energy), the chemical potential has to be negative. If it were not, the level at $E = 0$ would have infinite occupation. Thus $\mu < 0$, and hence the fugacity $z = e^{\beta \mu}$ must lie in the range $0 < z < 1$. But what value will the chemical potential take?

Equation 30.43 can be rearranged to give

$$\frac{n\lambda_{\text{th}}^3}{2S+1} = \text{Li}_{3/2}(z), \quad (30.51)$$

and here we hit an uncomfortable problem. The left-hand side can be increased if $n = N/V$ increases or if T decreases (because $\lambda_{\text{th}} \propto T^{-1/2}$). We can plug numbers for n and T into the left-hand side and then read off a value for z from the graph in Fig. 30.3, which shows the behaviour of the function $\text{Li}_{3/2}(z)$ (and also $\text{Li}_{5/2}(z)$). As we raise n or decrease T, we make the left-hand side of eqn 30.51 bigger and hence z bigger, so that μ becomes less negative, approaching 0 from below. However, if

$$\frac{n\lambda_{\text{th}}^3}{2S+1} > \zeta(\tfrac{3}{2}) = 2.612, \quad (30.52)$$

there is no solution to eqn 30.51. What has happened?

30.4 Bose–Einstein condensation (BEC)

The solution to the conundrum raised at the end of the previous section is remarkably subtle, but has far-reaching consequences. As the chemical potential has become closer and closer to zero energy, approaching this from below, the lowest energy level has become macroscopically occupied. The reason our mathematics has broken down is that our usual, normally perfectly reasonable, approximation in going from a sum to an integral in evaluating our grand partition function is no longer valid.

In fact, we can see when this fails using a rearranged[8] version of eqn 30.52. Failure occurs when we fall below a temperature T_c given by

[8] Recall that $\lambda_{\text{th}} = h/\sqrt{2\pi m k_B T}$.

$$k_B T_c = \frac{2\pi \hbar^2}{m} \left(\frac{n}{2.612(2S+1)} \right)^{2/3}. \quad (30.53)$$

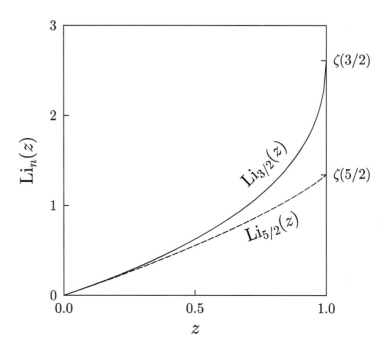

Fig. 30.3 The functions $\text{Li}_{3/2}(z)$ and $\text{Li}_{5/2}(z)$. For $z \ll 1$ (the classical regime), $\text{Li}_n(z) \approx z$. Also, $\text{Li}_n(1) = \zeta(n)$.

We can perform a corrected analysis of the problem as follows. We separate N into two terms:

$$N = N_0 + N_1, \tag{30.54}$$

where N_0 is the expected number of particles in the ground state,[9]

$$N_0 = \frac{1}{z^{-1} - 1} = \frac{1}{e^{-\beta\mu} - 1}, \tag{30.55}$$

[9] We will set the zero of energy to be the energy of the ground state.

(see Fig. 30.4) and N_1 is our original integral representing the expected number of bosons in all the other states. Thus above T_c, $z < 1$ and there are N_0 bosons in the ground state (see Fig. 30.4), a number much less than the total number N of bosons, which is

$$N = N_1 = \frac{(2S+1)V}{[\lambda_{\text{th}}(T)]^3} \text{Li}_{3/2}(z). \tag{30.56}$$

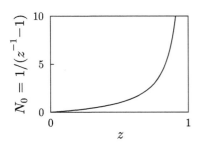

Fig. 30.4 The expected number of particles in the ground state as a function of the fugacity z according to eqn 30.55. Until z gets close to 1, $N_0 \ll N$.

In particular, at T_c, we can write the total number density n of bosons as

$$n \equiv \frac{N}{V} = \frac{(2S+1)\text{Li}_{3/2}(1)}{[\lambda_{\text{th}}(T_c)]^3} = \frac{(2S+1)\zeta(\tfrac{3}{2})}{[\lambda_{\text{th}}(T_c)]^3}. \tag{30.57}$$

Below T_c, z is extremely close to unity and so we will set $z = 1$. Later we will critically examine how good an approximation this actually is. Under this assumption n_1, the concentration of particles in the excited state is given by

$$n_1 \equiv \frac{N_1}{V} = \frac{(2S+1)\text{Li}_{3/2}(1)}{[\lambda_{\text{th}}(T)]^3} = \frac{(2S+1)\zeta(\tfrac{3}{2})}{[\lambda_{\text{th}}(T)]^3}. \quad (30.58)$$

Any remaining particles must be in the ground state, so that[10]

$$\frac{n_0}{n} = \frac{n - n_1}{n} = 1 - \left(\frac{T}{T_c}\right)^{3/2}. \quad (30.59)$$

This function is plotted in Fig. 30.5 and shows how the number of particles in the ground state grows as the temperature is lowered below T_c. At $T = 0$ *all* the bosons are in the ground state. For $0 < T < T_c$ the ground state has an appreciable fraction of the total number of bosons.[11] This macroscopic occupation of the ground state is known as **Bose–Einstein condensation** (often abbreviated to BEC). Note that this transition is not driven by interactions between particles (as we had for the liquid–gas transition); we have so far only considered non-interacting particles; the transition is driven purely by the requirements of exchange symmetry on the quantum statistics of the bosons.

The term "condensation" often implies a condensation in space, as when liquid water condenses on a cold window in a steamy bathroom. However, for Bose–Einstein condensation it is a condensation in k-space, with a macroscopic occupation of the lowest energy state occurring below T_c.

[10] Remember $\lambda_{\text{th}} \propto T^{-1/2}$, and so

$$\frac{n_1}{n} = \frac{[\lambda_{\text{th}}(T_c)]^3}{[\lambda_{\text{th}}(T)]^3} = \left(\frac{T}{T_c}\right)^{3/2}.$$

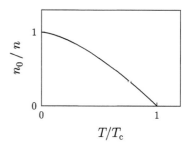

Fig. 30.5 The number of particles in the ground state as a function of temperature, after eqn 30.59.

[11] Note that using eqn 30.55, we can write $z^{-1} = 1 + N_0^{-1}$. Thus below T_c, $z \approx 1 - \frac{1}{N_0}$ and is therefore different from 1 by an amount which is about one over the number of particles in the system. Thus our earlier approximation of $z = 1$ below T_c is an extremely good one for reasonably large systems.

Example 30.8

Find the internal energy $U(T)$ at temperature T for the Bose gas.
Solution:
The internal energy of the system only depends on the excited states since, by assumption, the ground state (which is macroscopically occupied) has zero energy. Since $z = 1$ for $T \leq T_c$, we have that

$$\begin{aligned} U &= \frac{3}{2} N_1 k_B T \frac{\zeta(\tfrac{5}{2})}{\zeta(\tfrac{3}{2})} \\ &= \frac{3}{2} N k_B T \frac{\zeta(\tfrac{5}{2})}{\zeta(\tfrac{3}{2})} \left(\frac{T}{T_c}\right)^{3/2} \\ &= 0.77 N k_B T_c \left(\frac{T}{T_c}\right)^{5/2}. \quad (30.60) \end{aligned}$$

For $T > T_c$ we have (from eqn 30.46)

$$U = \frac{3}{2} N k_B T \frac{\text{Li}_{5/2}(z)}{\text{Li}_{3/2}(z)}. \quad (30.61)$$

This example gives the high-temperature internal energy as a function of the fugacity, but z is temperature dependent. For a system with a fixed number N of bosons, we can extract z using the equation $N/V = (2S+1)\mathrm{Li}_{3/2}(z)/\lambda_{\mathrm{th}}^3$ and equating this with eqn 30.57 yields

$$\frac{T}{T_\mathrm{c}} = \left[\frac{\zeta(\tfrac{3}{2})}{\mathrm{Li}_{3/2}(z)}\right]^{2/3}, \qquad (30.62)$$

which, although it cannot be straightforwardly inverted to make z the subject, does show how z is related to T above T_c. (As we have already noted, below T_c the fugacity z is extremely close to unity.)

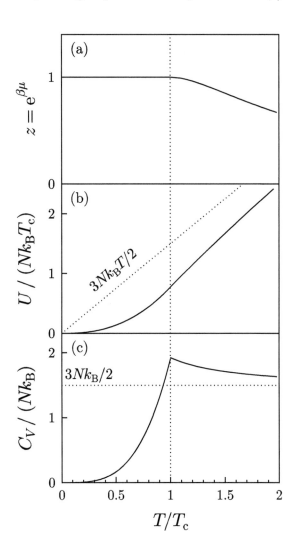

Fig. 30.6 The (a) fugacity, (b) internal energy, and (c) heat capacity for a system of bosons as a function of temperature.

The fugacity z, internal energy U, and heat capacity C_V, calculated for non-interacting bosons, are plotted in Fig. 30.6. The fugacity is obtained by numerical inversion of eqn 30.62; it rises up towards unity as you cool, and below T_c is not actually one but very close to it. The internal energy U in Fig. 30.6(b) is obtained from eqn 30.61, while the heat capacity C_V is plotted in Fig. 30.6(c) from eqn 30.65, to be proved in the exercises at the end of this chapter.

The Indian physicist S. N. Bose wrote to Einstein in 1924 describing his work on the statistical mechanics of photons. Einstein appreciated the significance of this work and used Bose's approach to predict what is now called Bose–Einstein condensation.

In the late 1930's, it was discovered that liquid ^{4}He becomes a **superfluid** when cooled below about 2.2 K. Superfluidity is a quantum-mechanical state of matter with very unusual properties, such as the ability to flow through very small capillaries with no measurable viscosity. Speculation arose as to whether this state of matter was connected with Bose–Einstein condensation, for the reason outlined in the following example.

Example 30.9

Estimate the Bose–Einstein condensation temperature for liquid ^{4}He, given that $m_{He} \approx 4m_p$ and that the density $\rho \approx 145 \, \text{kg m}^{-3}$.
Solution:
Using $n = \rho/m$, eqn 30.53 yields $T_c \approx 3.1 \, \text{K}$, which is remarkably close to the experimental value of the superfluid transition temperature.

Despite the agreement between this estimate and the experimental value, things are a bit more complicated. The particle density of ^{4}He is very high and interactions between helium atoms cannot be ignored; ^{4}He is a strongly interacting Bose gas, and therefore the predictions of the theory outlined in this chapter have to be modified.

A more suitable example of Bose–Einstein condensation is provided by the very dilute gases of alkali metal atoms[12] that can be prepared inside magnetic ion traps. The atoms, usually about 10^4–10^6 of them, can be trapped and cooled using the newly developed techniques of laser cooling. These alkali atoms have a single electronic spin due to their one valence electron and this can couple with the non-zero nuclear spin. Each atom therefore has a magnetic moment and thus can be trapped inside local minima of magnetic field. The densities of these **ultracold atomic gases** inside the traps are very low, more than seven orders of magnitude lower than that in ^{4}He, though their masses are higher. The Bose–Einstein condensation temperature is therefore also very low, typically 10^{-8}–10^{-6} K, but these temperatures can be reached using laser cooling. The low density precludes significant three-body collisions (in which two atoms bind with the third, taking away the excess kinetic energy, thus causing clustering), but two-body collisions do occur, which allow the cloud of atoms to thermalize. Example data are shown in Fig. 30.7 from one such experiment, which clearly shows that below a critical temperature Bose–Einstein condensation is taking place.[13]

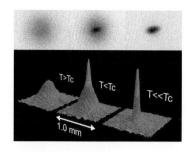

Fig. 30.7 Observation of Bose–Einstein condensation by absorption imaging. The data are shown as shadow pictures (upper panel) and as a three-dimensional plot (lower panel); the blackness of the shadow in the upper panel is here represented by height in the lower panel. These pictures measure the slow expansion of the trapped atoms observed after a 0.006 s time of flight, and thus measure the momentum distribution inside the cloud. The left-hand picture shows an expanding cloud cooled to just above the transition point. In the right-hand picture we see the velocity distribution well below T_c where almost all the atoms are condensed into the zero-velocity peak. (Image courtesy of W. Ketterle.)

[12] Alkali atoms are in Group I of the periodic table and include Li, Na, K, Rb, and Cs.

[13] The 2001 Nobel Prize was awarded to Eric Cornell and Carl Wieman (who did the experiment with rubidium atoms) and to Wolfgang Ketterle (who did it with sodium atoms).

Superfluidity is also found in these ultracold atomic gases; it turns out that the very weak interactions that exist between the alkali atoms are important for this to occur (a non-interacting Bose gas does not show superfluidity). Other experiments have explored the intriguing consequences of **macroscopic quantum coherence**, a term which means that in the condensed state all the atoms exist in a coherent quantum superposition.

Electrons do not exhibit Bose–Einstein condensation because they are fermions, not bosons, but they can show other condensation effects such as **superconductivity**. In a superconductor, a weak attractive interaction (which can be mediated by phonons) allows pairs of electrons to form **Cooper pairs**. A Cooper pair is a boson, and the Cooper pairs themselves can form a coherent state below the superconducting transition temperature. Many common superconductors can be described in this way using the **BCS theory of superconductivity**,[14] though many newly discovered superconductors, such as the **high-temperature superconductors**, which are ceramics, do not seem to be described by this model.

[14] BCS is named after its discoverers, John Bardeen, Leon Cooper, and Robert Schrieffer.

Chapter summary

- Non-interacting fermions and bosons can be described using

$$N = \frac{(2S+1)V}{\lambda_{th}^3}[\mp \text{Li}_{3/2}(\mp z)],$$

$$U = \frac{3}{2}Nk_BT\frac{\text{Li}_{5/2}(\mp z)}{\text{Li}_{3/2}(\mp z)},$$

$$\Phi_G = -\frac{2}{3}U.$$

- In a Fermi gas (a gas of fermions), fermions fill states up to E_F at absolute zero. The Pauli exclusion principle ensures that fermions only singly occupy states.

- The results for a Fermi gas can be applied to the electrons in a metal. At non-zero temperature, electrons with energies within k_BT of E_F are important in determining the properties.

- In a Bose gas, Bose–Einstein condensation can occur below a temperature given by

$$k_B T_c = \frac{2\pi\hbar^2}{m}\left(\frac{n}{2.612(2S+1)}\right)^{2/3}.$$

Bose–Einstein condensation is a macroscopic occupation of the ground state.

- The results for a Bose gas can be applied to liquid ^{4}He and dilute ultracold atomic gases.

Further reading

For further information, see Ashcroft and Mermin (1976), Annett (2004), Foot (2004), Ketterle (2002), and Pethick and Smith (2002).

Exercises

(30.1) Show that in the classical limit, when the fugacity $z = e^{\beta\mu} \ll 1$, z is the ratio of the thermal volume to the volume per particle of a single-spin excitation.

(30.2) Show that the pressure p exerted by a Fermi gas at absolute zero is

$$p = \frac{2}{5}nE_F, \qquad (30.63)$$

where n is the number density of particles.

(30.3) Show that for a gas of fermions with density of states $g(E)$, the chemical potential is given by

$$\mu(T) = E_F - \frac{\pi^2}{6}(k_B T)^2 \frac{g'(E_F)}{g(E_F)} + \ldots \qquad (30.64)$$

(30.4) Show that the heat capacity of a system of non-interacting bosons is given by

$$C_V = \frac{15}{4}\frac{\zeta(\frac{5}{2})}{\zeta(\frac{3}{2})} N k_B \left(\frac{T}{T_c}\right)^{3/2}, \quad T < T_c,$$

$$C_V = \frac{3}{2} N k_B \left(\frac{5}{2}\frac{\mathrm{Li}_{5/2}(z)}{\mathrm{Li}_{3/2}(z)} - \frac{3}{2}\frac{\mathrm{Li}_{3/2}(z)}{\mathrm{Li}_{1/2}(z)}\right),$$

$$T > T_c. \qquad (30.65)$$

(30.5) Show that Bose–Einstein condensation does not occur in two dimensions.

(30.6) In Bose–Einstein condensation, the ground state becomes macroscopically occupied. What about the first excited state, which might be only a small energy above the ground state; is it also macroscopically occupied?

Part IX

Special topics

In this final part, we apply some of the material presented earlier in this book to some specialized topics. This part is structured as follows:

- In Chapter 31 we describe *sound waves* and show that these are adiabatic. We derive an expression for the speed of sound in a fluid.
- A particular type of sound wave is the *shock wave*, and we consider such waves in Chapter 32. We define the *Mach number* and derive the *Rankine–Hugoniot* conditions, which allow us to consider the changes in density and pressure at a shock front.
- In Chapter 33, we examine how *fluctuations* can be studied in thermodynamics and lead to effects such as *Brownian motion*. We consider the *linear response* of a system to a generalized force and derive the *fluctuation–dissipation theorem*.
- In Chapter 34, we discuss *non-equilibrium thermodynamics* and show how fluctuations lead to the *Onsager reciprocal relations*, which connect certain *kinetic coefficients*. We apply these ideas to *thermoelectric* phenomena and briefly discuss *time-reversal symmetry*.
- In Chapter 35, we consider the physics of *stars* and study how gravity, nuclear reactions, convection, and conduction all lead to the observed properties of stellar material.
- In Chapter 36 we discuss what happens to stars when they run out of fuel, and consider the properties of *white dwarfs*, *neutron stars*, and *black holes*.
- In Chapter 37, we apply thermal physics to the *atmosphere*, attempting to understand how solar energy keeps the Earth at a certain temperature, the rôle played by the *greenhouse effect* and how humans may be causing *climate change*.

31 Sound waves

- 31.1 Sound waves under isothermal conditions   377
- 31.2 Sound waves under adiabatic conditions   377
- 31.3 Are sound waves in general adiabatic or isothermal?   378
- 31.4 Derivation of the speed of sound within fluids   379
- Chapter summary   381
- Further reading   382
- Exercises   382

Sound waves can be propagated in various fluids, such as liquids or gases, and consist of oscillations in the local pressure and density of the fluid. They are **longitudinal waves** (in which the displacement of molecules from their equilibrium positions is in the same direction as the wave motion) and can be described by alternating regions of compression and rarefaction (see Fig. 31.1). The speed at which sound travels through a material is therefore related to the material's compressibility (measured by its bulk modulus, see below) as well as to its inertia (represented by its density). In this chapter, we will show that the speed of sound v_s is given by

$$v_s = \sqrt{\frac{B}{\rho}}, \tag{31.1}$$

where v_s is the speed of sound and B is the **bulk modulus** of the material. The bulk modulus describes how much the volume of the fluid will change with changing pressure, so it is defined as the pressure increment dP divided by the fractional volume increment dV/V; since a pressure increase usually results in a volume decrease the definition is therefore

$$B = -V\frac{\partial p}{\partial V}, \tag{31.2}$$

to ensure that $B > 0$. It is also helpful to write the bulk modulus in terms of density rather than volume. Density ρ and volume V are related, for a fixed mass of material M,

$$\rho = \frac{M}{V}, \tag{31.3}$$

which means that fractional changes in density and in pressure are related by[1]

$$B = \rho\frac{\partial p}{\partial \rho}. \tag{31.4}$$

Later in this chapter we will see how to derive the equation for the speed of sound, which is quoted in eqn 31.1, but first we are going to see how it works for two different possible constraints: adiabatic and isothermal. These constraints determine the way in which we evaluate the partial differential in eqn 31.4.

Fig. 31.1 A sound wave in a fluid is a longitudinal wave consisting of compressions and rarefactions.

[1]Notice that
$$d\rho = -M\,dV/V^2 = -\rho\,dV/V.$$

31.1 Sound waves under isothermal conditions

We first begin by supposing that sound waves propagate under isothermal conditions. Simple differentiation of the ideal gas equation (eqn 6.20) at constant temperature gives that[2]

$$B_T = -V\left(\frac{\partial p}{\partial V}\right)_T = p, \qquad (31.5)$$

where the subscript T indicates that it is the temperature that is held constant (isothermal conditions).

Thus, using eqn 31.1, and then substituting in eqn 6.15 and writing the density as $\rho = nm$, we may write

$$v_s = \sqrt{\frac{B_T}{\rho}} = \sqrt{\frac{p}{\rho}} = \sqrt{\frac{\frac{1}{3}nm\langle v^2\rangle}{\rho}} = \sqrt{\frac{\langle v^2\rangle}{3}}. \qquad (31.6)$$

This implies that we can write

$$v_s = \sqrt{\langle v_x^2\rangle}, \qquad (31.7)$$

where v_x is as defined in eqn 5.15. This implies that the sound speed is very similar to the mean molecular speed in a given direction and is consistent with molecular interactions being the mediator of bulk sound waves.

[2] For an ideal gas at constant temperature, pV is a constant and hence $p \propto V^{-1}$. This implies that

$$\mathrm{d}p/p = -\mathrm{d}V/V,$$

and hence

$$-V\left(\frac{\partial p}{\partial V}\right)_T = p.$$

31.2 Sound waves under adiabatic conditions

A gas under adiabatic conditions obeys eqn 12.16 (pV^γ is constant) and hence $p \propto V^{-\gamma}$ so that

$$\frac{\mathrm{d}p}{p} = -\gamma \frac{\mathrm{d}V}{V}, \qquad (31.8)$$

and hence the adiabatic[3] bulk modulus B_S is

$$B_S = -V\left(\frac{\partial p}{\partial V}\right)_S = \gamma p. \qquad (31.9)$$

[3] The subscript S is because the entropy S is constant in an adiabatic process.

Hence the equation for sound speed under these conditions then becomes

$$v_s = \sqrt{\frac{\gamma p}{\rho}} = \sqrt{\frac{\gamma\langle v^2\rangle}{3}}. \qquad (31.10)$$

Comparison of the sound speed under isothermal and adiabatic conditions (i.e., eqns 31.6 and 31.10) tell us that the speed under adiabatic conditions is $\gamma^{1/2}$ times faster than it would be under isothermal conditions.

Example 31.1

What is the temperature dependence of the speed of sound assuming adiabatic conditions?

Solution:

The relationship between the sound speed v_s and the mean square speed of molecules in air $\langle v^2 \rangle$ given in eqn 31.10 enables us to relate the sound speed in air to its temperature. Using $\langle v^2 \rangle = 3k_\mathrm{B}T/m$, we have that

$$v_\mathrm{s} = \sqrt{\frac{\gamma \langle v^2 \rangle}{3}} = \sqrt{\frac{\gamma k_\mathrm{B} T}{m}}. \tag{31.11}$$

This shows that the speed of sound is a function of temperature and mass alone.[4] It is unsurprising that the speed of sound, i.e., the speed at which a pressure disturbance can be propagated, follows the same temperature dependence as the mean molecular speed since the molecular collision rates that govern the propagation of disturbances are proportional to the mean molecular speed.

[4] Note that γ can be weakly temperature dependent.

31.3 Are sound waves in general adiabatic or isothermal?

Because of the ideal gas law, one would expect that at the compressions in a sound wave the temperature rises, while at the rarefactions there is cooling. If there were sufficient time for thermal equilibration to take place as the sound wave passes (i.e., as the compressions and rarefactions reverse positions) then the wave would be isothermal. However, if there is insufficient time, then the wave is said to be adiabatic since there is no time for heat to flow.

To establish whether sound waves are usually likely to be adiabatic or isothermal, we are going to consider how far thermal changes can propagate in comparison with the length scale of a sound wave. The latter is given by the wavelength[5] λ of the sound wave, which is related to the angular frequency ω in a medium with sound speed v_s, by

$$\lambda = \frac{2\pi v_\mathrm{s}}{\omega}. \tag{31.12}$$

[5] Do not confuse λ as wavelength with λ as mean free path. The context should indicate which is meant.

The distance over which a thermal wave can propagate is the skin depth δ which we met in eqn 10.22. Thus the characteristic depth to which heat diffuses in a certain time T (using $T = 2\pi/\omega$ for the "thermal wave" which is driven at frequency ω) is given by

$$\delta^2 = \frac{2D}{\omega} = \frac{DT}{\pi}. \tag{31.13}$$

The frequency dependence of these two length scales, the wavelength of the sound wave and the skin depth or propagation distance of the heat wave driven at the same frequency, is shown in Fig. 31.2. In different frequency ranges, either λ or δ will be larger because they have a different frequency dependence ($\lambda \propto \omega^{-1}$ and $\delta \propto \omega^{-1/2}$). In the high-frequency regime, for which $\lambda < \delta$, the heat wave propagates over many wavelengths so any sound waves would be isothermal. In the low-frequency regime, for which $\lambda > \delta$, the sound waves would be adiabatic.

In fact, it turns out that the latter situation is usually satisfied in practice and sound waves are *adiabatic*. You can demonstrate this by substituting typical values for D and ω into eqn 31.13 to estimate δ and show that the wavelength of a sound wave will exceed the skin depth. In fact, for these typical values of δ, the wavelengths required to be in the isothermal regime are so tiny that they are smaller than the mean free path of the molecules in the gas (see Exercise 31.3).

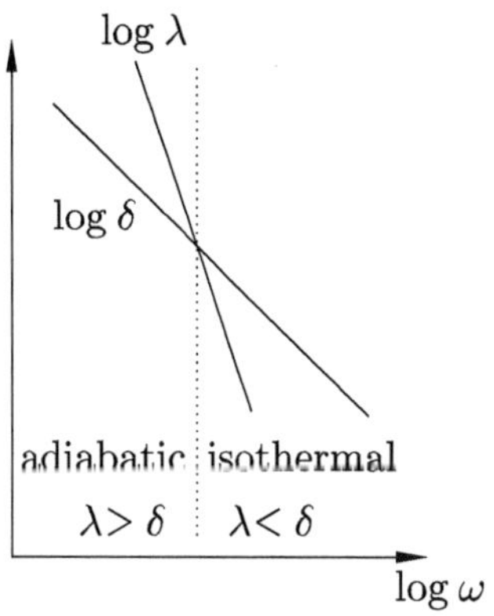

Fig. 31.2 Propagation distance of a sound wave and of a thermal wave as a function of frequency. In the region where $\lambda < \delta$ the sound waves would be isothermal and in the region where $\lambda > \delta$ the sound waves would be adiabatic.

Example 31.2

What is the speed of sound in a relativistic gas?
Solution:
For a non-relativistic gas we have from eqn 6.15 that $p = \frac{1}{3} nm \langle v^2 \rangle$. Using $\rho = nm$, we can write this as $p = \frac{1}{3} \rho \langle v^2 \rangle$. For a relativistic gas, this should be replaced by

$$p = \frac{1}{3} \rho c^2, \tag{31.14}$$

where c is the speed of light. Since $\rho \propto 1/V$, we have that $B = p$ and hence

$$v_s = \sqrt{B/\rho} = \frac{c}{\sqrt{3}}. \tag{31.15}$$

31.4 Derivation of the speed of sound within fluids

The speed of sound formula in eqn 31.1 can be derived by combining two equations, the continuity equation and the Euler equation (see boxes on page 380), to give a wave equation whose speed can be clearly identified. These equations are fully three dimensional, and a derivation for three dimensions is straightforward. However, fluids such as air cannot transmit shear and so no transverse waves can be propagated, only longitudinal waves. For this reason, we will just present the one-dimensional derivation appropriate for longitudinal waves; this is illustrative and perfectly analogous to the three-dimensional version.

The continuity equation in one dimension (see box on page 380) is given by

$$\frac{\partial(\rho u)}{\partial x} = -\frac{\partial \rho}{\partial t}. \tag{31.16}$$

The continuity equation

The continuity equation for a fluid (that is, for a liquid or a gas) can be derived in a similar manner to the diffusion equation, eqn 9.35. The mass flux out of a closed surface S is

$$\int_S \rho \boldsymbol{u} \cdot \mathrm{d}\boldsymbol{S}, \tag{31.17}$$

where ρ is the density and $\boldsymbol{u}$ is the local fluid velocity. This flux must be balanced by the rate of decrease of fluid concentration inside the volume:

$$\int_S \rho \boldsymbol{u} \cdot \mathrm{d}\boldsymbol{S} = -\frac{\partial}{\partial t} \int_V \rho \mathrm{d}V. \tag{31.18}$$

The divergence theorem then implies that

$$\int_V \nabla \cdot (\rho \boldsymbol{u}) \mathrm{d}V = -\int_V \frac{\partial \rho}{\partial t} \mathrm{d}V \tag{31.19}$$

and hence

$$\nabla \cdot (\rho \boldsymbol{u}) = -\frac{\partial \rho}{\partial t}, \tag{31.20}$$

or in one dimension that

$$\frac{\partial (\rho u)}{\partial x} = -\frac{\partial \rho}{\partial t}. \tag{31.21}$$

The Euler equation

The force per unit mass on an element of fluid owing to a pressure gradient ∇p is $-(1/\rho)\nabla p$. This leads to the Euler equation:

$$-\frac{1}{\rho}\nabla p = \frac{\mathrm{D}\boldsymbol{u}}{\mathrm{D}t}, \tag{31.22}$$

where $\mathrm{D}\boldsymbol{u}/\mathrm{D}t$ is the local acceleration of the fluid, described in the co-moving frame of the fluid via the convective derivative

$$\frac{\mathrm{D}\boldsymbol{X}}{\mathrm{D}t} \equiv \frac{\partial \boldsymbol{X}}{\partial t} + (\boldsymbol{u}\cdot\nabla)\boldsymbol{X}. \tag{31.23}$$

Here, $\mathrm{D}\boldsymbol{X}/\mathrm{D}t$ is the rate of change of property $\boldsymbol{X}$ with time following the fluid. Thus, eqn 31.22 becomes

$$-\frac{1}{\rho}\nabla p = \frac{\partial \boldsymbol{u}}{\partial t} + (\boldsymbol{u}\cdot\nabla)\boldsymbol{u}, \tag{31.24}$$

or in one dimension

$$-\frac{1}{\rho}\frac{\partial p}{\partial x} = \frac{\partial u}{\partial t} + u\frac{\partial u}{\partial x}. \tag{31.25}$$

Euler's equation for a fluid in one dimension (see box on page 380) is
$$-\frac{1}{\rho}\frac{\partial p}{\partial x} = \frac{\partial u}{\partial t} + u\frac{\partial u}{\partial x}. \tag{31.26}$$

Equation 31.16 may be expanded as
$$\frac{\partial(\rho u)}{\partial x} = u\frac{\partial \rho}{\partial x} + \rho\frac{\partial u}{\partial x} = -\frac{\partial \rho}{\partial t}. \tag{31.27}$$

Dividing through by ρ and writing $\mathrm{d}s = \mathrm{d}\rho/\rho$ yields
$$u\frac{\partial s}{\partial x} + \frac{\partial u}{\partial x} = -\frac{\partial s}{\partial t}. \tag{31.28}$$

For small-amplitude sound waves, any terms that are second order in u, such as $u\partial s/\partial x$, may be neglected so that eqn 31.27 becomes
$$\frac{\partial u}{\partial x} = -\frac{\partial s}{\partial t}. \tag{31.29}$$

Again neglecting terms that are second order in u, one finds that eqn 31.26 becomes
$$\frac{\partial u}{\partial t} = -\frac{1}{\rho}\frac{\partial p}{\partial x}. \tag{31.30}$$

In terms of a bulk modulus defined in eqn 31.4, we may rewrite eqn 31.30 as
$$\frac{\partial u}{\partial t} = -\frac{B}{\rho}\frac{\partial s}{\partial x}, \tag{31.31}$$

and then eliminating u from this equation and from eqn 31.29 we have a one-dimensional wave equation:
$$\frac{\partial^2 s}{\partial x^2} = \frac{\rho}{B}\frac{\partial^2 s}{\partial t^2}. \tag{31.32}$$

This has solutions that may be recognized as travelling waves of the form
$$s \propto e^{i(kx - \omega t)}, \tag{31.33}$$

for which the wave speed is then given by substituting eqn 31.33 into eqn 31.32 and obtaining
$$v_s = \frac{\omega}{k} = \sqrt{\frac{B}{\rho}}. \tag{31.34}$$

Chapter summary

- The speed of sound is defined by $v_s = \sqrt{B/\rho}$, where B is given by $B = -V\partial p/\partial V$.
- For adiabatic sound waves the speed of sound is given by $v_s = \sqrt{\gamma \langle v^2 \rangle/3} = \sqrt{\gamma k_B T/m}$.
- In a relativistic gas, the speed of sound is given by $v_s = c/\sqrt{3}$.

Further reading

Faber (1995) has a good discussion of sound waves in gases and liquids and is a useful primer on fluid dynamics in general.

Exercises

(31.1) The speed of sound in air at $0\,°C$ is $331.5\,\mathrm{m\,s^{-1}}$. Estimate the speed of sound at an aircraft's cruising altitude where the temperature is $-60\,°C$.

(31.2) Calculate the speed of sound in nitrogen at $200\,°C$.

(31.3) For sound waves in air of frequency (a) 1 Hz and (b) 20 kHz estimate both the wavelength λ of the sound wave and the skin depth δ (the characteristic depth to which a thermal wave of this frequency will diffuse). Hence show that sound waves are invariably adiabatic and not isothermal. For what frequency would $\delta = \lambda$?

(31.4) The speed of sound in air, hydrogen, and carbon dioxide at $0\,°C$ is $331.5\,\mathrm{m\,s^{-1}}$, $1270\,\mathrm{m\,s^{-1}}$, and $258\,\mathrm{m\,s^{-1}}$ respectively. Explain the relative magnitude of these values.

(31.5) Breathing helium gas can result in your voice sounding higher (do not try this as asphyxiation is a serious risk); explain this effect (and note that the actual pitch of the voice is not higher).

(31.6) *Estimate* the time taken for a sound wave to cross the Sun using eqn 31.11, assuming that the average temperature of the Sun is $6\times10^6\,\mathrm{K}$. [Assume that the Sun is mostly ionized hydrogen (protons plus electrons) so that the average mass per particle is about $m_\mathrm{p}/2$. The radius $R_\odot$ of the Sun is $6.96\times10^8\,\mathrm{m}$.]

Shock waves

32

Shock waves (known for short as shocks) occur when a disturbance is propagating through a medium faster than the sound speed of the medium. In this chapter we are going to consider the nature of shocks in gases and the thermodynamic properties of the gas on either side of such a shock.

32.1 The Mach number	383
32.2 Structure of shock waves	383
32.3 Shock conservation laws	385
32.4 The Rankine–Hugoniot conditions	386
Chapter summary	388
Further reading	389
Exercises	389

32.1 The Mach number

The **Mach number** M of a disturbance is defined to be the ratio of the speed w at which the disturbance is passing through a medium to the sound speed v_s of the medium. Thus we have

$$M = \frac{w}{v_\mathrm{s}}. \tag{32.1}$$

When $M > 1$, the disturbance is called a **shock front** and the speed of the disturbance is **supersonic**. The development of a shock wave can be seen in Fig. 32.1, which shows wavefronts from a moving point source. The point source, moving at speed w, emits circular wavefronts and these wavefronts overlap constructively to form a single conical wavefront when $w > v_\mathrm{s}$, i.e., when $M \geq 1$ (the cone looks like the two sides of a triangle in the figure, which is necessarily printed in two dimensions!). The semi-angle of the cone decreases as M increases. This shock wave is responsible for the sonic "boom" which can be heard when a supersonic aircraft passes overhead (a double boom is often heard owing to the fact that shock waves originate from both the nose and the tail of the aircraft). Because the semi-angle of the cone decreases for very high speeds, a very fast aircraft at high altitude does not produce a boom at ground level because the cone does not intersect the ground.

32.2 Structure of shock waves

What is actually going on at a shock front? To establish the thermodynamic properties either side of a shock front, it is helpful to treat it as a mathematical discontinuity across which there is an abrupt change in the values of the properties because of the motion of the shock. In reality, the width of the shock front is finite but its detailed structure does not matter for our purposes although we will discuss it briefly in Section 32.4. Figure 32.2 illustrates the velocities of unshocked and shocked

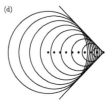

Fig. 32.1 The propagation of a shock wave for subsonic and supersonic flows. (a) $M = 0.8$, (b) $M = 1$, (c) $M = 1.2$, (d) $M = 1.4$.

gas with respect to a shock front (illustrated as a grey rectangle in each frame). This is shown for the two frames of reference in which it is convenient to work for these situations: the rest frame of the unshocked gas and the rest frame of the shock front (which we shall call the shock frame).

In the rest frame of the undisturbed gas, the shock front moves at velocity w while the gas through which the shock has already passed moves at velocity w_2 (where $w_2 < w$). There is a shock because the shock front propagates at speed $w > v_{s1}$, where v_{s1} is the sound speed in the unshocked gas. If $w \gg v_{s1}$ then there is said to be a **strong shock** whereas if w is just a little above v_{s1} then there is said to be a **weak shock**.

In the shock frame, the gas through which the shock front has passed moves away from the shock at velocity v_2 while the as-yet undisturbed gas moves towards it at velocity v_1. Therefore, $v_1 = w$, since this is the speed at which the undisturbed gas enters the shock front. In the same frame, the speed at which the shocked gas leaves the back of the shock is given by

$$v_2 = w - w_2. \qquad (32.2)$$

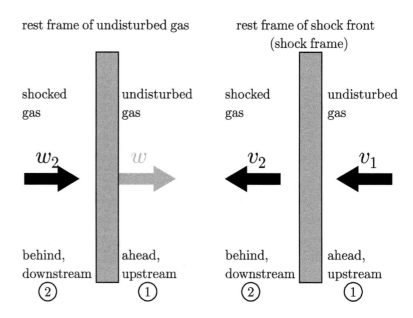

Fig. 32.2 Structure of a shock front in the rest frame of the undisturbed gas and in the rest frame of the shock front (which we call the shock frame). The terms "upstream" and "downstream" are best understood in the rest frame of the shock: in this frame, the shock is stationary and high-velocity gas (velocity v_1), which is yet to be disturbed by the shock front, streams towards the shock front (from "upstream") from region 1, while slower (velocity v_2) shocked gas moves away ("downstream") in region 2. Region 1 contains gas with *lower* internal energy, temperature, entropy, pressure, and density but *higher* velocity and hence bulk kinetic energy than region 2.

32.3 Shock conservation laws

To establish the physical properties of the gas before and after the passage of the shock, we have to think about the conservations laws, of mass, momentum, and energy, either side of the shock front. It is most convenient at this point to work in the shock frame (right panel of Fig. 32.2). We then have the following three conservation equations:

- The *conservation of mass* is applied by stating that the mass flux Φ_m, that is the mass crossing unit area in unit time, is equal on either side of the shock. Denoting the upstream region by 1 and the downstream region by 2, we may write

$$\rho_2 v_2 = \rho_1 v_1 = \Phi_m. \tag{32.3}$$

- The *conservation of momentum* requires that the momentum flux should be continuous; this means that the force per unit area plus the rate at which momentum is transported across unit area should be matched on either side of the shock front, giving

$$p_2 + \rho_2 v_2^2 = p_1 + \rho_1 v_1^2. \tag{32.4}$$

- The *conservation of energy* requires that the rate at which gas pressure does work per unit area (given by pv) and the rate of transport of internal and kinetic energy per unit area (($\rho\tilde{u}+\frac{1}{2}\rho v^2)v$, where $\tilde{u}$ is the *internal energy per unit mass*) is constant across a shock, which gives

$$p_2 v_2 + \left(\rho_2\tilde{u}_2 + \frac{1}{2}\rho_2 v_2^2\right) v_2 = p_1 v_1 + \left(\rho_1\tilde{u}_1 + \frac{1}{2}\rho_1 v_1^2\right) v_1. \tag{32.5}$$

The following example illustrates a simple algebraic manipulation of two of the conservation laws.

Example 32.1

Rearrange eqn 32.3 and eqn 32.4 to show that

$$\Phi_m^2 = (p_2 - p_1)/(\rho_1^{-1} - \rho_2^{-1}), \tag{32.6}$$

and hence find an expression for $v_1^2 - v_2^2$ in terms of pressures and densities.

Solution:
Equation 32.3 implies that $v_i = \rho_i^{-1}\Phi_m$. This, together with eqn 32.4, can be simply rearranged to give

$$p_2 - p_1 = \rho_1 v_1^2 - \rho_2 v_2^2 = \Phi_m^2(\rho_1^{-1} - \rho_2^{-1}), \tag{32.7}$$

and the desired result follows. The final step can be achieved by writing

$$v_1^2 - v_2^2 = (v_1 - v_2)(v_1 + v_2) = \Phi_m^2(\rho_1^{-1} - \rho_2^{-1})(\rho_1^{-1} + \rho_2^{-1}), \tag{32.8}$$

and substitution of eqn 32.7 yields

$$v_1^2 - v_2^2 = (p_2 - p_1)(\rho_1^{-1} + \rho_2^{-1}). \tag{32.9}$$

32.4 The Rankine–Hugoniot conditions

Having written down the conservation laws, we now wish to solve these simultaneously to find the pressures, densities, and temperatures on either side of the shock front.[1] We will treat the gas as an ideal gas, so that the internal energy per unit mass, $\tilde{u}$, is given by (see eqn 11.35)

$$\tilde{u} = \frac{p}{(\gamma-1)\rho}. \tag{32.10}$$

[1] The derivation in this section is nothing more than algebraic manipulations following from the conservation laws, but we give it in full since it is somewhat fiddly. If you are not concerned with these details, you can skip straight to equation 32.19.

Rearranging this gives $p = (\gamma-1)\rho\tilde{u}$ and substituting this into eqn 32.5 gives

$$\gamma\rho_2 v_2 \tilde{u}_2 + \frac{1}{2}\rho_2 v_2^3 = \gamma\rho_1 v_1 \tilde{u}_1 + \frac{1}{2}\rho_1 v_1^3. \tag{32.11}$$

Dividing the left-hand side by $\rho_2 v_2$ and the right-hand side by $\rho_1 v_1$ (and eqn 32.3 implies that these two factors are equal) and using eqn 32.10 yields

$$\frac{\gamma p_2}{(\gamma-1)\rho_2} + \frac{1}{2}v_2^2 = \frac{\gamma p_1}{(\gamma-1)\rho_1} + \frac{1}{2}v_1^2. \tag{32.12}$$

Using eqn 32.9, and multiplying by $\gamma-1$, this can be rearranged to give

$$2\gamma(p_1\rho_1^{-1} - p_2\rho_2^{-1}) + (\gamma-1)(p_2 - p_1)(\rho_1^{-1} + \rho_2^{-1}) = 0. \tag{32.13}$$

Hence, we have that

$$\frac{\rho_2^{-1}}{\rho_1^{-1}} = \frac{(\gamma+1)p_1 + (\gamma-1)p_2}{(\gamma-1)p_1 + (\gamma+1)p_2}. \tag{32.14}$$

Substitution into eqn 32.6 gives

$$\Phi_m^2 = \frac{p_2 - p_1}{\rho_1^{-1}[1 - \rho_2^{-1}/\rho_1^{-1}]} = \frac{1}{2}\rho_1[(\gamma-1)p_1 + (\gamma+1)p_2], \tag{32.15}$$

and hence

$$v_1^2 = \Phi_m^2 \rho_1^{-2} = \frac{1}{2}\rho_1^{-1}[(\gamma-1)p_1 + (\gamma+1)p_2]. \tag{32.16}$$

We would like to express everything in terms of the Mach number M_1 of the shock, and recalling that $M_1 = v_1/v_{s1}$ and $v_{s1} = \sqrt{\gamma p_1/\rho_1}$, we have that

$$M_1^2 = \frac{\rho_1 v_1^2}{\gamma p_1}. \tag{32.17}$$

Susbtitution of eqn 32.17 into eqn 32.16 gives

$$\rho v_1^2 = M_1^2 \gamma p_1 = \frac{1}{2}[(\gamma-1)p_1 + (\gamma+1)p_2], \tag{32.18}$$

and rearranging gives our desired equation relating the pressure on either side of the shock front:

$$\boxed{\frac{p_2}{p_1} = \frac{2\gamma M_1^2 - (\gamma-1)}{\gamma+1}.} \tag{32.19}$$

Substitution of eqn 32.19 into eqn 32.14, and using eqn 32.3, gives an equation for the ratio of the densities (and velocities) on either side of the shock:

$$\boxed{\frac{\rho_2}{\rho_1} = \frac{v_1}{v_2} = \frac{(\gamma+1)M_1^2}{2+(\gamma-1)M_1^2}.} \qquad (32.20)$$

Equations 32.19 and 32.20 are known as the **Rankine–Hugoniot conditions** and describe the physical properties of material on other side of the shock front. The results are plotted in Fig. 32.3.

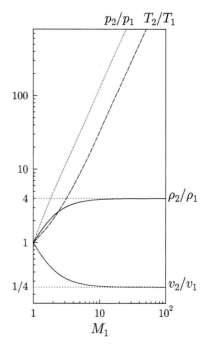

Fig. 32.3 The Rankine–Hugoniot conditions for a shock front as a function of Mach number M_1, where γ is assumed to take the value 5/3 (this is the value γ takes for a non-relativistic, monatomic gas).

Example 32.2

What are the ranges of values that can be taken by the following quantities for a shock front? (i) ρ_2/ρ_1, (ii) v_2/v_1, and (iii) p_2/p_1.

Solution:

When $M_1 = 1$, each of these quantities takes the value unity. In the limit as $M_1 \to \infty$, we find that

$$\frac{\rho_2}{\rho_1} \to \frac{\gamma+1}{\gamma-1}, \qquad (32.21)$$

$$\frac{v_2}{v_1} \to \frac{\gamma-1}{\gamma+1}, \qquad (32.22)$$

$$\frac{p_2}{p_1} \to \frac{2\gamma M_1^2}{\gamma+1}, \qquad (32.23)$$

so that ρ_2/ρ_1 and v_2/v_1 both saturate (at values of 4 and 1/4 respectively in the case of $\gamma = 5/3$) but p_2/p_1 can increase without limit. This is demonstrated in Fig. 32.3.

Example 32.3

Show that for a monatomic gas, the ratio ρ_2/ρ_1 can never exceed 4 and v_2/v_1 can never be lower than $\frac{1}{4}$.

Solution:

Equation 32.21, together with $\gamma = 5/3$ for a monatomic gas, shows that ρ_2/ρ_1 can never exceed $(\gamma+1)/(\gamma-1) = 4$. Since $v_2/v_1 = \rho_1/\rho_2$, this ratio can never be lower than $\frac{1}{4}$.

The Rankine–Hugoniot conditions, eqns 32.19 and 32.20 together with eqn 32.29, as they stand permit **expansive shocks**, that is with a reversal of roles for the two regions pictured in Fig. 32.2. The physical picture here would be that subsonically moving hot gas expands at a shock front and accelerates to become supersonic cool gas, i.e., internal energy would convert to bulk kinetic energy at the shock front. Such a situation is forbidden by the second law of thermodynamics (Chapter 14), which says that entropy can only increase. The second law, together with the Rankine–Hugoniot conditions only permit **compressive shocks** in which the shock speed (w) exceeds the sound speed v_{s1}, i.e., the Mach number $M_1 > 1$. In the shock frame, the flow ahead of the shock ("upstream") is *supersonic* and the flow behind the shock ("downstream") is *subsonic*.

For a shock to be compressive means that, $p_2 > p_1$ and $\rho_2 > \rho_1$ (which is of course consistent with $v_2 < v_1$). The ideal gas equation implies that $p/\rho \propto T$ and hence that

$$\frac{T_2}{T_1} = \frac{p_2/\rho_2}{p_1/\rho_1}. \tag{32.24}$$

This can be used to show that

$$T_2 > T_1, \tag{32.25}$$

so that a shock wave not only slows the gas but also heats it up, thus converting kinetic energy into thermal energy. The conversion of ordered energy into random motion occurs via collisions. The thickness of a shock front is thus usually of the order of the collisional mean free path.

We would expect from this that entropy increases as kinetic energy is converted into heat. The entropy increase to the gas downstream of the shock compared with that upstream is straightforwardly computed by using the relationship we established in eqn 16.90, namely

$$S = C_V \ln\left(\frac{p}{\rho^\gamma}\right) + \text{constant}. \tag{32.26}$$

Hence, the difference in entropy ΔS between the two regions is given by

$$\Delta S = S_2 - S_1 = C_V \ln\left[\frac{p_2}{p_1}\left(\frac{\rho_1}{\rho_2}\right)^\gamma\right]. \tag{32.27}$$

When we substitute eqns 32.19 and 32.20 into eqn 32.27 we obtain the following expression for the entropy difference across a shock:

$$\Delta S = C_V \ln\left[\frac{2\gamma M_1^2 - (\gamma - 1)}{\gamma + 1}\right]\left[\frac{2 + (\gamma - 1)M_1^2}{(\gamma + 1)M_1^2}\right]^\gamma. \tag{32.28}$$

This equation can be used to show that $\Delta S > 0$, so that entropy always increases as gas is shocked. Equation 32.28 is plotted in Fig. 32.4.

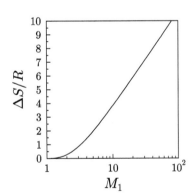

Fig. 32.4 The entropy change ΔS in units of R for one mole of gas, as a function of Mach number M_1.

Chapter summary

- Shock waves occur when a disturbance is propagating through a medium at a speed w, which is faster than the sound speed of the medium v_s.
- The Mach number $M = w/v_s$.
- Shocks convert kinetic energy into thermal energy.

Further reading

Faber (1995) contains useful information on shocks in fluids.

Exercises

(32.1) Show that the semi-angle of the cone of the shock waves shown in Fig. 32.1 is given by $\sin^{-1}(1/M)$, where M is the Mach number.

(32.2) Use eqn 32.24 to show that

$$\frac{T_2}{T_1} = \frac{[2\gamma M_1^2 - (\gamma - 1)][2 + (\gamma - 1)M_1^2]}{(\gamma + 1)^2 M_1^2}, \quad (32.29)$$

and hence for $M_1 \gg 1$ we have that

$$\frac{T_2}{T_1} \to \frac{2\gamma(\gamma - 1)M_1^2}{(\gamma + 1)^2}. \quad (32.30)$$

(32.3) For a shock wave in a monatomic gas show that

$$\frac{\rho_2}{\rho_1} \to 4, \quad \frac{p_2}{p_1} \to \frac{5}{4}M_1^2, \quad \frac{T_2}{T_1} \to \frac{5}{32}M_1^2, \quad (32.31)$$

in the limit $M_1 \gg 1$.

(32.4) Air is mostly nitrogen (N_2) and oxygen (O_2), which are both diatomic gases and for which $\gamma = 7/5$. Show that in this case, in the limit $M_1 \gg 1$ we have that

$$\frac{\rho_2}{\rho_1} \to 6, \quad \frac{p_2}{p_1} \to \frac{7}{6}M_1^2, \quad \frac{T_2}{T_1} \to \frac{7}{36}M_1^2. \quad (32.32)$$

(32.5) Show that in the limit as the Mach number of a shock becomes large, the increase in entropy from the upstream material flowing into the shock to the downstream material flowing away from it is given by

$$\Delta S = C_V \ln\left[\frac{2\gamma M_1^2}{\gamma + 1}\right]\left(\frac{\gamma - 1}{\gamma + 1}\right)^\gamma. \quad (32.33)$$

33
Brownian motion and fluctuations

33.1 Brownian motion 390
33.2 Johnson noise 393
33.3 Fluctuations 394
33.4 Fluctuations and the availability 395
33.5 Linear response 397
33.6 Correlation functions 400
Chapter summary 406
Further reading 407
Exercises 407

Our treatment of the thermodynamic properties of thermal systems has assumed that we can replace quantities such as pressure by their average values. Even though the molecules in a gas hit the walls of their container stochastically, there are so many of them that the pressure does not *appear* to fluctuate. But with very small systems, these fluctuations can become important. In this chapter, we consider these fluctuations in detail. A useful insight comes from the *fluctuation-dissipation theorem*, which is derived from the assumption that the response of a system in thermodynamic equilibrium to a small external perturbation is the same as its response to a spontaneous fluctuation. This implies that there is a direct relation between the fluctuation properties of a thermal system and what are known as its *linear response* properties.

33.1 Brownian motion

We introduced Brownian motion in Section 19.4. There we showed that the equipartition theorem implies that the translational motion of particles at temperature T fluctuates since each particle must have mean kinetic energy given by $\frac{1}{2}m\langle v^2 \rangle = \frac{3}{2}k_B T$. Einstein, in his 1905 paper on Brownian motion, noted that the same random forces that cause Brownian motion of a particle would also cause drag if the particle were pulled through the fluid.

Example 33.1

Find the solution to the equation of motion (known as the **Langevin equation**) for the velocity v of a particle of mass m, which is given by

$$m\dot{v} = -\alpha v + F(t), \tag{33.1}$$

where α is a damping constant (arising from friction), $F(t)$ is a random force whose average value over a long time period, $\langle F \rangle$, is zero.
Solution:
Note first that in the absence of the random force, eqn 33.1 becomes

$$m\dot{v} = -\alpha v, \tag{33.2}$$

which has solution
$$v(t) = v(0)\exp[-t/(m\alpha^{-1})], \tag{33.3}$$
so that any velocity component dies away with a time constant given by m/α. The random force $F(t)$ is necessary to give a model in which the particle's motion does not die away.

To solve eqn 33.1, write $v = \dot{x}$ and premultiply both sides by x. This leads to $m x \ddot{x} = -\alpha x \dot{x} + x F(t)$. A useful identity is
$$\frac{\mathrm{d}}{\mathrm{d}t}(x\dot{x}) = x\ddot{x} + \dot{x}^2, \tag{33.4}$$
and using this we can express our equation of motion as
$$m\frac{\mathrm{d}}{\mathrm{d}t}(x\dot{x}) = m\dot{x}^2 - \alpha x\dot{x} + xF(t). \tag{33.5}$$
We now average this result over time. We note that x and F are uncorrelated, and hence $\langle xF\rangle = \langle x\rangle\langle F\rangle = 0$. We can also use the equipartition theorem, which here states that
$$\frac{1}{2}m\langle \dot{x}^2\rangle = \frac{1}{2}k_\mathrm{B}T. \tag{33.6}$$
Hence, using eqn 33.6 in eqn 33.5, we have
$$m\frac{\mathrm{d}}{\mathrm{d}t}\langle x\dot{x}\rangle = k_\mathrm{B}T - \alpha\langle x\dot{x}\rangle, \tag{33.7}$$
or equivalently
$$\left(\frac{\mathrm{d}}{\mathrm{d}t} + \frac{\alpha}{m}\right)\langle x\dot{x}\rangle = \frac{k_\mathrm{B}T}{m}, \tag{33.8}$$
which has a solution
$$\langle x\dot{x}\rangle = C\mathrm{e}^{-\alpha t/m} + \frac{k_\mathrm{B}T}{\alpha}. \tag{33.9}$$
Putting the boundary condition that $x = 0$ when $t = 0$, one can find that the constant $C = -k_\mathrm{B}T/\alpha$, and hence
$$\langle x\dot{x}\rangle = \frac{k_\mathrm{B}T}{\alpha}(1 - \mathrm{e}^{-\alpha t/m}). \tag{33.10}$$
Using the identity
$$\frac{1}{2}\frac{\mathrm{d}}{\mathrm{d}t}\langle x^2\rangle = \langle x\dot{x}\rangle, \tag{33.11}$$
we then have
$$\langle x^2\rangle = \frac{2k_\mathrm{B}T}{\alpha}\left[t - \frac{m}{\alpha}(1 - \mathrm{e}^{-\alpha t/m})\right]. \tag{33.12}$$
When $t \ll m/\alpha$,
$$\langle x^2\rangle = \frac{k_\mathrm{B}Tt^2}{m}, \tag{33.13}$$
while for $t \gg m/\alpha$,
$$\langle x^2\rangle = \frac{2k_\mathrm{B}Tt}{\alpha}. \tag{33.14}$$
Writing[1] $\langle x^2\rangle = 2Dt$, where D is the diffusion constant, yields
$$D = k_\mathrm{B}T/\alpha. \tag{33.15}$$

[1] See Appendix C.12.

If a steady force F had been applied instead of a random one, then the terminal velocity (the velocity achieved in the steady state, with $\dot{v} = 0$) of the particle could have been obtained from

$$m\dot{v} = -\alpha v + F = 0, \qquad (33.16)$$

yielding $v = \alpha^{-1} F$, and so α^{-1} plays the rôle of a **mobility** (the ratio of velocity to force). It is easy to understand that the terminal velocity should be limited by frictional forces, and hence depends on α. However, the previous example shows that the diffusion constant D is proportional to $k_B T$ and also to the mobility α^{-1}. Note that the diffusion constant $D = k_B T/\alpha$ is independent of mass. The mass only enters in the transient term in eqn 33.12 (see also eqn 33.13) that disappears at long times.

Remarkably, we have found that the diffusion rate D, describing the random fluctuations of the particle's position, is related to the frictional damping α. The formula $D = k_B T/\alpha$ is an example of the fluctuation-dissipation theorem, which we will prove later in the chapter (Section 33.6).

As a prelude to what will come later, the following example considers the correlation function for the Brownian motion problem.

Example 33.2

Derive an expression for the velocity **correlation function** $\langle v(0)v(t) \rangle$ for the Brownian motion problem.

Solution:

The rate of change of v is given by

$$\dot{v}(t) = \frac{v(t+\tau) - v(t)}{\tau} \qquad (33.17)$$

in the limit in which $\tau \to 0$. Inserting this into eqn 33.1 and premultiplying by $v(0)$ gives

$$\frac{v(0)v(t+\tau) - v(0)v(t)}{\tau} = -\frac{\alpha}{m} v(0)v(t) + \frac{v(0)F(t)}{m}. \qquad (33.18)$$

Averaging this equation, and noting that $\langle v(0)F(t) \rangle = 0$ because v and F are uncorrelated, yields

$$\frac{\langle v(0)v(t+\tau) \rangle - \langle v(0)v(t) \rangle}{\tau} = -\frac{\alpha}{m} \langle v(0)v(t) \rangle, \qquad (33.19)$$

and taking the limit in which $\tau \to 0$ yields

$$\frac{d}{dt} \langle v(0)v(t) \rangle = -\frac{\alpha}{m} \langle v(0)v(t) \rangle, \qquad (33.20)$$

and hence

$$\langle v(0)v(t) \rangle = \langle v(0)^2 \rangle e^{-\alpha t/m}. \qquad (33.21)$$

Correlation functions are discussed in more detail in Section 33.6. The velocity correlation function $\langle v(0)v(t) \rangle$ is defined by

$$\lim_{T \to \infty} \frac{1}{T} \int_{-T/2}^{T/2} dt'\, v(t')v(t+t'),$$

and describes how well, on average, the velocity at a certain time is correlated with the velocity at a later time.

This example shows that the velocity correlation function decays to zero as time increases at exactly the same rate that the velocity itself relaxes (see eqn 33.3).

33.2 Johnson noise

We now consider another fluctuating system: the noise voltage that is generated across a resistor of resistance R by thermal fluctuations. Let us suppose that the resistor is connected to a transmission line of length L, which is correctly terminated at each end, as shown in Fig. 33.1.[2] Because the transmission line is matched, it should not matter whether it is connected or not. The transmission line can support modes of wave vector $k = n\pi/L$ and frequency $\omega = ck$, and therefore there is one mode per frequency interval $\Delta\omega$ given by

$$\Delta\omega = \frac{c\pi}{L}. \tag{33.22}$$

By the equipartition theorem, each mode has mean energy $k_B T$, and hence the energy per unit length of transmission line, in an interval $\Delta\omega$, is given by

$$k_B T \frac{\Delta\omega}{c\pi}. \tag{33.23}$$

Half this energy is travelling from left to right, and half from right to left. Hence, the mean power incident on the resistor is given by

$$\frac{1}{2\pi} k_B T \Delta\omega, \tag{33.24}$$

and in equilibrium this must equal the mean power dissipated by the resistor, which is given by

$$\langle I^2 R \rangle. \tag{33.25}$$

In the circuit, we have $I = V/(2R)$ and hence

$$\frac{\langle V^2 \rangle}{4R} = \langle I^2 R \rangle = \frac{1}{2\pi} k_B T \Delta\omega, \tag{33.26}$$

and hence

$$\langle V^2 \rangle = \frac{2}{\pi} k_B T R \Delta\omega, \tag{33.27}$$

which, using $\Delta\omega = 2\pi \Delta f$, can be written in the form

$$\boxed{\langle V^2 \rangle = 4 k_B T R \Delta f.} \tag{33.28}$$

This expression is known as the **Johnson noise** produced across a resistor in a frequency interval Δf. It is another example of the connection between fluctuations and dissipation, since it relates fluctuating noise power ($\langle V^2 \rangle$) to the dissipation in the circuit (R).

We can derive a quantum mechanical version of the Johnson noise formula by replacing $k_B T$ by $\hbar\omega/(e^{\beta\hbar\omega} - 1)$, which yields

$$\langle V^2 \rangle = \frac{2R}{\pi} \frac{\hbar\omega \Delta\omega}{e^{\beta\hbar\omega} - 1}. \tag{33.29}$$

[2] We will give a method of calculating the noise voltage that may seem a little artificial at first, but provides a convenient way of calculating how the resistor can exchange energy with a thermal reservoir. A more elegant approach will be made in Example 33.9.

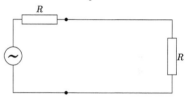

Fig. 33.1 The equivalent circuit to consider the Johnson noise across a resistor. The resistor is connected to a matched transmission line which is correctly terminated, hence the presence of the second resistor; one can consider the noise voltage as being an alternating voltage source connected in series with the second resistor.

33.3 Fluctuations

In this section, we will consider the origin of fluctuations and show how much freedom a system has to allow the functions of state to fluctuate. We will focus on one such function of state, which we will call x, and ask the question: if the system is in equilibrium, what is the probability distribution of x? Let us suppose that the number of microstates associated with a system characterized by this parameter x and having energy E (which we will consider fixed[3]) is given by

[3] This part of the argument assumes that we are working in the microcanonical ensemble (see Section 4.6).

$$\Omega(x, E). \qquad (33.30)$$

If x were constrained to this value, the entropy S of the system would be

$$S(x, E) = k_B \ln \Omega(x, E), \qquad (33.31)$$

which we could write equivalently as $\Omega(x, E) = e^{S(x,E)/k_B}$. If x were not constrained, its probability distribution function would then follow the function $p(x)$, where

$$p(x) \propto \Omega(x, E) = e^{S(x,E)/k_B}. \qquad (33.32)$$

At equilibrium the system will maximize its entropy, and let us suppose that this occurs when $x = x_0$. Hence

$$\left(\frac{\partial S(x, E)}{\partial x}\right) = 0 \quad \text{when } x = x_0. \qquad (33.33)$$

Let us now write a Taylor expansion of $S(x, E)$ around the equilibrium point $x = x_0$:

$$S(x, E) = S(x_0, E) + \left(\frac{\partial S}{\partial x}\right)_{x=x_0} (x - x_0) + \frac{1}{2}\left(\frac{\partial^2 S}{\partial x^2}\right)_{x=x_0} (x - x_0)^2 + \cdots \qquad (33.34)$$

which with eqn 33.33 implies that

$$S(x) = S(x_0) + \frac{1}{2}\left(\frac{\partial^2 S}{\partial x^2}\right)_{x=x_0} (x - x_0)^2 + \cdots \qquad (33.35)$$

Hence, defining $\Delta x = x - x_0$, we can write the probability function as a Gaussian,

$$p(x) \propto \exp\left(-\frac{(\Delta x)^2}{2\langle(\Delta x)^2\rangle}\right), \qquad (33.36)$$

where

$$\boxed{\langle(\Delta x)^2\rangle = -\frac{k_B}{\left(\frac{\partial^2 S}{\partial x^2}\right)_E}.} \qquad (33.37)$$

This equation shows that if the entropy S changes rapidly as a function of x, we are more likely to find the system with x close to x_0, as one might expect.

Example 33.3

Let x be the internal energy U for a system with fixed volume. Using $T = (\partial U/\partial S)_V$, we have that

$$\left(\frac{\partial^2 S}{\partial U^2}\right)_V = \left(\frac{\partial (1/T)}{\partial U}\right)_V = -\frac{1}{T^2 C_V}, \qquad (33.38)$$

and hence

$$\langle (\Delta U)^2 \rangle = -\frac{k_B}{\left(\frac{\partial^2 S}{\partial U^2}\right)_V} = k_B T^2 C_V. \qquad (33.39)$$

So if a system is in contact with a bath at temperature T, there is a non-zero probability that we may find the system away from the equilibrium internal energy: thus U can fluctuate. The size of the fluctuations is larger if the heat capacity is larger.

This result in eqn 33.39 is equivalent to that found using statistical mechanics in eqn 28.65.

Both the heat capacity C_V and the internal energy U are extensive parameters and therefore they scale with the size of the system. The rms fluctuations of U scale with the square root of the size of the system, so the *fractional* rms fluctuations scale with the size of the system to the power $-\frac{1}{2}$. Thus if the system has N atoms, then

$$C \propto N, \quad U \propto N \quad \sqrt{\langle(\Delta U)^2\rangle} \propto \sqrt{N}, \qquad (33.40)$$

and

$$\frac{\sqrt{\langle(\Delta U)^2\rangle}}{U} \propto \frac{1}{\sqrt{N}}. \qquad (33.41)$$

Hence as $N \to \infty$, we can ignore fluctuations. Fluctuations are more important in small systems. However, note that at a critical point for a first-order phase transition, $C \to \infty$ and hence

$$\frac{\langle(\Delta U)^2\rangle}{U} \to \infty. \qquad (33.42)$$

Hence fluctuations become divergent at the critical point and cannot be ignored, even for large systems.

33.4 Fluctuations and the availability

We now generalize an argument presented in Section 16.5 to the case in which numbers of particles can fluctuate. Consider a system in contact with a reservoir. The reservoir has temperature T_0, pressure p_0, and chemical potential μ_0. Let us consider what happens when we transfer energy dU, volume dV, and dN particles from the reservoir to the system. The internal energy of the reservoir changes by dU_0, where

$$dU_0 = -dU = T_0\, dS_0 - p_0(-dV) + \mu_0(-dN), \qquad (33.43)$$

where the minus signs express the fact that the energy, volume, and number of particles in the reservoir are *decreasing*. We can rearrange this expression to give the change of entropy in the reservoir as

$$dS_0 = \frac{-dU - p_0 dV + \mu_0 dN}{T_0}. \tag{33.44}$$

If the entropy of the system changes by dS, then the total change of entropy dS_{tot} is

$$dS_{\text{tot}} = dS + dS_0, \tag{33.45}$$

and the second law of thermodynamics implies that $dS_{\text{tot}} \geq 0$. Using eqn 33.44, we have that

$$dS_{\text{tot}} = -\frac{dU - T_0 dS + p_0 dV - \mu_0 dN}{T_0}, \tag{33.46}$$

which can be written as

$$dS_{\text{tot}} = -\frac{dA}{T_0}, \tag{33.47}$$

where $A = U - T_0 S + p_0 V - \mu_0 N$ is the **availability** (this generalizes eqn 16.32).

We now apply the concept of availability to fluctuations. Let us suppose that the availability depends on some variable x, so that we can write a function $A(x)$. Equilibrium will be achieved when $A(x)$ is minimized (so that S_{tot} is maximized, see eqn 33.47) and let us suppose that this occurs when $x = x_0$. Hence we can similarly write $A(x)$ in a Taylor expansion around the equilibrium point and hence

$$A(x) = A(x_0) + \frac{1}{2}\left(\frac{\partial^2 A}{\partial x^2}\right)_{x=x_0}(\Delta x)^2 + \cdots, \tag{33.48}$$

so that we can recover the probability distribution in eqn 33.36 with

$$\langle(\Delta x)^2\rangle = \frac{k_B T_0}{\left(\frac{\partial^2 A}{\partial x^2}\right)}. \tag{33.49}$$

Example 33.4

A system with a fixed number N of particles is in thermal contact with a reservoir at temperature T. It is surrounded by a tensionless membrane so that its volume is able to fluctuate. Calculate the mean square volume fluctuations. For the special case of an ideal gas, show that $\langle(\Delta V)^2\rangle = V^2/N$.

Solution:
Fixing T and N means that U can fluctuate. Fixing N implies that $dN = 0$ and hence we have that

$$dU = TdS - pdV. \tag{33.50}$$

Changes in the availability therefore follow:

$$dA = dU - T_0 dS + p_0 dV = (T - T_0)dS + (p_0 - p)dV, \quad (33.51)$$

and hence

$$\left(\frac{\partial A}{\partial V}\right)_{T,N} = p_0 - p \quad (33.52)$$

and

$$\left(\frac{\partial^2 A}{\partial V^2}\right)_{T,N} = -\left(\frac{\partial p}{\partial V}\right)_{T,N}. \quad (33.53)$$

Hence

$$\langle (\Delta V)^2 \rangle = -k_B T_0 \left(\frac{\partial V}{\partial p}\right)_{T,N}. \quad (33.54)$$

For an ideal gas, $(\partial V/\partial p)_{T,N} = -Nk_B T/p^2 = -V/p$, and hence

$$\langle (\Delta V)^2 \rangle = \frac{V^2}{N}. \quad (33.55)$$

Equation 33.55 implies that the *fractional* volume fluctuations follow

$$\frac{\sqrt{\langle (\Delta V)^2 \rangle}}{V} = \frac{1}{N^{1/2}}. \quad (33.56)$$

Thus for a box containing 10^{24} molecules of gas (a little over a mole of gas), the fractional volume fluctuations are at the level of one part in 10^{12}.

We can derive other similar expressions for other fluctuating variables, including

$$\langle (\Delta T)^2 \rangle = \frac{k_B T^2}{C_V}, \quad (33.57)$$

$$\langle (\Delta S)^2 \rangle = k_B C_p, \quad (33.58)$$

$$\langle (\Delta p)^2 \rangle = \frac{k_B T \kappa_S}{C_V}, \quad (33.59)$$

where κ_S is the adiabatic compressibility (see eqn 16.69).

33.5 Linear response

To understand in more detail the relationship between fluctuations and dissipation, it is necessary to consider how systems respond to external forces in a rather more general way. We consider a displacement variable $x(t)$ that is the result of some force $f(t)$, and require that the product xf has the dimensions of energy. (We will say that x and f are **conjugate variables** if their product has the dimensions of energy.) We assume that the response of x to a force f is linear (so that, for example, doubling the force doubles the response), but there could be some delay in the way in which the system responds. The most general way of writing this down is as follows: we say that the average value of x at time t is denoted by $\langle x(t)\rangle_f$ (the subscript f reminds us that a force f has been applied) and is given by

$$\langle x(t)\rangle_f = \int_{-\infty}^{\infty} \chi(t-t')f(t')\,\mathrm{d}t', \tag{33.60}$$

where $\chi(t-t')$ is a **response function**. This relates the value of $x(t)$ to a sum over values of the force $f(t')$ at all other times. Now it makes sense to sum over past values of the force, but not to sum over future values of the force. This will force the response function $\chi(t-t')$ to be zero if $t < t'$. Before seeing what effect this has, we need to Fourier transform eqn 33.60 to make it simpler to deal with. The **Fourier transform** of $x(t)$ is given by the function $\tilde{x}(\omega)$ given by

$$\tilde{x}(\omega) = \int_{-\infty}^{\infty} \mathrm{d}t\, \mathrm{e}^{-\mathrm{i}\omega t} x(t). \tag{33.61}$$

The inverse transform is then given by

$$x(t) = \frac{1}{2\pi} \int_{-\infty}^{\infty} \mathrm{d}\omega\, \mathrm{e}^{\mathrm{i}\omega t} \tilde{x}(\omega). \tag{33.62}$$

The expression in eqn 33.60 is a convolution of the functions χ and f, and hence by the convolution theorem we can write this equation in Fourier transform form as

$$\langle \tilde{x}(\omega)\rangle_f = \tilde{\chi}(\omega)\tilde{f}(\omega). \tag{33.63}$$

This is much simpler than eqn 33.60 as it is a product, rather than a convolution. Note that the response function $\tilde{\chi}(\omega)$ can be complex. The real part of the response function gives the part of the displacement that is in phase with the force. The imaginary part of the response function gives a displacement with is $\frac{\pi}{2}$ out of phase with the force. It corresponds to dissipation because the external force does work on the system at a rate given by the force multiplied by the velocity, i.e., $f(t)\dot{x}(t)$, and this work is dissipated as heat. For $f(t)$ and $\dot{x}(t)$ to be in phase, and hence give a non-zero average, $f(t)$ and $x(t)$ have to be $\frac{\pi}{2}$ out of phase (see Exercise 33.2).

We can build causality into our problem by writing the response function as

$$\chi(t) = y(t)\theta(t), \tag{33.64}$$

where $\theta(t)$ is a Heaviside step function (see Fig. 30.1) and $y(t)$ is a function, which equals $\chi(t)$ when $t > 0$ and can equal anything at all when $t < 0$. For the convenience of the following derivation, we will set $y(t) = -\chi(|t|)$ when $t < 0$, making $y(t)$ an odd function (and, importantly, making $\tilde{y}(\omega)$ purely imaginary). By the inverse convolution theorem, the Fourier transform of $\chi(t)$ is given by the convolution

$$\tilde{\chi}(\omega) = \frac{1}{2\pi} \int_{-\infty}^{\infty} \mathrm{d}\omega'\, \tilde{\theta}(\omega'-\omega)\tilde{y}(\omega'). \tag{33.65}$$

Writing the Heaviside step function as

$$\theta(t) = \lim_{\epsilon \to 0} \begin{cases} e^{-\epsilon t} & t > 0 \\ 0 & t < 0 \end{cases}, \quad (33.66)$$

its Fourier transform is given by

$$\tilde{\theta}(\omega) = \int_0^\infty dt\, e^{-i\omega t} e^{-\epsilon t} = \frac{1}{i\omega + \epsilon} = \frac{\epsilon}{\omega^2 + \epsilon^2} - \frac{i\omega}{\omega^2 + \epsilon^2}. \quad (33.67)$$

Thus, taking the limit $\epsilon \to 0$, we have that

$$\tilde{\theta}(\omega) = \pi \delta(\omega) - \frac{i}{\omega}. \quad (33.68)$$

Substituting this into eqn 33.65 yields[4]

$$\tilde{\chi}(\omega) = \frac{1}{2}\tilde{y}(\omega) - \frac{i}{2\pi} \mathcal{P} \int_{-\infty}^{\infty} \frac{\tilde{y}(\omega')\, d\omega'}{\omega' - \omega}. \quad (33.69)$$

We now write $\tilde{\chi}(\omega)$ in terms of its real and imaginary parts:

$$\tilde{\chi}(\omega) = \tilde{\chi}'(\omega) + i\tilde{\chi}''(\omega), \quad (33.70)$$

and since $\tilde{y}(\omega)$ is purely imaginary, eqn 33.69 yields

$$i\tilde{\chi}''(\omega) = \frac{1}{2}\tilde{y}(\omega), \quad (33.71)$$

and hence

$$\boxed{\tilde{\chi}'(\omega) = \mathcal{P} \int_{-\infty}^{\infty} \frac{d\omega'}{\pi} \frac{\tilde{\chi}''(\omega')}{\omega' - \omega}.} \quad (33.72)$$

This is one of the **Kramers–Kronig relations**, which connects the real and imaginary parts of the response function.[5] Note that our derivation has only assumed that the response is linear (eqn 33.60) and causal, so that the Kramers–Kronig relations are very general.

By putting $\omega = 0$ into eqn 33.72, we obtain another very useful result:

$$\boxed{\tilde{\chi}'(0) = \mathcal{P} \int_{-\infty}^{\infty} \frac{d\omega'}{\pi} \frac{\tilde{\chi}''(\omega')}{\omega'}.} \quad (33.73)$$

Sometimes the response function is called a **generalized susceptibility**, and the zero frequency real part, $\tilde{\chi}'(0)$, is called the **static susceptibility**. As discussed above, the imaginary part of the response function, $\tilde{\chi}''(\omega)$, corresponds to the dissipation of the system. Equation 33.73 therefore shows that the static susceptibility (the response at zero frequency) is related to an integral of the total dissipation of the system.

---

[4]The symbol $\mathcal{P}$ denotes the **Cauchy principal value** of the integral. This means that an integral whose integrand blows up at some value is evaluated using an appropriate limit. For example, $\int_{-1}^{1} dx/x$ is undefined since $1/x \to \infty$ at $x = 0$, but

$$\mathcal{P}\int_{-1}^{1} \frac{dx}{x} = \lim_{\epsilon \to 0^+} \left(\int_{-1}^{-\epsilon} \frac{dx}{x} + \int_{\epsilon}^{1} \frac{dx}{x} \right)$$
$$= 0.$$

[5]The other Kramers–Kronig relation is derived in Exercise 33.3.

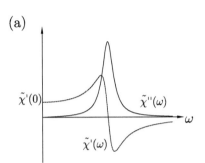

Example 33.5

Find the response function for the damped harmonic oscillator (mass m, spring constant k, damping α) whose equation of motion is given by

$$m\ddot{x} + \alpha\dot{x} + kx = f \tag{33.74}$$

and show that eqn 33.73 holds.

Solution:
Writing the resonant frequency $\omega_0^2 = k/m$, and writing the damping $\gamma = \alpha/m$, we have

$$\ddot{x} + \gamma\dot{x} + \omega_0^2 x = \frac{f}{m}, \tag{33.75}$$

and Fourier transforming this gives immediately that

$$\tilde{\chi}(\omega) = \frac{\tilde{x}(\omega)}{\tilde{f}(\omega)} = \frac{1}{m}\left[\frac{1}{\omega_0^2 - \omega^2 - i\omega\gamma}\right]. \tag{33.76}$$

Hence, the imaginary part of the response function is

$$\tilde{\chi}''(\omega) = \frac{1}{m}\left[\frac{\omega\gamma}{(\omega^2 - \omega_0^2)^2 + (\omega\gamma)^2}\right], \tag{33.77}$$

and the static susceptibility is

$$\tilde{\chi}'(0) = \frac{1}{m\omega_0^2} = \frac{1}{k}. \tag{33.78}$$

The real and imaginary parts of $\tilde{\chi}(\omega)$ are plotted in Fig. 33.2(a). The imaginary part shows a peak near ω_0. Equation 33.77 shows that $\tilde{\chi}''(\omega)/\omega = (\gamma/m)/[(\omega^2 - \omega_0^2) + (\omega\gamma)^2]$ and straightforward integration shows that $\int_{-\infty}^{\infty}(\tilde{\chi}''(\omega)/\omega)\,d\omega = \pi/(m\omega_0^2) = \pi\tilde{\chi}'(0)$ and hence that eqn 33.73 holds. This is illustrated in Fig. 33.2(b).

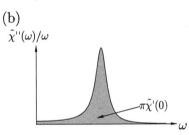

Fig. 33.2 (a) The real and imaginary parts of $\tilde{\chi}(\omega)$ as a function of ω. (b) An illustration of eqn 33.73 for the damped harmonic oscillator.

[6] See Appendix C.11.

33.6 Correlation functions

Consider a function $x(t)$. Its Fourier transform[6] is given by

$$\tilde{x}(\omega) = \int_{-\infty}^{\infty} dt\, e^{-i\omega t} x(t), \tag{33.79}$$

as before, and we define the **power spectral density** as $\langle|\tilde{x}(\omega)|^2\rangle$. This function shows how much power is associated with different parts of the frequency spectrum. We now define the **autocorrelation function** $C_{xx}(t)$ by

$$C_{xx}(t) = \langle x(0)x(t)\rangle = \int_{-\infty}^{\infty} x^*(t')x(t'+t)\,dt'. \tag{33.80}$$

The notation here is that the double subscript means we are measuring how much x at one time is correlated with x at another time. (We could also define a cross-correlation function $C_{xy}(t) = \langle x(0)y(t) \rangle$, which measures how much x at one time is correlated with a different variable y at another time.) The autocorrelation function is connected to the power spectral density by the **Wiener–Khinchin theorem**,[7] which states that the power spectral density is given by the Fourier transform of the autocorrelation function:

[7] Norbert Wiener (1894–1964); Alexsandr Y. Khinchin (1894–1959). The proof of this theorem is given in Appendix C.11.

$$\langle |\tilde{x}(\omega)|^2 \rangle = \tilde{C}_{xx}(\omega) = \int_{-\infty}^{\infty} e^{-i\omega t} \langle x(0)x(t) \rangle \, dt. \qquad (33.81)$$

The inverse relation also must hold:

$$\langle x(0)x(t) \rangle = \frac{1}{2\pi} \int_{-\infty}^{\infty} e^{i\omega t} \langle |\tilde{x}(\omega)|^2 \rangle \, d\omega, \qquad (33.82)$$

and hence for $t = 0$ we have that

$$\langle x(0)x(0) \rangle = \frac{1}{2\pi} \int_{-\infty}^{\infty} \langle |\tilde{x}(\omega)|^2 \rangle \, d\omega, \qquad (33.83)$$

or, more succinctly,

$$\langle x^2 \rangle = \frac{1}{2\pi} \int_{-\infty}^{\infty} \tilde{C}_{xx}(\omega) \, d\omega. \qquad (33.84)$$

This is a form of **Parseval's theorem** that states that the integrated power is the same whether you integrate over time or over frequency.[8]

[8] Parseval's theorem is actually nothing more than Pythagoras' theorem in an infinite-dimensional vector space. If you think of the function $x(t)$, or its transform $\tilde{x}(\omega)$, as a single vector in such a space, then the square of the length of the vector is equal to the sum of the squares on the "other sides", which in this case is the sum of the squares of the components (i.e., an integral of the squares of the values of the function).

Example 33.6

A random force $F(t)$ has average value given by

$$\langle F(t) \rangle = 0 \qquad (33.85)$$

and its autocorrelation function is given by

$$\langle F(t)F(t') \rangle = A\delta(t - t'), \qquad (33.86)$$

where $\delta(t - t')$ is a Dirac delta function.[9] Find the power spectrum.
Solution:
By the Wiener–Khinchin theorem, the power spectrum is simply the Fourier transform of the autocorrelation function, and hence

[9] See Appendix C.10.

$$\langle |F(\omega)|^2 \rangle = A, \qquad (33.87)$$

i.e., the power spectrum is completely flat (independent of ω).

This demonstrates that if the random force $F(t)$ has zero autocorrelation, it must have infinite frequency content.

Example 33.7

Find the velocity autocorrelation for the Brownian motion particle governed by eqn 33.1 where the random force $F(t)$ is as described in the previous example, i.e., with $\langle F(t)F(t')\rangle = A\delta(t-t')$. Hence relate the constant A to the temperature T.

Solution:
Equation 33.1 states that
$$m\dot{v} = -\alpha v + F(t), \tag{33.88}$$
and the Fourier transform of this equation is
$$\tilde{v}(\omega) = \frac{\tilde{F}(\omega)}{\alpha - im\omega}. \tag{33.89}$$
This implies that the Fourier transform of the velocity autocorrelation function is
$$\tilde{C}_{vv}(\omega) = \langle|v(\omega)|^2\rangle = \frac{A}{\alpha^2 + m^2\omega^2}, \tag{33.90}$$
using the result of eqn 33.87. The Wiener–Khinchin theorem states that
$$\tilde{C}_{vv}(\omega) = \int e^{-i\omega t} \langle v(0)v(t)\rangle \, dt, \tag{33.91}$$
and hence
$$\tilde{C}_{vv}(t) = \langle v(0)v(t)\rangle = \langle v^2\rangle e^{-\alpha t/m}, \tag{33.92}$$
in agreement with eqn 33.21 derived earlier using another method. Parseval's theorem (eqn 33.84) implies that
$$\langle v^2\rangle = \int_{-\infty}^{\infty} \frac{d\omega}{2\pi} \tilde{C}_{vv}(\omega) = \frac{A}{2m\alpha}. \tag{33.93}$$
Equipartition, which gives that $\frac{1}{2}m\langle v^2\rangle = \frac{1}{2}k_\mathrm{B}T$, leads immediately to
$$A = 2\alpha k_\mathrm{B}T. \tag{33.94}$$

Let us next suppose that the energy E of a harmonic system is given by $E = \frac{1}{2}\alpha x^2$ (as in Chapter 19). The probability $P(x)$ of the system taking the value x is given by a Boltzmann factor $e^{-\beta E}$ and hence
$$P(x) = \mathcal{N}e^{-\beta\alpha x^2/2}, \tag{33.95}$$
where $\mathcal{N}$ is a normalization constant. Now we apply a force f, which is conjugate to x, so that the energy E is lowered by xf. The probability $P(x)$ becomes
$$P(x) = \mathcal{N}'e^{-\beta(\alpha x^2/2 - xf)}, \tag{33.96}$$

where $\mathcal{N}'$ is a different normalization constant. By completing the square, this can be rewritten as

$$P(x) = \mathcal{N}'' e^{-\frac{\beta\alpha}{2}(x-\frac{f}{\alpha})^2}, \qquad (33.97)$$

where $\mathcal{N}''$ is yet another normalization constant. This equation is of the usual Gaussian form

$$P(x) = \mathcal{N}'' e^{-(x-\langle x \rangle_f)^2/2\langle x^2 \rangle}, \qquad (33.98)$$

where $\langle x \rangle_f = f/\alpha$ and $\langle x^2 \rangle = 1/\beta\alpha$. Notice that $\langle x \rangle_f$ is telling us about the average value of x in response to the force f, while $\langle x^2 \rangle = k_\mathrm{B}T/\alpha$ is telling us about fluctuations in x. The ratio of these two quantities is given by

$$\frac{\langle x \rangle_f}{\langle x^2 \rangle} = \beta f. \qquad (33.99)$$

Now $\langle x \rangle_f$ is the average value x takes when a force f is applied, and we know that $\langle x \rangle_f$ is related[10] to f by the static susceptibility, i.e. by

$$\frac{\langle x \rangle_f}{f} = \tilde{\chi}'(0). \qquad (33.100)$$

[10] Here we are making the assumption that the linear response function $\tilde{\chi}(\omega)$ governs both fluctuations *and* the usual response to perturbations.

Thus eqn 33.99 can be rewritten as

$$\boxed{\langle x^2 \rangle = k_\mathrm{B} T \tilde{\chi}'(0).} \qquad (33.101)$$

Equation 33.101 thus relates $\langle x^2 \rangle$ to the static susceptibility of the system. Using eqn 33.73, we can express this relationship as

$$\langle x^2 \rangle = k_\mathrm{B} T \int_{-\infty}^{\infty} \frac{\mathrm{d}\omega'}{\pi} \frac{\tilde{\chi}''(\omega')}{\omega'}, \qquad (33.102)$$

and together with eqn 33.84, this motivates

$$\boxed{\tilde{C}_{xx}(\omega) = 2k_\mathrm{B} T \frac{\tilde{\chi}''(\omega)}{\omega},} \qquad (33.103)$$

which is a statement of the **fluctuation-dissipation theorem**. This shows that there is a direct connection between the autocorrelation function of the *fluctuations*, $\tilde{C}_{xx}(\omega)$, and the imaginary part $\tilde{\chi}''(\omega)$ of the response function which is associated with *dissipation*.

Example 33.8

Show that eqn 33.103 holds for the problem considered in Example 33.5.
Solution:
Using the Wiener–Khinchin theorem, we have

$$\tilde{C}_{xx}(\omega) = \int e^{-\mathrm{i}\omega t} \langle x(0)x(t)\rangle \, \mathrm{d}t = \langle |\tilde{x}(\omega)|^2 \rangle = A|\chi(\omega)|^2, \qquad (33.104)$$

and hence using $\tilde{\chi}(\omega)$ from eqn 33.76 and A from eqn 33.94, we have that
$$\tilde{C}_{xx}(\omega) = \frac{2\gamma k_B T}{m}\left[\frac{1}{(\omega^2-\omega_0^2)^2+(\omega\gamma)^2}\right]. \qquad (33.105)$$

Equation 33.77 shows that
$$2k_B T\frac{\tilde{\chi}''(\omega)}{\omega} = \frac{2\gamma k_B T}{m}\left[\frac{1}{(\omega^2-\omega_0^2)^2+(\omega\gamma)^2}\right], \qquad (33.106)$$

and hence eqn 33.103 holds.

Example 33.9

Derive an expression for the Johnson noise across a resistor R using the circuit in Fig. 33.3 (which includes the small capacitance across the ends of the resistor).

Solution:
Simple circuit theory yields
$$V + IR = \frac{Q}{C}. \qquad (33.107)$$

The charge Q and voltage V are conjugate variables (their product has dimensions of energy) and so we write
$$\tilde{Q}(\omega) = \tilde{\chi}(\omega)\tilde{V}(\omega), \qquad (33.108)$$

where the response function $\tilde{\chi}(\omega)$ is given for this circuit by
$$\tilde{\chi}(\omega) = \frac{1}{C^{-1} - i\omega R}. \qquad (33.109)$$

Hence $\tilde{\chi}''(\omega)$ is given by
$$\tilde{\chi}''(\omega) = \frac{\omega R}{C^{-2} + \omega^2 R^2}. \qquad (33.110)$$

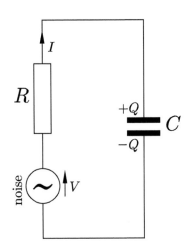

Fig. 33.3 Circuit for analysing Johnson noise across a resistor.

At low frequency ($\omega \ll 1/RC$, and since the capacitance will be small, $1/RC$ will be very high so that this is not a severe restriction) we have that $\tilde{\chi}''(\omega) \to \omega RC^2$. Thus the fluctuation-dissipation theorem (eqn 33.103) gives
$$\tilde{C}_{QQ}(\omega) = 2k_B T\frac{\tilde{\chi}''(\omega)}{\omega} = 2k_B TRC^2. \qquad (33.111)$$

Because $Q = CV$ for a capacitor, correlations in Q and V are related by
$$\tilde{C}_{VV}(\omega) = \frac{\tilde{C}_{QQ}(\omega)}{C^2}, \qquad (33.112)$$

and hence
$$\tilde{C}_{VV}(\omega) = 2k_B TR. \tag{33.113}$$

Equation 33.84 implies that
$$\langle V^2 \rangle = \frac{1}{2\pi} \int_{-\infty}^{\infty} \tilde{C}_{VV}(\omega)\, d\omega, \tag{33.114}$$

and hence if this integral is carried out, not over all frequencies, but only in a small interval $\Delta f = \Delta\omega/(2\pi)$ about some frequency $\pm\omega_0$ (see Fig. 33.4),
$$\langle V^2 \rangle = 2C_{VV}(\omega)\Delta f = 4k_B TR \Delta f, \tag{33.115}$$

in agreement with eqn 33.28.

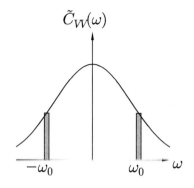

Fig. 33.4 The voltage fluctuations $\langle V^2 \rangle$ in a small frequency interval $\Delta f = \Delta\omega/(2\pi)$ centred on $\pm\omega_0$ are due to the part of the $\tilde{C}_{VV}(\omega)$ shown by the shaded boxes. One can imagine that the noise is examined through a filter, which only allows these frequencies through, so that the integral in eqn 33.114 only picks up the regions shown by the shaded boxes.

We close this chapter by remarking that our treatment so far applies only to classical systems. The quantum mechanical version of the fluctuation-dissipation theorem can be evaluated by replacing $k_B T$, the mean thermal energy in a classical system, by
$$\hbar\omega\left(n(\omega) + \frac{1}{2}\right) \equiv \frac{\hbar\omega}{2}\coth\frac{\beta\hbar\omega}{2}, \tag{33.116}$$

which is the mean energy in a quantum harmonic oscillator. In eqn 33.116,
$$n(\omega) = \frac{1}{e^{\beta\hbar\omega} - 1} \tag{33.117}$$

is the Bose factor, which is the mean number of quanta in the harmonic oscillator at temperature T. Hence, in the quantum mechanical case, eqn 33.103 is replaced by
$$\tilde{C}_{xx}(\omega) = \hbar\tilde{\chi}''(\omega)\coth\frac{\beta\hbar\omega}{2}. \tag{33.118}$$

At high temperature, $\coth(\beta\hbar\omega/2) \to 2/(\beta\hbar\omega)$ and we recover eqn 33.103. The quantum mechanical version of eqn 33.102 is
$$\langle x^2 \rangle = \frac{\hbar}{2}\int_{-\infty}^{\infty} d\omega'\, \tilde{\chi}''(\omega')\coth\frac{\beta\hbar\omega}{2}. \tag{33.119}$$

Chapter summary

- The fluctuation-dissipation theorem implies that there is a direct relation between the fluctuation properties of the thermal system (e.g., the diffusion constant) and its linear response properties (e.g., the mobility). If you've characterized one, you've characterized the other.
- Fluctuations are more important for small systems than for large systems, though they are always dominant near the critical point of a phase transition, even for large systems.
- Fluctuations in a variable x are given by

$$\langle (\Delta x)^2 \rangle = -k_B T_0 / (\partial^2 A / \partial x^2),$$

  where A is the availability.
- A response function is defined by

$$\langle x(t) \rangle_f = \int_{-\infty}^{\infty} \chi(t-t') f(t') \, dt',$$

  and causality implies the Kramers–Kronig relations.
- The Fourier transform of the correlation function gives the power spectrum. This allows us to show that

$$\langle x^2 \rangle = \frac{1}{2\pi} \int_{-\infty}^{\infty} \tilde{C}_{xx}(\omega) \, d\omega.$$

- The fluctuation-dissipation theorem states that

$$\tilde{C}_{xx}(\omega) = 2k_B T \frac{\tilde{\chi}''(\omega)}{\omega},$$

  and relates the autocorrelation function to the dissipations via the imaginary part of the response function.

Further reading

A good introduction to fluctuations and response functions can be found in Chaikin and Lubensky (1995). Another useful source of information is chapter XII of Landau and Lifshitz (1980) (particularly §110–114, §118–126).

Exercises

(33.1) If a system is held at fixed T, N and p, show that the fluctuations in a variable x are governed by the probability function

$$p(x) \propto e^{-G(x)/k_B T}, \qquad (33.120)$$

where $G(x)$ is the Gibbs function.

(33.2) A system has displacement $\tilde{x}(\omega) = \tilde{\chi}(\omega)\tilde{f}(\omega)$ in response to a force $\tilde{f}(\omega)$. Show that if the force is given by $f(t) = f_0 \cos \omega t$, the average power dissipated is $\frac{1}{2} f_0^2 \omega \tilde{\chi}''(\omega)$.

(33.3) Repeat the derivation that led to eqn 33.72 (one of the Kramers–Kronig relations), but this time set $y(-t) = y(t)$, so that $\tilde{y}(\omega)$ is purely real. In this case, prove the other Kramers–Kronig relation, which states that

$$\tilde{\chi}''(\omega) = -\mathcal{P} \int_{-\infty}^{\infty} \frac{d\omega'}{\pi} \frac{\tilde{\chi}'(\omega')}{\omega' - \omega}. \qquad (33.121)$$

34 Non-equilibrium thermodynamics

34.1 Entropy production 408
34.2 The kinetic coefficients 409
34.3 Proof of the Onsager reciprocal relations 410
34.4 Thermoelectricity 413
34.5 Time reversal and the arrow of time 417
Chapter summary 418
Further reading 419
Exercises 419

Much of the material in this book has been concerned with the properties of systems in thermodynamic equilibrium, in which the functions of state are time independent. However, we have also touched on transport properties (see Chapter 9) which deal with the flow of momentum, heat or particles from one place to another. Such processes are usually irreversible and result in entropy production. In this chapter, we will use the theory of fluctuations developed in Chapter 33 to derive a general relation concerning different transport processes, and apply it to thermoelectric effects. We conclude the chapter by discussing the asymmetry of time.

34.1 Entropy production

Changes in the internal energy density u of a system are related to entropy density s, number N_j of particles of type j, and electric charge density ρ_e by the combined first and second laws of thermodynamics, which states that

$$\mathrm{d}u = T\mathrm{d}s + \sum_j \mu_j \mathrm{d}N_j + \phi \mathrm{d}\rho, \qquad (34.1)$$

where μ_j is the chemical potential of atoms of type j and ϕ is the electric potential. Rearranging this equation to make entropy changes the subject gives

$$\mathrm{d}s = \frac{1}{T}\mathrm{d}u - \sum_j \left(\frac{\mu_j}{T}\right) \mathrm{d}N_j - \frac{\phi}{T}\mathrm{d}\rho_e, \qquad (34.2)$$

and this is of the form

$$\mathrm{d}s = \sum_k \phi_k \, \mathrm{d}\rho_k, \qquad (34.3)$$

where ρ_k is a generalized density and $\phi_k = \partial s/\partial \rho_k$ is the corresponding generalized potential. Possible values for these variables are listed in Table 34.1. Each of the generalized densities is subject to a continuity equation of the form

$$\frac{\partial \rho_k}{\partial t} + \nabla \cdot \boldsymbol{J}_k = 0, \qquad (34.4)$$

where $\boldsymbol{J}_k$ is a generalized current density. We can associate each of these currents with a flow of entropy, which is itself measured by the **entropy**

	ρ_k	ϕ_k	$\nabla \phi_k$
Energy	u	$1/T$	$\nabla(1/T)$
Number of particles of type j	N_j	μ_j/T	$\nabla(\mu_j/T)$
Charge density	ρ_e	$-\phi_e/T$	$-\nabla(\phi_e/T)$

Table 34.1 Terms in eqn 34.3.

current density $\boldsymbol{J}_s$. This will be subject to its own continuity equation, which states that the local entropy production rate Σ is given by

$$\Sigma = \frac{\partial s}{\partial t} + \nabla \cdot \boldsymbol{J}_s. \qquad (34.5)$$

We can relate the entropy current density $\boldsymbol{J}_s$ to the other current densities via the following equation:

$$\boldsymbol{J}_s = \sum_k \phi_k \boldsymbol{J}_k. \qquad (34.6)$$

Inserting this into eqn 34.5 yields

$$\Sigma = \sum_k \phi_k \dot{\rho}_k + \nabla \cdot \left(\sum_k \phi_k \boldsymbol{J}_k \right). \qquad (34.7)$$

Now some straightforward vector calculus, and use of eqn 34.4, yields

$$\begin{aligned}
\nabla \cdot \left(\sum_k \phi_k \boldsymbol{J}_k \right) &= \sum_k \nabla \phi_k \cdot \boldsymbol{J}_k + \sum_k \phi_k \nabla \cdot \boldsymbol{J}_k \\
&= \sum_k \nabla \phi_k \cdot \boldsymbol{J}_k + \sum_k \phi_k(-\dot{\rho}_k), \qquad (34.8)
\end{aligned}$$

and hence

$$\boxed{\Sigma = \sum_k \nabla \phi_k \cdot \boldsymbol{J}_k.} \qquad (34.9)$$

This equation relates the local entropy production rate Σ to the generalized current densities $\boldsymbol{J}_k$ and the generalized force fields $\nabla \phi_k$.

34.2 The kinetic coefficients

Very often the response of a system to a force is to produce a steady current. For example, a constant electric field applied to an electrical conductor produces an electric current; a constant temperature gradient applied to a thermal conductor produces a flow of heat. Assuming a linear response, one can write in general that the generalized current density $\boldsymbol{J}_i$ is related to the generalized force fields by the equation

$$\boldsymbol{J}_i = \sum_j L_{ij} \nabla \phi_j, \qquad (34.10)$$

where the coefficients L_{ij} are called **kinetic coefficients**.

Example 34.1

Recall the equation for heat flow (eqn 9.15):

$$\boldsymbol{J} = -\kappa \nabla T. \tag{34.11}$$

This can be cast into the form

$$\boldsymbol{J}_u = L_{uu} \nabla (1/T), \tag{34.12}$$

where $L_{uu} = \kappa T^2$, the subscript "u" denoting energy flow (see Table 34.1).

Equation 34.10 implies that the local entropy production Σ is given by

$$\Sigma = \sum_{ij} \nabla \phi_i \, L_{ij} \, \nabla \phi_j. \tag{34.13}$$

The second law of thermodynamics can be stated in the form that entropy must increase globally. However, entropy can go down in one place if it goes up at least as much somewhere else. Equation 34.5 relates the entropy produced locally in a small region to the entropy, which is transported into or out of that region (perhaps by matter, charge, heat, or some combination of these, being imported or exported). An even stronger statement can be made by insisting that not only is the global entropy change always positive but that so is the local equilibrium production rate: $\Sigma \geq 0$. Equation 34.13 then implies that L_{ij} must be a positive-definite matrix (all its eigenvalues must be positive). A further statement about L_{ij} can be made, which follows from **Onsager's reciprocal relations**, which state that

$$L_{ij} = L_{ji}. \tag{34.14}$$

We will prove these relations, first derived by Lars Onsager (Fig. 34.1) in 1929, in the following section.

Fig. 34.1 Lars Onsager (1903–1976).

34.3 Proof of the Onsager reciprocal relations

Near an equilibrium state, we define the variable $\alpha_k = \rho_k - \rho_k^{\text{eqm}}$, which measures the departure of the kth density variable from its equilibrium value. The probability of the system having density fluctuations given by $\boldsymbol{\alpha} = (\alpha_1, \alpha_2, \ldots, \alpha_m)$ can be written as

$$P(\boldsymbol{\alpha}) \propto e^{\Delta S/k_{\text{B}}}. \tag{34.15}$$

We assume that the probability P is suitably normalized, so that
$$\int P\,d\boldsymbol{\alpha} = 1. \tag{34.16}$$

The entropy change for a fluctuation ΔS is a function of $\boldsymbol{\alpha}$, which we can express using a Taylor expansion in $\boldsymbol{\alpha}$. There are no linear terms since we are measuring departures from equilibrium, where S is maximized, and hence we write
$$\Delta S = -\frac{1}{2}\sum_{ij} g_{ij}\alpha_i\alpha_j, \tag{34.17}$$

where $g_{ij} = \left(\frac{\partial^2 \Delta S}{\partial \alpha_i \partial \alpha_j}\right)_{\boldsymbol{\alpha}=0}$. Thus we can write the logarithm of the probability as
$$\ln P = \frac{\Delta S}{k_B} + \text{constant}, \tag{34.18}$$

and hence
$$\frac{\partial \ln P}{\partial \alpha_i} = \frac{1}{k_B}\frac{\partial \Delta S}{\partial \alpha_i}. \tag{34.19}$$

The next part of the proof involves working out a couple of averages of a fluctuation of one of the density variables with some other quantity.

(1) We begin by deriving an expression for $\langle (\partial S/\partial \alpha_i)\alpha_j \rangle$:
$$\begin{aligned}\left\langle \frac{\partial S}{\partial \alpha_i}\alpha_j \right\rangle &= k_B \left\langle \frac{\partial \ln P}{\partial \alpha_i}\alpha_j \right\rangle \\ &= k_B \int \frac{\partial \ln P}{\partial \alpha_i}\alpha_j P\,d\boldsymbol{\alpha} \\ &= k_B \int \frac{\partial P}{\partial \alpha_i}\alpha_j\,d\boldsymbol{\alpha} \\ &= k_B \left(\int d\boldsymbol{\alpha}'\,[P\alpha_j]_{-\infty}^{\infty} - \int \frac{\partial \alpha_j}{\partial \alpha_i}P\,d\boldsymbol{\alpha}\right).\end{aligned} \tag{34.20}$$

In this equation, $d\boldsymbol{\alpha}' = d\alpha_1\cdots d\alpha_{j-1}d\alpha_{j+1}\cdots d\alpha_m$, i.e., the product of all the $d\alpha_i$ except $d\alpha_j$. The term $[P\alpha_j]_{-\infty}^{\infty}$ is zero because $P \propto \exp[-\frac{1}{2k_B}\sum_{ij}g_{ij}\alpha_i\alpha_j]$ and hence goes to zero as $\alpha_j \to \pm\infty$. Using $\partial \alpha_j/\partial \alpha_i = \delta_{ij}$, we can therefore show that
$$\boxed{\left\langle \frac{\partial S}{\partial \alpha_i}\alpha_j \right\rangle = -k_B \delta_{ij}.} \tag{34.21}$$

(2) We now derive an expression for $\langle \alpha_i \alpha_j \rangle$:
$$\frac{\partial \Delta S}{\partial \alpha_i} = -\sum_k g_{ik}\alpha_k, \tag{34.22}$$

and hence
$$\sum_k g_{ik}\langle \alpha_k \alpha_j \rangle = -\left\langle \frac{\partial S}{\partial \alpha_i}\alpha_j \right\rangle = k_B \delta_{ij}. \tag{34.23}$$

Hence
$$\boxed{\langle \alpha_i \alpha_j \rangle = k_B (g^{-1})_{ij}.} \tag{34.24}$$

Example 34.2

Show that $\langle \Delta S \rangle = -mk_B/2$, explain the sign of the answer, and interpret the answer in terms of the equipartition theorem.
Solution:

$$\langle \Delta S \rangle = \left\langle -\frac{1}{2} \sum_{ij} g_{ij} \alpha_i \alpha_j \right\rangle = -\frac{1}{2} \sum_{ij} g_{ij} \langle \alpha_i \alpha_j \rangle = -\frac{k_B}{2} \sum_{i=1}^{m} \delta_{ii} = -\frac{mk_B}{2}. \tag{34.25}$$

The equilibrium configuration, $\boldsymbol{\alpha} = 0$, corresponds to maximum entropy, so $\langle \Delta S \rangle$ should be negative; a fluctuation corresponds to a statistically less likely state. If the system has m degrees of freedom, then its mean thermal energy is $mk_B T/2$, which is equal to $-T \langle \Delta S \rangle$.

We are now in a position to work out some correlation functions of the fluctuations. We now make the crucial assumption that any microscopic process and its reverse process take place on average with the same frequency. This is the **principle of microscopic reversibility**. This implies that

$$\begin{aligned}\langle \alpha_i(0) \alpha_j(t) \rangle &= \langle \alpha_i(0) \alpha_j(-t) \rangle \\ &= \langle \alpha_i(t) \alpha_j(0) \rangle.\end{aligned} \tag{34.26}$$

Subtracting $\langle \alpha_i(0) \alpha_j(0) \rangle$ from both sides of eqn 34.26 yields

$$\langle \alpha_i(0) \alpha_j(t) \rangle - \langle \alpha_i(0) \alpha_j(0) \rangle = \langle \alpha_i(t) \alpha_j(0) \rangle - \langle \alpha_i(0) \alpha_j(0) \rangle. \tag{34.27}$$

Dividing eqn 34.27 by t and factoring out common factors gives

$$\left\langle \alpha_i(0) \left[\frac{\alpha_j(t) - \alpha_j(0)}{t} \right] \right\rangle = \left\langle \left[\frac{\alpha_i(t) - \alpha_i(0)}{t} \right] \alpha_j(0) \right\rangle, \tag{34.28}$$

and in the limit as $t \to 0$, this can be written

$$\langle \alpha_i \dot{\alpha}_j \rangle = \langle \dot{\alpha}_i \alpha_j \rangle. \tag{34.29}$$

Now, assuming that the decay of fluctuations is governed by the same laws as the macroscopic flows as they respond to generalized forces, so that we can use the kinetic coefficients L_{ij} to describe the fluctuations, we have that

$$\dot{\alpha} = \sum_k L_{ik} \frac{\partial S}{\partial \alpha_k} \tag{34.30}$$

and hence substituting into eqn 34.29 yields

$$\left\langle \alpha_i \sum_k L_{jk} \frac{\partial S}{\partial \alpha_k} \right\rangle = \left\langle \sum_k L_{ik} \frac{\partial S}{\partial \alpha_k} \alpha_j \right\rangle, \tag{34.31}$$

which simplifies to

$$\sum_k L_{jk} \left\langle \alpha_i \frac{\partial S}{\partial \alpha_k} \right\rangle = \sum_k L_{ik} \left\langle \frac{\partial S}{\partial \alpha_k} \alpha_j \right\rangle. \qquad (34.32)$$

Using the relation in eqn 34.21, we have that

$$\sum_k L_{jk}(-k_B \delta_{ik}) = \sum_k L_{ik}(-k_B \delta_{jk}) \qquad (34.33)$$

and hence we have the Onsager reciprocal relations:

$$L_{ji} = L_{ij}. \qquad (34.34)$$

34.4 Thermoelectricity

In this section we apply the Onsager reciprocal relations and the other ideas developed in this chapter to the problem of **thermoelectricity**, which describes the relationship between flows of heat and electrical current. It is not surprising that heat current and electrical currents in metals should be related; both result from the flow of electrons and electrons carry both charge and energy.

Consider two dissimilar metals A and B, with different work functions[1] and chemical potentials, whose energy levels are shown schematically in Fig. 34.2(a). These two metals are connected together, as shown in Fig. 34.2(b), and both held at the same temperature T. Because initially $\mu_A \neq \mu_B$, some electrons will diffuse from A into B, resulting in a small build up of positive charge on A and a small build up of negative charge on B. This will lead to a small electric field across the junction of A and B that will oppose any further electrons moving into B. Once equilibrium is established, the chemical potential in A and B must equalize and hence $\mu_A = \mu_B$, see Section 22.2.

No voltage develops between the ends of the metals if they remain at the same temperature, but if the ends of A and B are at different temperatures, there will be a voltage difference. Electrons respond both to an applied electric field $\boldsymbol{E}$ and a gradient in the chemical potential $\nabla \mu$, the former producing a **drift current** and the latter a **diffusion current**. Near the junction between A and B shown in Fig. 34.2(b), these two currents coexist but are equal and opposite and therefore precisely cancel in equilibrium, if A and B are held at the same temperature. Thus a voltmeter responds not to the integrated electric field given by

$$\int \boldsymbol{E} \cdot \mathrm{d}\boldsymbol{l} \qquad (34.35)$$

around a circuit, but rather to

$$\int \boldsymbol{\mathcal{E}} \cdot \mathrm{d}\boldsymbol{l}, \qquad (34.36)$$

where

$$\boldsymbol{\mathcal{E}} = \boldsymbol{E} + \frac{1}{e}\nabla\mu \qquad (34.37)$$

[1] The **work function** of a metal is the minimum energy needed to remove an electron from the Fermi level to a point in the vacuum well away from the surface.

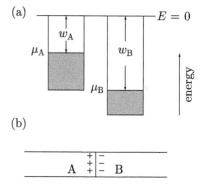

Fig. 34.2 (a) Two dissimilar metals with different work functions w_A and w_B and chemical potentials $\mu_A = -w_A$ and $\mu_B = -w_B$. (b) Metals A and B, held at the same temperature, are connected together.

is the **electromotive field**, which combines the effects of the fields driving the drift and diffusion currents. We thus write the current densities for charge and heat, $\boldsymbol{J}_e$ and $\boldsymbol{J}_Q$, in terms of the electromotive field and temperature gradient which drive them, in the following general way:

$$\boldsymbol{J}_e = \mathcal{L}_{\mathcal{E}\mathcal{E}}\boldsymbol{\mathcal{E}} + \mathcal{L}_{\mathcal{E}T}\nabla T, \tag{34.38}$$

$$\boldsymbol{J}_Q = \mathcal{L}_{T\mathcal{E}}\boldsymbol{\mathcal{E}} + \mathcal{L}_{TT}\nabla T. \tag{34.39}$$

Here the kinetic coefficients $\mathcal{L}_{\mathcal{E}\mathcal{E}}$, $\mathcal{L}_{\mathcal{E}T}$, $\mathcal{L}_{T\mathcal{E}}$, and $\mathcal{L}_{TT}$ are written using the symbol $\mathcal{L}$ rather than L because we haven't yet written them in the form of eqn 34.10. To work out what these coefficients are, let us examine some special cases:

- *No temperature gradient*
  If $\nabla T = 0$, then we expect that

$$\boldsymbol{J}_e = \sigma \boldsymbol{\mathcal{E}}, \tag{34.40}$$

where σ is the electrical conductivity, and hence we identify $\mathcal{L}_{\mathcal{E}\mathcal{E}} = \sigma$ from eqn 34.38. In this case, the heat current density is given by eqn 34.39 and hence

$$\boldsymbol{J}_Q = \mathcal{L}_{T\mathcal{E}}\boldsymbol{\mathcal{E}} = \frac{\mathcal{L}_{T\mathcal{E}}}{\mathcal{L}_{\mathcal{E}\mathcal{E}}}\boldsymbol{J}_e = \Pi \boldsymbol{J}_e, \tag{34.41}$$

where $\Pi = \mathcal{L}_{T\mathcal{E}}/\mathcal{L}_{\mathcal{E}\mathcal{E}}$ is known as the **Peltier coefficient**. (The Peltier coefficient has dimensions of energy/charge, and so is measured in volts.) Thus, an electrical current is associated with a heat current, and this is known as the **Peltier effect**.[2]

[2] J. C. A. Peltier (1785–1845) first observed the effect in 1834.

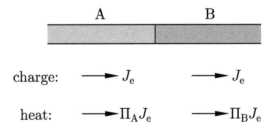

Fig. 34.3 A junction between wires of different metals, A and B, carrying an electrical current has a discontinuous jump in its heat current. This is the Peltier effect.

Consider an electrical current $\boldsymbol{J}_e$ flowing along a wire of metal A and then via a junction to a wire of metal B, as shown in Fig. 34.3. The electrical current must be the same in both wires, so the heat current must exhibit a discontinuous jump at the junction. This jump is given by $(\Pi_A - \Pi_B)\boldsymbol{J}_e$, and this results in the liberation of heat

$$\Pi_{AB}\boldsymbol{J}_e \tag{34.42}$$

at the junction, where $\Pi_{AB} = \Pi_A - \Pi_B$. If $\Pi_{AB} < 0$ this results in cooling, and this is the principle behind **Peltier cooling** in which heat is removed from a region by putting it in thermal contact with a junction between dissimilar wires and passing a current along the wires (see Fig. 34.4). Of course, the heat removed is simultaneously liberated elsewhere in the circuit, as shown in Fig. 34.4. The Peltier heat flow is reversible and so if the electrical currents are reversed, then so are the heat flows.

- *No electrical current*
  If $\boldsymbol{J}_e = 0$, then
  $$\boldsymbol{J}_Q = -\kappa \nabla T, \tag{34.43}$$
  where κ is the thermal conductivity. However, we also have an electric field $\boldsymbol{\mathcal{E}}$ given by
  $$\boldsymbol{\mathcal{E}} = \epsilon \nabla T \tag{34.44}$$
  where ϵ is the **Seebeck coefficient**[3] or the **thermopower** (units $\mathrm{V\,K^{-1}}$). Thus a thermal gradient is associated with an electric field: this is called the **Seebeck effect**. Equations 34.38 and 34.44 imply that
  $$\epsilon = -\frac{\mathcal{L}_{\mathcal{E}T}}{\mathcal{L}_{\mathcal{E}\mathcal{E}}}. \tag{34.45}$$
  A circuit consisting of a single material with a temperature gradient around it would produce a voltage given by
  $$\oint \boldsymbol{\mathcal{E}} \cdot \mathrm{d}\boldsymbol{l} = \epsilon \oint \nabla T \cdot \mathrm{d}\boldsymbol{l} = 0. \tag{34.46}$$

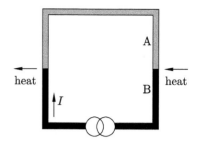

Fig. 34.4 In this circuit, $\Pi_{AB} < 0$, so that for the current direction shown, the junction on the right-hand side absorbs heat while the junction on the left-hand side liberates heat.

[3]T. J. Seebeck (1770–1831) discovered this in 1821.

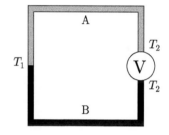

Fig. 34.5 Thermocouple circuit for measuring the differences between thermoelectric voltages.

To observe the thermopower, one needs a circuit containing two different metals: this is known as a **thermocouple** and such a circuit is shown in Fig. 34.5. An equivalent circuit is shown in Fig. 34.6. Thus the Seebeck emf (electromotive force) $\Delta \phi_S$ measured by the voltmeter in the circuit in Fig. 34.6 is given by

$$\begin{aligned}
\Delta \phi_S &= -\int \boldsymbol{\mathcal{E}} \cdot \mathrm{d}\boldsymbol{l} \\
&= \int_{T_0}^{T_1} \epsilon_B \mathrm{d}T + \int_{T_1}^{T_2} \epsilon_A \mathrm{d}T + \int_{T_2}^{T_0} \epsilon_B \mathrm{d}T \\
&= \int_{T_1}^{T_2} (\epsilon_A - \epsilon_B) \mathrm{d}T, \tag{34.47}
\end{aligned}$$

and we write
$$\epsilon_A - \epsilon_B = \frac{\mathrm{d}\phi_S}{\mathrm{d}T}. \tag{34.48}$$

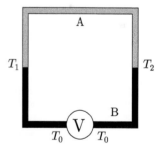

Fig. 34.6 Equivalent thermocouple circuit.

Example 34.3

Derive an expression for κ in terms of the kinetic coefficients.
Solution:
Substitution of eqn 34.45 into eqn 34.44 yields
$$\boldsymbol{\mathcal{E}} = -\frac{\mathcal{L}_{\mathcal{E}T}}{\mathcal{L}_{\mathcal{E}\mathcal{E}}} \nabla T. \tag{34.49}$$

Putting this into eqn 34.39 implies that

$$\boldsymbol{J}_Q = \left(\frac{\mathcal{L}_{\mathcal{E}\mathcal{E}}\mathcal{L}_{TT} - \mathcal{L}_{T\mathcal{E}}\mathcal{L}_{\mathcal{E}T}}{\mathcal{L}_{\mathcal{E}\mathcal{E}}}\right)\nabla T, \qquad (34.50)$$

and hence comparison with eqn 34.43 yields

$$\kappa = -\left[\frac{\mathcal{L}_{\mathcal{E}\mathcal{E}}\mathcal{L}_{TT} - \mathcal{L}_{T\mathcal{E}}\mathcal{L}_{\mathcal{E}T}}{\mathcal{L}_{\mathcal{E}\mathcal{E}}}\right]. \qquad (34.51)$$

Putting eqns 34.38 and 34.39 into the form of eqn 34.10, we have

$$\begin{aligned}\boldsymbol{J}_e &= L_{\mathcal{E}\mathcal{E}}\nabla(-\phi/T) + L_{\mathcal{E}T}\nabla(1/T), & (34.52)\\ \boldsymbol{J}_Q &= L_{T\mathcal{E}}\nabla(-\phi/T) + L_{TT}\nabla(1/T), & (34.53)\end{aligned}$$

where

$$\begin{aligned}L_{\mathcal{E}\mathcal{E}} &= T\mathcal{L}_{\mathcal{E}\mathcal{E}},\\ L_{T\mathcal{E}} &= T\mathcal{L}_{T\mathcal{E}},\\ L_{\mathcal{E}T} &= -T^2\mathcal{L}_{\mathcal{E}T},\\ L_{TT} &= -T^2\mathcal{L}_{TT}. \end{aligned} \qquad (34.54)$$

The Onsager reciprocal relation in this case thus implies that

$$L_{T\mathcal{E}} = L_{\mathcal{E}T}, \qquad (34.55)$$

and hence $\mathcal{L}_{T\mathcal{E}} = -T\mathcal{L}_{\mathcal{E}T}$, so that

$$\Pi = T\epsilon. \qquad (34.56)$$

This yields

$$\Pi_{AB} = T(\epsilon_A - \epsilon_B), \qquad (34.57)$$

[4] William Thomson, also known as Lord Kelvin (1824–1907). Thomson's proof was, of course, not based on the Onsager reciprocal relations and is somewhat suspect.

which is known as **Thomson's second relation**.[4] It is a very good example of the power of Onsager's approach: we have been able to relate the Peltier and Seebeck coefficients on the basis of Onsager's general theorem concerning the symmetry of general kinetic coefficients.

There is one other thermoelectric effect that we wish to consider. If the thermopower ϵ is temperature dependent, then there will even be heat liberated by an electric current which flows in a single metal. This heat is known as **Thomson heat**.[5] An electrical current $\boldsymbol{J}_e$ corresponds to a heat current $\boldsymbol{J}_Q = \Pi\boldsymbol{J}_e$ (by eqn 34.41). The heat liberated at a particular point per second is therefore given by the divergence of $\boldsymbol{J}_Q$ and hence

[5] Lord Kelvin again!

$$\nabla \cdot \boldsymbol{J}_Q = \nabla \cdot (\epsilon T \boldsymbol{J}_e), \qquad (34.58)$$

using eqn 34.56. If no charges build up, then $\boldsymbol{J}_e$ is divergenceless and hence

$$\nabla \cdot \boldsymbol{J}_Q = \boldsymbol{J}_e \cdot \nabla(\epsilon T) = \boldsymbol{J}_e \cdot \epsilon \nabla T + \boldsymbol{J}_e \cdot T\nabla\epsilon. \qquad (34.59)$$

Writing $\nabla \epsilon = (\mathrm{d}\epsilon/\mathrm{d}T)\nabla T$ and using eqn 34.44, we have finally that

$$\nabla \cdot \boldsymbol{J}_Q = \boldsymbol{J}_e \cdot \boldsymbol{\mathcal{E}} + \tau \boldsymbol{J}_e \cdot \nabla T, \qquad (34.60)$$

which is the sum of a resistive heating term ($\boldsymbol{J}_e \cdot \boldsymbol{\mathcal{E}}$) and a thermal gradient term ($\tau \boldsymbol{J}_e \cdot \nabla T$). In this equation, the **Thomson coefficient** τ is given by

$$\tau = T \frac{\mathrm{d}\epsilon}{\mathrm{d}T}. \qquad (34.61)$$

The Thomson coefficient is the heat generated per second per unit current per unit temperature gradient.

Equation 34.57 implies that

$$T \frac{\mathrm{d}}{\mathrm{d}T} \left(\frac{\Pi_{AB}}{T} \right) = \tau_A - \tau_B, \qquad (34.62)$$

and this implies that

$$\frac{\mathrm{d}\Pi_{AB}}{\mathrm{d}T} + \epsilon_A - \epsilon_B = \tau_A - \tau_B, \qquad (34.63)$$

which is known as **Thomson's first relation**.

34.5 Time reversal and the arrow of time

The proof of the Onsager reciprocal relations rested on the hypothesis of microscopic reversibility. This makes some degree of sense since molecular collisions and processes are based on laws of motion, which are themselves symmetric under time reversal. The heat produced in the Peltier effect considered in the previous section is reversible (one simply has to reverse the current) and this adds to our feeling that we are dealing with underlying reversible processes. But of course, that is not the whole story. The second law of thermodynamics insists that entropy never decreases and in fact increases in irreversible processes. This presents us with a dilemma since to explain this we have to understand why microscopic time-symmetric laws generate a Universe in which time is definitely asymmetric: eggs break but do not unbreak; heat flows from hot to cold and never the reverse; we remember the past but not the future. In our Universe, $+t$ is manifestly different from $-t$.

This problem afflicted Boltzmann when he tried to prove the second law on the basis of classical mechanics and derived his famous **H-theorem**, which showed how the Maxwell–Boltzmann distribution of velocities in a gas would emerge as a function of time on the basis of molecular collisions. One hypothesis which had gone into his proof was the innocent looking principle of molecular chaos (the **stoßzahlansatz**), which states that the velocities of molecules undergoing a collision are statistically indistinguishable from those of any pair of molecules in the gas selected at random. However, this cannot be right; Boltzmann's approach showed how molecules retain correlations in their motion *following* a collision and this "memory" of the collision is progressively redistributed among the molecular velocities until they adopt the most likely Maxwell–Boltzmann distribution. However, because the underlying dynamics are time symmetric, before a collision molecules must possess pre-collisional correlations, which are "harbingers of collisions to come".[6] This makes a nonsense of the stoßzahlansatz.

It seems more likely that the source of the time asymmetry is not in the dynamics but in the boundary conditions. If we watch an irreversible process, we are watching how a system prepared in a low-entropy state evolves to a state of higher entropy. For example, in a Joule expansion, it is the experimenter who prepares the two chambers appropriately in a low-entropy state (by producing entropy elsewhere in the Universe, pumping gas out from the second chamber). There is a boundary condition at the start, putting all the gas in one chamber in a non-equilibrium state, but not at the end. This lopsided nature of the boundary conditions results in asymmetric time evolution. Thus the operation of the second law of thermodynamics in our Universe may come about because the Universe was *prepared* in a low-entropy state; in this view, the boundary condition of the Universe is therefore the driving force for the asymmetry in time. Or is it that it is something to do with the operation of the microscopic laws, which leads to the asymmetry in the flow of time? Not perhaps as Boltzmann attempted, using classical mechanics, but at the quantum mechanical (or possibly at the quantum-gravity[7]) level? These questions are far from being resolved. We are so familiar with the arrow of time that we are not often struck how odd it is and how much it is out of alignment with our present understanding of the reversible, microscopic laws of physics.

[6]This wonderful phrase is used by Lockwood (2005).

[7]Wild speculations about quantum gravity are possible since we have, at present, no adequate theory of quantum gravity.

Chapter summary

- The local entropy production Σ is given by
$$\Sigma = \sum_k \nabla\phi_k \cdot \boldsymbol{J}_k = \sum_{ij} \nabla\phi_i L_{ij} \nabla\phi_j \geq 0.$$

- The Onsager reciprocal relations state that $L_{ij} = L_{ji}$.
- The Peltier effect is the liberation of heat at a junction owing to the flow of electrical current.
- The Seebeck effect is the development of a voltage in a circuit containing a junction between two metals owing to the presence of a temperature gradient between them.
- Onsager's reciprocal relations rest on the principle of microscopic reversibility. The arrow of time, which shows the direction in which irreversible processes occur, may result from the asymmetry in the boundary conditions.

Further reading

- A good introduction to non-equilibrium thermodynamics may be found in Kondepudi and Prigogine (1998), Plischke and Bergersen (1989) and chapter XII of Landau and Lifshitz (1980) (particularly §118–126).
- The problem of the arrow of time is discussed in a highly readable and thought-provoking fashion in Lockwood (2005).

Exercises

(34.1) If a system is held at fixed T, N, and V, show that the fluctuations in a variable x are governed by the probability function
$$p(x) \propto e^{-F(x)/k_B T}, \qquad (34.64)$$
where $F(x)$ is the Helmholtz function.

(34.2) For the thermoelectric problem considered in Section 34.4, show that
$$L_{EE} = T\sigma, \qquad (34.65)$$
$$L_{ET} = T^2 \epsilon \sigma, \qquad (34.66)$$
$$L_{TE} = T^2 \epsilon \sigma, \qquad (34.67)$$
$$L_{TT} = \kappa T^2 + \epsilon^2 T^3 \sigma. \qquad (34.68)$$

(34.3) At 0 °C the measured Peltier coefficient for a Cu–Ni thermocouple is 5.08 mV. Hence estimate the Seebeck coefficient at this temperature and compare your answer with the measured value of 20.0 μV K^{-1}.

(34.4) (a) Explain why the thermopower is a measure of the entropy per carrier.

(b) Consider a classical gas of charged particles and explain why the thermopower ϵ should be of the order of $k_B/e = 87\,\mu$V K^{-1} and be independent of temperature T.

(c) In a metal, the measured thermopower is much less than 87 μV K^{-1} and decreases as the metal is cooled. Give an argument for why one might expect the thermopower to behave as
$$\epsilon \approx \frac{k_B}{e} \frac{k_B T}{T_F}, \qquad (34.69)$$
where T_F is the Fermi temperature.

(d) In a semiconductor, the measured thermopower is much larger than 87 μV K^{-1} and increases as the semiconductor is cooled. Give an argument for why one might expect the thermopower to behave as
$$\epsilon \approx \frac{k_B}{e} \frac{E_g}{2k_B T}, \qquad (34.70)$$
where E_g is the energy gap of the semiconductor.

(e) Since thermopower is a function of the entropy of the carriers, the third law of thermodynamics leads one to expect that it should go to zero as $T \to 0$. Is this a problem for the semiconductor considered in (d)?

35 Stars

35.1 Gravitational interaction 421
35.2 Nuclear reactions 426
35.3 Heat transfer 427
Chapter summary 434
Further reading 434
Exercises 434

[1] There are thought to be at least 10^{12} galaxies in the observable Universe. On average, a galaxy might contain 10^{11} stars.

[2] The interstellar medium is the dilute gas, dust, and plasma, which exists between the stars within a galaxy.

[3] Luminosity is a term used to mean energy radiated per unit time, i.e., power, and has the units of watts. In astrophysics, one often uses **spectral luminosity** (which is often what astrophysicists mean when they say luminosity), which is the power radiated per unit energy band or per wavelength interval or per frequency interval, and so in the latter case has units W Hz^{-1}.

[4] The adjective *Galactic* pertains to our own Galaxy, the **Milky Way**, while the adjective *galactic* pertains to galaxies in general.

In this chapter we apply some of the concepts of thermal physics developed earlier in this book to **stellar astrophysics**. Astrophysics is the study of the physical properties of the Universe and the objects therein. In this field, we make the fundamental assumption that the laws of physics, including those governing the properties of atoms and gravitational and electromagnetic fields, which are all largely obtained from experiment on Earth, are valid throughout the entire Universe, way beyond the confines of the Solar System where they have been well tested. It is further assumed that the fundamental constants do not vary in time and space.

The Universe contains a great many galaxies.[1] Each of these galaxies contain a great many stars, which are born out of the condensation of the denser gas in the **interstellar medium**.[2] Gravitational collapse produces extremely high temperatures permitting fusion to take place and hence the radiation of energy. Stars live and evolve, seeming to follow the laws of physics with impressive obedience, changing size, temperature, and **luminosity**.[3] Ultimately stars die, some exploding as supernovae and returning their mass (at least partially) to the Galactic[4] interstellar medium.

The star about which we know the most is the Sun. It seems to be a rather average star in our Galaxy, and some of its properties are summarized in the following box. The first three properties are measured while the remaining ones are model dependent.

Solar quantities:			
	Mass	$M_\odot$	1.99×10^{30} kg
	Radius	$R_\odot$	6.96×10^8 m
	Luminosity	$L_\odot$	3.83×10^{26} W
	Effective temperature	T_{eff}	5780 K
	Age	$t_\odot$	4.55×10^9 yr
	Central density	ρ_c	1.45×10^5 kg m^{-3}
	Central temperature	T_c	15.6×10^6 K
	Central pressure	p_c	2.29×10^{16} Pa

Stellar astrophysics, the subject of this chapter, is a very interesting field because using fairly simple physics, we can make predictions that can be tested observationally. We will consider the main processes that determine the properties of stars (Section 35.1 gravity, Section 35.2 nuclear reactions and Section 35.3 heat transfer) and, importantly, derive the main equations of stellar structure used to model stars. We will not, however, address more complicated issues such as magnetic fields in stars or detailed particle physics. In the following chapter, we will consider what happens to stars at the ends of their lives.

35.1 Gravitational interaction

The fundamental force, which causes stars to form and which produces huge pressures and temperatures in the centre of stars, is gravity. In this section, we explore how the effect of gravity governs the behaviour of stars.

35.1.1 Gravitational collapse and the Jeans criterion

How do stars form in the first place? For a gas cloud to condense into stars, the cloud must be sufficiently dense that attractive gravitational forces predominate over the pressure (which is proportional to the internal energy) otherwise the cloud would expand and disperse. The critical condition for condensation, i.e., for a gas cloud to be gravitationally bound, is that the total energy E must be less than zero. Now $E = U + \Omega$, where U is the kinetic energy and Ω is the gravitational potential energy. To be gravitationally bound requires $E < 0$ and hence $-\Omega > U$. The gravitational potential energy is negative, and hence the condition for condensation is

$$|\Omega| > U. \tag{35.1}$$

Now consider a spherical gas cloud of radius R and mass M, which is in thermal equilibrium at temperature T. The cloud consists of N particles, each assumed to be of the same type and of mass $m = M/N$. The gravitational potential energy of this cloud is given by

$$\Omega = -f\frac{GM^2}{R}, \tag{35.2}$$

where G is the gravitational constant and f is a factor of order unity, which reflects the density profile within the cloud.[5] For simplicity, we will set $f = 1$ in what follows. The thermal kinetic energy U of the cloud is found by assuming that each particle contributes $\frac{3}{2}k_\mathrm{B}T$, so that

$$U = \frac{3}{2}Nk_\mathrm{B}T. \tag{35.3}$$

Thus making use of eqn 35.1, a gas cloud will collapse if its mass M exceeds the **Jeans mass**[6] M_J given by

$$M_\mathrm{J} = \frac{3k_\mathrm{B}T}{2Gm}R. \tag{35.4}$$

Thus the Jeans mass is the minimum mass of a gas cloud that will collapse under gravity. Increasing the temperature T causes the particles

[5] For a spherical cloud of uniform density, $f = \frac{3}{5}$. For a spherical shell, $f = 1$.

[6] Sir James Jeans (1877–1946).

to move faster and thus makes it harder for the cloud to collapse; increasing the mass m of each particle favours gravitational collapse. The condition

$$M > M_J \tag{35.5}$$

is known as the **Jeans criterion**. It is often helpful to write the Jeans mass in terms of the density ρ of the cloud given by

$$\rho = \frac{M}{\frac{4}{3}\pi R^3}, \tag{35.6}$$

assuming spherical symmetry. This can be rearranged to give

$$R = \left(\frac{3M}{4\pi\rho}\right)^{\frac{1}{3}}, \tag{35.7}$$

and hence the Jeans criterion can be written as

$$R > R_J = \left(\frac{9k_BT}{8\pi Gm\rho}\right)^{\frac{1}{2}}, \tag{35.8}$$

where R_J is the **Jeans length**. Substitution of eqn 35.8 into eqn 35.4 yields another expression for the Jeans mass:

$$M_J = \left(\frac{3k_BT}{2Gm}\right)^{\frac{3}{2}}\left(\frac{3}{4\pi\rho}\right)^{\frac{1}{2}}. \tag{35.9}$$

Equivalently, one may also write that a cloud of mass M will condense if its average density exceeds

$$\rho_J = \frac{3}{4\pi M^2}\left[\frac{3k_BT}{2Gm}\right]^3, \tag{35.10}$$

where ρ_J is known as the **Jeans density**.

Example 35.1

What is the Jeans density of a cloud composed of hydrogen atoms and with total mass $M_\odot$ at $10\,\text{K}$?
Solution:
Using eqn 35.10,

$$\rho_J = \frac{3}{4\pi M_\odot^2}\left[\frac{3k_B \times 10}{2Gm_H}\right]^3 \approx 5 \times 10^{-17}\,\text{kg}\,\text{m}^{-3}, \tag{35.11}$$

which corresponds to about 3×10^{10} particles per cubic metre.

35.1.2 Hydrostatic equilibrium

As we have seen, gravity is responsible for gas clouds condensing into stars. It also contributes to the pressure inside a star. Consider a spherical body of gas of mass M and radius R, in which the only forces acting are due to gravity and internal pressure (see Fig. 35.1). The mass enclosed by a spherical shell of radius r is

$$m(r) = \int_0^r \rho(r') 4\pi r'^2 dr', \qquad (35.12)$$

where $\rho(r)$ is the density of the star at radius r, and so is responsible for a gravitational acceleration given by

$$g(r) = \frac{Gm(r)}{r^2}. \qquad (35.13)$$

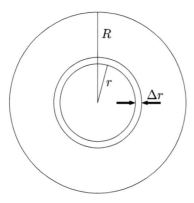

Fig. 35.1 Schematic illustration of a star of mass M and total radius R. Consider a small element at radius r from the centre having area ΔA perpendicular to the radius. We denote the pressure on the inner surface of the element at radius r as p and that at radius $r + \Delta r$ as $p + (dp/dr)\Delta r$.

In equilibrium, this is balanced by the internal pressure p of the star. Consider a small volume element at radius r extending to $r + \Delta r$ and having cross-sectional area ΔA. The force on the element due to the pressure either side is given by

$$\left[p(r) + \frac{dp}{dr} \Delta r \right] \Delta A - p(r) \Delta A = \frac{dp}{dr} \Delta r \Delta A. \qquad (35.14)$$

The gravitational attraction of the mass $m(r)$ within radius r is equal to $g(r)\rho(r)\Delta r \Delta A = g(r)\Delta M$. Since the mass of the element ΔM is given by $\rho(r)\Delta r \Delta A$, the inwards acceleration of any element of mass at distance r from the centre due to gravity and pressure is

$$-\frac{d^2 r}{dt^2} = g(r) + \frac{1}{\rho(r)} \frac{dp}{dr}. \qquad (35.15)$$

If the star is gravitationally stable it is said to be in **hydrostatic equilibrium** and an elemental volume will undergo no acceleration towards the centre of the star since the gravitational acceleration $g(r) =$

$Gm(r)/r^2$ will be balanced by the increased pressure on the inner surface compared with that on the outer surface. If this is true for all r then the left-hand side of eqn 35.15 will be zero, enabling us to rewrite this in a form known as the **equation of hydrostatic equilibrium**.

$$\frac{dp}{dr} = -\frac{Gm(r)\rho(r)}{r^2}. \tag{35.16}$$

The equation of hydrostatic equilibrium is one of the fundamental equations satisfied by static stellar structures.

35.1.3 The virial theorem

The **virial theorem** relates the average pressure (related to the internal kinetic energy) needed to support a self-gravitating system, thus balancing the gravitational potential energy with the kinetic energy. To derive this we first need to relate pressure to internal kinetic energy. Recall from Section 11.3 that the adiabatic index γ is used to describe the relation between the pressure and the volume of a gas during adiabatic compression or expansion, i.e., when the internal energy changes solely because of the work done on it. For such a process, pV^γ is a constant, and so we may write

$$0 = \gamma \frac{dV}{V} + \frac{dp}{p}. \tag{35.17}$$

Hence we can write

$$d(pV) = p\,dV + V\,dp = -(\gamma - 1)p\,dV. \tag{35.18}$$

If we denote the internal energy due to translational kinetic energy by dU then

$$dU = -p\,dV, \tag{35.19}$$

and hence

$$dU = \frac{1}{\gamma - 1}d(pV). \tag{35.20}$$

If the adiabatic index is a constant (which is not the case if different energy levels, for example rotational and vibrational, become excited) then this equation simply integrates to

$$U = \frac{pV}{\gamma - 1}. \tag{35.21}$$

Hence the internal energy density $u = U/V$ is given by

$$u = \frac{p}{\gamma - 1}. \tag{35.22}$$

Example 35.2

Use eqn 35.22 to derive the energy density of a gas of (i) non-relativistic particles ($\gamma = \frac{5}{3}$) and (ii) relativistic particles ($\gamma = \frac{4}{3}$).

Solution:
Straightforward substitution into eqn 35.22 yields

$$u = \frac{3}{2}p \quad \text{for } \gamma = \frac{5}{3}, \quad (35.23)$$

$$u = 3p \quad \text{for } \gamma = \frac{4}{3}, \quad (35.24)$$

in agreement with eqns 6.25 and 25.21.

The next part of the derivation of the virial theorem proceeds by multiplying both sides of the hydrostatic equation (eqn 35.16) by $4\pi r^3$ and integrating with respect to r from $r = 0$ to $r = R$. This leads to

$$\int_0^R 4\pi r^3 \frac{dp}{dr} dr = -\int_0^R \frac{Gm(r)\rho(r)}{r} 4\pi r^2 dr, \quad (35.25)$$

which becomes

$$\left[p(r)4\pi r^3\right]_0^R - 3\int_0^R p(r)\,4\pi r^2 dr = -\int_{m=0}^{m=M} \frac{Gm(r)}{r} dm. \quad (35.26)$$

The first term on the left-hand side is zero, because the surface of a star is defined to be at the radius where the pressure has fallen to zero. The second term on the left-hand side is equal to $-3\langle p\rangle V$, where V is the star's entire volume and $\langle p\rangle$ is the average pressure. The right-hand side is the gravitational potential energy of the star, Ω, so eqn 35.26, which came from the equation of hydrostatic equilibrium, leads us to

$$\langle p\rangle V = -\frac{\Omega}{3}, \quad (35.27)$$

which is a statement of the virial theorem. Equation 35.27 substituted into eqn 35.21 [which implies that $\langle p\rangle V = (\gamma - 1)U$] yields

$$3(\gamma - 1)U + \Omega = 0, \quad (35.28)$$

which is another statement of the virial theorem. The total energy E is the sum of the potential energy Ω and the kinetic energy U, i.e.,

$$E = U + \Omega. \quad (35.29)$$

Putting together eqns 35.28 and 35.29 gives

$$E = (4 - 3\gamma)U = \frac{3\gamma - 4}{3(\gamma - 1)}\Omega. \quad (35.30)$$

Example 35.3

Use eqns 35.28 and 35.30 to relate U, Ω and E for a gas of (i) non-relativistic particles ($\gamma = \frac{5}{3}$) and (ii) relativistic particles ($\gamma = \frac{4}{3}$).

Solution:

(i) For a gas of non-relativistic particles, we have (using $\gamma = \frac{5}{3}$ in eqns 35.28 and 35.30) that

$$2U + \Omega = 0, \qquad (35.31)$$

and hence

$$E = -U = \frac{\Omega}{2}. \qquad (35.32)$$

Since the kinetic energy U is positive, the total energy E is negative thus the system is bound. Moreover, this shows that if the total energy E of a star decreases, this corresponds to a decrease in the gravitational potential energy Ω, but an *increase* in the kinetic energy U. Since U is directly related to the temperature T, we conclude that a star has a "negative heat capacity": as a star radiates energy (E decreases), it contracts and heats up! This allows the nuclear heating process to be, to some extent, self-regulating. If a star loses energy from its surface, it contracts and heats up; therefore nuclear burning can increase, leading to an expansion, which cools the stellar core.

(ii) For a gas of relativistic particles, we have that (using $\gamma = \frac{4}{3}$ in eqns 35.28 and 35.30)

$$U + \Omega = 0, \qquad (35.33)$$

and hence

$$E = 0. \qquad (35.34)$$

Because the total energy is zero, a gravitationally bound state is not stable.

35.2 Nuclear reactions

The energy production in a star is dominated by **nuclear reactions**. These reactions are **fusion** processes, in which two or more nuclei combine and release energy. This is often called **nuclear burning**, though note that it is not burning in the conventional sense (we normally use the term burning to denote a chemical reaction with atmospheric oxygen; here we are talking about nuclear fusion reactions). Young stars are composed mainly of hydrogen, and the most important fusion reaction is **hydrogen burning** via the so-called **PP chain**, the first part of which is known as PP1 and is described in eqns 35.35. In these equations, ^{1}H is hydrogen, ^{2}D is deuterium, γ is a photon, and ^{3}He and ^{4}He are isotopes of helium.

$$\begin{aligned} ^1\text{H} + {^1\text{H}} &\rightarrow {^2\text{D}} + e^+ + \nu; \\ ^2\text{D} + {^1\text{H}} &\rightarrow {^3\text{He}} + \gamma; \\ ^3\text{He} + {^3\text{He}} &\rightarrow {^4\text{He}} + {^1\text{H}} + {^1\text{H}}. \end{aligned} \qquad (35.35)$$

This process releases 26.5 MeV of energy (of which about 0.3 MeV is carried away by the neutrino ν) by converting four ^{1}H to one ^{4}He. When helium becomes sufficiently abundant in the star, it too can burn in further cycles. Additional reactions can occur involving carbon and nitrogen to produce oxygen, which catalyse a further helium-burning reaction and is called the **CNO cycle**. This complex series of reactions, and other such cycles, which can produce elements as heavy as Fe, are now quite well understood, and can be used to understand the observed abundance of various chemical elements in the Universe, and to infer primordial abundances.

We will not examine the details of these reactions here, but suffice to say that when hydrogen is transmuted into iron via various complicated reaction pathways, the maximum possible energy release is equivalent to about 0.8% of the converted mass. In other words, the **mass defect** is 0.008. Hence the total energy available to the Sun can be estimated as $0.008 M_\odot c^2$, leading to an estimated solar lifetime $t_\odot^{\text{lifetime}}$ given by

$$t_\odot^{\text{lifetime}} \sim \frac{0.008 M_\odot c^2}{L_\odot} \sim 10^{11} \text{ years}. \tag{35.36}$$

The current age of the Sun is estimated to be 4.55×10^9 years, and hence the very rough estimate of the total solar lifetime is not obviously unrealistic. In fact, the long lifetimes of stars can only be explained by nuclear reactions.

35.3 Heat transfer

We have just seen that the release of nuclear energy is responsible for much of the energy radiated from stars. Energy is also released (or absorbed) owing to gravitational contraction (or expansion). A small mass dm makes a contribution dL to the total luminosity L of a star given by

$$\mathrm{d}L = \epsilon \mathrm{d}m, \tag{35.37}$$

where ϵ is the total power released per unit mass by nuclear reactions and gravity. For a spherically symmetric star, the luminosity d$L(r)$ of a thin shell of radius dr (and mass d$m = 4\pi r^2 \rho \mathrm{d}r$), and writing $\epsilon = \epsilon(r)$, is

$$\frac{\mathrm{d}L(r)}{\mathrm{d}r} = 4\pi r^2 \rho \epsilon(r). \tag{35.38}$$

How the luminosity varies with radius depends on how the heat is transported to the surface of the star, either by photon diffusion or by convection. We consider each of these in turn.

35.3.1 Heat transfer by photon diffusion

The passage of photons through a star towards its surface is a diffusive process and is precisely analogous to thermal conductivity via the free electrons in a metal. As such, we may use eqn 9.14 to describe the radial heat flux $J(r)$

$$J(r) = -\kappa_{\text{photon}}\left(\frac{\partial T}{\partial r}\right), \qquad (35.39)$$

where κ_{photon} is the thermal conductivity (see Section 9.2) due to photons in the star. Treating the photons as a classical gas, we can use a result from the kinetic theory of gases, eqn 9.18, to write

$$\kappa_{\text{photon}} = \frac{1}{3}Cl\langle c\rangle, \qquad (35.40)$$

where C here is the heat capacity of the photon gas per unit volume, l is the mean free path[7] for photons and $\langle c \rangle$ is the mean speed of the particles in the "gas", which here can be equated to the speed of light c. The heat capacity per unit volume C may be obtained from the energy density of a gas of photons in thermal equilibrium at temperature T, which is given by

$$u = \frac{4\sigma}{c}T^4 \qquad (35.41)$$

and from which we may derive the heat capacity per unit volume $C = \mathrm{d}u/\mathrm{d}T$ as

$$C = \frac{16\sigma}{c}T^3. \qquad (35.42)$$

[7] We have used the symbol l for mean free path in this section so as to keep the symbol λ for wavelength.

We next turn to the mean free path of the photons. This is determined by any process that results in photons being absorbed or scattered. Consider a beam of light with specific intensity[8] I_λ at wavelength λ. The change in specific intensity $\mathrm{d}I_\lambda$ of this beam as it travels through the stellar material is proportional to its specific intensity I_λ, the distance it has travelled $\mathrm{d}s$, and the density of the gas ρ. So we have

$$\mathrm{d}I_\lambda = -\kappa_\lambda \rho I_\lambda \mathrm{d}s, \qquad (35.43)$$

[8] The specific intensity in this section is the same thing as spectral radiance. Radiative transfer is considered in more detail in Section 37.3.

where the minus sign above shows that the specific intensity decreases with distance due to absorption. The constant κ_λ is called the **absorption coefficient** or **opacity**.[9] Equation 35.43 integrates to a dependence on distance of the form $I_\lambda(s) = I_\lambda(0)e^{-s/l}$ where $l = 1/(\kappa_\lambda \rho)$ is the mean free path. Hence we obtain a new and useful expression for the thermal conductivity of a gas of photons by substituting eqn 35.42 into eqn 35.40:

[9] It is also called the extinction coefficient, see Section 37.3, and sometimes the mass extinction cross-section.

$$\kappa = \frac{16}{3}\frac{\sigma T^3}{\kappa(r)\rho(r)}. \qquad (35.44)$$

The total radiative flux at radius r is $4\pi r^2 J(r)$ and this is equal to $L(r)$. Hence using eqn 35.39 and 35.44, we can write

$$L(r) = -4\pi r^2 \frac{16\sigma[T(r)]^3}{3\kappa(r)\rho(r)}\frac{\mathrm{d}T}{\mathrm{d}r}. \qquad (35.45)$$

For many stars, the dominant heat transfer mechanism is radiative diffusion, which crucially depends on the temperature gradient $\mathrm{d}T/\mathrm{d}r$.

We can now summarize the main equations of stellar structure that we have obtained so far.

Equations of stellar structure

$$\frac{dm(r)}{dr} = 4\pi r^2 \rho(r) \qquad (35.12)$$

$$\frac{dp(r)}{dr} = -\frac{Gm(r)\rho(r)}{r^2} \qquad (35.16)$$

$$\frac{dL(r)}{dr} = 4\pi r^2 \rho \epsilon(r) \qquad (35.38)$$

$$\frac{dT}{dr} = -\frac{3\kappa(r)\rho(r)L(r)}{64\pi r^2 \sigma [T(r)]^3} \qquad (35.45)$$

In these equations, the energy release due to nuclear reactions, $\epsilon(r)$, may need to be corrected for a term that includes the release of gravitational potential energy. Under certain circumstances, this term may in fact be dominant.[10] These equations ignore convection, which we will consider in the following section.

[10] One also has to consider the heat capacity of the stellar material whenever the stellar structure changes.

35.3.2 Heat transfer by convection

If the temperature gradient exceeds a certain critical value then the heat transfer in a star is governed by convection. The following analysis was first produced by Schwarzschild[11] in 1906.

[11] Karl Schwarzschild (1873–1916).

Consider a parcel of stellar material at radius r having initial values of density and pressure $\rho_*(r)$ and $p_*(r)$ respectively. The parcel subsequently rises by a distance dr through ambient material of density and pressure $\rho(r)$ and $p(r)$. Initially the parcel is in pressure equilibrium with its surroundings and so $p_*(r) = p(r)$; it initially has the same density as its surroundings, and hence $\rho_*(r) = \rho(r)$. We will assume that the parcel rises adiabatically, and hence $p_* \rho_*^{-\gamma}$ is constant[12] where γ is the adiabatic index. The parcel will be buoyant and will continue to rise if its density is lower than that of its surroundings (see Fig. 35.2), i.e., convection is possible if

[12] This follows from pV^γ being constant; see eqn 12.16.

$$\rho_* < \rho, \qquad (35.46)$$

which implies that

$$\frac{d\rho_*}{dr} < \frac{d\rho}{dr}. \qquad (35.47)$$

Because the parcel rises adiabatically, the constancy of $p_* \rho_*^{-\gamma}$ implies that

$$\frac{1}{p_*}\frac{dp_*}{dr} = \frac{\gamma}{\rho_*}\frac{d\rho_*}{dr}. \qquad (35.48)$$

We can treat the ambient material as an ideal gas (so that $p \propto \rho T$; see

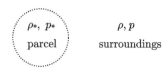

Fig. 35.2 A parcel of stellar material (of density ρ_*) will rise in its surroundings (density ρ) if $\rho_* < \rho$. This is the condition for **convection** to occur.

eqn 6.18) and hence
$$\frac{1}{p}\frac{dp}{dr} = \frac{1}{\rho}\frac{d\rho}{dr} + \frac{1}{T}\frac{dT}{dr}. \tag{35.49}$$

Substituting eqns 35.48 and 35.49 into eqn 35.47 leads to
$$\frac{1}{\gamma}\frac{\rho_*}{p_*}\frac{dp_*}{dr} < \frac{\rho}{p}\frac{dp}{dr} - \frac{\rho}{T}\frac{dT}{dr}. \tag{35.50}$$

Since pressure equilibration happens very rapidly, we can assume that $p(r) = p_*(r)$. Moreover, $\rho_* \approx \rho$ to first order, and hence
$$\left(\frac{1}{\gamma} - 1\right)\frac{dp}{dr} < -\frac{p}{T}\frac{dT}{dr}, \tag{35.51}$$

and thus
$$\frac{dT}{dr} < \left(1 - \frac{1}{\gamma}\right)\frac{T}{p}\frac{dp}{dr} \tag{35.52}$$

is the condition for convection to occur. In fact, both temperature and pressure decrease with increasing distance from the centre in a star; hence both the temperature and pressure gradients are negative. Thus, it is more convenient to write the condition for convection to occur as
$$\left|\frac{dT}{dr}\right| > \left(1 - \frac{1}{\gamma}\right)\frac{T}{p}\left|\frac{dp}{dr}\right|. \tag{35.53}$$

In this equation, the pressure gradient is governed by hydrostatic equilibrium, which we met in eqn 35.16. This equation shows that convection will occur if the temperature gradient is very large, or because γ becomes close to 1 (which makes the right-hand side of the equation small) and occurs when a gas becomes partially ionized (see Exercise 35.5).

35.3.3 Scaling relations

To find the detailed pressure and temperature dependences inside a star requires one to solve eqns 35.12, 35.16, 35.38 and 35.45 simultaneously (these equations are tabulated in the box at the end of Section 35.3.1) with a realistic form of the opacity $\kappa(r)$. This is very complicated and has to be performed numerically. However, we can gain considerable insight into general trends by deriving **scaling relations**. To do this, we assume that the principle of **homology** applies, which says that if a star of total mass M expands or contracts, its physical properties change by the same factor at all radii. This means that the radial profile as a function of the fractional mass is the same for all stars, the only difference being a constant factor, which depends on the mass M. For example, this implies that a pressure interval dp scales in exactly the same way as the central pressure p_c, and that the density profile $\rho(r)$ scales in the same way as the mean density ρ. The following example demonstrates the use of the principle of homology in deriving scaling relations for various stellar properties.

Example 35.4

Using the principle of homology for a star of total mass M and radius R, show that (a) $p(r) \propto R^{-4}$ and (b) $T(r) \propto R^{-1}$.

Solution:

(a) The equation for hydrostatic equilibrium, eqn 35.16, states that

$$\frac{dp}{dr} = \frac{Gm(r)\rho(r)}{r^2}, \qquad (35.54)$$

so using $\rho \propto MR^{-3}$ and writing $dp/dr = p_c/R$, we deduce that

$$\frac{p_c}{R} \propto M^2 R^{-5}. \qquad (35.55)$$

Equation 35.55 means that $p_c \propto M^2 R^{-4}$, and using the principle of homology,

$$p(r) \propto M^2 R^{-4}. \qquad (35.56)$$

(b) We next consider a relationship for scaling the temperature throughout a star. Our starting point this time is the ideal gas law, which we met in eqn 6.18, from which we may write the following:

$$T(r) \propto \frac{p(r)}{\rho(r)}. \qquad (35.57)$$

Using $\rho \propto MR^{-3}$ and eqn 35.56, we have

$$T(r) \propto MR^{-1}. \qquad (35.58)$$

Hence as the star shrinks, its central temperature increases. Note that this does not give information on the surface temperature $T(R)$, since this depends on the precise form of $T(r)$.

For a low–mass star, the opacity $\kappa(r)$ increases with density and decreases with temperature roughly according to

$$\kappa(r) \propto \rho(r) T(r)^{-3.5}, \qquad (35.59)$$

which is known as **Kramers opacity**.[13] In this case, scaling yields (via $\rho \propto MR^{-3}$ and eqn 35.58)

$$\kappa(r) \propto M^{-2.5} R^{0.5}. \qquad (35.60)$$

For a very massive star, in which electron scattering dominates the opacity, $\kappa(r)$ is a constant.

[13] H. A. Kramers (1894–1952).

Example 35.5

Determine the scaling of the luminosity L with M and R for (a) a low-mass star and (b) a high-mass star.

Solution: By the principle of homology, a temperature increment $\mathrm{d}T$ scales in the same way as T, which eqn 35.58 gives as $T(r) \propto MR^{-1}$. An increment in radius, however, scales with radius, i.e., $\mathrm{d}R \propto R$. Therefore the temperature gradient follows $\mathrm{d}T/\mathrm{d}r \propto MR^{-1}/R$, giving

$$\frac{\mathrm{d}T}{\mathrm{d}r} \propto MR^{-2}. \tag{35.61}$$

Equation 35.45 becomes

$$\frac{L(r)}{r^2} \propto -\frac{T(r)^3}{\rho(r)\kappa(r)}\frac{\mathrm{d}T}{\mathrm{d}r}, \tag{35.62}$$

and hence in case (a), for which $\kappa(r) \propto \rho(r)T(r)^{-3.5}$, we find

$$L(r) \propto \frac{M^{5.5}}{R^{0.5}}. \tag{35.63}$$

The assumption of homology means that if the luminosity at any radius r scales as $M^{5.5}R^{-0.5}$, then the surface luminosity scales in this way, so we may write

$$L \propto \frac{M^{5.5}}{R^{0.5}}. \tag{35.64}$$

For case (b), since $\kappa(r)$ is a constant, we find $L(r) \propto M^3$ and hence

$$L \propto M^3. \tag{35.65}$$

[14] Ejnar Hertzsprung 1873–1967, Henry Norris Russell 1877–1957.

The **Hertzsprung–Russell diagram**[14] is a plot for a collection of stars of the luminosity against effective surface temperature T_{eff}, where the latter quantity is obtained by measuring the colour of a star, and hence the wavelength of its peak emission which is inversely proportional to T_{eff} by Wien's law. Fig 35.3 shows a Hertzsprung–Russell diagram for a selection of stars in our Galaxy. The most striking feature of this diagram is the **main sequence**, which represents stars that are burning mainly hydrogen; this is how almost all stars spend most of their "active" life. The correlation between L and T_{eff} occurs because both quantities depend on the star's mass. Empirically it is found that, for main sequence stars, $L \propto M^a$, where a is a positive constant which takes a value of about 3.5 (which is intermediate between the value of 5.5 for low-mass stars and 3 for massive stars, as we found in Example 35.5). Note that the lifetime of a star must be proportional to M/L (since the total mass M measures how much "fuel" is "on board") and hence is proportional to M^{1-a}. Hence more massive stars burn up faster than less massive stars.

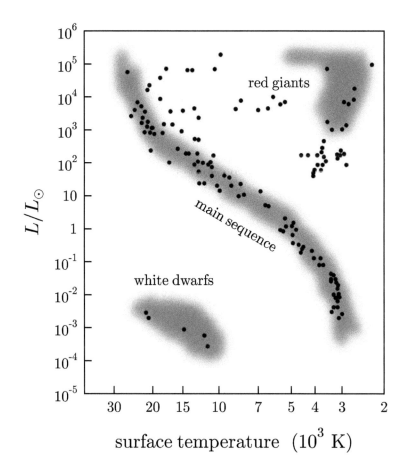

Fig. 35.3 A schematic Hertzsprung–Russell diagram.

The Hertzsprung–Russell diagram in Fig. 35.3 also shows various **red giants**, which are stars that have exhausted their supply of hydrogen in their cores. Red giants are very luminous due to a very hot inert helium core (far hotter than in a main-sequence star), which causes the hydrogen shell around it (which undergoes nuclear fusion) to greatly expand; the surface is very large and cooler, leading to a lower surface temperature. Eventually, the temperature in the helium core rises so high that beryllium and carbon can be formed; the outer part of the core can be ejected leading to the formation of a **nebula**, and the remaining core can collapse to form a white dwarf. White dwarfs, which are not very luminous but have a high surface temperature, will be described in the following chapter. It is expected that our own Sun will eventually pass through a red-giant phase, the core of which will ultimately become a white dwarf.

Chapter summary

- A gas cloud will condense if its density is above the Jeans density.
- The equation of hydrostatic equilibrium is

$$\frac{dp(r)}{dr} = -\frac{Gm(r)\rho(r)}{r^2}.$$

- The luminosity obeys

$$\frac{dL(r)}{dr} = 4\pi r^2 \rho \epsilon(r).$$

- The temperature profile inside a star obeys

$$\frac{dT}{dr} = -\frac{3\kappa(r)\rho(r)L(r)}{64\pi r^2 \sigma [T(r)]^3}.$$

- The virial theorem states that

$$\langle p \rangle V = -\frac{\Omega}{3} \quad \text{and} \quad 3(\gamma - 1)U + \Omega = 0.$$

Further reading

Recommended texts on stellar physics include Binney and Merrifield (1998), Prialnik (2000), Carroll and Ostlie (1996), and Zeilik and Gregory (1998).

Exercises

(35.1) Estimate the number of protons in the Sun.

(35.2) Find the critical density for condensation of a cloud of molecular hydrogen gas of total mass $1000 M_\odot$ at $20\,\text{K}$, expressing your answer in number of molecules per cubic metre. How would this answer change if (a) the mass of the cloud was only one solar mass, (b) the temperature was $100\,\text{K}$.

(35.3) Assume that the density of baryonic matter in the Universe is $3 \times 10^{-27}\,\text{kg m}^{-3}$ and that the distance to the edge of the Universe is given by $c\tau$ where τ is the age of the Universe, 13×10^9 years and c is the speed of light. Given that a typical galaxy has a mass $10^{11} M_\odot$, estimate the number of galaxies in the observable Universe. Estimate how many protons there are in the observable Universe, stating all your assumptions.

(35.4) Show that for a uniform density cloud in eqn 35.2, $f = 3/5$.

(35.5) Consider a gas consisting of neutral hydrogen atoms with number density n_0, protons of number density n_+, and electrons with number density $n_e = n_+$. The ionization potential is χ. Find C_V.

(35.6) Show that for low-mass stars, the luminosity L scales with the effective surface temperature T_{eff} and mass M according to $L \propto M^{22/5} T_{\text{eff}}^{4/5}$.

Compact objects

36

When a star is near the end of its lifetime, and all of its fuel is used up, there is no longer enough outwards pressure due to radiation to resist the inwards pull of gravity and the star starts to collapse again. However, there is another source of internal pressure. The electrons inside a star, being fermions, are subject to the Pauli exclusion principle and take unkindly to being squashed into a small space. They produce an outwards **electron degeneracy pressure** which we calculate in the following section. This concept leads to white dwarfs (Section 36.2) and, for the case of neutron degeneracy pressure, neutron stars (Section 36.3). More massive stars can turn into black holes (Section 36.4). We consider how mass can accrete onto such objects in Section 36.5 and consider the entropy of a black hole in Section 36.6.

36.1 Electron degeneracy pressure	435
36.2 White dwarfs	437
36.3 Neutron stars	438
36.4 Black holes	440
36.5 Accretion	441
36.6 Black holes and entropy	442
36.7 Life, the Universe, and entropy	443
Chapter summary	445
Further reading	445
Exercises	445

36.1 Electron degeneracy pressure

Using the results from Chapter 30 concerning fermion gases, we can write the Fermi momentum p_F as

$$p_F = \hbar(3\pi^2 n)^{1/3}, \qquad (36.1)$$

where n is the number density of electrons, so that equivalently n can be written as

$$n = \frac{1}{3\pi^2}\left(\frac{p_F}{\hbar}\right)^3. \qquad (36.2)$$

If we assume that the electrons behave non-relativistically, the Fermi energy is

$$E_F = \frac{p_F^2}{2m_e}, \qquad (36.3)$$

and the average internal energy density u is

$$u = \frac{3}{5} n E_F = \frac{3\hbar^2}{10 m_e}(3\pi^2)^{2/3} n^{5/3}. \qquad (36.4)$$

This gives an expression for the electron degeneracy pressure p_{electron} (using eqn 6.25) as

$$p_{\text{electron}} = \frac{2}{3}u = \frac{\hbar^2}{5m_e}(3\pi^2)^{2/3} n^{5/3}. \qquad (36.5)$$

We can relate the number density n of electrons to the density ρ of the star by the following argument. If the star contains nuclei with atomic

Note that our expression for the electron degeneracy pressure is inversely proportional to the electron mass m_e. This is why we have worried about electron degeneracy pressure, and not proton or neutron degeneracy pressure; the pressure produced by neutrons and protons is much smaller because they are more massive.

number Z and mass number A, each nucleus has mass Am_p and positive charge $+Ze$ (where $-e$ is the charge of an electron). For charge balance, for every nucleus there must be Z electrons. Hence, by ignoring the mass of the electrons themselves (which is much less than the mass of the nuclei), n is given by

$$n \approx \frac{Z\rho}{Am_p}. \tag{36.6}$$

Putting this into eqn 36.5, we find that the electron degeneracy pressure $p_{\text{electron}} \propto \rho^{5/3}$.

This outwards electron degeneracy pressure must balance the inwards pressure due to the gravitational force. This pressure, which we will here denote by p_{grav}, is related by eqn 35.27 to the gravitational potential energy Ω, which is given by

$$\Omega = -\frac{3GM^2}{5R}, \tag{36.7}$$

so that

$$p_{\text{grav}} = \frac{\Omega}{3V} = -\frac{G}{5}\left(\frac{4\pi}{3}\right)^{1/3} M^{2/3} \rho^{4/3}, \tag{36.8}$$

where we have used $\rho = M/V$ and $R^3 = 3M/(4\pi\rho)$ to obtain the final result.

Note the important results that, for non-relativistic electrons:

- The outwards pressure is $p_{\text{electron}} \propto \rho^{5/3}$.
- The inwards pressure is $p_{\text{grav}} \propto \rho^{4/3}$.

This leads to a stable situation since, if a star supported only by electron degeneracy pressure begins to shrink so that ρ begins to increase, the outwards pressure p_{electron} increases faster than p_{grav}, producing an outwards restoring force.

Example 36.1

What is the condition for balancing p_{electron} and p_{grav}?
Solution:
We set

$$p_{\text{electron}} = p_{\text{grav}}, \tag{36.9}$$

and using eqns 36.5 and 36.8 this implies that

$$\rho = \frac{4G^3 M^2 m_e^3}{27\pi^3 \hbar^6}\left(\frac{Am_p}{Z}\right)^5. \tag{36.10}$$

36.2 White dwarfs

A star supported from further collapse only by electron degeneracy pressure is called a **white dwarf**[1] and is the fate of many stars once they have exhausted their nuclear fuel. Equation 36.10 shows that

$$\rho \propto M^2, \quad (36.11)$$

which together with $\rho \propto M/R^3$ implies that

$$R \propto M^{-1/3}. \quad (36.12)$$

This implies that the radius of a white dwarf *decreases* as the mass *increases*.

[1] White dwarfs are called dwarfs because they are small and white because they are hot and luminous.

Example 36.2

What is the electron degeneracy pressure for relativistic electrons?
Solution:
The Fermi energy is now

$$E_F = p_F c, \quad (36.13)$$

and the average internal energy density is

$$u = \frac{3}{4} n E_F = \frac{3c\hbar}{4} (3\pi^2)^{1/3} n^{4/3}. \quad (36.14)$$

The pressure p_{electron} now follows from eqn 25.21 and is

$$p = \frac{u}{3} = \frac{c\hbar}{4} (3\pi^2)^{1/3} n^{4/3}. \quad (36.15)$$

Note the important result that, for relativistic electrons:

- The outwards pressure is $p_{\text{electron}} \propto \rho^{4/3}$.
- The inwards pressure is $p_{\text{grav}} \propto \rho^{4/3}$.

This leads to an unstable situation, since now if a star begins to shrink, so that ρ begins to increase, the outwards pressure p_{electron} increases at exactly the same rate as p_{grav}. Electron degeneracy pressure cannot halt further collapse.

We can estimate the mass above which the electrons in a white dwarf will behave relativistically. This will occur when

$$p_F \gtrsim m_e c, \quad (36.16)$$

and hence when

$$n \gtrsim \frac{1}{3\pi^2} \left(\frac{m_e c}{\hbar} \right)^3, \quad (36.17)$$

or equivalently

$$\rho \gtrsim \left(\frac{Am_{\rm p}}{Z}\right) \frac{1}{3\pi^2} \left(\frac{m_{\rm e}c}{\hbar}\right)^3. \tag{36.18}$$

Substituting in eqn 36.10 for ρ in this equation, and then rearranging, yields

$$M \gtrsim \left(\frac{Z}{Am_{\rm p}}\right)^2 \frac{3\sqrt{\pi}}{2} \left(\frac{\hbar c}{G}\right)^{3/2} \approx 1.2 M_\odot, \tag{36.19}$$

assuming that $Z/A = 0.5$ (appropriate for hydrogen). A more exact treatment leads to an estimate around $1.4 M_\odot$. This is known as the **Chandrasekhar limit**[2], and is the mass above which a white dwarf is no longer stable. Above the Chandrasekhar limit, the electron degeneracy pressure is no longer sufficient to support the star against gravitational collapse.

White dwarfs are fairly common and it is believed that most small and medium-size stars will end up in this state, often after going through a red-giant phase. The first-discovered white dwarf was Sirius B, the so-called dark companion of Sirius A (the brightest star visible in the night sky, to be found in the constellation of Canis Major), which is shown in an X-ray image in Fig. 36.1. Though Sirius B is much less bright in the visible region of the spectrum, it is a stronger emitter of X-rays because of its high temperature and thus appears as the brighter object in the X-ray image.

36.3 Neutron stars

Once a star is more massive than about $1.4 M_\odot$, electrons behave relativistically and cannot prevent further collapse. However, the star will contain neutrons and these will still be non-relativistic since the neutron mass is larger than the electron mass. Neutrons are fermions and their pressure, albeit lower than the electron pressure below the Chandrasekhar limit, will follow $\rho^{5/3}$ and can therefore balance the inwards gravitational pressure. Free neutrons decay with a mean lifetime of about 15 minutes, but in a star one has to consider the equilibrium

$$\rm n \rightleftharpoons p^+ + e^- + \nu_e. \tag{36.20}$$

Because the electrons are relativistic, their Fermi energy is proportional to $p_{\rm F} \propto n^{1/3}$, while the neutrons are non-relativistic and so their Fermi energy is proportional to $p_{\rm F}^2 \propto n^{2/3}$. Thus at high density, an equilibrium can be established in the reaction in eqn 36.20. This implies that the Fermi momentum of the electrons is much smaller than that of the neutrons, and hence the number density of electrons will be much smaller than that of the neutrons. This moves the equilibrium towards the left-hand side of eqn 36.20.

A compact object composed mainly of neutrons is called a **neutron star**. The first observational evidence of such an object came from the discovery of **pulsars** by Jocelyn Bell Burnell in 1967. These were soon

[2] Subrahmanyan Chandrasekhar 1910–1995

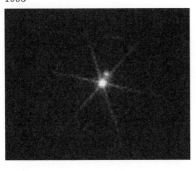

Fig. 36.1 Sirius is the brightest star in the night sky, but is actually a binary star. What you see with the naked eye is the bright normal star "Sirius A", but the small star in orbit around it, known as "Sirius B" (discovered by Alvan G. Clark in 1862), is a white dwarf. Because a white dwarf is so dense, it is very hot and can emit X-rays. The X-ray image shown in the figure was taken with the High Resolution Camera on the *Chandra* satellite. In this X-ray image the white dwarf Sirius B is much brighter than Sirius A. The bright "spokes" in the image are produced by X-rays scattered by the support structure of a diffraction grating, which was in the optical path for this observation. (Image courtesy of NASA.)

identified as rapidly rotating neutron stars emitting beams of radiation from their north and south magnetic poles. If their axis of rotation is not aligned with the poles, then lighthouse-type sweeping beams are produced as they rotate. When these intersect with the line of sight of an observer, pulses of radiation with a regular frequency are seen. The physical mechanism by which the radiation is emitted from pulsars is currently the subject of active research.

Neutron stars are thought to form from the collapsed remnant of a very massive star after a supernova explosion. Even though the mass of a neutron star is a few solar masses, they are very compact, having radii in the range 10–20 km (see Exercise 36.3). One such neutron star is found at the centre of the Crab Nebula (Fig. 36.2), in the constellation of Taurus. This object is 6500 light years from us and is the remnant of a supernova explosion, which was recorded by Chinese and Arab astronomers in 1054 as being visible during daylight for over three weeks. The neutron star at the centre currently rotates at a rate of thirty times per second.

Fig. 36.2 The Crab Nebula, as seen by the VLT telescope in Paranal, Chile. At the centre of the nebula is a neutron star. (Figure courtesy European Southern Observatory.)

Example 36.3

Estimate the minimum rotation period τ of a pulsar of radius R and mass M.

Solution:

For a neutron star rotating at $\omega = 2\pi/\tau$, the gravitational force at the equator GM/R^2 must be bigger than the centrifugal force $\omega^2 R$, so that

$$\tau = 2\pi\sqrt{\frac{R^3}{GM}}. \tag{36.21}$$

By analogy with a white dwarf, the mass M of a neutron star follows $M \propto R^{-1/3}$, so that more massive neutron stars are smaller than lighter ones. When the mass of a neutron star becomes very large, the neutrons behave relativistically and the neutron star becomes unstable.

Example 36.4

Above what mass will a neutron star become unstable?

Solution:

The high gravitational fields and compact nature of neutron stars mean that we really ought to include the effects of general relativity and the strong nuclear interactions. However, ignoring these, we can make an estimate on the basis that the neutron star will become unstable when the neutrons themselves become relativistic. By analogy with eqn 36.19, and taking $Z/A = 1$, we have the maximum mass[3] as

$$M \gtrsim \frac{3\sqrt{\pi}}{2m_p^2}\left(\frac{\hbar c}{G}\right)^{3/2} \approx 5M_\odot. \tag{36.22}$$

[3] Including general relativity reduces the maximum mass to about $0.7 M_\odot$, but including a more realistic equation of state raises the maximum mass up again, to somewhere around 2–$3 M_\odot$.

36.4 Black holes

If a neutron star undergoes gravitational collapse, there is no other pressure to balance the gravitational attraction and the gravitational collapse of the star is total. The result is a **black hole**. To treat black holes properly requires general relativity, but we can derive a few results about them using simple arguments. The escape velocity v_{esc} at the surface of a star can be obtained by equating kinetic energy $\frac{1}{2}mv_{esc}^2$ to the magnitude of the gravitational potential energy GMm/R so that

$$v_{esc} = \sqrt{\frac{2GM}{R}}. \tag{36.23}$$

For a black hole of mass M, the escape velocity reaches the speed of light, c, at the **Schwarzschild radius**[4] R_S given by

$$R_S = \frac{2GM}{c^2}. \tag{36.24}$$

[4]Karl Schwarzschild (1873–1916)

This result seems to imply that photons from a black hole cannot escape and the black hole appears black to an observer. Actually this is not quite true for two reasons, one practical and one esoteric.

(1) Matter falling into a black hole is ripped apart by the enormous gravitational tidal forces, well before it enters the **event horizon**[5] at the Schwarzschild radius. This results in powerful emission of X-rays and radiation at other wavelengths. Supermassive black holes at the centres of certain galaxies, the most luminous **active galactic nuclei** having masses $\gtrsim 10^8 M_\odot$, are responsible for the most powerful sustained electromagnetic radiation in the Universe.

[5]An event horizon is a mathematical, rather than physical, surface surrounding a black hole within which the escape velocity for a particle exceeds the speed of light, making escape impossible.

(2) Even neglecting this powerful observed emission, there is believed to be weak emission of radiation from black holes due to quantum fluctuations close to the event horizon. This **Hawking radiation** can be thought of as resulting from vacuum fluctuations which produce particle–antiparticle pairs in which one-half of the virtual pair falls into the black hole and the other half escapes. Hawking radiation has a black-body spectrum, implying that the black hole has a well-defined temperature. The **Hawking temperature** T_H of a black hole of mass M is given by

$$k_B T_H = \frac{\hbar c^3}{8\pi G M}, \tag{36.25}$$

so that as the black hole loses energy owing to Hawking radiation it becomes hotter. It also loses mass and this is termed **black-hole evaporation**. If we ignore all other processes, the lifetime of a black hole can be estimated using

$$\frac{dM}{dt} c^2 = -4\pi R_S^2 \sigma T_H^4, \tag{36.26}$$

which leads to a lifetime proportional to M^3. Thus small black holes evaporate due to Hawking radiation much faster than very massive ones.

36.5 Accretion

Black holes and neutron stars increase their mass as matter falls on to them. There is, however, a maximum rate of this **accretion** of mass onto any compact object.[6] This occurs because the higher the rate of accretion, the greater the luminosity due to the infalling matter, and hence a higher outwards radiation flux. Therefore the radiation pressure increases, pushing outwards on any further matter attempting to fall inwards and accrete. To analyse this situation, we will assume spherically

[6]Or, indeed, onto a normal star.

symmetric accretion and consider a very small piece of matter accreting onto a star at radius R. This piece of matter has density ρ, and volume $\mathrm{d}A\mathrm{d}R$. The gravitational force dragging it towards the star is

$$-\frac{GM}{R^2}\rho \mathrm{d}A\,\mathrm{d}R. \tag{36.27}$$

However, the radiation from the luminosity L of the stellar object produces a radiation pressure[7] on the falling matter, which results in an outwards force equal to

$$\frac{L}{4\pi R^2 c}\kappa\rho \mathrm{d}A\,\mathrm{d}R. \tag{36.28}$$

[7] The power per unit area is $L/4\pi R^2$ and the fraction of radiant energy absorbed by the piece of matter is given by eqn 35.43, i.e., $\kappa\rho \mathrm{d}R$. Dividing the product of these two terms by c gives the radiation pressure (see Section 23.4) and multiplying by $\mathrm{d}A$ gives a force.

The piece of matter will be able to accrete onto the star if the gravitational force dominates, so that

$$\frac{GM}{R^2}\rho \mathrm{d}A\,\mathrm{d}R > \frac{L\kappa\rho}{4\pi R^2 c}\mathrm{d}A\,\mathrm{d}R. \tag{36.29}$$

Hence a condition for accretion to occur is that the luminosity of the compact object is below a maximum limit, i.e.,

$$L < L_{\mathrm{edd}} = \frac{4\pi GMc}{\kappa}, \tag{36.30}$$

[8] Arthur Stanley Eddington (1882–1944)

[9] If the conversion of gravitational potential energy to luminosity, via accretion, is not perfectly efficient, then one needs to include a radiative efficiency factor. This depends on the specifics of the situation, such as whether one is dealing with a rotating or a stationary black hole.

where L_{edd} is the **Eddington luminosity**[8]. If the luminosity L of the compact object is itself entirely[9] produced by accreting matter, then the luminosity must be balanced by the change of gravitational potential energy and hence $L = GM\dot{M}/R$. This implies that there is a maximum rate of accretion given by

$$\dot{M}_{\mathrm{edd}} = \frac{4\pi c R}{\kappa}. \tag{36.31}$$

The treatment in this section assumes spherically symmetric accretion and luminosity. Many compact objects actually accrete mass at a rate above the **Eddington limit** given by eqn 36.31, by accreting near the object's equator but radiating photons from the object's polar regions.

36.6 Black holes and entropy

In this section we consider the entropy of black holes. If we ignore the quantum mechanical Hawking radiation, the mass of a black hole can only increase since mass can enter but not leave. This means that the event horizon expands and the area A of the horizon, given by $4\pi R_\mathrm{S}^2$, only increases. It turns out that the area of an event horizon can be associated with its entropy S according to

$$S = k_\mathrm{B}\frac{A}{4l_\mathrm{P}^2}, \tag{36.32}$$

where $l_\mathrm{P} = (G\hbar/c^3)^{1/2}$ is the **Planck length**, a result obtained by Hawking and Bekenstein. The entropy (and hence the area) of a black hole increases in all classical processes, as it should according to the second law of thermodynamics. Since all information concerning matter is lost when it falls into a black hole, the entropy of a black hole can be thought of as a large reservoir of missing information. Information can be measured in bits, and relating information to entropy (see Chapter 15) implies that for a black hole, one bit corresponds to four Planck areas (where the Planck area is l_P^2) on its surface. This is indicated schematically in Fig. 36.3.

The entropy of the black hole measures the uncertainty concerning which of its internal configurations are realized. We can speculate that a particular black hole may have been formed from a collapsing neutron star, the collapse of a normal star, or (somewhat improbably) the collapse of a giant cosmic spaghetti monster: we have no way of telling which, because all of this information has become completely *inaccessible* to us and all we can measure is the black hole's mass, charge, and angular momentum. Information about the black hole's past history or its current chemical composition is hidden from our eyes.

As the mass M of a black hole increases, so too does R_S and hence so does S. Therefore the maximal limit of entropy (and hence information) for any ordinary region of space is directly proportional not to the region's volume, but to its *area*. This is a counterexample to the usual rule that entropy is an extensive property, being proportional to volume. Although the entropy of a black hole increases in all classical processes, it decreases in the quantum mechanical black-hole evaporation due to Hawking radiation. Finding out what has happened to the information in black-hole evaporation, and whether information can ever escape from a black hole, is a current conundrum in black hole physics.

It is useful to consider what happens when a body containing ordinary entropy falls into a black hole. The ordinary body has entropy for the usual reasons, namely that it can exist in a wide variety of different configurations and its entropy expresses our uncertainty in knowledge of its precise configuration. All that entropy seems at first sight to have been lost when the body falls into the black hole, since it can now only exist in one single configuration: the state of being annihilated! It therefore appears that the entropy of the Universe has gone down. However, the increase in mass of the black hole leads to an increase in the black hole's area and hence in its entropy. It turns out that this more than compensates for any entropy apparently "lost" by matter falling into the black hole. This motivates Bekenstein's generalized second law of thermodynamics, which states that the sum of the usual entropy of matter in the Universe plus the entropy of the black holes never decreases.

Fig. 36.3 The entropy of a black hole is proportional to its area A. This corresponds to a quantity of information such that one bit is "stored" in four Planck areas across the surface of the black hole.

36.7 Life, the Universe, and entropy

We often hear it said that we receive our energy from the Sun. This is true, but though Earth receives about 1.5×10^{17} W of energy, mainly in ultraviolet and visible photons (radiation corresponding to the tem-

[10]See Chapter 37.

[11]See eqn 23.76.

perature on the surface of the Sun), the planet ultimately radiates it again as infrared photons (radiation corresponding to the temperature in Earth's atmosphere[10]). If we did not do this, our planet would get progressively warmer and warmer and so for the conditions on Earth to be approximately time independent, we require that the total solar energy arriving at Earth must balance the total energy leaving Earth. The crucial point is that the frequency of radiation coming in is higher than that going out; a visible or ultraviolet photon thus has more energy than an infrared photon so fewer photons arrive than leave. The entropy per photon[11] is a constant, independent of frequency, so that by having fewer high-energy photons coming in and a larger number of lower-energy photons leaving, the incoming energy is low-entropy energy while the energy that leaves is high entropy. Thus the Sun is, for planet Earth, a convenient low-entropy energy source and the planet benefits from this incoming flux of low-entropy energy. When we digest food, and our body builds new cells and tissue, we are extracting some low-entropy energy from the plant and animal matter we have eaten, all of which derives from the Sun. Similarly, the process of evolution over million of years, in which the complexity of life on Earth has increased with time, is driven by this flux of solar low-entropy energy.

Since the Universe is bathed in 2.7 K black-body radiation, the Sun, with its 6000 K surface temperature, is clearly in a non-equilibrium state. The "ultimate equilibrium state" of the Universe would be everything sitting at some uniform, low temperature, such as 2.7 K. During the Sun's lifetime, almost all its low-entropy energy will be dissipated, filling space with photons; they will travel through the Universe and eventually interact with matter. The resulting high-entropy energy will tend to feed eventually into the cosmic slush of the ultimate equilibrium state. However, it is in the process of these interacting with matter that fun can begin: life is a non-equilibrium state, and prospers on Earth through non-equilibrium states that are driven by the constant influx of low-entropy energy.

The origin of the Sun's low entropy is, of course, gravity. The Sun has gravitationally condensed from a uniform hydrogen cloud, which is a source of low entropy as far as gravity is concerned (the operation of gravity is to cause such a cloud to condense and the entropy increases as the particles clump together). The clouds of gas of course came from the matter dispersed in the big bang. A crucial insight is to realize that although the matter and electromagnetic degrees of freedom in the early Universe were in thermal equilibrium (i.e., in a thermalized, high-entropy state, and thus producing the almost perfectly uniform cosmic microwave background we see today), the gravitational degrees of freedom were not thermalized. These unthermalized gravitational degrees of freedom provided the reservoir of low entropy that could drive gravitational collapse, and hence lead to the emission of low-entropy energy from stars which can, in favourable circumstances, drive life itself.

Chapter summary

- Electron degeneracy pressure is proportional to $\rho^{5/3}$ for non-relativistic electrons and to $\rho^{4/3}$ for relativistic electrons. In the former case, it can balance the gravitational pressure, which is proportional to $\rho^{4/3}$.
- A white dwarf is stable up to $1.4 M_\odot$ and is supported by electron degeneracy pressure. Its radius R depends on mass M as $R \propto M^{-1/3}$.
- For $1.4 M_\odot < M \lesssim 5 M_\odot$, electrons behave relativistically, but a star can be supported by neutron degeneracy pressure, resulting in the formation of a neutron star. These are very compact and rotate with a period $\propto R^{3/2} M^{-1/2}$.
- The Schwarzschild radius R_S of a black hole is $2GM/c^2$.
- The maximum accretion rate for spherically symmetric accretion is given by the Eddington limit $\dot{M}_{\mathrm{edd}} = 4\pi cR/\kappa$.
- A black hole has entropy $S/k_\mathrm{B} = A/(4l_\mathrm{P}^2)$ so that one bit of information can be associated with four Planck areas.

Further reading

More information may be found in Carroll and Ostlie (1996), Cheng (2005), Prialnik (2000), Perkins (2003) and Zeilik and Gregory (1998).

Exercises

(36.1) Show that for a white dwarf MV is a constant.

(36.2) Estimate the radius of a white dwarf with mass $M_\odot$.

(36.3) Estimate the radius of a neutron star with mass $2M_\odot$ and calculate its minimum rotation period.

(36.4) What is the Schwarzschild radius of a black hole with mass (i) $10 M_\odot$, (ii) $10^8 M_\odot$, and (iii) $10^{-8} M_\odot$?

(36.5) For a black hole of mass $100 M_\odot$, estimate the Schwarzschild radius, the Hawking temperature, and the entropy.

37 Earth's atmosphere

37.1 Solar energy 446
37.2 The temperature profile in the atmosphere 447
37.3 Radiative transfer 449
37.4 The greenhouse effect 452
37.5 Global warming 456
Chapter summary 459
Further reading 460
Exercises 460

[1] The mass of the Earth is $M_\oplus = 5.97 \times 10^{24}$ kg.

The **atmosphere** is the layer of gases gravitationally bound to the Earth, composed of $\sim 78\%$ N_2, 21% O_2 and very small amounts of other gases. The Earth has radius $R_\oplus = 6378$ km, and atmospheric pressure at sea level is $p = 10^5$ Pa, and hence the mass M_{atmos} of the atmosphere is given by

$$M_{\mathrm{atmos}} = \frac{4\pi R_\oplus^2 p}{g} = 5 \times 10^{18} \text{ kg.} \tag{37.1}$$

Thus, $M_{\mathrm{atmos}}/M_\oplus \sim 10^{-6}$, where $M_\oplus$ is the mass of the Earth.[1] The atmosphere is able to exchange thermal energy with the ocean (whose mass is considerably larger ($\approx 10^{21}$ kg) than that of the atmosphere) and also with space (absorbing ultraviolet and visible radiation from the Sun, and emitting infrared radiation). In this chapter, we will outline a few of the thermodynamic properties of the atmosphere. More details on all of these issues may be found in the further reading.

37.1 Solar energy

Energy is continuously pumped into the atmosphere by the Sun in the form of solar radiation. The luminosity of the Sun is $L_\odot = 3.83 \times 10^{26}$ W and can be related to the Sun's effective surface temperature $T_\odot$ using eqn 23.12 so that

$$L_\odot = 4\pi R_\odot^2 \sigma T_\odot^4, \tag{37.2}$$

where $R_\odot = 6.96 \times 10^8$ m is the solar radius. This gives $T_\odot \approx 5800$ K. When the Sun is directly overhead, the power incident on unit area of the Earth's surface is

$$S = \frac{L_\odot}{4\pi R_{\mathrm{ES}}^2} = 1.36 \text{ kW m}^{-2}, \tag{37.3}$$

a quantity called the **solar constant**. R_{ES} is the distance of the Earth from the Sun,[2] equal to one **astronomical unit** (1.496×10^{11} m). The Earth absorbs energy (see Fig. 37.1(a)) at a rate $\pi R_\oplus^2 S(1-A)$ where $A \approx 0.3$ is the Earth's **albedo**, defined as the fraction of solar radiation reflected (so a fraction $1 - A$ of the incoming energy $\pi R_\oplus^2 S$ is absorbed). The Earth also emits radiation[3] (see Fig. 37.1(b)) at a rate given by $4\pi R_\oplus^2 \sigma T_{\mathrm{E}}^4$, where T_{E} is the **radiative temperature** of the Earth, sometimes called its **radiometric temperature**.

[2] The Earth is in an elliptical orbit about the Sun and so the Earth–Sun distance varies by about $\pm 2\%$.

[3] The emitted energy is principally in the form of infrared radiation, at much longer wavelength than the energy the Earth receives from the Sun, which has a large component in the visible and ultraviolet region of the spectrum.

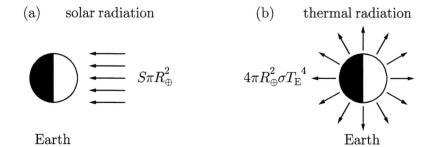

Fig. 37.1 Schematic illustration of (a) the solar power received on the Earth's surface and (b) the power radiated from the Earth.

Balancing the power absorbed with the power emitted yields

$$\pi R_\oplus^2 S(1-A) = 4\pi R_\oplus^2 \sigma T_E^4, \qquad (37.4)$$

and hence

$$T_E = T_\odot \left(\frac{R_\odot}{2R_{ES}}\right)^{1/2} (1-A)^{1/4}, \qquad (37.5)$$

and this leads to $T_E \approx 255\,\text{K}$, which is $\sim -20°\,\text{C}$. This is much lower than the mean surface temperature, which is $\sim 283\,\text{K}$. This is because most of the thermal radiation into space comes from high up in the atmosphere, where the temperature is lower than it is at the surface.

Example 37.1

How large a solar panel do you need to drive a television (which needs 100 W to run) on a sunny day, assuming that the solar panel operates at 15 % efficiency?

Solution:
Assuming that you have the full $S = 1.36\,\text{kW m}^{-2}$ at your disposal, the area needed is

$$\frac{100\,\text{W}}{0.15 \times 1.36 \times 10^3\,\text{W m}^{-2}} \approx 0.5\,\text{m}^2. \qquad (37.6)$$

37.2 The temperature profile in the atmosphere

In this section we wish to derive the dependence of the temperature T as a function of height z above the ground. In the lowest region of the atmosphere, the temperature profile is governed by the adiabatic lapse rate (see Section 12.4), whose derivation we will briefly review. Consider a fixed mass of dry air, which retains its identity as it rises. If it does not exchange heat with its surroundings ($đQ = 0$) it can be treated adiabatically. Its change of enthalpy dH is given by

$$dH = C_p dT = đQ + V dp, \qquad (37.7)$$

and hence

$$C_p dT = V dp. \qquad (37.8)$$

Pressure p can be related to height z using the hydrostatic equation which we met in eqn 4.23 ($dp = -\rho g dz$), and this leads to[4]

$$\frac{dT}{dz} = -\frac{\rho g V}{C_p} = -\frac{g}{c_p} \equiv -\Gamma, \qquad (37.9)$$

where $c_p = C_p/\rho V$ is the specific heat capacity of dry air at constant pressure. We define $\Gamma = g/c_p$ to be the **adiabatic lapse rate**.

Considerable heat transfer takes place within the lowest ≈ 10 km of the atmosphere, which is termed the **troposphere**. Air is warmed by contact with the Earth's surface. The heating of the air drives the temperature gradient $-dT/dz$ to be larger than Γ, making it unstable to these convection currents.[5] When the temperature gradient vertically upwards from the Earth becomes too great (so that air at low altitudes is too warm and air higher up is too cool) then convection will take place, just as we learned it takes place within the interior of stars (Section 35.3.2). As the air rises into lower-pressure regions, it cools owing to adiabatic expansion. This instability to convection is why this region of the atmosphere is given the name troposphere (the name comes from the Greek *tropos*, meaning "turning"). Moreover, if the temperature gradient as a function of latitude is similarly too great, the atmosphere exhibits **baroclinic instability**. When combined with the Coriolis[6] force due to the rotation of Earth, this gives rise to cyclones and anti-cyclones which can transport considerable energy between the equator and the poles.

At the top of the troposphere, there is an interface region called the **tropopause**, where there is no convection. Vertically above this is the next layer, called the **stratosphere**, and in the lowest part of this layer temperature is found to be roughly constant with height z (see Fig. 37.2). The atmosphere becomes "stratified" into layers which tend not to move up or down, but just hang there ("in much the same way that bricks don't", to borrow a phrase from Douglas Adams). The stratosphere is "optically thin" (see Section 37.3) and hence absorbs little energy from the incoming solar radiation. If the stratosphere has absorptivity ϵ (typically $\epsilon \approx 0.1$), it will absorb energy radiated at infrared wavelengths from the Earth's surface at the rate $\epsilon \sigma T_E^4$ per unit area, where T_E is the effective radiative temperature of the Earth (including the troposphere).

If the temperature of the stratosphere is T_strat, it will emit (mainly infrared) radiation at a rate $\epsilon \sigma T_\text{strat}^4$ from its upper surface and $\epsilon \sigma T_\text{strat}^4$ from its lower surface, i.e., at a total rate of $2\epsilon \sigma T_\text{strat}^4$, and hence

$$T_\text{strat} = \frac{T_E}{2^{1/4}}. \qquad (37.10)$$

[4] Remember that because temperature T falls with height z, the quantity dT/dz is negative.

[5] If a parcel of air is quickly displaced upwards by dz, eqn 37.9 implies that its temperature will change by $-\Gamma dz$ (the process is adiabatic because there is no time for heat to be exchanged with the surroundings). If the temperature of the surrounding air drops more rapidly with height ($-dT/dz > \Gamma$), the parcel will become less dense than its surroundings and it will *carry on* moving upwards (convection occurs). The parcel will return to its starting position (and hence be stable against convection) only if the surrounding air is less dense, and this occurs if $-dT/dz < \Gamma$.

[6] The Coriolis force arises because the Earth is rotating. A description of this may be found in Andrews (2000) and in books on mechanics.

The effective radiative temperature of the Earth is ≈ 255 K, and this yields $T_{\text{strat}} \approx 214$ K, a typical average temperature of the lower stratosphere (the roughly isothermal region just above the tropopause).

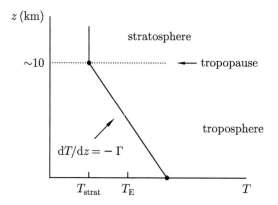

Fig. 37.2 Diagrammatic form of a very simple model of the troposphere and the stratosphere. For real data, see Taylor (2005). Note that T_E is lower than the surface temperature.

At higher altitudes in the stratosphere, the temperature starts rising with increasing height, owing to absorption of ultraviolet radiation in the ozone[7] layer, reaching around 270 K. At about 50 km is the **stratopause**, which is the interface between the stratosphere and the **mesosphere**. In the mesosphere, the temperature falls again owing to the absence of ozone, bottoming out below 200 K at ≈90 km, roughly the location of the **mesopause**. Above this is the **thermosphere**, where the temperature rises very high (to above 1000 °C) owing to very energetic solar photons and cosmic ray particles which cause dissociation of molecules in the upper atmosphere.

[7] Ozone is the name given to the O_3 molecule.

37.3 Radiative transfer

To understand how energy is exchanged between the various parts of the atmosphere, consider first the passage of energy through an absorbing medium. In a small distance dz, the mass of material per unit area is $\rho_a\, dz$ where ρ_a is the density of absorbing molecules. We can define an **extinction coefficient**[8] κ_ν as the fraction of photons of frequency ν that are absorbed or scattered per unit mass of absorber. Then if spectral radiance[9] I_ν passes through a distance dz, it will change by

$$dI_\nu = -\kappa_\nu \rho_a I_\nu\, dz. \tag{37.11}$$

This can be rearranged to give

$$\frac{dI_\nu}{dz} = -\kappa_\nu \rho_a I_\nu. \tag{37.12}$$

If ρ_a does not depend on z (appropriate for a uniform medium), then this can be integrated to give

$$I_\nu(z) = I_\nu(0) e^{-\kappa_\nu \rho_a z}, \tag{37.13}$$

[8] The quantity κ_ν has units $m^2\, kg^{-1}$ and is also called the **absorption coefficient** or **specific opacity**, particularly in astrophysics contexts, see Section 35.3.1. Sometimes κ_ν is split into a scattering component and an absorption component.

[9] Recall that the radiance I is the power per unit area per unit solid angle (units $W\, m^{-2}\, sr^{-1}$). The spectral radiance I_ν is the radiance in a particular frequency interval (units $W\, m^{-2}\, sr^{-1}\, Hz^{-1}$). Both quantities may depend on direction, so in general one could write $I_\nu(\theta, \phi)$ as the spectral radiance in a particular direction parameterized by angles θ and ϕ. In astrophysics, spectral radiance is commonly called **specific intensity**, see Section 35.3.1.

i.e., an exponential decrease of spectral radiance with distance, and this is known as the **Beer–Lambert law**. However, if the density ρ_a changes with distance then the solution to eqn 37.11 is

$$I_\nu(z) = I_\nu(0) e^{-\chi_\nu}, \tag{37.14}$$

where χ_ν is the optical path length[10] given by

$$\chi_\nu = \int_0^z \kappa_\nu \rho_a(s) \, \mathrm{d}s. \tag{37.15}$$

However, a missing ingredient in our treatment is that the absorbing medium is also capable of reradiating because it has a temperature. The amount of radiation it emits at frequency ν will be proportional to its density and also to κ_ν (because "good absorbers are good emitters") and so eqn 37.11 can be generalized to

$$\mathrm{d}I_\nu = -\kappa_\nu \rho_a I_\nu \, \mathrm{d}z + \kappa_\nu \rho_a J_\nu \, \mathrm{d}z, \tag{37.16}$$

where J_ν is the source function. In local thermal equilibrium, $J_\nu = B_\nu(T)$, the black-body radiance. Equation 37.16 leads immediately to the **radiative-transfer equation**, also known as the **Schwarzschild equation**, which is

$$\frac{\mathrm{d}I_\nu}{\mathrm{d}z} = -\kappa_\nu \rho_a(z)(I_\nu - J_\nu). \tag{37.17}$$

Since $\rho_a(z)$ depends on z, this equation is usefully expressed as

$$\frac{\mathrm{d}I_\nu}{\mathrm{d}\chi_\nu} + I_\nu = J_\nu. \tag{37.18}$$

For a proper calculation of the transfer of radiation in the atmosphere, you need to think about the power travelling in all directions. One therefore needs the **irradiance**[11] F, which is the power per unit area normal to a surface and can be obtained by taking the component of the radiance normal to the surface ($I \cos\theta$) and integrating over a hemisphere on one side of the surface.[12] There are two choices for this and therefore one can either calculate the upwards irradiance or the downwards irradiance. The units of irradiance are $\mathrm{W\,m^{-2}}$. One often deals with the spectral irradiance[13] F_ν, which is the irradiance in a particular frequency interval and has units $\mathrm{W\,m^{-2}\,Hz^{-1}}$. Spectral irradiance F_ν can be related to spectral radiance $I_\nu(\theta, \phi)$ by

$$F_\nu = \int I_\nu(\theta, \phi) \cos\theta \, \mathrm{d}\Omega = \int_0^{2\pi} \int_0^{\pi/2} I_\nu(\theta, \phi) \cos\theta \sin\theta \, \mathrm{d}\theta \, \mathrm{d}\phi, \tag{37.19}$$

where the integral over θ goes only from 0 to $\frac{\pi}{2}$ because we are only integrating over a hemisphere. For isotropic radiation, $I_\nu(\theta, \phi)$ does not depend on θ and ϕ and so

$$F_\nu = I_\nu \int_0^{2\pi} \int_0^{\pi/2} \cos\theta \sin\theta \, \mathrm{d}\theta \, \mathrm{d}\phi = \pi I_\nu. \tag{37.20}$$

---

[10] When $\chi_\nu \gg 1$, the region is said to be **optically thick** because

$$\frac{I_\nu(z)}{I_\nu(0)} = e^{-\chi_\nu}$$

is extremely small and so very little radiation can penetrate. When $\chi_\nu \ll 1$, the region is said to be **optically thin** because $I_\nu(z) \approx I_\nu(0)$ and so most of the radiation is transmitted. In many treatments of the atmosphere, it is conventional to measure z from the top of the atmosphere downwards. In this case χ_ν is called the **optical depth** and is zero at the top of the atmosphere and increases as you penetrate deeper into the atmosphere. In this case, we define the optical depth χ_ν as

$$\chi_\nu = \int_z^\infty \kappa_\nu \rho_a(s) \, \mathrm{d}s.$$

The optical depth is sometimes given the symbol τ_ν rather than χ_ν.

[11] The irradiance F is sometime called **radiative flux**.

[12] This is analogous to the calculation of molecular flux in Chapter 7, see Exercise 37.7.

[13] In astrophysics, the term **flux density** is commonly used for the spectral irradiance F_ν.

Example 37.2

Black-body radiation is isotropic and therefore eqn 37.20 holds. The irradiance $F(T)$ in black-body radiation is given by

$$F(T) = \frac{1}{4}uc = \pi B(T) = \sigma T^4. \tag{37.21}$$

The spectral irradiance $F_\nu(T)$ is given by

$$F_\nu(T) = \frac{1}{4}u_\nu c = \pi B_\nu(T). \tag{37.22}$$

The function $B(T)$ is defined by

$$B(T) \equiv \int_0^\infty B_\nu(T)\,d\nu = \frac{\sigma T^4}{\pi},$$

where the last equality follows from eqn 23.49. Recall from eqn 23.50 that $B_\nu(T)$ is

$$B_\nu(T) = \frac{2h}{c^2}\frac{\nu^3}{e^{\beta h\nu}-1}.$$

Irradiances are what we want to consider because they tell us how much energy is transported. However, the radiative-transfer equation (eqn 37.17) is formulated in terms of radiances, not irradiances. Although it is straightforward to convert the black-body radiance $B_\nu(T)$ into an irradiance $F_\nu(T)$ (by multiplying by π, since $F_\nu(T) = \pi B_\nu(T)$), it is not so easy in general. For example, radiation travelling vertically downwards through the atmosphere will be less absorbed than radiation travelling at some angle to the vertical, simply because the path length is shorter. In the atmosphere, the infrared radiation travels in many different directions and therefore one must consider a distribution of path lengths. To treat this as a one-dimensional problem one can often use the diffuse approximation, which states that eqn 37.18 can be replaced by

$$\frac{dF_\nu}{d\chi_\nu^*} + F_\nu = \pi B_\nu, \tag{37.23}$$

where $\chi_\nu^* \approx r\chi_\nu$ is the scaled optical depth and the scaling factor of r roughly accounts for the fact that we are averaging over different directions and therefore different amounts of absorbance.[14] Equation 37.23 is valid for radiation travelling downwards. For upwards-travelling radiation, $d\chi_\nu^*$ changes sign[15] and we have

$$-\frac{dF_\nu}{d\chi_\nu^*} + F_\nu = \pi B_\nu, \tag{37.24}$$

These differential equations can be solved for particular assumed density profiles for the atmosphere.

[14] It has been found that the value $r \approx 1.66$ gives reasonably good agreement with an exact treatment in the limit of strong absorption, though larger values work better for weak absorption, see Goody and Yang (1989).

[15] This is assuming the convention mentioned above, whereby χ_ν^* is zero at the top of the atmosphere and increases with decreasing altitude.

Example 37.3

If we multiply the equation for upwards irradiance, eqn 37.24, by the integrating factor $e^{-\chi^*}$ we have that

$$\frac{d}{d\chi_\nu^*}(e^{-\chi_\nu^*}F_\nu) = e^{-\chi_\nu^*}\frac{dF_\nu}{d\chi_\nu^*} - e^{-\chi_\nu^*}F_\nu = -\pi B_\nu e^{-\chi_\nu^*}. \tag{37.25}$$

Remember χ is defined in such a way that it is zero at the top of the Earth's atmosphere and non-zero on the Earth's surface, so when integrating quantities that are a function of χ you integrate "up to" the Earth's surface.

We now integrate eqn 37.25 up to the Earth's surface χ_s from some optical depth χ (dropping both the ν subscript and $*$ superscript from all variables for clarity), we then have

$$\left[e^{-\chi'}F\right]_\chi^{\chi_s} = -\pi \int_\chi^{\chi_s} Be^{-\chi'}\,d\chi', \qquad (37.26)$$

and so

$$F(\chi) = e^{-(\chi_s-\chi)}F(\chi_s) + \pi \int_\chi^{\chi_s} B(\chi')e^{-(\chi'-\chi)}\,d\chi'. \qquad (37.27)$$

The term on the left-hand side is the irradiance at optical depth χ and is equal to a contribution from the Earth's surface (the first term on the right-hand side) and a term due to radiation from the intervening layers (the second term on the right-hand side, which is an integral). Using the fact that $F(\chi_s) = \pi B(\chi_s)$ for black-body radiation, and integrating the final term in eqn 37.27 by parts, one has

$$F(\chi) = \pi B(\chi) + \pi \int_\chi^{\chi_s} \frac{dB(\chi')}{d\chi'} e^{-(\chi'-\chi)}\,d\chi'. \qquad (37.28)$$

This can be rewritten as

$$F(\chi) = \pi B(\chi_s) - \pi \int_\chi^{\chi_s} \frac{dB(\chi')}{d\chi'}\left[1 - e^{-(\chi'-\chi)}\right]d\chi', \qquad (37.29)$$

showing that the upwards irradiance $F(\chi)$ is simply the irradiance from the surface reduced by the colder atmosphere above it.[16]

[16] This means that in either an isothermal or a completely transparent atmosphere, $F(\chi) = \pi B(\chi_s)$, as it should be in each case.

Equation 37.29 can be used to calculate the upwards irradiance under an assumed model of the variation of density of absorber with height, which controls χ, and the temperature profile, which also feeds into $B(\chi)$. Note that the function $B_\nu(T)$ is not far off from being a linear function of temperature for the frequencies relevant to infrared radiation in the atmosphere, see Fig. 37.3, a fact we will use later. Equation 37.29 also shows that it is the function $B(\chi)$ and also the value of $F(\chi_s) = \pi B(\chi_s)$ (the irradiance from the surface of the Earth), which determine the outgoing irradiance at the top of the atmosphere. Rather than the main atmospheric constituents N_2 and O_2, it is gases such as CO_2 and H_2O, that contribute to the absorption of infrared radiation in the atmosphere. The concentrations of these gases therefore affect atmospheric radiative transfer and hence the outgoing irradiance, as described in the next section.

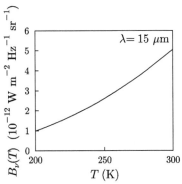

Fig. 37.3 The function $B_\nu(T)$ for a frequency $\nu = c/\lambda$ where $\lambda = 15\,\mu\mathrm{m}$, the frequency of the CO_2 bending mode.

37.4 The greenhouse effect

The Earth receives energy from the Sun in the form of short wavelength radiation (mainly in the visible and ultraviolet regions of the spectrum)

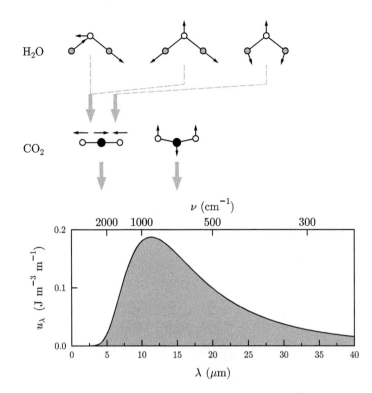

Fig. 37.4 This graph shows a blackbody spectrum at 255 K corresponding to the radiation emitted from the Earth. Shown above are cartoons of relevant normal modes of the CO_2 and H_2O molecules. The grey vertical arrows indicate the relevant vibrational wavelengths. (The third normal mode of CO_2 is a symmetric stretch mode and is not infrared active.)

because the Sun has a surface temperature $T_\odot \approx 5800\,\text{K}$. The Earth reradiates this energy in the form of long-wavelength radiation (mainly in the infrared region of the spectrum) because the Earth's temperature is about a factor of twenty lower than $T_\odot$ and so the outwards radiation has a wavelength that is about a factor of twenty longer. Some of the incoming radiation is reflected by the atmosphere but the rest is only weakly absorbed by the atmosphere and is then either absorbed or reflected by the land and oceans. Radiation emitted by the Earth's surface is absorbed by the atmosphere and then reradiated in a manner described by eqn 37.29. Reradiation occurs both upwards and downwards and can therefore lead to surface warming. Now, the atmosphere contains different constituents and it is important to understand which of these absorbs infrared radiation.

The different molecules found in the atmosphere do indeed respond differently to incident radiation from the Earth, which is at infrared wavelengths (see Fig. 37.4). The main constituents of air are N_2 and O_2. Both these molecules are composed of two identical atoms and are termed diatomic **homonuclear molecules**. They do not couple directly to infrared radiation because no vibrations of such molecules produce an **electric dipole moment**,[17] but rather they can only stretch along the bond. However, for **heteronuclear molecules** like CO_2, the situation is different. Two of the **vibrational modes** of CO_2, which

[17] A molecule is said to have an electric dipole moment if there is charge separation across the molecule. Electric dipole moment is a vector quantity, and if two charges $+q$ and $-q$ are separated by a distance D then it takes the value qD in the direction from the negative charge towards the positive charge. A molecule can possess a permanent electric dipole moment, or have one induced by a vibrational mode.

is a linear molecule, are the **asymmetric stretch** mode (at ≈5 μm) and the bending mode (at ≈15 μm). These are both **infrared active** because they correspond to a change in electric dipole moment when the vibration takes place. The **symmetric stretch** mode is not infrared active.

Water (H_2O) behaves similarly to CO_2, but because H_2O is a bent molecule with a permanent electric dipole moment, all three normal modes of vibration are infrared active, although the symmetric stretch and bending modes are at high frequencies (<3 μm). The antisymmetric stretch mode (at ≈3 μm) is relevant to atmospheric absorption. These vibrational modes are sketched in Fig. 37.4.

The strong infrared absorption of gases like CO_2 and H_2O (but not N_2 and O_2) gives rise to the **greenhouse effect**.[18] This effect depends on very small concentrations of these heteronuclear molecules, or **greenhouse gases**, in the atmosphere. Greenhouse gases are capable of absorbing radiation emitted by the Earth at certain frequencies, and produce strong absorption in the emitted spectrum as shown in Fig. 37.5.

[18] The term "greenhouse effect" was coined in 1827 by Jean Baptiste Joseph Fourier, whom we met on page 105.

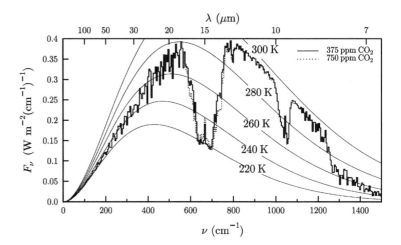

Fig. 37.5 Calculated irradiance from the top of the atmosphere, assuming a tropical clear-sky atmosphere. The atmosphere is not transparent around 9.5 microns or around 15 microns owing to absorption by ozone (O_3) and CO_2, respectively. The effect of doubling of CO_2 concentration is shown and is most pronounced around the 15 μm absorption (see also Exercise 37.4). The smooth curves show black-body curves for selected temperatures. Data provided by Anu Dudhia and Myles Allen.

What this means is that the radiation at these wavelengths, which would pass out of the atmosphere in the absence of the greenhouse gases, is retained in the atmosphere: the greenhouse gases act as a "winter coat" at these particular wavelengths and so contribute to increasing the temperature at the Earth's surface. To some extent, of course, this is a good thing as this planet would be a very cold place without any of the winter-coat effect of H_2O, CO_2, and the other greenhouse gases. However, too much winter coat when it is not needed will elevate the temperature of the planet, with potentially disastrous consequences. In the next example we will consider how the outgoing irradiance from the top of the atmosphere (which quantifies how the Earth radiates energy away into space) depends on the concentration of an absorber such as CO_2. Our approach will be a direct application of the ideas introduced in Section 37.3 concerning radiative transfer.

Example 37.4

This calculation is intended to give some insight into how greenhouse gases in the atmosphere reduce the outwards irradiance at the top of the atmosphere. Let the top of the tropopause have an optical depth χ_1, while at the Earth's surface the optical depth is $\chi_s \gg \chi_1$. Remember, optical depth is measured from the top of the atmosphere downwards. Both of these quantities depend on frequency ν. Equation 37.29 then gives the outgoing irradiance $F(\chi_1)$ to be

$$F(\chi_1) = \pi B(\chi_s) - \pi \int_{\chi_1}^{\chi_s} \frac{dB(\chi)}{d\chi}\left[1 - e^{-(\chi-\chi_1)}\right] d\chi. \quad (37.30)$$

The outgoing irradiance depends on frequency because the optical depth depends on frequency. At frequencies at which the atmosphere is optically thin, $\chi_s \to 0$ and hence eqn 37.30 reduces to

$$F(\chi_1) = \pi B(\chi_s), \quad (37.31)$$

which is simply black-body radiation with a temperature given by the Earth's surface temperature. At these frequencies the atmosphere has little effect. On the other hand, at frequencies at which the atmosphere is optically thick, $\chi_s \to \infty$ and, if also $\chi_1 \gg 1$, it is easy to show that eqn 37.30 reduces to

$$F(\chi_1) = \pi B(\chi_1), \quad (37.32)$$

which is simply black-body radiation with a temperature given by the temperature of the top of the atmosphere. The interesting case occurs for frequencies which lead to behaviour in between these two limits, when the atmosphere is neither so optically thin as to be negligible nor so optically thick that radiation cannot penetrate it.

To make the calculation tractable we will make some simplifying assumptions. We consider a very simple model of how the optical depth χ relates to the height z above the Earth's surface. The density ρ of the atmosphere can be assumed to vary roughly as $\rho \propto e^{-z/H}$ where $H = k_B T/mg$ is the **scale height** (see eqn 4.26). Assuming that the molecules in the atmosphere which absorb or scatter radiation at the relevant frequencies are well mixed with the rest of the atmosphere, their density ρ_a will be proportional to ρ. We will also assume that the extinction coefficient κ will follow the density[19] and hence $\kappa \propto e^{-z/H}$. Hence, the optical depth χ is

$$\chi = \int_z^\infty \kappa(z') \rho_a(z')\, dz' \propto e^{-z/h}, \quad (37.33)$$

where $h = H/2$, so χ also varies exponentially with height z, albeit with a different scale height (h rather than H). Since $\chi = \chi_s$ at $z = 0$, we can write $\chi = \chi_s e^{-z/h}$, or equivalently $z = -h \ln(\chi/\chi_s)$. Now, temperature $T(z)$ falls roughly linearly with height in the troposphere (see Fig. 37.2) and so $T(z) = T(0) - \Gamma z$, hence

$$T(z) = T(0) + \Gamma h \ln(\chi/\chi_s). \quad (37.34)$$

[19] Absorption occurs at certain frequencies corresponding to the vibrational spectral lines. These lines are broadened due to collisions; the width of the lines is proportional to the collision rate which scales with pressure. We assume therefore that κ scales roughly with pressure, and hence also with density. We are ignoring the temperature dependence of κ, which would be needed for a more realistic calculation.

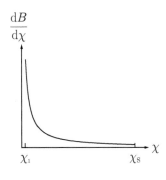

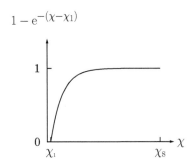

Fig. 37.6 The two factors that are integrated in eqn 37.30 in order to calculate the outgoing irradiance.

[20] Atmospheric physicists often refer to the concentration of absorbing molecules in the atmosphere in terms of the **mixing ratio**. This can be defined either in terms of a mass mixing ratio (the mass fraction of the air that consists of the absorber) or a volume mixing ratio (which is equivalent to the mole fraction). See Andrews (2000).

[21] CO_2 levels are rising at a rate unprecedented in the last 20 million years.

The function $B(T)$ is approximately linear with temperature at these frequencies (see Fig. 37.3) and so we can write $B(T) = B_0 + B_1 T$ where B_0 and B_1 are constants. Using this and eqn 37.34 we can evaluate $dB/d\chi$ as follows:

$$\frac{dB}{d\chi} = \frac{dT}{d\chi}\frac{dB}{dT} = \frac{\Gamma h B_1}{\chi}. \qquad (37.35)$$

The quantity $dB/d\chi$ has to be multiplied by $1 - e^{-(\chi - \chi_1)}$ and then integrated, as in eqn 37.30, to obtain the outgoing irradiance. These two factors are plotted in Fig. 37.6. When the concentration of the absorbing molecules (such as CO_2) is increased, the optical depth χ_s increases with it. Let us say that χ_s increases by a factor $\alpha > 1$. In this case, the outgoing irradiance increases by an amount

$$\begin{aligned}
\Delta F(\chi_1) &= -\pi \int_{\chi_s}^{\alpha \chi_s} \frac{dB(\chi)}{d\chi}[1 - e^{-(\chi - \chi_1)}]\, d\chi \qquad (37.36)\\
&\approx -\pi \int_{\chi_s}^{\alpha \chi_s} \frac{dB(\chi)}{d\chi}\, d\chi\\
&= -\pi \int_{\chi_s}^{\alpha \chi_s} \frac{\Gamma h B_1}{\chi}\\
&= -\pi \Gamma h B_1 \ln \alpha,
\end{aligned}$$

where the first approximation is made by setting $1 - e^{-(\chi - \chi_1)} \approx 1$ which is valid for large χ. Thus $F(\chi_1)$ *decreases* when $\alpha > 1$.

This example demonstrates that in this model the outwards irradiance depends logarithmically on α and hence on the concentration of the absorbing molecules in the atmosphere.[20] It is important to note that the outwards irradiance is still sensitive to the concentration of absorbing molecules, even in a relatively optically thick atmosphere. Of course, predicting *exactly* the effect of, say, doubling CO_2 requires us to know over how much of the spectrum CO_2 is partially absorbing in this way, which depends on other gases, cloud cover and many other variables, as well as a more accurate model of the temperature and density profiles of the atmosphere. Nevertheless, the calculation in the example above gives an indication of the mechanism involved.

37.5 Global warming

There is now much accumulated evidence from independent measurements that the concentrations of greenhouse gases and in particular CO_2 are changing[21] as a result of human activity and, further, are giving rise to **global warming** (i.e., the elevation of the average temperature of Earth's atmosphere, see Fig. 37.7) and consequent **climate change**. This is termed **anthropogenic climate change**, with anthropogenic meaning "having its origins in the activities of humans".

Over the last few *hundred* years, since the industrial revolution, we have released into the atmosphere the combustion products of fossil fuels

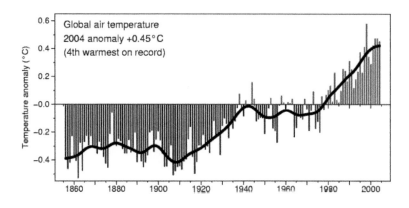

Fig. 37.7 Variations in the globally averaged near-surface air temperature over the last 140 years; reproduced by kind permission of P. Jones of the Climate Research Unit, University of East Anglia.

that were laid down over a few *hundred million* years. As we have explored in the last section, even the small changes in the chemical composition of the Earth's atmosphere this brings can have considerable influence on climate. The immense heat capacity of the oceans means that the full consequences of global warming and consequent climate change do not instantly become apparent.[22] Already however these are significant as can be seen in Fig. 37.7, which shows measurements of globally averaged temperatures since 1851.

Predictions of global warming are complicated because this is a multi-parameter problem, which depends on boundary conditions that themselves cannot be known exactly; for example, the details of cloud cover cannot be predicted with accuracy. Clouds play a part in the Earth's radiation balance because they reflect some of the incident radiation from the Sun but they also absorb and emit thermal radiation and have the same winter-coat insulating effect as greenhouse gases. In addition, the presence of water vapour (water in gaseous form as distinct from water droplets in a cloud) plays an important rôle as a greenhouse gas. Furthermore, as the atmosphere heats up, so it can hold more water vapour before it begins to condense out as liquid droplets.[23] This increased capacity further increases the winter-coat effect. The increasing presence of CO_2 in the atmosphere thus gives rise to a **positive feedback mechanism**: as the global temperature rises, the atmosphere can hold a greater amount of H_2O before saturation and precipitation[24] is reached. This leads to an even larger greenhouse effect from atmospheric H_2O.

These and other feedback effects will influence the future of this planet. There is a competition between positive feedback effects (i.e., warming triggering further warming, as ice cover on the planet is reduced, more land is exposed that, being less reflective, absorbs heat more quickly) and negative feedback effects (e.g., higher temperatures will tend to promote the growth rate of plants and trees, which will increase their intake of CO_2) but it seems that positive feedback effects have a much greater effect.

It is also difficult to forecast accurately future trends in the world's human population, especially in developing countries that are, in addi-

[22] This is explored further in Exercise 37.3.

[23] It is helpful to consider the capacity of the atmosphere to "hold" increasing water vapour as its temperature increases in terms of the phase diagram for water (shown in Fig. 28.7): on the phase boundary between gas and liquid, p increases with increasing T. Here, p should be interpreted as the partial pressure of water vapour, i.e., a measure of how much water vapour is present. Fig. 28.7 shows that as temperature increases, a larger partial pressure of water vapour can be attained before condensation takes place.

[24] Precipitation is any form of water that falls from the sky, such as rain, snow, sleet and hail. Analysis of the effect of water is greatly complicated by the fact that the atmosphere doesn't have a uniform temperature and water partial pressure. Hence, in practice it is necessary to consider the effect of the temperature increase on a huge ensemble of complicated air-parcel trajectories.

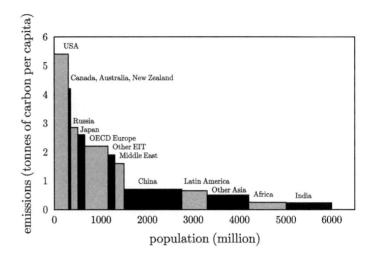

Fig. 37.8 CO_2 emissions in 2000 from different countries or groups of countries in tonnes per capita versus their population in millions. Data from M. Grubb, The economics of the Kyoto Protocol, *World Economics* **4**, 143 (2003).

[25] Carbon-neutral (sometimes referred to as "zero carbon") means that there is no net input of CO_2 into the atmosphere as a result of that particular energy supply.

tion, becoming increasingly industrialized. It is also difficult to make precise predictions about the economies of developed and developing worlds and their reliance on fossil fuels rather than **carbon-neutral**[25] energy supplies. Figure 37.8 gives some sense of the uncertainty in future CO_2 production: the width of each bar represents the population of each nation (or group of nations) in millions and the height represents the CO_2 emission per capita. Both the rate of change of width and the rate of change of height of each bar are uncertain, but it seems highly likely that increasing population and increasing industrialization will lead to increasing CO_2 production worldwide.

However, although many of these factors are uncertain, a very wide range of plausible input models for global warming (covering extremes such as a world with a continuously increasing population and one which has emphasis on local solutions to provide economic, social, and environmental sustainability) predict a temperature rise of at least two degrees in 2100 compared with that in the first half of the twentieth century (see Fig. 37.9). The centuries-long relaxation time of atmospheric CO_2 and the high heat capacity of the oceans imply that we are already committed to a substantial warming, even if emissions stop immediately. This motivates urgent action now.

Some consequences of global warming are already apparent: at the time of writing we have observed a $0.6\,°C$ rise in the annual average global temperature, a $1.8\,°C$ rise in the average Arctic temperature, 90% of the planet's glaciers have been retreating since 1850, and Arctic sea ice has reduced by 15–20 %. One of the predicted consequences of global warming is the rise by $2\,°C$ of the average global temperature (see Fig. 37.9) by the second half of this century. This will promote the melting of the Greenland ice and cause sea water to expand. Both effects will lead to a significant rise in sea level and consequent reduction of habitable land on the planet.

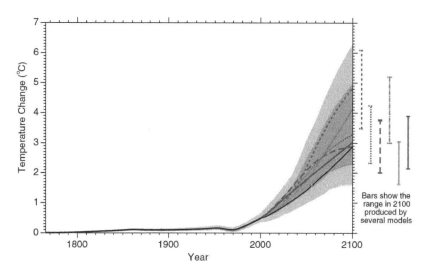

Fig. 37.9 Predictions of global warming from a wide range of different input models. Data from U. Cubasch et al., Projections of future climate change, in *Climate Change 2001: The Scientific Basis*, IPCC 2001 Report, CUP, Cambridge, chapter 9, and figure from Houghton (2005).

Chapter summary

- Earth receives about 1.4 kW per m^2 from the Sun as radiation.
- The Beer–Lambert law for the distance-dependence of the spectral radiance I_ν through an absorbing medium is $I_\nu(z) = I_\nu(0)\mathrm{e}^{-\kappa_\nu \rho_a z}$.
- For black-body radiation, the spectral irradiance F_ν is related to the black-body radiance $B_\nu(T)$ by $F_\nu = \pi B_\nu(T)$.
- Radiative transfer in the atmosphere can be considered using the radiative-transfer equations, which describe how radiation is both absorbed and reradiated from gases in the atmosphere.
- The presence of some CO_2 molecules in the atmosphere keeps Earth from being a much colder place to inhabit.
- The CO_2 concentrations in the atmosphere have increased significantly since the industrial revolution.
- Increasing CO_2 in the atmosphere catalyses the increasing temperature of the atmosphere, by promoting the presence in the atmosphere of another greenhouse gas, H_2O vapour.
- Although there are considerable uncertainties in the timescales over which global warming will take place, it seems hard to avoid the conclusion that significant and devastating global warming has begun.

Further reading

- The Intergovernmental Panel on Climate Change: http://www.ipcc.ch
- The Climate Research Unit: http://www.cru.uea.ac.uk
- Climate prediction for everyone: http://www.climateprediction.net
- Useful background reading and an introduction to the physics of atmospheres may be found in Andrews (2000), Taylor (2005), and Houghton (2005). For more on global warming, see Archer (2007).

Exercises

(37.1) What is the *average* power per unit area received from the Sun per year (a) on the equator, (b) at 35° latitude and (c) over all the Earth?

(37.2) Using the mass of Earth's atmosphere given at the start of this chapter, estimate its heat capacity.

(37.3) Using the mass of Earth's ocean given at the start of this chapter, find its heat capacity. Compare your answer with that for the previous question.

(37.4) Suppose that the Earth did not have any atmosphere, and neglecting any thermal conduction between the oceans and the land, estimate how long would it take for the power from the Sun to bring the ocean to the boil. State any further assumptions that you make.

(37.5) The total annual energy consumption at the start of the twenty-first century is about 13 TW (13×10^{12} W). If the efficiency of a solar panel is 15%, what area of land would you need to cover with solar panels (a) at the equator and (b) at 35° latitude, to supply the energy needs for the Earth's population?

(37.6) The Moon has no atmosphere. Assuming that the Sun's rays strike a small region on the Moon at angle θ from the normal to the surface, show that the temperature T at that region is given by $T = (S(1-A)\cos\theta/\sigma)^{1/4}$, stating your assumptions. The albedo of the Moon is $A \approx 0.15$ and the solar constant $S = 1.36\,\text{kW}\,\text{m}^{-2}$. Using these numbers, calculate T for various values of θ.

(37.7) In the kinetic theory of gases it is possible to define a quantity $I = n\langle v \rangle/4\pi$ as the number of molecules per second striking unit area per steradian of solid angle, where n is the number of molecules per unit volume and $\langle v \rangle$ is the mean speed. Note that I does not depend on direction. Show that the flux Φ of molecules is given by

$$\Phi = \int I \cos\theta \, d\Omega = \frac{1}{4}n\langle v\rangle, \qquad (37.37)$$

as in eqn 7.6, and relate this to eqn 37.20.

(37.8) This question explores a purely radiative equilibrium model, which grossly simplifies the radiative transfer in the atmosphere. Assume that the radiative transfer equations for upwards and downwards radiation can be written (following eqns 37.23 and 37.24 and dropping the frequency subscript)

$$-\frac{dF^\uparrow}{d\chi^*} + F^\uparrow = \pi B \qquad (37.38)$$

$$\frac{dF^\downarrow}{d\chi^*} + F^\downarrow = \pi B. \qquad (37.39)$$

In radiative equilibrium, $F^\uparrow - F^\downarrow$ is a constant, which we will call ϕ. Writing $\psi = F^\uparrow + F^\downarrow$, show that

$$\frac{d\psi}{d\chi^*} = \phi \qquad (37.40)$$

and

$$\frac{d\phi}{d\chi^*} = \psi - 2\pi B \qquad (37.41)$$

and hence that $\psi = 2\pi B$. At the top of the atmosphere, $\chi^* = 0$ and $F^\downarrow = 0$. Hence show that

$$B = \frac{\phi}{2\pi}(\chi^* + 1), \qquad (37.42)$$

and sketch $F^\uparrow$, $F^\downarrow$ and πB as a function of optical depth χ^* on the same graph. If the optical depth at the bottom of the atmosphere increases because of increasing CO_2, what happens to the surface temperature in this model? (Remember that B is a function of T.)

Fundamental constants

Bohr radius		a_0	5.292×10^{-11} m
speed of light in free space		c	2.9979×10^{8} m s^{-1}
Electronic charge		e	1.6022×10^{-19} C
Planck constant		h	6.626×10^{-34} J s
	$h/2\pi =$	$\hbar$	1.0546×10^{-34} J s
Boltzmann constant		k_B	1.3807×10^{-23} J K^{-1}
electron rest mass		m_e	9.109×10^{-31} kg
proton rest mass		m_p	1.6726×10^{-27} kg
Avogadro number		N_A	6.022×10^{23} mol^{-1}
standard molar volume			22.414×10^{-3} m^3 mol^{-1}
molar gas constant		R	8.314 J mol^{-1} K^{-1}
fine structure constant	$\dfrac{e^2}{4\pi\epsilon_0 \hbar c} =$	α	$(137.04)^{-1}$
permittivity of free space		ϵ_0	8.854×10^{-12} F m^{-1}
magnetic permeability of free space		μ_0	$4\pi \times 10^{-7}$ H m^{-1}
Bohr magneton		μ_B	9.274×10^{-24} A m^2 or J T^{-1}
nuclear magneton		μ_N	5.051×10^{-27} A m^2 or J T^{-1}
neutron magnetic moment		μ_n	$-1.9130 \mu_N$
proton magnetic moment		μ_p	$2.7928 \mu_N$
Rydberg constant		R_∞	1.0974×10^{7} m^{-1}
		$R_\infty hc$	13.606 eV
Stefan constant		σ	5.670×10^{-8} W m^{-2} K^{-4}
gravitational constant		G	6.674×10^{-11} N m^2 kg^{-2}
mass of the Sun		$M_\odot$	1.99×10^{30} kg
mass of the Earth		$M_\oplus$	5.97×10^{24} kg
radius of the Sun		$R_\odot$	6.96×10^{8} m
radius of the Earth		$R_\oplus$	6.378×10^{6} m
1 astronomical unit			1.496×10^{11} m
1 light year			9.460×10^{15} m
1 parsec			3.086×10^{16} m
Planck length	$\sqrt{\dfrac{\hbar G}{c^3}} =$	l_P	1.616×10^{-35} m
Planck mass	$\sqrt{\dfrac{\hbar c}{G}} =$	m_P	2.176×10^{-8} kg
Planck time	$l_P/c =$	t_P	5.391×10^{-44} s

B   Useful formulae

(1) **Trigonometry**

$$e^{i\theta} = \cos\theta + i\sin\theta$$

$$\sin\theta = \frac{e^{i\theta} - e^{-i\theta}}{2i}$$

$$\cos\theta = \frac{e^{i\theta} + e^{-i\theta}}{2}$$

$$\sin(\theta + \phi) = \sin\theta\cos\phi + \cos\theta\sin\phi$$
$$\cos(\theta + \phi) = \cos\theta\cos\phi - \sin\theta\sin\phi$$

$$\tan\theta = \sin\theta/\cos\theta$$
$$\cos^2\theta + \sin^2\theta = 1$$
$$\cos 2\theta = \cos^2\theta - \sin^2\theta$$
$$\sin 2\theta = 2\cos\theta\sin\theta$$

(2) **Hyperbolics**

$$\sinh x = \frac{e^x - e^{-x}}{2}$$

$$\cosh x = \frac{e^x + e^{-x}}{2}$$

$$\cosh^2 x - \sinh^2 x = 1$$
$$\cosh 2x = \cosh^2 x + \sinh^2 x$$
$$\sinh 2x = 2\cosh x \sinh x$$
$$\tanh x = \sinh x / \cosh x$$

(3) **Logarithms**

$$\log_b(xy) = \log_b(x) + \log_b(y)$$

$$\log_b(x/y) = \log_b(x) - \log_b(y)$$

$$\log_b(x) = \frac{\log_k(x)}{\log_k(b)}$$

$$\ln(x) \equiv \log_e(x) \text{ where } e = 2.718\,281\,828\,46\ldots$$

(4) **Geometric progression**

N-term series:

$$a + ar + ar^2 + \cdots + ar^{N-1} = a\sum_{n=0}^{N-1} r^n = \frac{a(1 - r^N)}{1 - r}.$$

∞-term series:

$$a + ar + ar^2 + \cdots = a\sum_{n=0}^{\infty} r^n = \frac{a}{1 - r}.$$

(5) **Taylor and Maclaurin series**
A Taylor series of a real function $f(x)$ about a point $x = a$ is given by

$$f(x) = f(a) + (x - a)\left(\frac{df}{dx}\right)_{x=a}$$
$$+ \frac{(x - a)^2}{2!}\left(\frac{d^2 f}{dx^2}\right)_{x=a} + \cdots$$

If $a = 0$, the expansion is a Maclaurin series

$$f(x) = f(0) + x\left(\frac{df}{dx}\right)_{x=0} + \frac{x^2}{2!}\left(\frac{d^2 f}{dx^2}\right)_{x=0} + \cdots$$

(6) **Some Maclaurin series** (valid for $|x| < 1$)

$$(1 + x)^n = 1 + nx + \frac{n(n-1)}{2!}x^2$$
$$+ \frac{n(n-1)(n-2)}{3!}x^3 + \cdots$$

$$(1 - x)^{-1} = 1 + x + x^2 + x^3 + \cdots$$

$$e^x = 1 + x + \frac{x^2}{2!} + \frac{x^3}{3!} + \frac{x^4}{4!} + \cdots$$

$$\sin x = x - \frac{x^3}{3!} + \frac{x^5}{5!} - \cdots$$

$$\cos x = 1 - \frac{x^2}{2!} + \frac{x^4}{4!} - \cdots$$

$$\tan x = x + \frac{x^3}{3} + \frac{2x^5}{15} + \cdots$$

$$\tanh x = x - \frac{x^3}{3} + \frac{2x^5}{15} - \cdots$$

$$\tanh^{-1} x = x + \frac{x^3}{3} + \frac{x^5}{5} + \frac{x^7}{7} + \cdots$$

$$\ln(1+x) = x - \frac{x^2}{2} + \frac{x^3}{3} - \cdots$$

(7) **Integrals** Indefinite (with $a > 0$):

$$\int \frac{dx}{x^2+a^2} = \frac{1}{a}\tan^{-1}\frac{x}{a}$$

$$\int \frac{dx}{x^2-a^2} = \frac{1}{2a}\ln\left|\frac{x-a}{x+a}\right|$$

$$\int \frac{dx}{\sqrt{x^2+a^2}} = \sinh^{-1}\frac{x}{a}$$

$$\int \frac{dx}{\sqrt{x^2-a^2}} = \begin{cases} \cosh^{-1}\frac{x}{a} & \text{if } x > a \\ -\cosh^{-1}\frac{x}{a} & \text{if } x < -a \end{cases}$$

$$\int \frac{dx}{\sqrt{a^2-x^2}} = \sin^{-1}\frac{x}{a}$$

(8) **Vector operators**

- grad acts on a scalar field to produce a vector field:
$$\text{grad}\,\phi = \nabla\phi = \left(\frac{\partial\phi}{\partial x}, \frac{\partial\phi}{\partial y}, \frac{\partial\phi}{\partial z}\right);$$

- div acts on a vector field to produce a scalar field:
$$\text{div}\,\boldsymbol{A} = \nabla\cdot\boldsymbol{A} = \frac{\partial A_x}{\partial x} + \frac{\partial A_y}{\partial y} + \frac{\partial A_z}{\partial z};$$

- curl acts on a vector field to produce another vector field:
$$\text{curl}\,\boldsymbol{A} = \nabla\times\boldsymbol{A} = \begin{vmatrix} \boldsymbol{i} & \boldsymbol{j} & \boldsymbol{k} \\ \partial/\partial x & \partial/\partial y & \partial/\partial z \\ A_x & A_y & A_z \end{vmatrix};$$

where $\phi(\boldsymbol{r})$ and $\boldsymbol{A}(\boldsymbol{r})$ are any given scalar and vector field respectively.

(9) **Vector identities:**

$$\begin{aligned}
\nabla\cdot(\nabla\phi) &= \nabla^2\phi \\
\nabla\times(\nabla\phi) &= 0 \\
\nabla\cdot(\nabla\times\boldsymbol{A}) &= 0 \\
\nabla\cdot(\phi\boldsymbol{A}) &= \boldsymbol{A}\cdot\nabla\phi + \phi\nabla\cdot\boldsymbol{A} \\
\nabla\times(\phi\boldsymbol{A}) &= \phi\nabla\times\boldsymbol{A} - \boldsymbol{A}\times\nabla\phi \\
\nabla\times(\nabla\times\boldsymbol{A}) &= \nabla(\nabla\cdot\boldsymbol{A}) - \nabla^2\boldsymbol{A} \\
\nabla\cdot(\boldsymbol{A}\times\boldsymbol{B}) &= \boldsymbol{B}\cdot\nabla\times\boldsymbol{A} - \boldsymbol{A}\cdot\nabla\times\boldsymbol{B} \\
\nabla(\boldsymbol{A}\cdot\boldsymbol{B}) &= (\boldsymbol{A}\cdot\nabla)\boldsymbol{B} + (\boldsymbol{B}\cdot\nabla)\boldsymbol{A} \\
&\quad + \boldsymbol{A}\times(\nabla\times\boldsymbol{B}) + \boldsymbol{B}\times(\nabla\times\boldsymbol{A}) \\
\nabla\times(\boldsymbol{A}\times\boldsymbol{B}) &= (\boldsymbol{B}\cdot\nabla)\boldsymbol{A} - (\boldsymbol{A}\cdot\nabla)\boldsymbol{B} \\
&\quad + \boldsymbol{A}(\nabla\cdot\boldsymbol{B}) - \boldsymbol{B}(\nabla\cdot\boldsymbol{A})
\end{aligned}$$

These identities can be easily proved by application of the alternating tensor and use of the summation convention. The alternating tensor ϵ_{ijk} is defined according to

$$\epsilon_{ijk} = \begin{cases} 1 & \text{if } ijk \text{ is an even permutation} \\ & \text{of 123} \\ -1 & \text{if } ijk \text{ is an odd permutation} \\ & \text{of 123} \\ 0 & \text{if any two of } i, j \text{ or } k \text{ are equal} \end{cases}$$

so that the vector product can be written

$$(\boldsymbol{A}\times\boldsymbol{B})_i = \epsilon_{ijk}A_j B_k.$$

The summation convention is used here, so that twice repeated indices are assumed summed. The scalar product is then

$$\boldsymbol{A}\cdot\boldsymbol{B} = A_i B_i.$$

Use can be made of the identity

$$\epsilon_{ijk}\epsilon_{ilm} = \delta_{jl}\delta_{km} - \delta_{jm}\delta_{kl}$$

where δ_{ij} is the Kronecker delta given by

$$\delta_{ij} = \begin{cases} 1 & i = j \\ 0 & i \neq j. \end{cases}$$

The vector triple product is given by

$$\boldsymbol{A}\times(\boldsymbol{B}\times\boldsymbol{C}) = (\boldsymbol{A}\cdot\boldsymbol{C})\boldsymbol{B} - (\boldsymbol{A}\cdot\boldsymbol{B})\boldsymbol{C}.$$

(10) **Cylindrical coordinates**

$$\nabla^2\psi = \frac{1}{r}\frac{\partial}{\partial r}\left(r\frac{\partial\psi}{\partial r}\right) + \frac{1}{r^2}\frac{\partial^2\psi}{\partial\phi^2} + \frac{\partial^2\psi}{\partial z^2}$$

$$\nabla\psi = \left(\frac{\partial\psi}{\partial r}, \frac{1}{r}\frac{\partial\psi}{\partial\phi}, \frac{\partial\psi}{\partial z}\right)$$

(11) **Spherical polar coordinates**

$$\nabla^2\psi = \frac{1}{r^2}\frac{\partial}{\partial r}\left(r^2\frac{\partial\psi}{\partial r}\right) + \frac{1}{r^2\sin\theta}\frac{\partial}{\partial\theta}\left(\sin\theta\frac{\partial\psi}{\partial\theta}\right)$$

$$+ \frac{1}{r^2\sin^2\theta}\frac{\partial^2\psi}{\partial\phi^2}$$

$$\nabla\psi = \left(\frac{\partial\psi}{\partial r}, \frac{1}{r}\frac{\partial\psi}{\partial\theta}, \frac{1}{r\sin\theta}\frac{\partial\psi}{\partial\phi}\right)$$

C Useful mathematics

C.1 The factorial integral — 464
C.2 The Gaussian integral — 464
C.3 Stirling's formula — 466
C.4 Riemann zeta function — 469
C.5 The polylogarithm — 470
C.6 Partial derivatives — 471
C.7 Exact differentials — 472
C.8 Volume of a hypersphere — 473
C.9 Jacobians — 473
C.10 The Dirac delta function — 475
C.11 Fourier transforms — 475
C.12 Solution of the diffusion equation — 476
C.13 Lagrange multipliers — 477

C.1 The factorial integral

One of the most useful integrals in thermodynamics problems is the following one (which is worth memorizing):

$$n! = \int_0^\infty x^n e^{-x}\, dx. \tag{C.1}$$

- This integral is simple to prove by induction as follows. First, show that it is true for the case $n = 0$. Then assume it is true for $n = k$ and prove it is true for $n = k+1$. (Hint: integrate $(k+1)! = \int_0^\infty x^{k+1} e^{-x}\, dx$ by parts.)
- It allows you to *define* the factorial of non-integer numbers. This is so useful that the integral is given a special name, the **gamma function**. The traditional definition of the gamma function is

$$\Gamma(n) = \int_0^\infty x^{n-1} e^{-x}\, dx \tag{C.2}$$

so that $\Gamma(n) = (n-1)!$, i.e., the factorial function and the gamma function are "out of step" with each other, a rather confusing feature. The gamma function is plotted in Fig. C.1 and has a surprisingly complicated structure for negative n. Selected values of the gamma function are listed in Table C.1. The gamma function will appear again in later integrals.

z	$-\frac{3}{2}$	$-\frac{1}{2}$	$\frac{1}{2}$	1	$\frac{3}{2}$	2	$\frac{5}{2}$	3	4
$\Gamma(z)$	$\frac{4\sqrt{\pi}}{3}$	$-2\sqrt{\pi}$	$\sqrt{\pi}$	1	$\frac{\sqrt{\pi}}{2}$	1	$\frac{3\sqrt{\pi}}{4}$	2	6

Table C.1 Selected values of the gamma function. Other values can be generated using $\Gamma(z+1) = z\Gamma(z)$.

C.2 The Gaussian integral

The **Gaussian** is a function of the form $e^{-\alpha x^2}$, and is plotted in Fig. C.2. It has a maximum at $x = 0$ and a shape that has been likened to that of

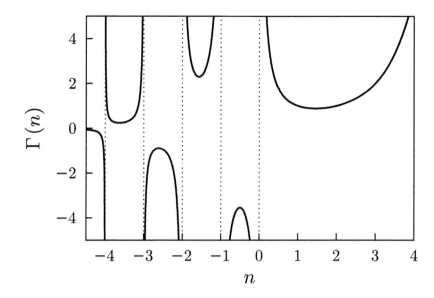

Fig. C.1 The gamma function $\Gamma(n)$ showing the singularities for integer values of $n \leq 0$. For positive, integer n, $\Gamma(n) = (n-1)!$.

a bell. It turns up in many statistical problems, often under the name of the **normal distribution**. The integral of a Gaussian is another extremely useful integral:

$$\boxed{\int_{-\infty}^{\infty} e^{-\alpha x^2} \, dx = \sqrt{\frac{\pi}{\alpha}}.} \tag{C.3}$$

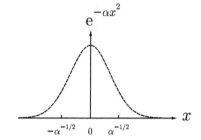

Fig. C.2 A Gaussian $e^{-\alpha x^2}$.

- It can be proved by evaluating the two-dimensional integral

$$\int_{-\infty}^{\infty} dx \int_{-\infty}^{\infty} dy \, e^{-\alpha(x^2+y^2)} = \left(\int_{-\infty}^{\infty} dx \, e^{-\alpha x^2}\right)\left(\int_{-\infty}^{\infty} dy \, e^{-\alpha y^2}\right)$$
$$= I^2, \tag{C.4}$$

where I is our desired integral. We can evaluate the left-hand side using polar coordinates, so that

$$I^2 = \int_0^{2\pi} d\theta \int_0^{\infty} dr \, r e^{-\alpha r^2}, \tag{C.5}$$

which with the substitution $z = \alpha r^2$ (and hence $dz = 2\alpha r \, dr$) gives

$$I^2 = 2\pi \times \frac{1}{2\alpha} \int_0^{\infty} dz \, e^{-z} = \frac{\pi}{\alpha}, \tag{C.6}$$

and hence $I = \sqrt{\pi/\alpha}$ is proved.

- Even more fun begins when we employ a cunning stratagem: we differentiate both sides of the equation with respect to α. Because x does not depend on α, this is easy to do. Hence $(d/d\alpha)e^{-\alpha x^2} = -x^2 e^{-\alpha x^2}$ and $(d/d\alpha)\sqrt{\pi/\alpha} = -\sqrt{\pi}/2\alpha^{3/2}$ so that

$$\boxed{\int_{-\infty}^{\infty} x^2 e^{-\alpha x^2} \, dx = \frac{1}{2}\sqrt{\frac{\pi}{\alpha^3}}.} \tag{C.7}$$

[1]A general formula is
$$\int_{-\infty}^{\infty} x^{2n} e^{-\alpha x^2}\, dx = \frac{(2n)!}{n! 2^{2n}} \sqrt{\frac{\pi}{\alpha^{2n+1}}},$$
for integer $n \geq 0$.

[2]A general formula is
$$\int_{0}^{\infty} x^{2n+1} e^{-\alpha x^2}\, dx = \frac{n!}{2\alpha^{n+1}},$$
for integer $n \geq 0$.

Another method of getting these integrals is to make the substitution $y = \alpha x^2$ and turn them into the factorial integrals considered above. This is all very well, but you need to know things like $(-\frac{1}{2})! = \sqrt{\pi}$ to proceed.

- This trick can be repeated with equal ease. Differentiating again gives
$$\boxed{\int_{-\infty}^{\infty} x^4 e^{-\alpha x^2}\, dx = \frac{3}{4}\sqrt{\frac{\pi}{\alpha^5}}.} \qquad (C.8)$$

- Therefore we have a way of generating the integrals between $-\infty$ and ∞ of $x^{2n} e^{-\alpha x^2}$, where $n \geq 0$ is an integer.[1] Because these functions are even, the integrals of the same functions between 0 and ∞ are just *half* of these results:
$$\int_0^\infty e^{-\alpha x^2}\, dx = \frac{1}{2}\sqrt{\frac{\pi}{\alpha}},$$
$$\int_0^\infty x^2 e^{-\alpha x^2}\, dx = \frac{1}{4}\sqrt{\frac{\pi}{\alpha^3}},$$
$$\int_0^\infty x^4 e^{-\alpha x^2}\, dx = \frac{3}{8}\sqrt{\frac{\pi}{\alpha^5}}.$$

- To integrate $x^{2n+1} e^{-\alpha x^2}$ between $-\infty$ and ∞ is easy: the functions are all odd and so the integrals are all zero. More interesting is to integrate between 0 and ∞, and for this you can start off with the integral $\int_0^\infty x e^{-\alpha x^2}\, dx = \left[(-1/2\alpha) e^{-\alpha x^2}\right]_0^\infty = 1/(2\alpha)$ (this integral is performed by noticing that differentiating $e^{-\alpha x^2}$ gives rise to something proportional to $x e^{-\alpha x^2}$) and after this you can then play the same trick as before. Thus, all the odd powers of x can now be obtained straightforwardly[2] by differentiating that integral with respect to α. Hence,
$$\int_0^\infty x e^{-\alpha x^2}\, dx = \frac{1}{2\alpha},$$
$$\int_0^\infty x^3 e^{-\alpha x^2}\, dx = \frac{1}{2\alpha^2},$$
$$\int_0^\infty x^5 e^{-\alpha x^2}\, dx = \frac{1}{\alpha^3}.$$

- A useful expression for a normalized Gaussian (one whose integral is unity) is
$$\frac{1}{\sqrt{2\pi\sigma^2}} e^{-(x-\mu)^2/2\sigma^2}. \qquad (C.9)$$

This has mean $\langle x \rangle = \mu$ and variance $\langle (x - \langle x \rangle)^2 \rangle = \sigma^2$. It is used extensively in mathematical statistics.

C.3 Stirling's formula

The derivation of Stirling's formula proceeds by using the integral expression for $n!$ in eqn C.1, namely
$$n! = \int_0^\infty x^n e^{-x}\, dx. \qquad (C.10)$$

We will play with the right-hand side of this integral and develop an approximation for it. We notice that the integrand $x^n e^{-x}$ consists of a function that increases with x (the function x^n) and a function that decreases with x (the function e^{-x}), and so it must have a maximum somewhere (see Fig. C.3(a)). Most of the integral is due to the bulge around this maximum, so we will try to approximate this region around the bulge. As we are eventually going to take logs of this integral, it is natural to work with the logarithm of this integrand, which we will call $f(x)$. Hence we define the function $f(x)$ by

$$e^{f(x)} = x^n e^{-x}. \tag{C.11}$$

This implies that $f(x)$ is given by

$$f(x) = n \ln x - x, \tag{C.12}$$

which is sketched in Fig. C.3(b). When the integrand has a maximum, so will $f(x)$. Hence the maximum of the integrand, and also the maximum of this function $f(x)$, can be found using

$$\frac{df}{dx} = \frac{n}{x} - 1 = 0, \tag{C.13}$$

which implies that the maximum in f is at $x = n$. We can differentiate again and get

$$\frac{d^2 f}{dx^2} = -\frac{n}{x^2}. \tag{C.14}$$

Now we can perform a Taylor expansion[3] around the maximum, so that

$$\begin{aligned} f(x) &= f(n) + \left(\frac{df}{dx}\right)_{x=n}(x-n) + \frac{1}{2!}\left(\frac{d^2 f}{dx^2}\right)_{x=n}(x-n)^2 + \cdots \\ &= n\ln n - n + 0 \times (x-n) - \frac{1}{2}\frac{n}{n^2}(x-n)^2 + \cdots \\ &= n\ln n - n - \frac{(x-n)^2}{2n} + \cdots \end{aligned} \tag{C.15}$$

The Taylor expansion approximates $f(x)$ by a quadratic (see the dotted line in Fig. C.3) and hence $e^{f(x)}$ approximates to a Gaussian.[4] Putting this as the integrand in eqn C.1, and removing from this integral the terms that do not depend on x, we have

$$n! = e^{n\ln n - n} \int_0^\infty e^{-(x-n)^2/2n + \cdots} \, dx. \tag{C.16}$$

The integral in this expression can be evaluated with the help of eqn C.3 to be

$$\int_0^\infty e^{-(x-n)^2/2n + \cdots} \, dx \approx \int_{-\infty}^\infty e^{-(x-n)^2/2n} \, dx = \sqrt{2\pi n}. \tag{C.17}$$

(Here we have used the fact that it doesn't matter if you put the lower limit of the integral as $-\infty$ rather than 0 since the integrand, $e^{-(x-n)^2/2n}$,

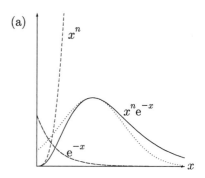

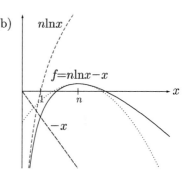

Fig. C.3 (a) The integrand $x^n e^{-x}$ (solid line) contains a maximum. (b) The function $f(x) = -x + n\ln x$ (solid line), which is the natural logarithm of the integrand. The dotted line is the Taylor expansion around the maximum (from eqn C.15). These curves have been plotted for $n = 3$, but the ability of the Taylor expansion to model the solid line improves as n increases. Note that (b) shows the natural logarithm of the curves in (a).

[3] See Appendix B.

[4] See Appendix C.2.

is a Gaussian centred at $x = n$ with a width that scales as $\sqrt{n}$ so that the contribution to the integral from the region between $-\infty$ and 0 is vanishingly small as n becomes large.) We have that

$$n! \approx e^{n \ln n - n} \sqrt{2\pi n}, \tag{C.18}$$

and hence

$$\boxed{\ln n! \approx n \ln n - n + \tfrac{1}{2} \ln 2\pi n,} \tag{C.19}$$

which is one version of **Stirling's formula**. When n is very large, this can be written

$$\boxed{\ln n! \approx n \ln n - n,} \tag{C.20}$$

which is another version of Stirling's formula.

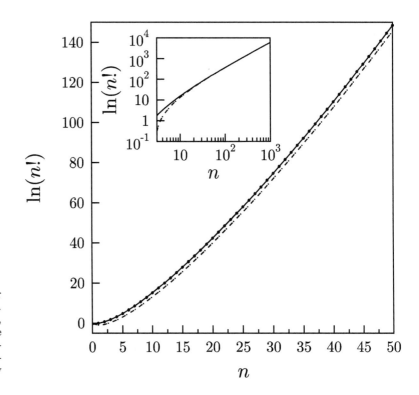

Fig. C.4 Stirling's approximation for $\ln n!$. The dots are the exact results. The solid line is according to eqn C.19, while the dashed line is eqn C.20. The inset shows the two lines for larger values of n and demonstrates that as n becomes large, eqn C.20 becomes a very good approximation.

The approximation in eqn C.19 is very good, as can be seen in Fig. C.4. The approximation in eqn C.20 (the dotted line in Fig. C.4) slightly underestimates the exact result when n is small, but as n becomes large (as is often the case in thermal physics problems) it becomes a very good approximation (as shown in the inset to Fig. C.4).

C.4 Riemann zeta function

The **Riemann zeta function** $\zeta(s)$ is usually defined by

$$\zeta(s) = \sum_{n=1}^{\infty} \frac{1}{n^s}, \qquad (C.21)$$

and converges for $s > 1$ (see Fig. C.5). For $s = 1$ it gives a divergent series. Some useful values are listed in Table C.2.

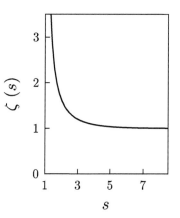

Fig. C.5 The Riemann zeta function $\zeta(s)$ for $s > 1$.

s	$\zeta(s)$
1	∞
$\frac{3}{2}$	≈ 2.612
2	$\pi^2/6 \approx 1.645$
$\frac{5}{2}$	≈ 1.341
3	≈ 1.20206
4	$\pi^4/90 \approx 1.0823$
5	≈ 1.0369
6	$\pi^6/945 \approx 1.017$

Table C.2 Selected values of the Riemann zeta function.

Our reason for introducing the Riemann zeta function is that it is involved in many useful integrals. One such is the **Bose integral** $I_B(n)$ defined by

$$I_B(n) = \int_0^\infty dx \, \frac{x^n}{e^x - 1}. \qquad (C.22)$$

We can evaluate this as follows:

$$\begin{aligned} I_B(n) &= \int_0^\infty dx \, \frac{x^n e^{-x}}{1 - e^{-x}} \\ &= \int_0^\infty dx \, x^n \sum_{k=0}^{\infty} e^{-(k+1)x} \\ &= \sum_{k=0}^{\infty} \frac{1}{(k+1)^{n+1}} \int_0^\infty dy \, y^n e^{-y} \\ &= \zeta(n+1)\, \Gamma(n+1). \end{aligned} \qquad (C.23)$$

Thus we have that

$$\boxed{I_B(n) = \int_0^\infty dx \, \frac{x^n}{e^x - 1} = \zeta(n+1)\, \Gamma(n+1).} \qquad (C.24)$$

So, for example,

$$\int_0^\infty dx \, \frac{x^3}{e^x - 1} = \zeta(4)\, \Gamma(4) = \frac{\pi^4}{90} \times 3! = \frac{\pi^4}{15}. \qquad (C.25)$$

Another useful integral can be derived as follows. Consider the integral

$$I = \int_0^\infty dx \frac{x^{n-1}}{e^{ax} - 1}. \tag{C.26}$$

This can be evaluated easily by making the substitution $y = ax$, yielding

$$I = \frac{1}{a^n} \int_0^\infty dy \frac{y^{n-1}}{e^y - 1}. \tag{C.27}$$

Now, differentiating I with respect to a using eqn C.26 gives

$$\frac{dI}{da} = -\int_0^\infty dx \frac{x^n e^{ax}}{(e^{ax} - 1)^2}, \tag{C.28}$$

while using eqn C.27 yields

$$\frac{dI}{da} = -\frac{n}{a^{n+1}} \int_0^\infty dy \frac{y^{n-1}}{e^y - 1}. \tag{C.29}$$

These two expressions should be the same, and hence equating them and putting $a = 1$ yields

$$\boxed{\int_0^\infty dx \frac{x^n e^x}{(e^x - 1)^2} = n\,\zeta(n)\,\Gamma(n).} \tag{C.30}$$

So, for example,

$$\int_0^\infty dx \frac{x^4 e^x}{(e^x - 1)^2} = 4\zeta(4)\,\Gamma(4) = 4 \times \frac{\pi^4}{90} \times 3! = \frac{4\pi^4}{15}. \tag{C.31}$$

C.5 The polylogarithm

The **polylogarithm** function $\text{Li}_n(z)$ (also known as **de Jonquière's function**) is defined as

$$\text{Li}_n(z) = \sum_{k=1}^\infty \frac{z^k}{k^n}, \tag{C.32}$$

where z is in the open unit disc in the complex plane, i.e., $|z| < 1$. The definition over the whole complex plane follows via the process of analytic continuation. The polylogarithm is useful in the evaluation of integrals of Bose–Einstein and Fermi–Dirac distribution functions. First note that we can write

$$\frac{1}{z^{-1}e^x - 1} = \frac{ze^{-x}}{1 - ze^{-x}} = \sum_{m=0}^\infty (ze^{-x})^{m+1}, \tag{C.33}$$

i.e., as a geometric progression. Hence we can evaluate the following integral:

$$\begin{aligned}
\int_0^\infty \frac{x^{n-1}\,dx}{z^{-1}e^x - 1} &= \sum_{m=0}^\infty \int_0^\infty dx\, x^{n-1}(ze^{-x})^{m+1} \\
&= \sum_{m=0}^\infty z^{m+1} \int_0^\infty dx\, x^{n-1}e^{-(m+1)x} \\
&= \sum_{m=0}^\infty \frac{z^{m+1}}{(m+1)^n} \int_0^\infty dy\, y^{n-1}e^{-y} \\
&= \Gamma(n) \sum_{m=0}^\infty \frac{z^{m+1}}{(m+1)^n} \\
&= \Gamma(n) \sum_{k=1}^\infty \frac{z^k}{k^n} \\
&= \Gamma(n)\mathrm{Li}_n(z).
\end{aligned} \qquad (C.34)$$

Similarly one can show that

$$\int_0^\infty \frac{x^{n-1}\,dx}{z^{-1}e^x + 1} = -\Gamma(n)\mathrm{Li}_n(-z). \qquad (C.35)$$

Combining these equations, one can write in general that

$$\boxed{\int_0^\infty \frac{x^{n-1}\,dx}{z^{-1}e^x \pm 1} = \mp\Gamma(n)\mathrm{Li}_n(\mp z)\,.} \qquad (C.36)$$

Note that when $|z| \ll 1$, only the first term in the series in eqn C.32 contributes, and

$$\mathrm{Li}_n(z) \approx z. \qquad (C.37)$$

Note also that

$$\mathrm{Li}_n(1) = \sum_{k=1}^\infty \frac{1}{k^n} = \zeta(n), \qquad (C.38)$$

where $\zeta(n)$ is the Riemann zeta function (eqn C.21).

C.6 Partial derivatives

Consider x as a function of two variables y and z. This can be written $x = x(y, z)$, and we have that

$$dx = \left(\frac{\partial x}{\partial y}\right)_z dy + \left(\frac{\partial x}{\partial z}\right)_y dz. \qquad (C.39)$$

But rearranging $x = x(y, z)$ can lead to having z as a function of x and y so that $z = z(x, y)$, in which case

$$dz = \left(\frac{\partial z}{\partial x}\right)_y dx + \left(\frac{\partial z}{\partial y}\right)_x dy. \qquad (C.40)$$

Substituting C.40 into C.39 gives

$$\mathrm{d}x = \left(\frac{\partial x}{\partial z}\right)_y \left(\frac{\partial z}{\partial x}\right)_y \mathrm{d}x + \left[\left(\frac{\partial x}{\partial y}\right)_z + \left(\frac{\partial x}{\partial z}\right)_y \left(\frac{\partial z}{\partial y}\right)_x\right] \mathrm{d}y.$$

The terms multiplying $\mathrm{d}x$ give the **reciprocal theorem**

$$\left(\frac{\partial x}{\partial z}\right)_y = \frac{1}{\left(\frac{\partial z}{\partial x}\right)_y}, \qquad (C.41)$$

and the terms multiplying $\mathrm{d}y$ give the **reciprocity theorem**

$$\left(\frac{\partial x}{\partial y}\right)_z \left(\frac{\partial y}{\partial z}\right)_x \left(\frac{\partial z}{\partial x}\right)_y = -1. \qquad (C.42)$$

This can be combined with the reciprocal theorem to write that

$$\left(\frac{\partial x}{\partial y}\right)_z = -\left(\frac{\partial x}{\partial z}\right)_y \left(\frac{\partial z}{\partial y}\right)_x, \qquad (C.43)$$

which is a very useful identity.

C.7 Exact differentials

An expression such as $F_1(x,y)\,\mathrm{d}x + F_2(x,y)\,\mathrm{d}y$ is known as an **exact differential** if it can be written as the differential

$$\mathrm{d}f = \left(\frac{\partial f}{\partial x}\right)\mathrm{d}x + \left(\frac{\partial f}{\partial y}\right)\mathrm{d}y, \qquad (C.44)$$

of a differentiable single-valued function $f(x,y)$. This implies that

$$F_1 = \left(\frac{\partial f}{\partial x}\right), \quad F_2 = \left(\frac{\partial f}{\partial y}\right), \qquad (C.45)$$

or in vector form, $\boldsymbol{F} = \nabla f$. Hence the integral of an exact differential is path independent, so that [where 1 and 2 are shorthands for (x_1, y_1) and (x_2, y_2)]

$$\int_1^2 F_1(x,y)\,\mathrm{d}x + F_2(x,y)\,\mathrm{d}y = \int_1^2 \boldsymbol{F}\cdot\mathrm{d}\boldsymbol{r} = \int_1^2 \mathrm{d}f = f(2)-f(1), \quad (C.46)$$

and the answer depends only on the initial and final states of the system. For an **inexact differential** this is not true and knowledge of the initial and final states is not sufficient to evaluate the integral: you have to know which path was taken.

For an exact differential the integral round a closed loop is zero:

$$\oint F_1(x,y)\,\mathrm{d}x + F_2(x,y)\,\mathrm{d}y = \oint \boldsymbol{F}\cdot\mathrm{d}\boldsymbol{r} = \oint \mathrm{d}f = 0, \qquad (C.47)$$

which implies that $\nabla \times \boldsymbol{F} = 0$ (by Stokes' theorem) and hence

$$\left(\frac{\partial F_2}{\partial x}\right) = \left(\frac{\partial F_1}{\partial y}\right) \quad \text{or} \quad \left(\frac{\partial^2 f}{\partial x \partial y}\right) = \left(\frac{\partial^2 f}{\partial y \partial x}\right). \qquad (C.48)$$

For thermal physics, a crucial point to remember is that *functions of state have exact differentials.*

C.8 Volume of a hypersphere

A **hypersphere** in D dimensions and with radius r is described by the equation

$$\sum_{i=1}^{D} x_i^2 = r^2. \tag{C.49}$$

It has volume V_D given

$$V_D = \alpha r^D, \tag{C.50}$$

where α is a numerical constant which we will now determine.
Consider the integral I given by

$$I = \int_{-\infty}^{\infty} dx_1 \cdots \int_{-\infty}^{\infty} dx_D \exp\left(-\sum_{i=1}^{D} x_i^2\right). \tag{C.51}$$

This can be evaluated as follows:

$$I = \left[\int_{-\infty}^{\infty} dx\, e^{-x^2}\right]^D = \pi^{D/2}. \tag{C.52}$$

Alternatively, we can evaluate it in hyperspherical polars as follows:

$$I = \int_0^{\infty} dV_D\, e^{-r^2}, \tag{C.53}$$

where the volume element is given by $dV_D = \alpha D r^{D-1}\, dr$. Hence, equating eqn C.52 and eqn C.53 we have that

$$\pi^{D/2} = \alpha D \int_0^{\infty} dr\, r^{D-1} e^{-r^2} = \frac{\alpha D \Gamma(\frac{D}{2})}{2}, \tag{C.54}$$

and hence

$$\alpha = \frac{2\pi^{D/2}}{D\Gamma(D/2)}. \tag{C.55}$$

Hence[5] we obtain the volume of a hypersphere in D dimensions as

[5] Using $\Gamma(\frac{D}{2} + 1) = \frac{D}{2}\Gamma(\frac{D}{2})$.

$$V_D = \frac{\pi^{D/2} r^D}{\Gamma(\frac{D}{2} + 1)}. \tag{C.56}$$

C.9 Jacobians

Let $x = g(u, v)$ and $y = h(u, v)$ be a transformation of the plane. Then the **Jacobian** of this transformation is

$$\frac{\partial(x, y)}{\partial(u, v)} = \begin{vmatrix} \frac{\partial x}{\partial u} & \frac{\partial x}{\partial v} \\ \frac{\partial y}{\partial u} & \frac{\partial y}{\partial v} \end{vmatrix} = \frac{\partial x}{\partial u}\frac{\partial y}{\partial v} - \frac{\partial x}{\partial v}\frac{\partial y}{\partial u}. \tag{C.57}$$

Example C.1

The Jacobian of the polar coordinate transformation $x(r,\theta) = r\cos\theta$ and $y(r,\theta) = r\sin\theta$ is

$$\frac{\partial(x,y)}{\partial(r,\theta)} = \begin{vmatrix} \frac{\partial x}{\partial r} & \frac{\partial x}{\partial \theta} \\ \frac{\partial y}{\partial r} & \frac{\partial y}{\partial \theta} \end{vmatrix} = \begin{vmatrix} \cos\theta & -r\sin\theta \\ \sin\theta & r\cos\theta \end{vmatrix} = r. \tag{C.58}$$

If g and h have continuous partial differentials such that the Jacobian is never zero, we then have

$$\int\int_R f(x,y)\,dx\,dy = \int\int_S f(g(u,v),h(u,v))\left|\frac{\partial(x,y)}{\partial(u,v)}\right|\,du\,dv. \tag{C.59}$$

So in our example, we would have

$$\int\int_R f(x,y)\,dx\,dy = \int\int_S f(g(r,\theta),h(r,\theta))r\,dr\,d\theta. \tag{C.60}$$

The Jacobian of the inverse transformation is the reciprocal of the Jacobian of the original transformation.

$$\left|\frac{\partial(x,y)}{\partial(u,v)}\right| = \frac{1}{\left|\frac{\partial(u,v)}{\partial(x,y)}\right|}, \tag{C.61}$$

which is a consequence of the fact that the determinant of the inverse of a matrix is the reciprocal of the determinant of the matrix. Other useful identities are

$$\frac{\partial(x,y)}{\partial(u,v)} = -\frac{\partial(y,x)}{\partial(u,v)} = \frac{\partial(y,x)}{\partial(v,u)}, \tag{C.62}$$

$$\frac{\partial(x,y)}{\partial(x,y)} = 1, \tag{C.63}$$

$$\frac{\partial(x,y)}{\partial(x,z)} = \left(\frac{\partial y}{\partial z}\right)_x, \tag{C.64}$$

and

$$\frac{\partial(x,y)}{\partial(u,v)} = \frac{\partial(x,y)}{\partial(a,b)}\frac{\partial(a,b)}{\partial(u,v)}. \tag{C.65}$$

Quick exercise:

The Jacobian can be generalized to three dimensions, as

$$\frac{\partial(x,y,z)}{\partial(u,v,w)} = \begin{vmatrix} \frac{\partial x}{\partial u} & \frac{\partial x}{\partial v} & \frac{\partial x}{\partial w} \\ \frac{\partial y}{\partial u} & \frac{\partial y}{\partial v} & \frac{\partial y}{\partial w} \\ \frac{\partial z}{\partial u} & \frac{\partial z}{\partial v} & \frac{\partial z}{\partial w} \end{vmatrix}. \tag{C.66}$$

Show that for the transformation of spherical polars $x = r\sin\theta\cos\phi$, $y = r\sin\theta\sin\phi$, $z = r\cos\theta$, the Jacobian is

$$\frac{\partial(x,y,z)}{\partial(r,\theta,\phi)} = r^2\sin\theta. \tag{C.67}$$

C.10 The Dirac delta function

The Dirac delta function $\delta(x-a)$ centred at $x=a$ is zero for all x not equal to a, but its area is 1. Hence

$$\int_{-\infty}^{\infty} \delta(x-a) = 1. \tag{C.68}$$

Because the Dirac delta function is such a narrow "spike", integrals of the Dirac delta function multiplied by any other function $f(x)$ are simple to evaluate:

$$\int_{-\infty}^{\infty} f(x)\delta(x-a) = f(a). \tag{C.69}$$

C.11 Fourier transforms

Consider a function $x(t)$. Its Fourier transform is defined by

$$\tilde{x}(\omega) = \int_{-\infty}^{\infty} dt\, e^{-i\omega t} x(t). \tag{C.70}$$

The inverse transform is

$$x(t) = \frac{1}{2\pi} \int_{-\infty}^{\infty} d\omega\, e^{i\omega t} \tilde{x}(\omega). \tag{C.71}$$

We now state some useful results concering Fourier transforms.

- The Fourier transform of a delta function $\delta(t-t')$ is given by

$$\int_{-\infty}^{\infty} dt\, e^{-i\omega t} \delta(t-t') = e^{-i\omega t'}, \tag{C.72}$$

and putting this into the inverse transform shows that

$$\int_{-\infty}^{\infty} d\omega\, e^{i(\omega-\omega')t} = 2\pi\delta(\omega-\omega'), \tag{C.73}$$

which is an identity that will be useful later.
- The Fourier transform of $\dot{x}(t)$ is $i\omega\tilde{x}(\omega)$, and so differential equations can be Fourier transformed into algebraic equations.
- The Fourier transform of $x^*(t)$ is $\tilde{x}^*(-\omega)$.
- Parseval's theorem states that

$$\int_{-\infty}^{\infty} dt\, |x(t)|^2 = \frac{1}{2\pi} \int_{-\infty}^{\infty} d\omega\, |\tilde{x}(\omega)|^2. \tag{C.74}$$

- The convolution $h(t)$ of two functions $f(t)$ and $g(t)$ is defined by

$$h(t) = \int_{-\infty}^{\infty} dt'\, f(t-t')g(t'). \tag{C.75}$$

The **convolution theorem** states that the Fourier transform of $h(t)$ is then given by the multiplication of the Fourier transforms of $f(t)$ and $g(t)$, i.e.,

$$\tilde{h}(\omega) = \tilde{f}(\omega)\tilde{g}(\omega). \tag{C.76}$$

- We now prove the Wiener–Khinchin theorem (mentioned in Section 33.6). Using the inverse Fourier transform, we can write the correlation function $C_{xx}(t)$ as

$$\begin{aligned}
C_{xx}(t) &= \int_{-\infty}^{\infty} x^*(t')x(t'+t)\,\mathrm{d}t' \\
&= \int_{-\infty}^{\infty} \mathrm{d}t' \left[\frac{1}{2\pi}\int_{-\infty}^{\infty} \mathrm{d}\omega'\, \mathrm{e}^{\mathrm{i}\omega't'}\tilde{x}^*(-\omega')\right] \\
&\qquad\qquad \left[\frac{1}{2\pi}\int_{-\infty}^{\infty} \mathrm{d}\omega\, \mathrm{e}^{\mathrm{i}\omega(t'+t)}\tilde{x}(\omega)\right] \\
&= \frac{1}{4\pi^2}\int_{-\infty}^{\infty} \mathrm{d}\omega\, \mathrm{e}^{\mathrm{i}\omega t} \int_{-\infty}^{\infty} \mathrm{d}\omega'\, \tilde{x}^*(-\omega')\tilde{x}(\omega) \int_{-\infty}^{\infty} \mathrm{d}t'\, \mathrm{e}^{\mathrm{i}(\omega+\omega')t'} \\
&= \frac{1}{2\pi}\int_{-\infty}^{\infty} \mathrm{d}\omega\, \mathrm{e}^{\mathrm{i}\omega t} \int_{-\infty}^{\infty} \mathrm{d}\omega'\, \tilde{x}^*(-\omega')\tilde{x}(\omega)\delta(\omega+\omega') \\
&= \frac{1}{2\pi}\int_{-\infty}^{\infty} \mathrm{d}\omega\, \mathrm{e}^{\mathrm{i}\omega t}\, \tilde{x}^*(\omega)\tilde{x}(\omega) \\
&= \frac{1}{2\pi}\int_{-\infty}^{\infty} \mathrm{d}\omega\, \mathrm{e}^{\mathrm{i}\omega t}\, |\tilde{x}(\omega)|^2,
\end{aligned} \qquad (\mathrm{C.77})$$

where use has been made of eqn C.73. This result is simply the inverse Fourier transform of $|\tilde{x}(\omega)|^2$, the power spectrum of the function.

C.12 Solution of the diffusion equation

The diffusion equation

$$\frac{\partial n}{\partial t} = D\frac{\partial^2 n}{\partial x^2} \qquad (\mathrm{C.78})$$

can be solved by Fourier transforming $n(x,t)$ using

$$\tilde{n}(k,t) = \int_{-\infty}^{\infty} \mathrm{d}x\, \mathrm{e}^{-\mathrm{i}kx} n(x,t), \qquad (\mathrm{C.79})$$

so that

$$-\mathrm{i}k\tilde{n}(k,t) = \int_{-\infty}^{\infty} \mathrm{d}x\, \mathrm{e}^{-\mathrm{i}kx}\frac{\partial n(x,t)}{\partial x}. \qquad (\mathrm{C.80})$$

Hence eqn C.78 becomes

$$\frac{\partial \tilde{n}(k,t)}{\partial t} = -Dk^2\tilde{n}(k,t), \qquad (\mathrm{C.81})$$

which is now a simple first-order differential equation whose solution is

$$\tilde{n}(k,t) = \tilde{n}(k,0)\,\mathrm{e}^{-Dk^2 t}. \qquad (\mathrm{C.82})$$

Inverse Fourier transforming then yields

$$n(x,t) = \frac{1}{2\pi}\int_{-\infty}^{\infty} \mathrm{d}k\, \mathrm{e}^{\mathrm{i}kx}\, \mathrm{e}^{-Dk^2 t}\tilde{n}(k,0). \qquad (\mathrm{C.83})$$

In particular, if the initial distribution of n is given by

$$n(x,0) = n_0 \delta(x), \qquad (C.84)$$

then

$$\tilde{n}(k,0) = n_0, \qquad (C.85)$$

and hence

$$n(x,t) = \frac{n_0}{\sqrt{4\pi Dt}} e^{-x^2/(4Dt)}. \qquad (C.86)$$

This equation is plotted in Fig. C.6 and describes a Gaussian whose width increases with time. Note that $\langle x^2 \rangle = 2Dt$.

Quick exercise

Repeat this in three dimensions for the diffusion equation

$$\frac{\partial n}{\partial t} = D \nabla^2 n \qquad (C.87)$$

and show that if $n(\mathbf{r}, 0) = n_0 \delta(\mathbf{r})$ then

$$n(\mathbf{r},t) = \frac{n_0}{\sqrt{4\pi Dt}} e^{-r^2/(4Dt)}. \qquad (C.88)$$

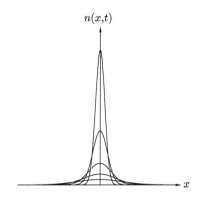

Fig. C.6 Equation C.86 plotted for various values of t. At $t = 0$, $n(x,t)$ is a delta function at the origin, i.e., $n(x,0) = n_0 \delta(x)$. As t increases, $n(x,t)$ becomes broader and the distribution spreads out.

C.13 Lagrange multipliers

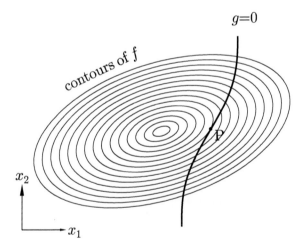

Fig. C.7 We wish to find the maximum of the function f subject to the constraint that $g = 0$. This occurs at the point P at which one of the contours of f and the curve $g = 0$ touch tangentially.

The method of **Lagrange multipliers**[6] is used to find the extrema of a function of several variables subject to one or more constraints. Suppose we wish to maximize (or minimize) a function $f(\mathbf{x})$ subject to the constraint $g(\mathbf{x}) = 0$. Both f and g are functions of the N variables $\mathbf{x} = (x_1, x_2, \ldots, x_N)$. The maximum (or minimum) will occur when one of the contours of f and the curve $g = 0$ touch tangentially; let us call the set of points at which this occurs P (this is shown in Fig. C.7 for a two-dimensional case). Now ∇f is a vector normal to the contours of f

[6] Joseph-Louis, Comte de Lagrange (1736–1813).

and ∇g is a vector normal to the curve $g = 0$, and these two vectors will be parallel to each other at P. Hence

$$\nabla[f + \lambda g] = 0, \tag{C.89}$$

where λ is a constant, called the Lagrange multiplier. Thus we have N equations to solve:

$$\frac{\partial F}{\partial x_k} = 0, \tag{C.90}$$

where $F = f + \lambda g$ and $k = 1, \ldots, N$. This allows us to find λ and hence identify the $(N-2)$-dimensional surface on which f is extremized subject to the constraint $g = 0$.

If there are M constraints, so that for example $g_i(x) = 0$ where $i = 1, \ldots, M$, then we solve eqn C.90 with

$$F = f + \sum_{i=1}^{M} \lambda_i g_i, \tag{C.91}$$

where $\lambda_1, \ldots, \lambda_M$ are Lagrange multipliers.

Example C.2

Find the ratio of the radius r to the height h of a cylinder, which maximizes its total surface area subject to the constraint that its volume is constant.

Solution:
The volume $V = \pi r^2 h$ and area $A = 2\pi rh + 2\pi r^2$, so we consider the function F given by

$$F = A + \lambda V, \tag{C.92}$$

and solve

$$\frac{\partial F}{\partial h} = 2\pi r + \lambda \pi r^2 = 0, \tag{C.93}$$

$$\frac{\partial F}{\partial r} = 2\pi h + 4\pi r + 2\lambda \pi r h = 0, \tag{C.94}$$

which yields $\lambda = -2/r$ and hence $h = 2r$.

The electromagnetic spectrum

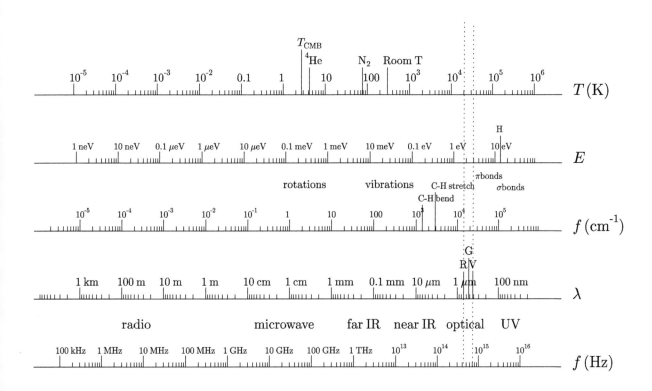

Fig. D.1 The electromagnetic spectrum. The energy of a photon is shown as a temperature $T = E/k_B$ in K and as an energy E in eV. The corresponding frequency f is shown in Hz and, because the unit is often quoted in spectroscopy, in cm^{-1}. The cm^{-1} scale is marked with some common molecular transitions and excitations (the typical ranges for molecular rotations and vibrations are shown, together with the C–H bending and stretching modes). The energy of typical π and σ bonds is also shown. The wavelength $\lambda = c/f$ of the photon is shown (where c is the speed of light). The particular temperatures marked on the temperature scale are $T_{\rm CMB}$ (the temperature of the cosmic microwave background), the boiling points of liquid helium (^{4}He) and nitrogen (N$_2$), both at atmospheric pressure, and also the value of room temperature. Other abbreviations on this diagram are IR = infrared, UV = ultraviolet, R = red, G = green, V = violet. The letter H marks 13.6 eV, the magnitude of the energy of the 1s electron in hydrogen. The frequency axis also contains descriptions of the main regions of the electromagnetic spectrum: radio, microwave, infrared (both "near" and "far"), optical and UV.

E Some thermodynamical definitions

- **System** = whatever part of the Universe we select.
- **Open** systems can exchange particles with their surroundings.
- **Closed** systems cannot.
- An **isolated** system is not influenced from outside its boundaries.
- **Adiathermal** = without flow of heat. A system bounded by adiathermal walls is **thermally isolated**. Any work done on such a system produces adiathermal change.
- **Diathermal** walls allow flow of heat. Two systems separated by diathermal walls are said to be **in thermal contact**.
- **Adiabatic** = adiathermal and reversible (often used synonymously with adiathermal).
- Put a system in thermal contact with some new surroundings. Heat flows or work is done. Eventually no further change takes place: the system is said to be in a state of **thermal equilibrium**.
- A **quasistatic** process is one carried out so slowly that the system passes through a series of equilibrium states so is always in equilibrium. A process that is quasistatic and has no hysteresis is said to be **reversible**.
- **Isobaric** = at constant pressure.
- **Isochoric** = at constant volume.
- **Isenthalpic** = at constant enthalpy.
- **Isentropic** = at constant entropy.
- **Isothermal** = at constant temperature.

Thermodynamic expansion formulae

	$(*)_T$	$(*)_P$	$(*)_V$	$(*)_S$	$(*)_U$	$(*)_H$	$(*)_F$
(∂G)	-1	$-S/V$	$\kappa S - \alpha V$	$\alpha S - C_p/T$	$S(T\alpha - P\kappa)$ $-C_p + PV\alpha$	$S(T\alpha - 1)$ $-C_p$	$S - P(\kappa S - V\alpha)$
(∂F)	$-\kappa P$	$-(S/V) - P\alpha$	κS	$\alpha S - p\kappa C_V/T$	$S(T\alpha - P\kappa)$ $-P\kappa C_V$	$S(T\alpha - 1)$ $-P(\kappa C_V + V\alpha)$	0
(∂H)	$T\alpha - 1$	C_p/V	$-\kappa C_V - V\alpha$	$-C_p/T$	$P(\kappa C_V + V\alpha)$ $-C_p$	0	
(∂U)	$T\alpha - p\kappa$	$(C_p/V) - P\alpha$	$-\kappa C_V$	$-P\kappa C_V/T$	0		
(∂S)	α	C_p/TV	$-\kappa C_V/T$	0			
(∂V)	κ	α	0				
(∂P)	$-1/V$	0					

Table F.1 Expansion formulae for first-order partial derivatives of thermal variables. (After E. W. Dearden, *Eur. J. Phys.* **16** 76 (1995).)

Table F.1 contains a listing of various partial derivatives, some of which have been derived in this book. To evaluate a partial differential, one has to take the ratio of two terms in this table using the equation

$$\left(\frac{\partial x}{\partial y}\right)_z \equiv \frac{(\partial x)_z}{(\partial y)_z}. \tag{F.1}$$

Note that $(\partial A)_B \equiv -(\partial B)_A$.

Example F.1

To evaluate the Joule–Kelvin coefficient:

$$\mu_{\text{JK}} = \left(\frac{\partial T}{\partial P}\right)_H = \frac{(\partial T)_H}{(\partial P)_H} = \frac{(\partial H)_T}{(\partial H)_p} = \frac{V(T\alpha - 1)}{C_p}. \tag{F.2}$$

G Reduced mass

Consider two particles with masses m_1 and m_2 located at positions $\mathbf{r}_1$ and $\mathbf{r}_2$ and held together by a force $\mathbf{F}(r)$ that depends only on the distance $r = |\mathbf{r}| = |\mathbf{r}_1 - \mathbf{r}_2|$ (see Fig. G.1).

Thus we have

$$m_1 \ddot{\mathbf{r}}_1 = \mathbf{F}(r), \tag{G.1}$$
$$m_2 \ddot{\mathbf{r}}_2 = -\mathbf{F}(r), \tag{G.2}$$

and hence

$$\ddot{\mathbf{r}} = (m_1^{-1} + m_2^{-1})\mathbf{F}(r), \tag{G.3}$$

which can be written

$$\mu \ddot{\mathbf{r}} = \mathbf{F}(r), \tag{G.4}$$

where μ is the **reduced mass** given by

$$\frac{1}{\mu} = \frac{1}{m_1} + \frac{1}{m_2}, \tag{G.5}$$

or equivalently

$$\mu = \frac{m_1 m_2}{m_1 + m_2}. \tag{G.6}$$

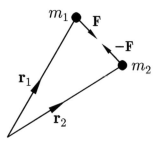

Fig. G.1 The forces exerted by two particles on one another.

Glossary of main symbols

α damping constant
α_λ spectral absorptivity
$\beta = 1/(k_B T)$
$\Gamma(n)$ gamma function
γ adiabatic index
γ surface tension
δ skin depth
ϵ Seebeck coefficient
ϵ_0 permittivity of free space
$\zeta(s)$ Riemann zeta function
η viscosity
$\theta(x)$ Heaviside step function
κ thermal conductivity
κ_ν extinction coefficient
Λ relativistic thermal wavelength
λ mean free path
λ wavelength
λ_{th} thermal wavelength
μ chemical potential
μ_0 permeability of free space
$\mu^\ominus$ chemical potential at STP
μ_J Joule coefficient
μ_{JK} Joule–Kelvin coefficient
ν frequency
$\pi = 3.1415926535\ldots$
Π momentum flux
Π Peltier coefficient
Π Osmotic pressure
ρ density
ρ resistivity
ρ_J Jeans density
Σ local entropy production
σ standard deviation
σ collision cross-section
σ_p Prandtl number
τ mean scattering time

τ_{xy} shear stress across xy plane
Φ_G grand potential
Φ flux
χ magnetic suseptibility
$\chi(t-t')$ response function
$\chi(t)$ response function
χ_ν optical depth at frequency ν
$\psi(\boldsymbol{r})$ wave function
Ω solid angle
Ω potential energy
$\Omega(E)$ number of microstates with energy E
ω angular frequency
A availability
A area
A albedo
A_{21} Einstein coefficient
A_{12} Einstein coefficient
B_{12} Einstein coefficient
B bulk modulus
B magnetic field
B_λ radiance or surface brightness in a wavelength interval
B_ν radiance or surface brightness in a frequency interval
B_S bulk modulus at constant entropy
B_T bulk modulus at constant temperature
$B(T)$ virial coefficient as a function of T
C heat capacity
C number of chemically distinct constituents
C capacitance
CMB cosmic microwave background
c speed of light
c specific heat capacity
D coefficient of self-diffusion
$\boldsymbol{E}$ electric field
E energy
E_F Fermi energy
$\boldsymbol{\mathcal{E}}$ electromotive field
$e = 2.7182818\ldots$

e_λ spectral emissive power
F Helmholtz function
F number of degrees of freedom
F_ν spectral irradiance
f frequency
$f(v)$ speed distribution function
$f(E)$ distribution function, Fermi function
G gravitational constant
G Gibbs function
g gravitational acceleration on Earth's surface
g degeneracy
$g(k)$ density of states as a function of wave vector
$g(E)$ density of states as a function of energy
H enthalpy
H magnetic field strength
I current
I moment of inertia
I_ν spectral radiance
J heat flux
J_ν source function
K equilibrium constant
K_b ebullioscopic constant
K_f cryoscopic constant
k wave vector
k_B Boltzmann constant
k_F Fermi wave vector
L latent heat
L luminosity
$L_\odot$ luminosity of the Sun
L_{edd} Eddington luminosity
L_{ij} kinetic coefficients
$\text{Li}_n(z)$ polylogarithm function
l_P Planck length
M magnetization
M Mach number
$M_\odot$ mass of the Sun
$M_\oplus$ mass of the Earth
M_J Jeans mass
m magnetic moment
m mass of particle or system
N number of particles
N_A Avogadro number

n number density (number per unit volume)
n_m number of moles
n_Q quantum concentration
P number of phases present
$P(x)$ probability of x
$\hat{P}_{12}$ exchange operator
$\mathcal{P}$ Cauchy principal value
p_F Fermi momentum
p pressure
$p^{\ominus}$ standard pressure (1 atmosphere)
Q heat
q phonon wave vector
R gas constant
R resistance
$R_\odot$ radius of the Sun
$R_\oplus$ radius of the Earth
S spin
S entropy
STP standard temperature and pressure
T temperature
T_B Boyle temperature
T_b temperature at boiling point
T_C Curie temperature
T_c critical temperature
T_F Fermi temperature
t time
U internal energy
u internal energy per unit volume
$\tilde{u}$ internal energy per unit mass
u_λ spectral energy density
V speed of particle
v speed of particle
$\langle v \rangle$ mean speed of particle
$\langle v^2 \rangle$ mean squared speed of particle
$\sqrt{\langle v^2 \rangle}$ root mean squared (rms) speed of particle
v_s speed of sound
W work
Z partition function
Z_1 partition function for single-particle state
$\mathcal{Z}$ grand partition function
z fugacity

Bibliography

Thermal physics

- C. J. Adkins, *Equilibrium Thermodynamics*, CUP, Cambridge (1983).
- P. W. Anderson, *Basic Notions of Condensed Matter Physics*, Addison-Wesley (1984).
- D. G. Andrews, *An Introduction to Atmospheric Physics*, CUP, Cambridge (2000).
- J. F. Annett, *Superconductivity, Superfluids and Condensates*, OUP, Oxford (2004).
- D. Archer, *Global Warming: Understanding the Forecast*, Blackwell, Malden (2007).
- N. W. Ashcroft and N. D. Mermin, *Solid State Physics*, Thomson Learning, London (1976).
- P. W. Atkins and J. de Paulo, *Physical Chemistry*, (8th edition), OUP, Oxford (2006).
- R. Baierlein, *Thermal Physics*, CUP, Cambridge (1999).
- R. Baierlein, The elusive chemical potential, *Am. J. Phys.* **69**, 423 (2001).
- J. J. Binney, N. J. Dowrick, A. J. Fisher and M. E. J. Newman, *The Theory of Critical Phenomena*, OUP, Oxford (1992).
- J. J. Binney and M. Merrifield, *Galactic Astronomy*, Princeton University Press, Princeton, New Jersey (1998).
- S. Blundell, *Magnetism in Condensed Mattter*, OUP, Oxford (2001).
- M. L. Boas, *Mathematical Methods in the Physical Sciences*, (2nd edition), Wiley, New York(1983).
- M. G. Bowler, *Lectures on Statistical Mechanics*, Pergamon, Oxford (1982).
- L. Brillouin, *Wave propagation in Periodic Structures*, (2nd edition), Dover, New York (1953) [reissued (2003)].
- H. B. Callen, *Thermodynamics and an Introduction to Thermostatics*, Wiley, (1985).
- G. S. Canright and S. M. Girvin, Fractional statistics: quantum possibilities in two dimensions *Science* **247**, 1197 (1990).
- B. W. Carroll and D. A. Ostlie, *An Introduction to Modern Astrophysics*, Addison-Wesley, Reading, Massachusetts (1996).
- P. M. Chaikin and T. C. Lubensky, *Principles of Condensed Matter Physics*, CUP, Cambridge (1995).
- S. Chapman and T. G. Cowling, *The Mathematical Theory of Non-uniform Gases*, (3rd edition), CUP, Cambridge (1970).
- T.-P. Cheng, *Relativity, Gravitation and Cosmology*, OUP, Oxford (2005).
- G. Cook and R. H. Dickerson, Understanding the chemical potential, *Am. J. Phys.* **63**, 737 (1995).
- M. T. Dove, *Structure and Dynamics*, OUP, Oxford (2003).

- T. E. Faber, *Fluid Dynamics for Physicists*, CUP, Cambridge (1995).
- R. P. Feynman, *Lectures on Physics* Vol I, Chapters 44–46, Addison-Wesley, Reading, Massachusetts (1970).
- R. P. Feynman, *Lectures on Computation*, Perseus, (1996).
- C. J. Foot, *Atomic Physics*, OUP, Oxford (2004).
- M. Glazer and J. Wark, *Statistical Mechanics: A Survival Guide*, OUP, Oxford (2001).
- R. M. Goody and Y. L. Yang, *Atmospheric Radiation*, 2nd edn., OUP, Oxford (1989).
- D. J. Griffiths, *Introduction to Electrodynamics*, Prentice Hall, Upper Sadler River, New Jersey (2003).
- J. Houghton, Global warming, *Rep. Prog. Phys.* **68**, 1343 (2005).
- K. Huang, *An Introduction to Statistical Physics*, Taylor and Francis, London (2001).
- W. Ketterle, Nobel lecture. When atoms behave as wave: Bose–Einstein condensation and the atom laser, *Rev. Mod. Phys.* **74**, 1131 (2002).
- D. Kondepudi and I. Prigogine, *Modern Thermodynamics*, Wiley, Chichester (1998).
- W. Krauth, *Statistical Mechanics: Algorithms and Computations*, OUP, Oxford (2006).
- L. D. Landau and E. M. Lifshitz, *Statistical Physics Part 1*, Pergamon, Oxford (1980).
- M. Le Bellac, F. Mortessagne and G. G. Batrouni, *Equilibrium and Non-equilibrium Statistical Thermodynamics*, CUP, Cambridge (2004).
- H. Leff and R. Rex, *Maxwell's Demon 2: Entropy, Classical and Quantum Information, Computing*, IOP Publishing (2003).
- A. Liddle, *An Introduction to Modern Cosmology*, Wiley, (2003).
- E. M. Lifshitz and L. P. Pitaevskii, *Statistical Physics Part 2*, Pergamon, Oxford (1980).
- M. Lockwood, *The Labyrinth of Time*, OUP, Oxford (2005).
- M. S. Longair, *Theoretical Concepts in Physics*, CUP, Cambridge (1991).
- D. J. C. Mackay, *Information Theory, Inference and Learning Algorithms*, CUP, Cambridge (2003).
- M. A. Nielsen and I. L. Chuang, *Quantum Computation and Quantum Information*, CUP, Cambridge (2000).
- A. Papoulis, *Probability, Random Variables and Stochastic Processes*, (2nd edition), McGraw-Hill (1984).
- D. Perkins, *Particle Astrophysics*, OUP, Oxford (2003).
- C. J. Pethick and H. Smith, *Bose–Einstein Condensation in Dilute Gases*, CUP, Cambridge (2002).
- A. C. Phillips, *The physics of Stars*, Wiley, Chichester (1999).
- M. Plischke and B. Bergersen, *Equilibrium Statistical Physics*, Prentice-Hall, Englewood Cliffs, New Jersey (1989).
- F. Pobell, *Matter and Methods at Low Temperatures*, (2nd edition), Springer, Berlin (1996).
- D. Prialnik, *An Introduction to the Theory of Stellar Structure and evolution*, CUP, Cambridge (2000).
- S. Rao, *An anyon primer*, `arXiv:hep-th/9209066` (1992).
- F. Reif, *Fundamentals of Statistical and Thermal Physics*, McGraw Hill, New York (1965).

- K. F. Riley, M. P. Hobson and S. J. Bence, *Mathematical Methods for Physics and Engineering: a Comprehensive Guide*, CUP, Cambridge (2006).
- P. Saha, *Principles of Data Analysis*, Capella Archive, Great Malvern (2003).
- F. W. Sears and G. L. Salinger, *Thermodynamics, Kinetic Theory and Statistical Thermodynamics*, (3rd edition), Addison-Wesley, Reading Massachusetts (1975).
- P. W. B. Semmens and A. J. Goldfinch, *How Steam Locomotives Really Work*, OUP, Oxford (2000).
- A. Shapere and F. Wilczek (editors), *Geometric Phases in Physics*, World-Scientific, Singapore (1989).
- J. Singleton, *Band Theory and Electronic Properties of Solids*, OUP, Oxford (2001).
- D. Sivia and J. Skilling, *Data Analysis a Bayesian Tutorial*, (2nd edition), OUP, Oxford (2006).
- F. W. Taylor, *Elementary Climate Physics*, OUP, Oxford (2005).
- J. R. Waldram, *The Theory of Thermodynamics*, CUP, Cambridge (1985).
- J. V. Wall and C. R. Jenkins, *Practical Statistics for Astronomers*, CUP, Cambridge (2003).
- G. H. Wannier, *Statistical Physics*, Dover, New York (1987).
- G. K. White and P. J. Meeson, *Experimental Techniques in Low-temperature Physics*, (4th edition), OUP, Oxford (2002).
- J. M. Yeomans, *Statistical Mechanics of Phase Transitions*, OUP, Oxford (1992).
- M. Zeilik and S. A. Gregory, *Introductory Astronomy and Astrophysics*, (4th edition), Thomson Learning, London (1998).

Thermal physicists

- H. R. Brown, *Physical Relativity*, OUP, Oxford (2005) [on Einstein].
- S. Carnot, *Reflections on the Motive Power of Fire*, Dover, New York (1988) [this translation also contains a short biography of Carnot and papers by Clapeyron and Clausius].
- C. Cercignani, *Ludwig Boltzmann: The Man Who Trusted Atoms*, OUP, Oxford (2006).
- W. H. Cropper, *Great Physicists*, OUP, Oxford (2001).
- G. Farmelo, *The Strangest Man: The Hidden Life of Paul Dirac, Quantum Genius*, Faber and Faber, London (2009).
- S. Inwood, *The Man Who Knew Too Much*, Macmillan, London (2002) [on R. Hooke].
- H. Kragh, *Dirac: A Scientific Biography*, CUP, Cambridge (2005).
- B. Mahon, *The Man Who Changed Everything*, Wiley, Chichester (2003) [on J. C. Maxwell].
- B. Marsden, *Watt's Perfect Engine*, Icon, Cambridge (2002).
- A. Pais, *Inward Bound*, OUP, Oxford (1986) [on the development of quantum mechanics].
- A. Pais, *Subtle is the Lord*, OUP, Oxford (1982) [on Einstein].
- E. Segre, *Enrico Fermi, Physicist*, University of Chicago Press, Chicago (1970).
- S. Shapin, *The Social History of Truth*, University of Chicago Press, Chicago (2002) [on R. Boyle].
- M. White, *Rivals: Conflict as the Fuel of Science*, Secker & Warburg, London (2001) [on Lavoisier].

Index

absolute zero
    unattainability of, 207
absorption, **275**
absorption coefficient, **428**, **449**
accretion, **441**
acoustic branch, **286**
acoustic modes, **286**
activation energy, **179**
active galactic nuclei, **441**
adiabatic, **121**, **480**
adiabatic atmosphere, 121
adiabatic compressibility, **184**
adiabatic demagnetization, **197**, **198**
adiabatic expansion, 121, 294
adiabatic expansivity, **183**
adiabatic exponent, **114**
adiabatic index, **114**, 424
adiabatic lapse rate, **123**, **448**
adiabatic process, 140
adiabats, **121**
adiathermal, **121**, **480**
albedo, 278, **446**
alternating tensor, 463
anharmonic terms, **216**
anthropogenic climate change, **456**
antiferromagnetic, **338**
anyons, **347**
arrow of time, 126, 417
astronomical unit, **446**
asymmetric stretch, **454**
atmosphere, 331, **446**
    adiabatic, 121
    isothermal, 43
atom trapping, 371
autocorrelation function, **400**
availability, **178**, 395, **396**
average, **19**
Avogadro number, **3**

baroclinic instability, **448**
Bayes' theorem, **165**
BCS theory of superconductivity, **372**
Beer–Lambert law, **450**
Bell Burnell, Jocelyn, **438**
Bernoulli, Daniel, 56
Bernoulli, Jacob, 26
Bernoulli trial, **26**, 159
Berthelot equation, **305**
binomial distribution, **26**
binomial theorem, **27**

bits, **158**
black body, **267**
black hole, **440**, 440–443
black-body cavity, **267**
black-body distribution, **271**
black-body radiation, **267**
black-hole evaporation, **441**
Bohr magneton, **229**
boiling point
    elevation of, 334
Boltzmann constant, **6**, **38**
Boltzmann distribution, **39**
Boltzmann factor, **39**, 39–45
Boltzmann, Ludwig, **31**, 418
bond enthalpy, 257
Bose–Einstein condensation, **369**, 367–373
Bose–Einstein distribution function, **352**
Bose factor, **352**
Bose gas, **366**
Bose integral, **469**
Bose, Satyendranath, **356**
bosons, **345**
Boyle, Robert, 63, 89
Boyle temperature, **306**
Boyle's law, **6**
British thermal unit (Btu), 110
broken symmetry, 337
Brown, Robert, 217
Brownian motion, **217**, 390
bubble chamber, **331**
bubbles
    formation of, 331
bulk modulus, 363, **376**
burning, 426

caloric, 110, 117
calorie, 110
canonical distribution, **39**
canonical ensemble, **38**, **39**, 38–43, 248
carbon-neutral, **458**
Carnot cycle, **127**
Carnot engine, 34, **127**
Carnot, Sadi, **139**
Cauchy principal value, **399**
causality, 398
cavity, 264
Celsius, Anders, 34
Chandrasekhar limit, **438**
Chapman and Enskog theory, 87

Charles' law, **6**
chemical potential, **244**, 244–263
    and chemical reactions, 252
    and particle number conservation laws, 251
    and phase changes, 324
    as a function of pressure, 329
    as a function of temperature, 329
    generalization to many types of particle, 250
    Gibbs function per particle, 250
    meaning, 245
chemical reactions, 44
chicken, spherical, 95
classical thermodynamics, **5**
Clausius–Clapeyron equation, 325
Clausius inequality, **136**, 141
climate change, **456**
closed system, **480**
CNO cycle, **427**
coefficient of self-diffusion, **83**
coefficient of viscosity, **76**
colligative properties, **334**
collision cross-section, **72**, 70–72
collisional broadening, **53**
collisions, 70
combination, **8**
combinatorial problems, 7–9
compressibility, 302, 312
compressive shocks, **388**
conditional probability, **165**
conjugate variables, **181**, **397**
constraint, **16**
continuity equation, 380
continuous phase transition, **336**
continuous probability distribution, 20
continuous random variable, **20**
contractile vacuole, **258**
convection, **100**, **429**, 429, 448
    forced, 100
    free, 100
convective derivative, 380
convolution theorem, **475**
Cooper pairs, **372**
correlation function, **392**
correlation length, **340**
cosmic microwave background, **273**, 479
counter-current heat exchanger, 318
C_p, 16, 112–115
Crab nebula, 439

490 Index

critical isotherm, **299**
critical opalescence, **336**
critical point, **299**, **328**
critical pressure, **300**, 305
critical slowing down, **342**
critical temperature, **300**, 305
critical volume, **300**, 305
cross-section, 71–72
cryoscopic constant, **334**
Curie's law, **196**, 206
curl, 463
C_V, 16, 112–115
cycle
    Carnot, 127
    Otto, 133, 138

Dalton's law, **60**
data compression, 160
de Jonquière's function, **470**
Debye frequency, **281**
Debye model, 281
Debye temperature, **282**
degenerate limit, **362**
degrees of freedom, 212
    number of, 214
delta function, 475
density matrices, **162**
density matrix, 163
    thermal, 163
density of states, **234**
derivative, convective, 380
deviation, **22**
Dewar, James, 319
diathermal, 480
diatomic gas
    heat capacity, 241
    rotational energy levels, 221, 226
    rotational motion, 213
    vibrational motion, 214
Dieterici equation, **305**
Dieterici gas, 304–305, 312
diffusion, 76, 83–90
diffusion constant, 83, 90, 391
diffusion current, **413**
diffusion equation, 84, **84**, 476
Dirac delta function, 475
Dirac, Paul, **357**
discrete probability distribution, 19
discrete random variable, **19**
dispersion relation, **269**, **290**
distinguishability, 236–237, 345–349
distribution function, **352**
div, 463
Donne, John, 207
Doppler broadening, **53**
drift current, **413**
droplets
    condensation of, 332
Dulong–Petit rule, 215, **281**
duvets, 94

dynamic viscosity, **76**

early Universe, 337
ebullioscopic constant, **334**
Eddington limit, **442**
Eddington luminosity, **442**
efficiency, **129**
effusion, **64**, 64–70
effusion rate, **66**
Ehrenfest, Paul, 335
Einstein A and B coefficients, **276**
Einstein, Albert, **355**
Einstein model, **279**
elastic rod, 191
electric dipole moment, **453**
electromagnetic spectrum, 479
electromotive field, **414**
electron degeneracy pressure, **435**
electroweak force, 337
endothermic, **179**, 256
energy, 108–116
engine, **127**
    Hero, 131
    Newcomen, 131
    Stirling, 132
    Watt, 132
ensemble, **38**
entanglement, **164**, **348**
enthalpy, 173, 175, 223, 292
entropic force, **257**
entropy, **38**, **140**, 140–154, 222, 223, 292
    black-hole, 442
    Boltzmann formula, 147
    connection with probability, 150
    current density, 408
    discontinuity at phase boundary, 322
    Gibbs expression, 152, 159
    in adiabatic demagnetization, 200
    in Joule expansion, 147
    in shock waves, 388
    life, the Universe and, 443
    of ideal gas, 239
    of mixing, 147
    of rubber, 193
    of van der Waals gas, 315
    per photon, 278, 444
    production, 408
    Shannon, 157–160
    statistical basis for, 146
    thermodynamic definition of, 140
    various systems, 200
equation of hydrostatic equilibrium, **424**
equation of state, **56**, 110
equations of stellar structure, 429
equilibrium constant, **253**
equilibrium state, **108**
equilibrium, thermal, 108
equipartition theorem, **212**, 391
erg, 110
Eucken's formula, **86**

Euler equation, 380
event horizon, **441**
exact differential, **109**, **472**
exchange gas, 198
exchange symmetry, **346**
exothermic, **179**, 256
expansive shocks, **387**
expected value, **19**
exponential distribution, **29**
extensive, **110**
extensive variables, **5**
extinction coefficient, 428, 449

factorial integral, 464
Fahrenheit, Daniel Gabriel, 33
Fermi–Dirac distribution function, **352**
Fermi energy, **361**, 435
Fermi, Enrico, **356**
Fermi factor, **352**
Fermi gas, **361**
Fermi level, **361**
Fermi momentum, 435
Fermi surface, **365**
Fermi temperature, **362**
Fermi wave vector, **361**
fermions, **345**
ferromagnetic, **338**
Fick's law, **83**
first law of thermodynamics, 110
    $dU = TdS - pdV$, 143, 144
    energy is conserved, 111
    energy of Universe, 141
first-order phase transitions, 335
fluctuation-dissipation theorem, 392, **403**, 405
fluctuations, 4, 336, 390–407, 410–413
flux, **64**
flux density, **450**
forced convection, **100**
Fourier, Jean Baptiste Joseph, **105**
Fourier transform, **398**
Fourier transforms, 475
fractional quantum Hall effect, **347**
fractional statistics, **347**
fractional width, **27**
free convection, **100**
free energy, 174, **176**
freezing point
    depression of, 334
fugacity, **262**, **359**, 367, 370, 373
functions of state, **108**
    in terms of Z, 223
    of ideal gas, 237
fusion, **426**

gain, **276**
gamma function, **464**
gas constant, **59**
Gaussian, **21**, **464**, 465
Gaussian integral, 465

Gay-Lussac's law, **6**
generalized force, **183**
generalized susceptibility, **183**, **399**
geometric progression, 462
Gibbs distribution, **247**
Gibbs function, 175, 178, 223, 292
Gibbs, Josiah Willard, **190**
Gibbs paradox, **240**
Gibbs phase rule, **333**
Gibbs' expression for the entropy, **152**
Gibbs–Helmholtz equations, **176**
global warming, **456**
grad, 463
Graham's law of effusion, **64**, 65
Graham, Thomas, 64
grand canonical ensemble, **38**, **247**, **248**
grand partition function, **247**, 349
grand potential, **248**, 358
gravitational collapse, 421
gravity, 421
greenhouse effect, **454**
greenhouse gases, **454**

H-theorem, **417**
hard-sphere potential, **71**
harmonic approximation, **216**
Hawking radiation, **441**
Hawking temperature, **441**
heat, **13**, 13–18, **81**, 111
heat bath, **38**
heat capacity, **15**, 14–17, 112–115, 183
  negative, **426**
  of a Bose gas, 373
  of a diatomic gas, 241
  of a metal, 365
  of a solid, 215, 280, 282
heat engines, 126–139
heat flow, 32, 90, 245
heat flux, 64, 81
heat pump, 133
heat transfer in stars, 427
Heaviside step function, 361, 399
helium, 34
  liquefaction of, 319
Helmholtz free energy, 177, 178
Helmholtz function, 174, 175, 178, 222, 223, 292
Helmholtz, Hermann von, **189**
Hero's engine, 131
Hertzsprung–Russell diagram, **432**
heteronuclear molecules, **453**
Higgs mechanism, **337**
high-temperature superconductors, **372**
homology, **430**
homonuclear molecules, **453**
hydrogen
  liquefaction of, 319
hydrogen bonding, **328**
hydrogen burning, **426**
hydrogen molecule

bond enthalpy, 257
hydrostatic equation, **44**, 121, 425
hydrostatic equilibrium, **423**
hypersphere, **473**
hypersphere, volume of, 473
hypertonic, **258**
hypotonic, **258**

ice, skating on thin, 344
ideal gas, 6–7, **56**, 120, 121, 206
  functions of state of, 237
ideal gas equation, **6**, **59**, 296
ideal gas law, 59
identical particles, 345
impact parameter, **72**
independent random variables, **24**
indistinguishability, 8, 236–237, 345–349
inexact differential, **109**, **472**
information, **158**, 157–169
infrared active, **454**
intensive, **110**
intensive variables, **5**
internal combustion engine, **132**
internal energy, **111**, 172, 175, 221, 223, 292
interstellar medium, **420**
inversion curve, **318**
inversion temperature, 318
irradiance, **450**
irreversible change, 140
isenthalpic, **480**
isentropic, **140**, 480
Ising model, **338**
island, no man is an, 207
isobaric, **173**, 480
isobaric expansivity, **183**
isochoric, **173**, 480
isolated system, 480
isothermal, **120**, 480
isothermal compressibility, **184**
isothermal expansion
  of ideal gas, 120
  of non-ideal gas, 315
isothermal magnetization, **197**, **198**
isothermal Young's modulus, **192**
isotherms, **121**
isotonic, **258**
isotopes, separation of, 64

Jacobian, **473**
Jeans criterion, **422**
Jeans density, **422**
Jeans length, **422**
Jeans mass, **421**
Johnson noise, **393**, 404
joint probability, **165**
joule, 14, 110
Joule coefficient, **313**
Joule expansion, **144**
  and time asymmetry, 418

apparent paradox refuted, 146
Gibbs paradox, 240
of van der Waals gas, 313
Joule heating, **155**
Joule–Kelvin expansion, **316**
Joule–Thomson expansion, **316**

Kelvin Lord, **189**, 416
Kelvin's formula, **330**
kilocalorie (kcal), 110
kinematic viscosity, **76**
kinetic coefficients, **410**
kinetic theory of gases, **5**, **47**
Kirchhoff's law, **267**
Knudsen effect, **67**
Knudsen flow, **68**
Knudsen gas, **78**
Knudsen method, 66
Kramers–Kronig relations, **399**, 407
Kramers opacity, **431**
Kronecker delta, 463
k-space, 233
kurtosis, 22

Lagrange multipliers, 477, **477**
Landauer, Rolf, 159
Langevin equation, **390**
Laplace's equation, 94
large numbers, 2–3
laser, **277**
laser cooling, 371
latent heat, **321**
lattice, **279**
Lavoisier, Antoine, 110, **117**
law of corresponding states, **310**
Le Chatelier's principle, **256**
life, 443
Linde process, **319**
Linde, Karl von, 319
linear expansivity at constant tension, **192**
linear response, 397
liquefaction of gas, 318
localized particles, 237
logarithms, 9, 462
longitudinal waves, **376**
low-pass filter, 98
luminosity, **268**, 420

Mach number, **383**
machine, sausage, 221
Maclaurin series, 462
macroscopic quantum coherence, **372**
macrostate, **35**
magnetic field, 196, **196**
magnetic field strength, **196**
magnetic flux density, **196**
magnetic induction, **196**
magnetic moment, **196**
magnetic susceptibility, **196**

magnetization, 196, **196**
magnetocaloric effect, **197**
main sequence, **432**
mass defect, **427**
mass extinction cross-section, 428
matter dominated, **294**
maximum entropy estimate, **168**
Maxwell–Boltzmann distribution, **50**, 48–55
Maxwell construction, **304**
Maxwell, James Clerk, **55**
Maxwell's demon, **149**, 149–160
Maxwell's relations, **180**, 181–183, 187
Maxwellian distribution, **50**
mean, **19**, 20
mean free path, 67, **73**, 70–74, 78
mean scattering time, **71**
mean squared deviation, 22
mean squared value, 19
mesopause, **449**
mesosphere, **449**
metals, 362
metastability, 329
metastable, **302**
Metropolis algorithm, **341**
microcanonical ensemble, **38**, **248**, **354**, 394
microstate, **35**
Milky Way, **420**
mixed state, **163**
mixing ratio, **456**
mobility, **392**
mode, **212**
molar heat capacity, **16**
molar mass, **3**
molar volume, **60**
mole, **3**
mole fraction, **332**
molecular flux, 64
moment, **22**
moment generating function, **30**
momentum flux, 77
monatomic gas
  translational motion, 213
Monte-Carlo method, **340**

natural radiative lifetime, **275**
natural variables, **143**
nebula, **433**
negative heat capacity, 426
neutron star, **438**
Newcomen's engine, 131
Newton's law of cooling, **99**
Newtonian fluids, **76**
no-cloning theorem, **164**
non-equilibrium, 76
non-equilibrium state, **108**
non-equilibrium thermodynamics, 408–419
non-interacting quantum fluid, 358

non-Newtonian fluids, 76
non-relativistic limit, **290**
normal distribution, **465**
nuclear burning, **426**
nuclear reactions, **426**
number of degrees of freedom, 214

occupation number, **350**
omelette
  exceedingly large, 3
Onnes, Heike Kamerlingh, 319
Onsager's reciprocal relations, **410**
opacity, **428**
open system, 480
optic branch, **286**
optic modes, **286**
optical depth, **450**
optically thick, **450**
optically thin, **450**
order, **335**
ortho-hydrogen, **354**
osmometry, **260**
osmosis, **257**
osmotic flow, **257**
osmotic pressure, **257**
Otto cycle, **138**

para-hydrogen, **354**
paramagnet, **230**
paramagnetism, **196**
parcel, 122
Parseval's theorem, **401**, 475
partial derivative, **471**
partial pressure, **60**
particle flow, 246
partition function, **40**, **219**, 219–233
  combination of several, 228
  single–particle, 220
Pauli exclusion principle, **345**, 435
Peltier coefficient, **414**
Peltier cooling, **414**
Peltier effect, **414**
perpetual motion, **137**
phase diagram, 327, 328
phase equilibrium, 324
phase transitions, **321**, 321–344
phonon dispersion relation, 284
phonons, **279**, 279–288
photons, **263**, 263–279
Pirani gauge, **83**
Planck area, 443
Planck length, **442**
poetry, badly paraphrased, 207
Poisson distribution, **29**
polylogarithm, **360**, **470**
population inversion, **277**
positive feedback mechanism, **457**
posterior probability, **166**
power spectral density, **400**
PP chain, **426**

Prandtl number, **100**
pressure, 4, 56–62, 223
  partial, 60
  units, 56
pressure broadening, **53**
principle of microscopic reversibility, **412**
prior probability, **166**
probability, 18–31
pulsars, **438**
pure state, **163**

quanta, **7**
quantum concentration, **235**
quantum dots, **244**
quantum gravity, 418
quantum information, 162–164
quantum theory, **5**
quasistatic, **119**, 480
qubits, **164**

R value, **94**
radiance, **272**
radiation dominated, **294**
radiation field, **275**
radiative flux, **450**
radiative temperature, **446**
radiative-transfer equation, **450**
radiometric temperature, **446**
random walks, **26**, 28
Rankine–Hugoniot conditions, **387**
Raoult's law, **259**
Rayleigh–Jeans law, **272**
reciprocal theorem, **472**
reciprocity theorem, **472**
red giants, **433**
reduced coordinates, **310**
reduced mass, **482**
refrigerator, 133
relative entropy, **169**
relativistic gases, 290–295
reservoir, **38**, **126**
response function, **398**
rest mass, **290**
reverse osmosis, **258**
reversibility, 118–120
reversible, 480
reversible engine, 130
Riemann zeta function, **469**
Rinkel's method, 124
r.m.s., 23
root mean squared value, 23
rubber, 192
Rüchhardt's method, 124
Rumford, Count, 110, **117**

Sackur–Tetrode equation, **240**, 243
Saha equation, **262**
sausage machine, 221
scale height, **455**
scaling relations, **430**

Schottky anomaly, **225**
Schwarzschild equation, **450**
Schwarzschild radius, **441**
second law of thermodynamics
  Carnot's theorem, 130
  Clausius' statement, 126
  Clausius' theorem, 136
  $dS \geq 0$, 141
  entropy of Universe, 141
  Kelvin's statement, 127
second-order phase transition, **335**
Seebeck coefficient, **415**
Seebeck effect, **415**
self-diffusion, 85
semipermeable membrane, **257**
Shannon entropy, 157–160
Shannon's noiseless channel coding theorem, **161**
shock front, **383**
shock waves, 383–389
simple harmonic oscillator, 220, 226
single–particle partition function, **220**
Sirius A and B, 438
skewness, 22
skin depth, **93**
solar constant, **446**
solid angle, **57**
solubility gap, **333**
solute, **257**, **334**
solvent, **257**, **334**
Sommerfeld formula, **364**
sound waves, 376–382
  adiabatic, 377
  isothermal, 377
  relativistic gas, 379
  speed of, 379
specific heat capacity, **15**
specific intensity, **449**
specific opacity, **449**
spectral absorptivity, **266**
spectral emissive power, **266**
spectral energy density, **266**
spectral luminosity, **420**
spectral radiance, 428
spin, 359
spin-statistics theorem, **345**
spontaneous emission, **275**
stability, 329
standard deviation, **22**
standard temperature and pressure, **60**
static susceptibility, **399**
statistical mechanics, **5**
statistics, **345**
steady state, 76, **94**
Stefan's law, **265**
Stefan–Boltzmann constant, **265**, 270
Stefan–Boltzmann law, **265**
stellar astrophysics, **420**
stellar structure, equations of, 429
steradians, **57**

stimulated emission, **276**
Stirling's engine, **132**
Stirling's formula, **9**, **468**
stoßzahlansatz, **417**
strain, **192**
stratopause, **449**
stratosphere, **448**
stress, **192**
strong force, 337
strong shock, **384**
sublimation, **328**
Sun, 45, 278
  main properties, 420
superconductivity, **372**
supercooled vapour, **329**
superfluid, **371**
superheated liquid, **329**
supersonic, **383**
surface brightness, **272**
surface tension, **194**
surroundings, **108**
sycophantic applause, 169
symmetric stretch, **454**
symmetry breaking, **336**
system, **38**, **108**, **480**
systematic errors, 26

Taylor series, 462
tea, good cup of, 100, 343
temperature, **32**, **37**, 36–38
temperature gradient, 81
terminal velocity, 392
thermal conductance, **94**
thermal conductivity, 76, **81**, 81–83, 86–90
  measurement of, 83
thermal contact, **32**
thermal density matrix, **163**
thermal diffusion equation, **91**, 90–105
thermal diffusivity, **91**
thermal equilibrium, **32**, 32–33, 108, 480
thermal expansion, 206
thermal radiation, **263**
thermal resistance, **94**
thermal wavelength, **235**
thermalization, **33**
thermally isolated system, **111**
thermocouple, **415**
thermodynamic limit, **4**, 6
thermodynamic potentials, **172**
thermodynamics, **108**
thermoelectricity, **413**, 413–419
thermometer, **33**
  platinum, 34
  RuO_2, 34
thermopower, **415**
thermosphere, **449**
third law of thermodynamics, 203–208
  consequences, 205
  Nernst statement, 204
  Planck statement, 204

Simon statement, 205
  summary, 208
Thompson, Benjamin, 110, **117**
Thomson coefficient, **417**
Thomson heat, **416**
Thomson, William, **189**, 416
Thomson's first relation, **417**
Thomson's second relation, **416**
time asymmetry, 417
tog, 94
torr, 56
torsion constant, **80**
trace, **162**
translational motion, 213
transpiration, **258**
transport properties, **76**
trigonometry, 462
triple point, **328**
tropopause, **448**
troposphere, **448**
Trouton's rule, **323**
turgor pressure, **258**
two-level system, 220, 224

U value, **94**
ultracold atomic gases, **371**
ultrarelativistic gas, 290–292
ultrarelativistic limit, **290**
ultraviolet catastrophe, **272**

van der Waals gas, **296**, 296–320
van 't Hoff equation, **256**
vapour, 323
vapour pressure, **330**
variables of state, **108**
variance, **22**
vector operators, 463
velocity distribution function, **48**
vibrational modes, **453**
virial coefficients, **306**
virial expansion, **306**
virial theorem, **424**
viscosity, **76**, 76–81, 86–90
  dynamic, 76
  kinematic, 76
  measurement of, 80
von Neumann entropy, **162**

water, 328, 344
watt, 14
wave vector, **233**
weak shock, **384**
white dwarf, 433, **437**
Wien's law, **272**
Wiener–Khinchin theorem, **401**
work, 111
work function, **413**

zeroth law of thermodynamics, **33**
Zustandssumme, 219